Biological Control of Vertebrate Pests

The History of Myxomatosis, an Experiment in Evolution

Biological Control of Vertebrate Pests

The History of Myxomatosis, an Experiment in Evolution

Frank Fenner
John Curtin School of Medical Research
Australian National University
Canberra
Australia

and

Bernardino Fantini
Louis Jeantet Institute for the History of Medicine
University of Geneva
Geneva
Switzerland

CABI *Publishing*

CABI *Publishing* is a division of CAB *International*

CABI Publishing
CAB International
Wallingford
Oxon OX10 8DE
UK

Tel: +44 (0)1491 832111
Fax: +44 (0)1491 833508
E-mail: cabi@cabi.org

CABI Publishing
10 E 40th Street
Suite 3203
New York, NY 10016
USA

Tel: +1 212 481 7018
Fax: +1 212 686 7993
E-mail: cabi-nao@cabi.org

A catalogue record for this book is available from the British Library, London, UK.

Library of Congress Cataloging-in-Publication Data
Fenner, Frank, 1914–
Biological control of vertebrate pests : the history of myxomatosis ; an experiment in evolution / by Frank Fenner and Bernardino Fantini.
p. cm.
Includes bibliographical references and indexes.
ISBN 0-85199-323-0 (alk. paper)
1. Myxomatosis--History. 2. Vertebrate pests--Biological control. I. Fantini, Bernardino. II. Title.
SF997.5.R2F38 1999
632′.66--dc21 98-52951
CIP

ISBN 0 85199 323 0

Typeset in Melior by Columns Design Ltd, Reading
Printed and bound in the UK by Biddles Ltd, Guildford and King's Lynn

Contents

Preface

Biological control is a popular and very cost-effective way of controlling a large number of insect pests and weeds. However, it has been difficult to apply to vertebrate pests, for several reasons. Early attempts to introduce predators to control pest animals such as rats or rabbits were dismal failures, the predators themselves becoming pests because they preyed on native animals. Attempts to use microbes meet with other problems. Thus, many animals regarded as pests are domestic animals or pets which have escaped to become wild again, for example, feral pigs and cats in Australia and other parts of the world. Use of a transmissible agent to control the pest would inevitably endanger its valued relatives. In other cases the wild animals are so closely related to domestic animals, for example foxes or coyotes and domestic dogs, that it is impossible to find a bacterium or virus that is pest specific. Another major difficulty is that animals such as rodents, rabbits and squirrels are so fecund that the control agent must be extremely lethal to produce more than a very shortlived effect on pest numbers.

By chance, the European rabbit, which is a major pest in Australia, New Zealand and Chile, has been found to suffer from two very different and highly lethal virus diseases, myxomatosis, caused by a poxvirus, and rabbit haemorrhagic disease, caused by a calicivirus. Both viruses are highly specific for the European rabbit, although myxoma virus causes a trivial infection in certain species of rabbits in the Americas, which are, indeed, its natural hosts. Myxomatosis burst onto the world scene in 1951, when, after proposals to introduce biological control dating back to 1918, it spread with amazing speed throughout the vast numbers of wild rabbits living in the south-eastern part of Australia. Then, in June 1952, it was illegally introduced into France, and soon spread among both wild and domestic rabbits. Within a few years it had spread throughout Europe, to the delight of many farmers but the consternation of rabbit breeders and chasseurs.

In 1984 another very lethal disease of European rabbits was recognized in commercial rabbitries in China and soon spread to Europe and other parts of the world. This was rabbit haemorrhagic disease, caused by a virus of the family *Caliciviridae*. In 1995, after extensive testing for its specificity, it was introduced in Australia as a biological control agent, and has spread naturally since then and caused high mortalities in some areas. In 1997 it was introduced illegally into New Zealand and has spread extensively there also.

This book is primarily a history of these two diseases, which were deliberately introduced into Australia to assist in the control of that country's major agricultural pest, the European rabbit. The history of myxomatosis is almost as long as the history of virology, since it was first described in Uruguay in 1898. The history of rabbit haemorrhagic disease is very short, since the disease did not exist before 1984. It is truly a new disease, evolved by mutation from a virus of the European rabbit that causes a completely subclinical infection. These two very different diseases are the only methods of biological control of vertebrate pests ever to have met with any success. To set their histories into context, we have provided a brief history of the concept of biological control as applied to other pest species, primarily insects and weeds, achieved by the use of insects, nematodes or microbes, and we recount earlier unsuccessful attempts at the biological control of vertebrate animals.

After chapters describing the biology of the European rabbit and of myxoma virus, we recount, in five chapters, the story of the introduction and spread of myxomatosis in Australia and in Europe. The biology of rabbit haemorrhagic disease virus and the history of the disease it causes are described in a single chapter. Since these diseases had an unprecedented effect on the size of wild rabbit populations in Australia and in some countries of Europe, we devote a chapter to the description of the dramatic ecological and environmental effects of the reductions they produced in rabbit numbers.

The penultimate chapter deals with a number of interesting theoretical questions raised in the earlier chapters, such as the effects of the age of the host and ambient temperature on the severity of infectious diseases, different methods of transmission of viral diseases by arthropods, problems of host range and possible changes in host range due to mutations and the role of infectious diseases in controlling animal populations. The final chapter considers at length a problem of great biological interest for which myxomatosis provides the best available example, namely the coevolution of viruses and hosts in infectious diseases.

Frank Fenner
Bernardino Fantini
October 1998

Acknowledgements

The production of this book has been made possible only with the help of many organizations and individuals. One of us (FF) is deeply indebted to the John Curtin School of Medical Research, in the Australian National University, for providing him with an office and supporting facilities as a Visiting Fellow since 1980. We are particularly grateful to Mr S.R. Butterworth and his colleagues in the Photographic Services section of the John Curtin School for their help with preparation of the many figures that appear in this book. Mrs V. Lyon of the Department of Geography of the Australian National University kindly prepared the maps. We have also been greatly assisted by the staff and the library services of the CSIRO Division of Wildlife and Ecology, which has responsibility for continuing research on rabbit control, and Ms S. Thomas of the Bureau of Resource Sciences, which is producing an excellent series of monographs on the control of vertebrate pests in Australia. Staff of the Services des Archives de l'Institut Pasteur, CSIRO Archives, the Manuscript Section of the National Library of Australia, the National Archives of Australia, the Archives Section of CSL Ltd and the Basser Library of the Australian Academy of Science have provided help in tracing archival material.

Dr J.H. Calaby, Dr B.D. Cooke, Mr J.W. Edmonds, Mr B.V. Fennessy, Dr P.J. Kerr, Dr I.D. Marshall, Dr K. Myers, Mr N. Newland, Dr A. Newsome, Mr I. Parer, Dr A.J. Robinson, Dr J. Ross, Ms R.C.H. Shepherd, Dr W.R. Sobey, Mr H.V. Thompson and Dr C.K. Williams have read through and provided valuable comments on one or more chapters. Others who have provided us with useful data include Dr A. Brun, Dr T. Berke, Dr J.J. Burdon, Professor W. Bynum, Dr L. Capucci, Dr B.J. Coman, Mr D. Demelier, Dr A.L. Dyce, Mr A. Girard, Mr M. Harper, Dr M.K. Holland, Dr G. Hood, Dr J. Kovalski, Mme C. Lardy, Professor J. Lederberg, Dr I. Lugton, Dr A.R. Mead-Briggs, Ms M.C. O'Dea, Mme D. Ogilvie, Dr D.C. Regnery, the Honourable Miriam Rothschild, Professor C.B. Schedvin, Dr R. Soriguer, Dr D.S. Strayer, Professor M.J. Studdert, Ms S. Thomas, Dr L.E. Twigg, Dr C.H. Tyndale-Biscoe, Dr B.H. Walker, Dr H.C. Westbury, Dr R.W. Wichmann, Dr R.T. Williams, and Dr D.H. Wood.

The staff of CABI *Publishing* have been most cooperative throughout the production of this book. We would like to thank all of those involved, especially Emma Critchley and Tim Hardwick.

1
Pest Animals and Plants

Overview

The idea that an animal or plant is a pest is anthropocentric; a living thing is regarded as a pest if it is troublesome to humans. Most pests fall into the category of animals (particularly insects) or plants which reduce agricultural or pastoral productivity. More recently the definition of pests has been enlarged to include plants, animals and microorganisms which threaten the natural environment. In several countries of Europe, and in Australia, New Zealand and Chile, the rabbit is an important pest, affecting both agricultural productivity and the environment.

Some animals are pests in their native habitats, but many of the most serious pest problems have followed the introduction of animals and plants into new habitats. In the 19th century introductions of vertebrate animals (but not insects) were often the result of a deliberate policy of 'acclimatization', and Acclimatization Societies were set up in Europe and in many British colonies.

From ancient times affected human communities have tried to control pests. A wide range of measures has been used against animal pests: scarecrows and sonic deterrents, barriers to movement, killing by hunting, trapping, poisoning or predatory animals, and control by introduced diseases. The principal methods used to control pest rabbits, particularly in Australia, have been barrier fencing, hunting, trapping, poisoning, habitat destruction by ripping warrens, and control by introduced diseases.

What is a Pest?

The word pest is derived from the Latin *pestis* = plague, and is usually defined as a troublesome or destructive animal. It is important to differentiate between animals that have been traditional enemies of humans since their days as hunter-gatherers; large carnivores such as tigers and wolves were regarded as dangerous predators, but not as pests. The idea of pest animals dates from the agricultural revolution, when animals that were not dangerous, such as locusts, caterpillars, rats and mice, became pests because of their numbers and their interference with crops or stored food. Pest plants, i.e. plants that grow where they are not wanted, are usually called weeds, whereas troublesome or destructive microorganisms, whether protozoa, fungi, bacteria or viruses, are generally called parasites. During recent decades, with the increase in environmental awareness, the definition of pests has been enlarged to include plants, animals and microorganisms that threaten ecological equilibria and biodiversity.

Within human populations, the definitions of which particular animals or plants are pests or weeds depend very much on the interests of the observer. With the development of agriculture some 10,000 years ago, man's major pests became, and remain, animals that feed on or otherwise damage his crops: vertebrates such as rabbits, rats, mice and large wild herbivores, and invertebrates such as insects and helminths. Many insects carry viruses or protozoa that cause infectious diseases of humans, domestic animals and plants, and insects that are persistent and annoying, like mosquitoes and flies, are also regarded as pests. The major weeds are plant species that compete with crop species or with pasture or garden plants, or foreign plants that spread into native woodlands or savanna.

Every country has indigenous animals or plants that are at some time regarded as pests or weeds, their numbers, achieved as a result of evolution and natural selection, being greater than humans regard as desirable. When land uses change, animals or plants that were previously tolerated may come to be regarded as pests. However, the most troublesome pests and weeds are animals and plants introduced into new environments, either deliberately or accidentally. Here they intrude on ecosystems in which they escape from the indigenous predators and diseases that controlled their numbers in their home countries. Further, they may introduce new parasites; for example, exotic birds introduced bird malaria into Hawaii, devastating several indigenous species of birds.

The process of introduction of animals and plants into new environments has been going on for millions of years, the invasion of new habitats being one of the major engines driving evolution. The scale and rate of new introductions increased over the last few thousand years, as humans moved around the world. Historically, it was seen on the largest scale during the period between 1500 and 1900, when peoples of European descent occupied other continents and islands and brought with them animals and plants from their home countries, either deliberately or accidentally. By and large, the importation of vertebrate pests is now controlled by quarantine, but with the vastly increased international trade and commerce that has occurred during the second half of the 20th century, such introductions are still occurring. Garden plants that 'escape', insects brought in from other countries with cut flowers, and unwanted invertebrates and microbes in ships' bilgewater are examples of a continuing problem.

Changing perceptions of what is a pest animal

The way in which perceptions of pests change with changing circumstances is well illustrated by a study of the occupation by Europeans of the Bega district, on the south coast of New South Wales (Lunney and Leary, 1988). Now a dairying district, the area was occupied by a few hundred Aboriginals before European settlement began in the 1830s. The new settlers used river flats for grazing cattle and sheep for meat and wool production and then the forests were cleared and cattle numbers were greatly increased. Initially native animals such as possums, bandicoots and various macropods were viewed as the most important pests (Fig. 1.1), and between 1880 and 1898 bounties were paid for their scalps. It is possible that the peaks in numbers of several native animals were due to the decline in hunting by Aboriginals after Europeans took over their land. European hares were introduced into Australia between 1859 and 1865. They reached the Bega district in the 1880s and in the early 20th century they briefly peaked at 'super-plague' levels. Rabbits were first seen in the district about 1900 and by 1910 they had reached super-plague levels, replacing hares as the major vertebrate pest. Unlike the hares, they have remained relatively common ever since, fluctuating since 1950 in response to outbreaks of myxomatosis. Due to destruction of their habitat, most species of native mammals since about 1950 have become uncommon or rare; six species are now extinct and another four species are endangered.

Why do some species become pests?

There are numerous reasons why animals or plants may be regarded as pests or weeds. Often they may be defined thus

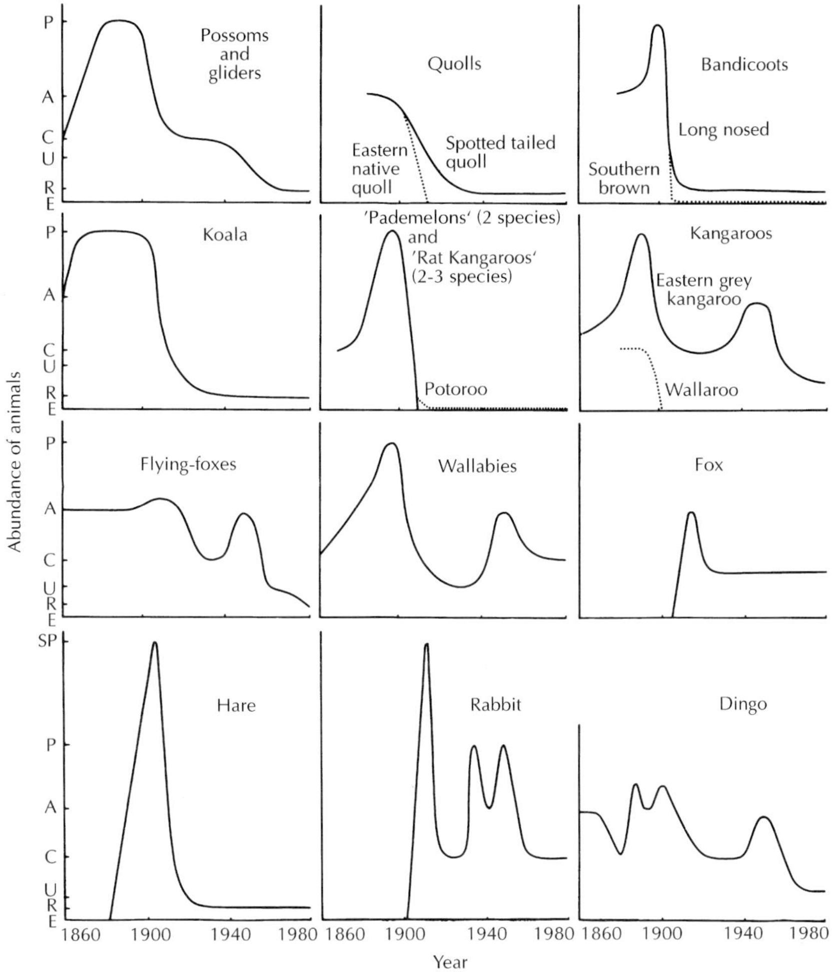

Fig. 1.1. Stylized curves showing the fluctuations in the numbers of native and exotic mammals and the changing perceptions of their importance as pests by people in Bega district, on the south coast of New South Wales. Lettering indicates perceptions by local farmers: SP = super-plague, P = plague, A = abundant, C = common, U = uncommon, R = rare, E = extinct. Since they preyed on lambs and poultry, foxes and dingoes were always regarded as pests; other animals only when they were present in plague or super-plague numbers. From Lunney and Leary (1988), with permission.

when human activities change so that species that were previously tolerated come to be seen as competing with human wishes concerning land use, or when the number of individuals in a tolerated species becomes too great. Sometimes introduced animals may find an ecological niche in which they compete very successfully with the indigenous animals, a classic situation in places previously long isolated from the Eurasian landmass, the Americas and Africa, such as Australia, New Zealand and Hawaii.

Clearly, what is tolerated or enjoyed as a desirable plant or animal in some countries or situations may be regarded as a pest in others. In relation to vertebrate pests, Hone (1994) used statistical, economic and modelling analyses to help identify their effects in particular situations, examine the economic and ecological impacts of these effects, and analyse the influence of various control measures on pest populations.

The Acclimatization of Animals and Plants

The development of ocean-going ships in the 15th century initiated several centuries of exploration and colonization in all parts of the world by the major European powers. Traditionally, explorers and their immediate successors, who in the southern hemisphere were whalers and sealers, deliberately released rabbits and goats on oceanic islands to provide a food source for shipwrecked mariners, and unintentionally, they often introduced rodents and pet animals such as cats and dogs. On small islands such animals often became extinct by eating themselves 'out of house and home'. Sometimes humans also suffered this fate; Easter Island in the South Pacific is the classic example.

The last countries in the world to be colonized were those most distant from Europe: Australia in 1788, and New Zealand in 1840. The colonizers, from Britain, found two different but alien environments, populated by Polynesians in New Zealand and by primitive hunter-gatherers, the Australian Aborigines, in Australia, each living in country dominated by trees, shrubs, grasses, animals and birds of types unknown anywhere else in the world. The early settlers brought their domestic animals and plants with them, initially to produce food for their sustenance. Once settlement had been firmly established, they arranged for other animals and plants with which they were familiar to be brought out, so that they could establish at least a domestic environment in which they would feel more comfortable. With increasing affluence, the more prosperous colonists wished to introduce field sports such as fox hunting, rabbit shoots, deer hunting and the like and brought out the animals needed. Deer were imported into Sydney in 1803, and deer, partridges and hares into Tasmania by 1830. Although hutch rabbits were introduced with other domestic animals with the First Fleet in 1788, and periodically thereafter, they did not become common until Thomas Austin imported wild rabbits from England in 1859 (see p. 17). To further foster such introductions, acclimatization societies were established in both Australia (Rolls, 1984) and New Zealand (Thomson, 1922; Wodzicki, 1950), and in both countries these societies enjoyed widespread support, especially from the more prosperous colonists.

Acclimatization societies were also established in several European countries, stimulated by curiosity about exotic species and the possibility of the commercial exploitation of new plants and animals. The first such society in the world was set up in Paris in 1854, subscribers including no less than fourteen crowned heads and almost all of the nobility of Europe. A similar society was set up in London in October 1860, stimulated by a letter to the London *Times* by Edward Wilson of Melbourne. Many exotic wild animals, including rabbits, foxes and deer, and several exotic plants, including prickly pear, were already established in Australia when the Acclimatization Society of Victoria was established in 1862, with Wilson as president. Similar societies were set up in New South Wales, Queensland and South Australia (Francis, 1862; Fig. 1.2) soon afterwards, and they organized a wide range of importations.

In New Zealand legislative acts were passed by the Colonial Parliament in 1861 'to encourage the importation of these animals and birds, not native to New Zealand, which would contribute to the pleasure and profit of the inhabitants, when they became acclimatized and spread over the country in sufficient numbers' (Wodzicki, 1950). In 1867 provision was made for the registration of acclimatization societies at the Colonial Secretary's office, and by the early years of the 20th century some 48 species of mammal and 30 species of bird had been introduced into New Zealand. Prominent among the reasons for these introductions was their value for sport, listed as a reason for some 45% of the mammals introduced.

THE ACCLIMATISATION

OF

HARMLESS, USEFUL, INTERESTING, AND ORNAMENTAL,

ANIMALS AND PLANTS,

BEING

A Paper Read before the Philosophical Society,

ADELAIDE, SOUTH AUSTRALIA,

ON MAY 13TH, 1862,

BY

GEORGE WILLIAM FRANCIS, ESQ., F.H.S.,

DIRECTOR OF THE BOTANIC GARDEN.

"The earth was made so various, that the mind
Of desultory man, studious of change
And pleased with variety, might be indulged."

COWPER.

Published by the Philosophical Society for Gratuitous Distribution.

ADELAIDE:
PRINTED AT THE REGISTER AND OBSERVER GENERAL PRINTING OFFICES, GRENFELL STREET.

1862.

Fig. 1.2. Title page of an early Australian address on acclimatization societies.

The fate of introduced animals and plants

By far the largest number of species imported under the patronange of the acclimatization societies were garden plants. Many of these were, and still remain, valued garden plants. Others spread outside of gardens and several have become major pests in forests and farmlands, for example the shrubs *Lantana* (from South America) in Queensland and

Pyracantha (from Europe) in parts of New South Wales. The same happened in Europe, with some species of *Eucalyptus* (from Australia) and with *Bougainvillea* (from South America), and in South Africa, with *Acacia* and *Hakea* from Australia. One factor contributing to the distribution of such plants outside gardens is the production of large numbers of seeds in fruit that are eaten by local birds.

In Australia and New Zealand, respectively, 16 and 25 introduced mammals are currently viewed as pests, as well as a further 17 native mammals in Australia (Cowan and Tyndale-Biscoe, 1997). Freed from the pressure of competitors, predators and disease that controlled their numbers in the original habitats, rabbits, foxes, rats, mice, cats, goats, pigs and horses and also, in places, red deer and camels became pests in Australia (Ramsay, 1994). In New Zealand also several of the animals introduced by acclimatization societies became pests – these included opossums from Australia, and rabbits, rats, red deer, cats, goats, pigs, stoats and weasels, mainly from England. Of all of these, by the 1860s the European rabbit was by far the most important pest animal in both Australia and New Zealand.

Rabbits as a pest in Australia and New Zealand

In southern Australia rabbits encountered a favourable climatic environment, a country with few effective native predators, and an ecological niche occupied by a variety of small marsupials, none of which could match the reproductive capacity or the aggressive behaviour of the rabbit. In New Zealand, there were no predators and their only ecological competitors were flightless birds. In both countries rabbits soon became agricultural pests (Wodzinski, 1950; Rolls, 1984).

Measures to Counteract Pests

A variety of measures can be undertaken to reduce the impact of particular pests, varying from a minimum of excluding them from contact with the threatened plants or animals through an intermediate level designated as control to the extreme response of eradication of the pest. For all pests, microbial, plant or animal, country-wide eradication is almost always the most cost-effective goal, but this is rarely possible with plant and animal pests unless started very early. Eradication campaigns are the usual response of veterinary authorities to the importation of exotic viruses. Thus, during the latter part of the 20th century, introductions of foot-and-mouth disease, rinderpest and avian influenza viruses into Europe, North America and Australia have been followed by vigorous and successful eradication campaigns, often involving the slaughter of large numbers of domestic animals. Likewise, when myxomatosis was first introduced into England in 1953 the initial but unsuccessful response was to attempt to eradicate the disease, and when rabbit haemorrhagic disease virus was introduced into Mexico in 1988 it was eradicated from that country by 1991.

There are only a few examples of the eradication of introduced vertebrate animals which escaped from farms and became feral. Muskrats (*Ondatra zibethicus*) were introduced into Britain in the 1920s to be farmed for pelts. Over 80 muskrat farms were established, from which escapes occurred and feral populations were established. Since the environmental damage they caused was well known from experience in Europe, the farms were phased out in 1933, and an energetic campaign to eradicate muskrats, started at that time, was brought to a successful conclusion in 1938, although in the process there was a large toll of non-target species (Gosling *et al.*, 1988; Gosling and Baker, 1989a).

Coypus (*Myocastor coypus*), a semi-aquatic rodent introduced into Britain from South America at about the same time as muskrats, also escaped from the farms and established feral populations in wetlands in East Anglia. They were not recognized as being environmentally damaging until the late 1950s, and in 1962 a limited, unsuccessful trapping campaign was begun. After a preliminary investigation indicated its feasibility (Gosling *et al.*,

1988), an eradication campaign was initiated in 1981, based on the results of a long-term investigation of coypu population ecology. One important feature of coypu behaviour is that when females are rare and dispersed, a female sex ratio of at least 50% is required for the maintenance of population fecundity (Gosling and Baker, 1989b). By cage-trapping the more widely ranging males, increasing numbers of females failed to conceive, and coypus have now been eradicated from Britain.

Traditional Methods of Pest Control

There are a great number of methods of controlling pests and parasites; some of those with a wide applicability are discussed below.

Barriers against pest species

One obvious method that applies to a few pests is to keep them out of the environment in which they may harm humans or human activities. Large animals may be kept out by fencing, small animals such as carnivorous or frugiverous birds by netting, and insects by screening.

Fencing

Fencing is traditionally used to keep livestock inside designated areas, but fencing of an appropriate type may also be used to exclude wild animals that may damage crops. Thus high fences are used to exclude animals such as deer from pasture or cropland, and rabbit-proof fences have been extensively used on individual farms in Australia and New Zealand to exclude rabbits from properties from which they have been eliminated or their numbers greatly reduced. In certain circumstances electric fencing may be effective and relatively cheap, but its maintenance may present problems.

On a larger scale, 'barrier fences' have been erected to prevent the movement of vertebrate pests from one region to another. In the latter part of the 19th century and the early part of the 20th century very long, nominally 'rabbit-proof', fences were erected beween and within several Australian states in an endeavour to stop the migration of rabbits from one part of Australia to another. They failed with rabbits (see p. 33), but the same or similar fences are still in use in some parts of Australia to control the movement of dingoes (the Australian wild dog).

Screening of fowl runs and fruit trees

On a still smaller scale than the individual farm, chickens and other domestic birds are usually kept in and carnivorous pests kept out by appropriate netting wire screens, and birds in zoological gardens are often kept within large wire-screened buildings. The same principle, usually in the form of screening erected temporarily at the time of fruiting, is sometimes used to exclude birds and fruit bats from orchards.

Screening of houses

In many tropical and warm temperate areas, insects such as flies and mosquitoes may be so numerous as to be a severe nuisance, or, as carriers of disease organisms such as malaria parasites, they may be dangerous to human health. An extreme response of human settlers to such risks has been to abandon settlements in tropical regions; this happened to the first three attempts by Europeans to settle in the Northern Territory of Australia (Price, 1930). A less extreme response was, and is, to screen dwellings against troublesome or dangerous insects with insect-proof wire mesh. This is the rule in houses for the more well-to-do residents in many parts of the tropics, as is the use of mosquito nets and the control of pest insects by other means. Screening may also be used to protect farmed small vertebrates against mosquito-borne diseases, such as domestic rabbits in parts of California where the local *Sylvilagus* rabbits are infected with myxoma virus, and since the introduction of myxomatosis into France, to protect domestic rabbits against that disease.

Scarecrows and sonic deterrents

Crude effigies of the human form, or of raptors such as eagles or owls, have long been used to frighten birds away from crops of fruit or cereals, with little effect unless they are associated with hostile sounds such as may be produced by guns, firecrackers or the like (Salmon and Marsh, 1991). In recent years attempts have been made to control animal pests, especially birds, in a non-lethal, non-toxic and humane way, by the use of sonic and ultrasonic devices. There are many such devices on the market but very few of them have been tested in a controlled way. The main problem appears to be habituation, a process that is likely to render any apparent successes transient. In a review of sonic deterrents, Bomford and O'Brien (1990) comment that: '... Sonic pest control devices should be viewed with considerable scepticism by legislators, pest controllers, and consumers. Few conclusive tests have been conducted of their efficacy, and even those devices that work may not be cost-effective ...'.

Use of predators

Predators have been used from time immemorial to control small vertebrate pests. Dogs were used by hunter-gatherers to warn and protect nomads against wolves and other large carnivores, and to help the nomads hunt. At the domestic level, cats have long been used to control rats and mice. In countries in which they were not endemic, predators have been released into the wild to control noxious animals, for example, the mongoose to control snakes and stoats and weasels to control rabbits (see pp. 31 and 48). The great reproductive capacity of rodents and rabbits suggests that few predatory species are likely to have a substantial effect on their numbers.

Use of poisons

Agrochemicals is a general term covering all chemicals used in agriculture. Poisonous chemicals (pesticides) have long been used for the control of all kinds of pests, being called insecticides when used for the control of insects, herbicides when used for the control of weeds and, for vertebrate pests, rodenticides (for rats and mice) or just poisons (for larger animals, such as foxes and rabbits).

Insecticides

The origins of chemical insecticides date back to the dawn of agriculture, when it became essential to preserve stored grains between seasons. Sulphur was used by the Sumerians about 2500 BC; in China, chalk and wood ash, and botanical products, were used for the treatment of stored grain from about 1200 BC, and arsenic was used as an insecticide in the second century BC. After centuries of use as elements of traditional folklore, the insecticidal properties of certain botanical products such as pyrethrum, nicotine and derris were recognized from about the 16th century. The early 20th century saw the standardization of petroleum oils and botanical products and the beginnings of the exploration of relationships between chemical structure and biological activity. The explosive development of chemical insecticides dates from about 1940, when the insecticidal properties of DDT (1,1,1,-dichlorodiphenyltrichloroethane) and BHC (benzene hexachloride) were discovered. A large number of different chemicals were tested for their insecticidal properties and many of these came to be used on a worldwide scale.

In the early 1950s the toxicity of many of these chemicals for vertebrates, including humans, and the presence of pesticide residues in food became matters of concern. At about the same time resistance of target species to the effects of some of the more widely used insecticides became a problem. Since the 1970s the use of a number of very effective agricultural insecticides, including DDT and the organophosphates, has been phased out, primarily because of concern for their danger to humans. They also became less effective, because of the development of resistance by many of the target species, and because they were only slowly degradable, they posed threats to fish and wildlife. The disillusionment with many chemical insecticides increased pressure for biological control, often conducted as part of a system of integrated pest management (see p. 52).

Herbicides

For as long as humans have practised agriculture, they have struggled to control weeds, by hand weeding or by mechanical methods such as hoeing or ploughing. The first chemical herbicide, 2,4-D, was introduced for large-scale use in 1947 and within 10 years some 90% of farmers in advanced farming areas depended on this and other herbicides for weed control. Herbicides do not pose as serious a risk to the health of humans or other animals as do insecticides, but they are not without danger, and for the most part they are non-specific in their effects and pose a threat to plants other than the target species.

Control of vertebrate pests by the use of poisons

Perhaps reflecting the minor importance to agriculture of vertebrate pests compared with insects and weeds, there is no general term to cover this category of chemical pest control; the term rodenticides reflects the universal human concern with rats and mice as pests. Early in the 20th century, in both New Zealand and Australia, strychnine was used in baits for rabbits and rats, phosphorized raspberry jam or pollard baits for rabbit control, and later warrens were gassed with carbon bisulphide, calcium cyanide or chloropicrin (Gibb and Williams, 1994). More recently warfarin (an anticoagulant) has been widely used for the control of rodents and sodium fluoroacetate ('1080') for the control of rabbits and foxes, both being administered in baits. However, these poisons are non-specific, and the widespread use of 1080 in the field poses serious risks to other animals that may take the baits.

Biological Control

If it works, one of the most cost effective methods of control of pests or parasites is biological control, i.e. control by means of a predator, an insect, a disease or, for some microbial parasites, a vaccine, because this method offers the possibility, although not the certainty, of being effective indefinitely. There are many examples of the successful biological control of insect pests and weeds, and vaccines are widely used to provide protection of humans and domestic animals against bacterial and viral diseases. Many bacterial diseases can be cured by antibiotics, that is by substances that are produced by fungi or bacteria and are poisonous for bacterial pathogens. Predators have been used, unsuccessfully, to control vertebrate pests. Although the bacterium *Salmonella* spp. has been used for rodent control (see p. 49), the only examples of effective biological control agents for vertebrate pests are myxoma virus and rabbit haemorrhagic disease virus for control of the rabbit. Most of this book is concerned with these diseases.

Evaluation of Pest Control Strategies

All methods of pest control involve costs as well as benefits. Not only does the design, performance and monitoring of the control programme cost time and money, but many vertebrate pests are commercially exploited. For example, in Australia the annual wholesale value of industries based on rabbits and feral horses, pigs and goats is estimated to be $A80 million. For this reason the role of the commercial use of vertebrate pests needs to be integrated into pest management strategy. In Australia, the Bureau of Resource Sciences has produced a series of books on managing vertebrate pests, comprising an overview volume, *Managing Vertebrate Pests. Principles and Strategies* (Braysher, 1993), and specific books on rabbits (Williams *et al.*, 1995), foxes (Saunders *et al.*, 1995) and various feral animals that are now pests in Australia. These provide valuable analyses of the costs and benefits of various methods of pest control. Details of the analysis of costs and benefits of the biological control of vertebrate pests are provided in Chapter 3 (see p. 51).

History of Methods of Control of Rabbits

In many parts of Europe rabbits were initially maintained as semi-domesticated animals in artificially constructed warrens, a word derived from the French *garenne,* now used in France to designate wild rabbits. Nevertheless, trapping was developed very early as a way of catching rabbits. In countries like Australia and New Zealand, after the initial enthusiasm of the wealthy landholders for the sport of hunting rabbits, they and ordinary farmers whose crops were being damaged sought to control their numbers. Initially they tried to do this by trapping, so that they could sell the carcasses and skins; later this was supplemented by methods aimed at destroying the pest, often salvaging the skins but not the carcasses. Methods of controlling rabbits are discussed at length in Chapter 2 (pp. 29–34), but the history of common methods of rabbit control will be described here.

Trapping

Bronze Age portable wooden traps have been found in peat bogs in Ireland and Wales (Lloyd, 1962) and iron traps identical to modern gin traps were in use by the mid-17th century (Thompson, 1994). The gin trap most commonly used in England and Australia for mammals varying in size from weasels to foxes had a flat spring set under tension, so that when an animal stepped on a treadle plate and released the spring, two 10-cm hinged jaws were clamped round the animal's leg. They were inhumane, since they captured the animal by the leg and it was not killed until the trap was inspected, many hours later. They were efficient, and could be used to eradicate rabbits locally, but more often they were used for 'rabbit farming', a system in which the trapper caught some 40% of the available rabbits and left the rest to breed up for next season's trapping. The trapper then moved on to another property.

In England there was a great deal of discussion about the inhumanity of the gin trap, and after passage of the *Pests Act, 1954* a Humane Traps Advisory Committee was appointed (Sheail, 1991). Several effective rabbit traps were developed and approved, the important feature of which was that they had two arms that struck the neck or head of the rabbit and killed it (Lloyd, 1963). In 1991 the European Community decided that as from 1995, the Community would ban the importation of fur from countries still using gin traps, which then included Australia and New Zealand (European Communities, 1991). Currently the situation differs in different Australian states, but as of April 1998 they are not yet banned in the Northern Territory, Western Australia and Queensland[1].

Hunting and shooting

Although in France rabbits are sometimes hunted with small dogs or ferrets by those aiming to destroy pests, hunting for sport, by shooting, is a major pastime and is not designed to control rabbits. Shooting is also practised as a sport on a large scale in Spain.

Shooting is a relatively 'humane' way of killing rabbits, and where rabbits are numerous in Australia and New Zealand it has long been practised by farmers in their spare time as a way of killing rabbits and providing the family table, and their dogs, with a meal. Since the use of gin traps was banned in many jurisdictions in Australia in the 1980s, wild rabbits for commercial use are usually field-shot and eviscerated by hunters and delivered to field chillers (Ramsay, 1994). When rabbits are few in number, shooting is sometimes practised, together with the use of dogs, as a preliminary to the ripping of warrens.

Biological control

The rabbit is unique amongst vertebrate pests in that during the last half century not one but two effective biological control tools have been exploited in Australia, where it is the most important agricultural and environmental pest animal. Following early attempts to use bacteria, the virus disease myxomatosis was suggested as a method of biological control by the Brazilian scientist H. de Beaurepaire Aragão in 1918 and eventually introduced into the Australian wild rabbit population in 1950 (see Chapter 6). Its initial effect was dramatic, but eventually its efficacy was reduced because of the development of genetic resistance in the rabbits. In 1984 a new lethal disease of rabbits, rabbit haemorrhagic disease, was reported in China and

Europe, and five years later investigations were commenced in Australia to see whether the causative virus could be used for biological control of rabbits in Australia and New Zealand. It was introduced in Australia in 1995 and very high kills were reported in the arid rangelands in some parts of Australia (see Chapter 11).

Endnote

[1]Basser Library Archives MS 143/25/5A. Letter from R.J. Downward to Fenner, 20 April 1998.

References

Bomford, M. and O'Brien, P.H. (1990) Sonic deterrents in animal damage control: a review of device tests and effectiveness. *Wildlife Society Bulletin* 18, 411–422.

Braysher, M. (1993) *Managing Vertebrate Pests: Principles and Strategies.* Australian Government Publishing Service, Canberra, 58 pp.

Cowan, P.E. and Tyndale-Biscoe, C.H. (1997) Australian and New Zealand mammal species considered to be pests or problems. *Reproduction, Fertility and Development* 9, 27–36.

European Communities (1991) Council Regulation (EEC) No. 3254/91. *Official Journal* 34, L 308/1.

Francis, G.W. (1862) *The Acclimatisation of Harmless, Useful, Interesting, and Ornamental Animals and Plants.* The Philosophical Society, Adelaide, 22 pp.

Gibb, J.A. and Williams, J.M. (1994) The rabbit in New Zealand. In: Thompson, H.V. and King, C.M. (eds) *The European Rabbit. The History and Biology of a Successful Colonizer.* Oxford University Press, Oxford, pp. 158–204.

Gosling, L.M. and Baker, S.J. (1989a) The eradication of muskrats and coypus from Britain. *Biological Journal of the Linnean Society* 38, 39–51.

Gosling, L.M. and Baker, S.J. (1989b) Demographic consequences of differences in the ranging behaviour of male and female coypus. In: Putman, R.J. (ed.) *Mammals as Pests.* Chapman & Hall, London. pp. 155–167.

Gosling, L.M., Baker, S.J. and Clarke, C.N. (1988) An attempt to remove coypus (*Myocastor coypus*) from a wetland habitat in East Anglia. *Journal of Applied Ecology* 25, 49–62.

Hone, J. (1994) *Analysis of Vertebrate Pest Control.* Cambridge University Press, Cambridge, 258 pp.

Lloyd, H.G. (1962) Humane traps. *The Review (June), Royal Agricultural Society of England,* pp. 15–16.

Lloyd, H.G. (1963) Spring traps and their development. *Journal of Applied Biology* 51, 329–333.

Lunney, D. and Leary, T. (1988) The impact on native mammals of land-use changes and exotic species in the Bega district, New South Wales, since settlement. *Australian Journal of Ecology* 13, 67–92.

Price, A.G. (1930) *The History and Problems of the Northern Territory, Australia.* A.E. Acott, Adelaide, 22 pp.

Ramsay, B.J. (1994) *Commercial Use of Wild Animals in Australia.* Australian Government Publishing Service, Canberra.

Rolls, E.C. (1984) *They All Ran Wild.* Angus & Robertson, Sydney. (An annotated and illustrated version of a book of the same name published in 1969).

Salmon, T.P. and Marsh, R.E. (1991) Bird hazing and frightening methods and techniques (with emphasis on containment ponds). Department of Wildlife and Fisheries Biology, University of California, Davis.

Saunders, G., Coman, B., Kinnear, J. and Braysher, M. (1995) *Managing Vertebrate Pests. Foxes.* Australian Government Publishing Service, Canberra, 141 pp.

Sheail, J. (1991) The management of an animal population: changing attitudes towards the wild rabbit in Britain. *Journal of Environmental Management* 33, 189–203.

Thompson, H.V. (1994) The rabbit in Britain. In: Thompson, H.V. and King, C.M. (eds) *The European Rabbit. The History and Biology of a Successful Colonizer.* Oxford University Press, Oxford, pp. 64–107.

Thomson, G.M. (1922) *The Naturalization of Animals and Plants in New Zealand.* Cambridge University Press, Cambridge, 607 pp.
Williams, K., Parer, I., Coman, B., Burley, J. and Braysher, M. (1995) *Managing Vertebrate Pests: Rabbits.* Australian Government Publishing Service, Canberra, 284 pp.
Wodzicki, K.A. (1950) *Introduced Mammals of New Zealand.* Bulletin No. 98, Department of Scientific and Industrial Research, Wellington, New Zealand, 255 pp.

2

The Rabbit

Overview

Since the development of agriculture some 10,000 years ago, man's major vertebrate pests have been animals that feed on crops or pasture plants. Initially rats and mice were the villains, principally because of damage to stored products, but with the changes in ground cover in western Europe and their introduction into new habitats like Australia, European rabbits (*Oryctolagus cuniculus*) have become a major agricultural pest[1].

The family Leporidae contains 11 genera and 43 species. Only five species, belonging to three genera, are of interest for this history: *Oryctolagus cuniculus,* the European rabbit, *Lepus europaeus,* the brown hare, and three American species belonging to the genus *Sylvilagus,* namely *S. brasiliensis* and *S. bachmani,* the natural hosts of myxoma virus, and *S. floridanus,* the natural host of Shope's fibroma virus.

Until the middle of the 19th century the European rabbit was prized as a game animal more than it was cursed as an agricultural pest. That picture changed when cropping and plantation forestry were developed more intensively in Britain and France and especially during the Second World War. However, wild rabbits have remained an important game animal in many European countries, especially France and Spain. In addition, there is a large commercial rabbit industry in several European countries and in China.

Besides being placed on many uninhabited islands to provide emergency food for shipwrecked sailors, rabbits were introduced to novel and highly favourable environments in Australia, New Zealand and Chile during the mid-19th century. Within 30 years they had spread over most of temperate eastern Australia, and 20 years later they inhabited every part of Australia in which climate and soil were suitable. They became established in many parts of New Zealand, but they failed to become common in mainland Chile until about 1950. In all three countries they were regarded as a pest by pastoralists, but a trade based on their meat and skins grew up, and rabbit trapping became an important industry.

Laws to control rabbits were enacted within 15 years of their introduction into Australia and a great variety of methods were used to control them: trapping, shooting, poisoning, destruction of burrows, and the erection of rabbit-proof fencing. None of these measures was more than palliative; it was not until myxomatosis spread in south-eastern Australia in 1951 that a cheap and effective control measure seemed to have been found. However, it was important that the good kills produced by successful biological control should be followed up by conventional methods of control, especially warren destruction.

The Family Leporidae

The family Leporidae, comprising the rabbits and hares, is one of two families in the Order Lagomorpha (Corbet, 1983, 1994). It contains 11 genera, only three of which are significant as far as myxomatosis is concerned, namely *Oryctolagus, Sylvilagus* and *Lepus* (Fig. 2.1). The genus *Oryctolagus* contains only one species, *Oryctolagus cuniculus.* Before it was more widely distributed around the world during the European colonial expansion (Flux, 1994), it was confined to Europe and north-western Africa. All 12 species of the genus *Sylvilagus* occur only in the Americas; three species are significant in the natural history of myxomatosis: *S. brasiliensis* and *S. bachmani* are natural hosts of myxoma virus and *S. floridanus* is the host of the related leporipoxvirus, Shope's fibroma virus. Only one of the 20 species in the genus *Lepus*, the European hare (*Lepus europaeus*), is relevant to the story of myxomatosis, and only marginally so.

Oryctolagus cuniculus

The European rabbit (*O. cuniculus*), which henceforth we shall call 'the rabbit', is the only member of that genus and the only rabbit found in Europe. It is also the only leporid that lives in burrows. From fossil records it appears to have evolved in south-western Europe, bones from Spain, southern France and Swanscombe, in Kent, dating back to the Middle Pleistocene.

Fig. 2.1. Leporids of importance as hosts of myxoma and rabbit fibroma viruses. (a) European hare (*Lepus europaeus*). (b) European rabbit (*Oryctolagus cuniculus*). (c) Eastern cottontail (*Sylvilagus floridanus*). (d) Tapeti or tropical forest rabbit (*Sylvilagus brasiliensis*). (e) Brush rabbit (*Sylvilagus bachmani*). From Fenner and Ratcliffe (1965), with permission.

Rabbits died out in the northerly part of their distribution during the Pleistocene glaciations, but subsequently extended progressively northwards from a relict population on the coast of the Mediterranean. The Phoenicians, sailing to the Iberian Peninsula in 1100 BC, noted large numbers of rabbits, and gave the country their name for rabbits, 'Sphania', which translated to Hispania and thus Spain (Rogers *et al.*, 1994).

Later rabbits were domesticated and bred in large numbers as a source of food. Many 'fancy' breeds were produced (Fox, 1974), and when laboratory medicine was developed in the mid-19th century certain breeds, notably the 'New Zealand White' became favoured experimental animals, especially for producing antisera (Weisbroth *et al.*, 1974). Laboratory rabbits taken to South America towards the end of the 19th century contracted a previously unknown and very severe infection, almost invariably lethal, which was called 'myxomatosis' and shown to be caused by a novel virus which was called myxoma virus (*myx-oma* = mucinous tumour).

Sylvilagus brasiliensis, S. bachmani, *and* S. floridanus

The appearance of these leporids is illustrated in Fig. 2.1 and their distribution in the Americas in Fig. 2.2. *S. brasiliensis,* the tapeti or tropical forest rabbit, is the natural host of the strains of myxoma virus that first brought myxomatosis to the attention of scientists and that were later used in releases of the virus in Australia, Europe and Chile. *S. bachmani,* the brush rabbit of the western United States, is the natural host to another subtype of myxoma virus, some strains of which are more rapidly lethal for European rabbits than the South American strains. In both *Sylvilagus* species myxoma virus produces merely a small fibroma in the skin. *S. floridanus,* the Eastern cottontail, is the natural host of a related virus, rabbit fibroma virus, which produces benign fibromas in the skin of both cottontail and European rabbits. Inoculation of European rabbits with fibroma virus provides substantial protection against myxomatosis.

Lepus europaeus

The European hare is much larger than any of the foregoing rabbits and is a surface-dwelling animal highly adapted for running. It is of peripheral importance in relation to myxomatosis, in that it is the only animal outside of the Americas from which a leporipoxvirus has been isolated (see p. 70), and very occasionally cases of myxomatosis have occurred in this species (see p. 73). Although hares were difficult to maintain on the long sea voyage, they were successfully introduced into areas near Melbourne and Adelaide in the 1860s (Rolls, 1984), both for coursing and as a game animal. Over the next 40 years they spread through the temperate south-eastern corner of Australia to reach their present distribution (see Fig. 2.7c, p. 22) by about 1900, spreading at a rate of about 60 km a year (Myers *et al.*, 1989). Hares are valued as game animals in Europe, but most Australians find the strong gamey flavour of the meat unacceptable, and they are regarded as pests, but of minor significance compared with rabbits.

The Spread of the Rabbit

The spread of rabbits in Europe and around the world is well described, for Europe, by Rogers *et al.* (1994) and Thompson (1994), and worldwide, by Flux (1994). The worldwide spread of rabbits coincided with European colonization of lands with temperate climates. Early introductions consisted of domestic rabbits, which usually failed to become established in large land areas. However, liberations of these rabbits on islands, as food for shipwrecked sailors or for sport, were often successful, and there are now rabbit colonies on some 800 islands in all the major oceans (Fig. 2.3).

Europe

Phoenicians, and later the Romans, translocated rabbits to various places around the Mediterranean, and they were a favourite food of the Romans, who kept them in enclosures, roofed to make them 'impenetrable to cats, badgers, wolves and eagles' (Barrett-Hamilton, 1912). In about

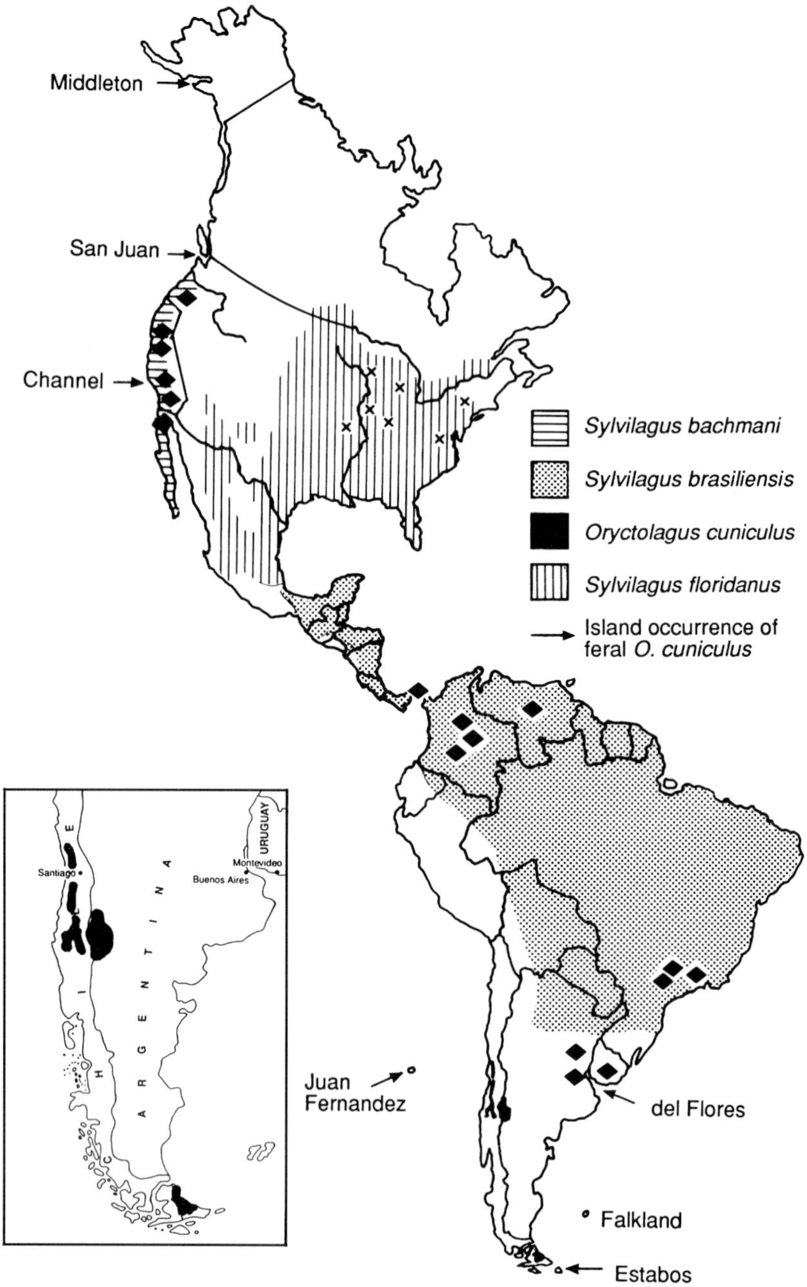

Fig. 2.2. Distribution in the Americas of wild rabbits relevant to leporipoxvirus infections, showing locations where myxoma virus (diamonds) and fibroma virus (crosses) have been isolated. The species concerned are feral *Oryctolagus cuniculus* (see Fig. 2.3 also, for occurrence on islands around the Americas); *Sylvilagus brasiliensis,* the natural host of myxoma virus in South America; *Sylvilagus bachmani,* the natural host of myxoma virus in North America; and *Sylvilagus floridanus,* the natural host of rabbit fibroma virus in North America. From Marshall[12] updated with data from J.P. Fullagar, personal communication (1978). Below left, enlargement of distribution of feral *Oryctolagus cuniculus* in Chile and Argentina. From Flux (1994), with permission.

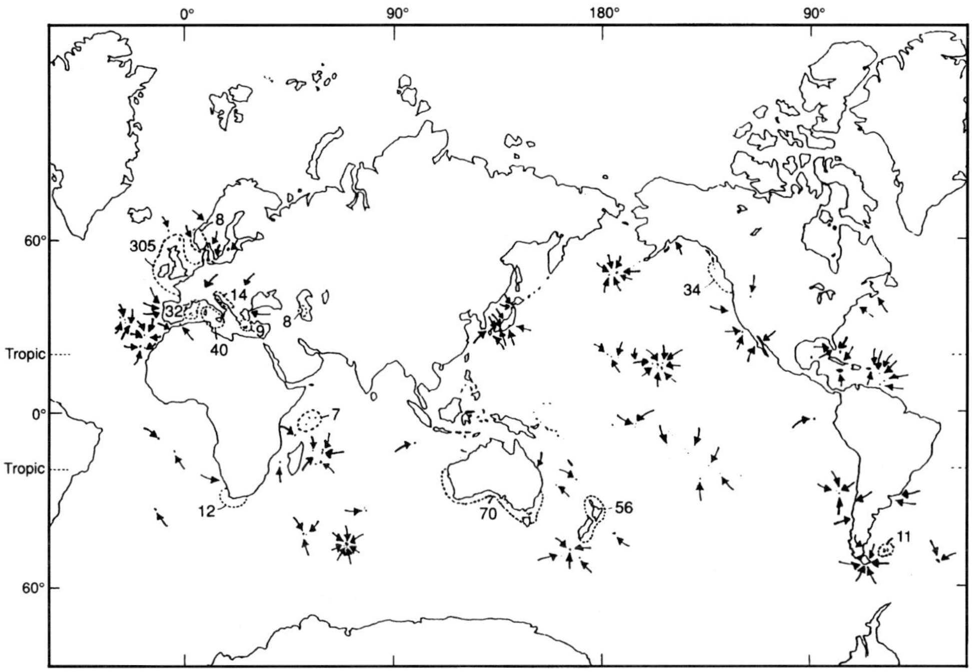

Fig. 2.3. Distribution of rabbits (*Oryctolagus cuniculus*) on islands around the world. Numbers indicate total number if islands within the areas indicated. Arrows point to separate islands. From Flux (1994), with permission.

30 BC Strabo wrote of a plague of rabbits in the Balearic Islands that caused so much damage that the settlers there petitioned the Emperor Augustus for help (Schwenk, 1986).

Rabbits were domesticated in the Middle Ages by monks in France, for the quaint reason that newborn rabbits ('laurices') were considered to be aquatic, and could therefore be eaten during Lent. They were introduced into Britain in the 12th and 13th centuries and became a favoured item of diet for feasts. In general, they were looked after carefully in warrens, which were often protected by stone walls (Fig. 2.4). By the late 18th century, however, there were also wild rabbits in most counties in Britain except Wales, and by this time they had become somewhat of a nuisance to farmers although still valued as a game animal. The present distribution of wild rabbits in Europe and north Africa is shown in Fig. 2.5.

Rabbits are now an important domestic animal in many countries in Europe, especially in France and Italy, and many distinctive breeds have been developed. They are a source of meat and fur, of a kind that is especially useful for making felt hats. As well as being a widely used laboratory animal, many distinctive breeds have been developed for the pet trade. Wild rabbits are greatly valued by sporting shooters (chasseurs), especially in France, where they are also regarded as a pest by agriculturalists and foresters, hence the mixed reactions to the release of myxoma virus in France in 1952 (see p. 213).

Australia

The introduction of rabbits into Australia and their spread around the continent has been described in detail, with full references, by Rolls (1984) and Stodart and Parer (1988), from which this abridged account is derived. Early introductions

Fig. 2.4. Ditsworthy Warren, near Plymouth, showing the artificial mound erected to provide a location for a warren, and part of the wall around the warren. Until relatively recently, rabbits were kept within such warrens and periodically harvested; annual crops of up to 100 rabbits an acre were said to be possible. From Thompson (1994), with permission.

consisted of small numbers of domesticated rabbits, which usually died out if they escaped. Five such rabbits were brought out with the First Fleet in 1788, and unrecorded importations were made on many subsequent convict or supply ships. In 1825 domestic rabbits were bred around houses in Sydney, and small numbers were brought from England or occasionally from other parts of Australia into Hobart (Tasmania), Perth (Western Australia), Portland (Victoria) and Adelaide (South Australia) shortly after each of those towns were established. Although there were few effective predators, these importations failed to establish rabbits in the wild, except near Sydney and in southern Tasmania, where feral domestic rabbits were common in the 1820s, but did not become a significant pest until 1870,

Fig. 2.5. Distribution of wild rabbits in Europe and North Africa. From Flux (1994), with permission.

possibly as a result of an importation of a genetically distinct wild strain at that time (Richardson *et al.*, 1980).

Even before the establishment of acclimatization societies in the Australian colonies and in New Zealand in the 1850s and 1860s, there was much enthusiasm for introducing game animals, including rabbits. Among the keenest was Thomas Austin of Barwon Park[2], a sheep station near Winchelsea, in western Victoria. After several unsuccessful attempts to establish domestic rabbits imported from England, he introduced two dozen pairs of wild-caught rabbits (possibly reduced to 13 animals by the time of their arrival) on the brig *Lightning,* which berthed in Melbourne on Christmas Day 1859. These rabbits were first housed in a warren at Barwon Park, and a few years later, when they had become established and were breeding well, animals were sent to friends in other country districts. By 1865 they had multiplied sufficiently at Barwon Park for 6000 to be harvested in eight months; and in 1864 Austin sent consignments to New Zealand. He often held shooting parties, and in 1867 Prince Alfred, Duke of

Edinburgh, participated in a successful shoot there (Fig. 2.6). Aided by the opening up of land for sheep farming, by this time they had spread 55 kilometres to the east and 20 kilometres to the north-west of Austin's property. Most of the landed gentry continued to encourage further liberations, but a few people foresaw what might happen. Thus in 1869 a member of the Victorian Parliament, Mr Connor, prophesied that 'the rabbit nuisance in this colony promised to be as great as that of the locusts in the land of Egypt', and unsuccessfully moved to introduce control measures into the Local Government Bill.

In 1870 rabbits, possibly of a different stock from those introduced by Austin, were released near Kapunda, in South Australia, and the main spread stemmed from Barwon Park and Kapunda in the years following 1875 (Fig. 2.7a). By 1880 they had crossed the River Murray into New South Wales and joined up with those spreading from Sydney; ten years later they had entered Queensland, moving at a rate of about 150 km a year across New South Wales. Movement westwards across the very arid country of northern South Australia was slower, but they had entered Western Australia near Eucla by 1894, and reached the good agricultural lands in the south-west corner by 1910. Their current distribution in Australia is illustrated in Fig. 2.7b. The extent of the infestation and the damage to crops and pasture done by the rabbits can be gauged by the number of books dealing specifically with the rabbit pest that were published in the late 19th

Fig. 2.6. Rabbits in Australia before they were recognized as a pest. Prince Alfred, Duke of Edinburgh, at a rabbit shoot at Barwon Park in December 1867. Just seven years after Thomas Austin of Barwon Park had introduced the wild rabbit into Australia, Prince Alfred, the second son of Queen Victoria, shot 416 rabbits in three and a half hours. From *The Illustrated Australian News*, 27 December 1867. National Library of Australia.

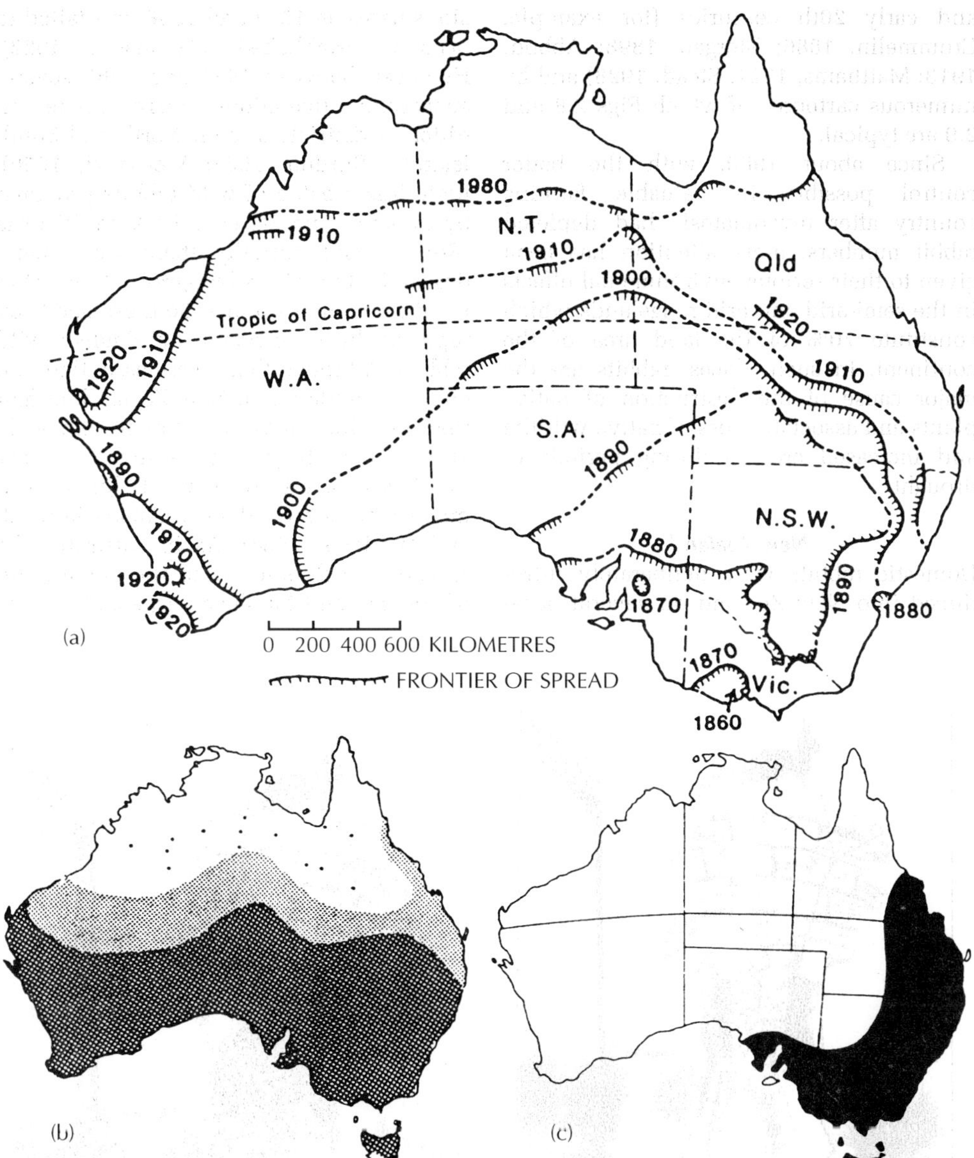

Fig. 2.7. Rabbits and hares in Australia. (a) The spread of the European rabbit over the mainland of Australia after the introduction of wild rabbits from England to Barwon Park in Victoria in December 1859. The arrow above '1860' indicates the locality of Barwon Park; the ring above '1870' in South Australia indicates the locality of Kapunda, the other significant centre from which spread occurred. (b) Present distribution of rabbits in Australia. (c) Present distribution of hares in Australia. (a) from Stodart and Parer (1988), with permission. (b) and (c) from Myers *et al.* (1989), Commonwealth of Australia copyright reproduced by permission.

and early 20th centuries (for example, Crommelin, 1886; Morgan, 1898; Abbott, 1913; Matthams, 1921; Stead, 1928) and by numerous cartoons, of which Figs 2.8 and 2.9 are typical.

Since about 1980, with the better control possible in valuable farming country after myxomatosis had depleted rabbit numbers, more attention has been given to their serious environmental effects in the semi-arid and arid rangelands which constitute 70% of the land area of the continent. In such places, rabbits are the major cause of the destruction of native plants and associated loss of native wildlife and increased erosion during periods of drought.

New Zealand

Domestic rabbits were deliberately introduced into New Zealand on several occasions between 1838 and 1858, but failed to become established (Thomson, 1922). However, between 1864 and 1867 several successful liberations were made in different districts of both North and South Islands (Burdon, 1938; Wodzicki, 1950), including a batch of wild rabbits provided by Austin from Barwon Park in Victoria (Rolls, 1984). Initially there were many descendants of silver-grey (domestic) rabbits, but after about 1880 these had been replaced by descendants of English wild rabbits. Although there were essentially no effective predators in New Zealand, at first they remained localized, but by the early 1870s they began spreading to other districts at a rate of about 16 km a year, much more slowly than in Australia (Gibb and Williams, 1994). As in Australia, the spread of rabbits followed the opening up of grazing land for sheep, especially in the

Fig. 2.8. 'Backcountry squatter, A.D. 1892'. Cartoon drawn by Alexander Campbell and published in the Supplement to the *Australasian Pastoralists Review*, 15 August 1892. It was produced during a major drought and a period of great economic depression. The original of this particular copy was annotated in French, probably by Loir or one of his colleagues, to explain the meaning of the burdens on the squatter's back. As illustrated on the right, the cartoon was republished in France, with a legend explaining the numbers[13].

Fig. 2.9. 'The King is dead; Long live the King'. Cartoon in the Supplement to the *Australasian Pastoralists Review,* 15 March 1893.

dry tussock grasslands of Otago. In 1874 '... their enormous numbers burst upon man's realization ...' (Burdon, 1938). Although a large export trade in skins developed rapidly (33,000 in 1873, about 1,000,000 in 1877 and over 9,000,000 in 1882), their pest potential was soon realized and their damage caused the abandonment of many sheep runs in Otago. In 1876 Parliament appointed a Select Committee to inquire into the Rabbit Nuisance and various steps were taken to bring them under control. As in Australia, cessation of control measures during the Second World War led to another explosion in their numbers, such that the value of rabbit carcasses and fur exported from New Zealand in 1946 was one-third of the total value of those exported from the whole of Australia (Fennessy, 1958).

Passage of the Rabbit Nuisance Amendment Act 1947 led to 'decommercialization' of the rabbit and eventually to its effective control in all but a few areas, primarily in hilly country in Central Otago, without recourse to the use of myxoma virus (Gibb and Williams, 1994).

North America

Besides the setting up of hutch rabbits for the production of meat and fur and the use of rabbits in biomedical laboratories, many attempts were made to introduce rabbits into the wild in the United States, but the vast majority were unsuccessful, except for liberations on small islands along the coasts and in the Caribbean (see Fig. 2.3, p. 16). The only mainland populations are in San Clemente Canyon Natural Park, San Diego, where there is a small feral stock of wild-type and piebald rabbits, near Marblemount and Seattle, Washington, and in some islands in the Snake River, Idaho (Flux, 1994). In response to the concerns of wildlife administrators and biologists in the United States, Thompson (1955) warned against attempts of sportsmen's clubs to introduce onto the mainland the 'San Juan' rabbits (*Oryctolagus cuniculus*), which had been released on the island of San Juan, off the coast of the State of Washington, about 1900 (Kirkpatrick, 1959). Attempts to establish European wild rabbits on the mainland continued, especially by beaglers in Indiana, and in 1959 the Indiana Department of Conservation issued a long report recommending that their importation into Indiana should be banned (Kirkpatrick, 1960).

South America

Rabbits were introduced from France via the Falkland Islands into the Beagle Channel islands in 1880. The infestation of Tierra del Fuego is said to have originated by the release of two pairs of rabbits near the port city of Porvenir in 1936 (Arentsen, 1953). By 1940 the rabbit was established near Porvenir and on the Chilean mainland near Punta Arenas. In 1950 rabbits were introduced near Ushuaia, near the Beagle Channel islands, and quickly became a pest, such that an area of 1 million hectares supported 30 million rabbits: '... there were so many rabbits that the whole countryside seemed to get up and move as one drove along' (Goodall, 1979).

Attempts were made to introduce rabbits from Spain into central Chile from the mid-19th century, but as late as 1940 they were not abundant. However, by 1960 they were present throughout Malleco province in Central Chile (Greer, 1966) and quickly increased to plague numbers. Between 1945 and 1950 they moved across the Andes into Argentina (Howard and Amaya, 1975), where their range has been expanding at about the rate of 15–20 km a year to occupy 45,000 km^2 by 1984 (Bonino and Amaya, 1984; see Fig. 2.2, p. 16), despite the use of myxomatosis (which was illegal in Argentina but not in Chile) by Argentinian landholders in 1971 and 1972. There seems no reason to doubt that they will become a pest in Argentina wherever there are favourable habitats, although it is likely that enzootic myxomatosis in *S. brasiliensis* will prevent their establishment where that animal occurs in north-east Argentina.

Comparing rabbits from central Chile and those from their place of origin in Spain (Housse, 1953), Jaksic and Fuentes (1991) noted that the rabbits in central Chile were larger in size and had larger litters and a longer lifespan than those in Spain. They ascribed their increased life expectancy and greater density in Chile to the low levels of predation compared with the situation in Spain and the absence of myxomatosis (the host range of *Sylvilagus brasiliensis* does not include Chile).

Africa and Asia

Rabbits were probably introduced into north-western Africa by the Romans and are currently found in Algeria and Morocco (Flux, 1994), but they do not occur elsewhere in Africa. Although pikas (*Ochotona* spp.) are common in some parts of China and have been domesticated, the only European rabbits found in Asian countries are domestic rabbits of various breeds. The People's Republic of China is now the world's largest exporter of meat from domestic rabbits, exports rising from 308 tonnes in 1957 to 53,200 tonnes in 1983, since when exports have diminished somewhat because of increased domestic consumption (Feng-Yi, 1990). In 1984 Angora rabbits in China came into the news as the source of a virulent virus that caused a highly lethal haemorrhagic disease, later identified as being caused by a calicivirus, rabbit haemorrhagic disease virus, which was accidentally exported to Europe in 1986 and to Mexico in 1989 (see p. 237).

Wild Rabbits as a Resource

One factor mitigating against rabbit control has been the pressure from groups exploiting wild rabbits for sport or commercial purposes. In France, particularly, there were controversies after the introduction of myxomatosis in 1952. The powerful hunting organizations tried hard to protect wild rabbits against the disease, even to the extent of vaccinating them, whereas foresters and most farmers welcomed the destruction of a major pest. However, some farmers found it more profitable to maintain wild rabbits for shooting organizations than to use their farms for agricultural production. There was also a large rabbit breeding industry, on a commercial and a family scale, and rabbit breeders naturally wished to minimize the impact of myxomatosis on their animals, so they vaccinated them.

In Australia the position was different. Rabbits were originally introduced for sport and were spread from one property to another by wealthy landowners (who were called squatters because they had 'squatted' on the land that they occupied). However, the squatters soon realized their pest potential and they led the pressure groups calling for rabbit control. But rabbits were also a resource for meat and fur, especially fur for felt hats, and pressure from the carcass and fur trade played a role (although not the deciding role) in preventing the introduction of myxomatosis in 1919–1920 (see p. 117). During the Great Depression of the 1930s, rabbits were a valued source of meat for many families and rabbiting was a common occupation for the unemployed.

In the years immediately following the Second World War, some 100 million rabbits passed through human hands each year in the form of carcasses or skins for export alone, with unknown but considerable numbers entering the domestic market or being killed but not recovered. It was therefore not surprising that rabbiters physically threatened CSIRO[3] field workers in the early days of the spread of myxomatosis. With the advent of myxomatosis the supply of wild rabbits for commercial use was severely reduced. Concurrently, improvements in husbandry practices in commercial rabbitries in Europe and China introduced strong competition in the export market. The export of wild rabbit products (meat and skin) was variable because of the fluctuation of rabbit numbers due to droughts, and was greatly reduced after the spread of myxomatosis in the early 1950s (Fig. 2.10).

The position on the commercial use of wild rabbits in Australia in the 1980s was reviewed by Ramsay (1994), and in the 1990s by Foster and Telford (1996). In 1989 about 3 million wild rabbits were harvested, 2.5 million of which were used for the domestic market, amounting to 1,800–2,000 tonnes of meat, valued at $A5–5.6 million, figures that remained much the same through the 1990s. The remaining 400 tonnes used for export compares with a world production of meat from wild and farmed rabbits of about 1,250,000 tonnes. About 200 tonnes of dried skins, valued at about $A1 million, were used for the manufacture of felt hats.

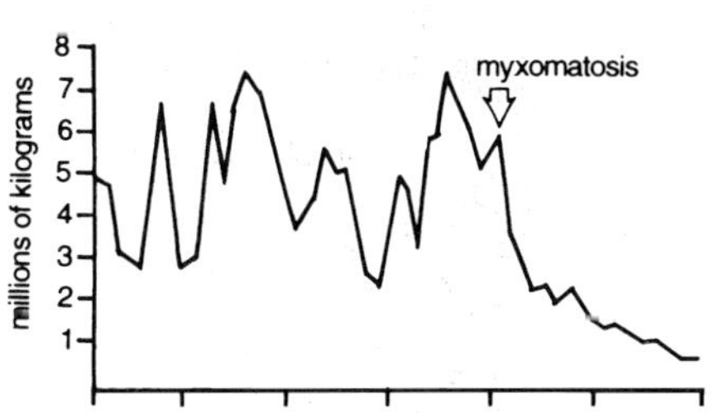

Fig. 2.10. Rabbit skins exported from Australia, in millions of kilograms, between 1910 and 1970, showing the abrupt fall after the spread of myxomatosis in the early 1950s.

More recently attention has been directed to the importance of wild rabbits to Aboriginal communities in the arid outback. Many Aboriginals living in Central Australia hunt wild rabbits for food (Hetzel, 1978), and they also sell rabbits to their community stores and to commercial processors supplying domestic and export markets (Wilson *et al.*, 1992). Currently, rabbits are relatively well controlled in the well-watered parts of the continent and occur in their greatest numbers in the arid rangelands (Fig. 2.11a). These are the areas of greatest importance to outback Aboriginals, and also constitute some of the important sources of rabbit harvesting for commercial purposes (Fig. 2.11b). When it was suggested that rabbit haemorrhagic disease virus should be introduced for rabbit control (see Chapter 11), Central Australian Aboriginals[4] and the fur and carcass trades were active in opposing its use. The success of rabbit haemorrhagic disease virus in the outback areas in 1996 led to a collapse of the trade in wild rabbit carcasses and fur, and within months there was an explosion of applications from struggling farmers in Western Australia and New South Wales (the only states where rabbit farming is legal) to set up rabbit breeding farms.

Rabbit Control in New Zealand

Although rabbits became a major agricultural pest in New Zealand at much the same time and for the same reasons as they did in Australia, the pattern of control in New Zealand was quite different from that in Australia. The following summary is drawn from the detailed account of the rabbit in New Zealand by Gibb and Williams (1994). From early days New Zealanders had tried control by predators (stoats, weasels, ferrets and cats) on a larger scale than in Australia, but there is no evidence that these animals played an important part in reducing rabbit numbers. Rabbit-proof barrier fences were erected in

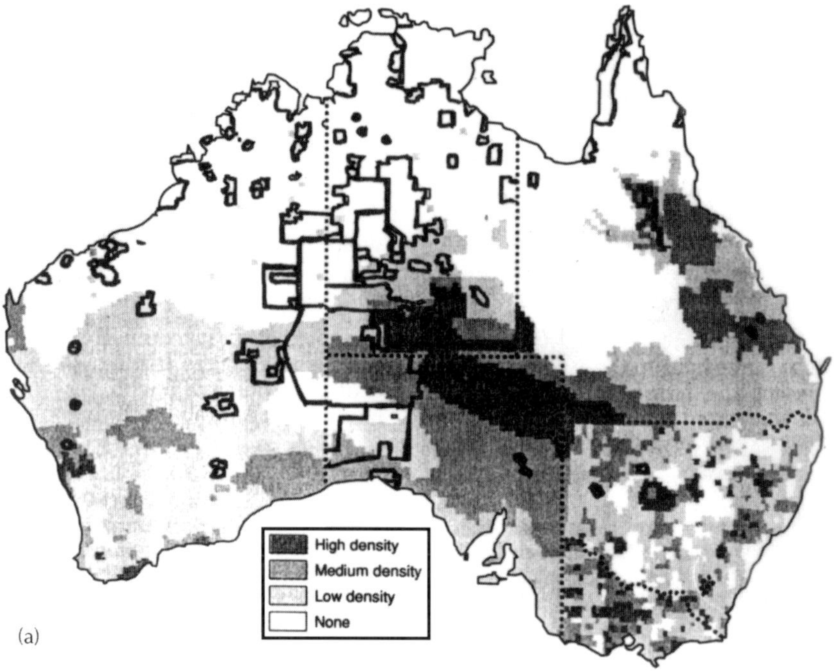

(a)

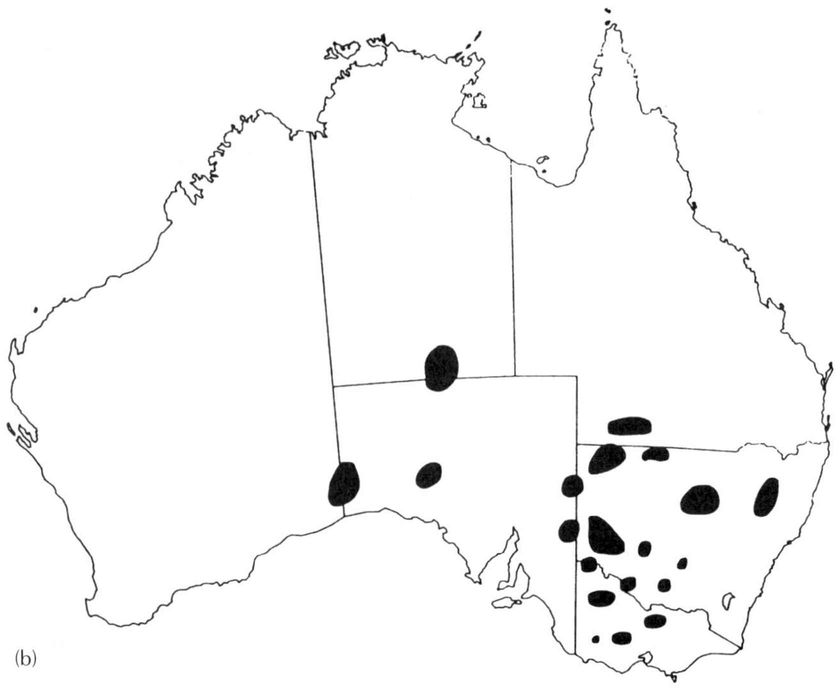

(b)

Fig. 2.11. Maps of mainland Australia showing (a) the present (1990) density of wild rabbit infestation in relation to Aboriginal lands, which are indicated by the rectangular lines; (b) regions where commercial rabbit harvesting occurs, as assessed by rabbit processors and field agents. (a) From Wilson *et al.* (1992), Commonwealth of Australia copyright reproduced by permission; (b) from Ramsay (1994), Commonwealth of Australia copyright reproduced by permission.

both North and South Island, but were ineffective. The first Rabbit Nuisance Act was passed in 1871 but few Rabbit Boards were set up, partly because of the extent of commercial farming of wild rabbits. Initially, when the sale of skins gave some cash return, poisoning was the most popular control method and strychnine the most popular poison. However, by the middle 1940s the rabbit problem was as bad as it ever had been; rabbits were out of control in many areas, extensively 'farmed' for their skins in some, but held in check in a few areas. Over 95% of the export trade in rabbit products was in rabbit skins (Wodzicki, 1950), and in the 1940s it grew to be quite substantial, but dropped to almost zero after the introduction of decommercialization.

'Decommercialization' of rabbits

The number of Rabbit Boards increased substantially after the Second World War and in 1947, as a result of representation from the North and South Islands Rabbit Boards' Associations, the Rabbit Nuisance Amendment Act 1947 was passed. Under this Act a Rabbit Destruction Council was appointed and all Boards except two were required to adopt a 'killer' policy. This required the dramatic and enforced abolition (decommercialization) of the rabbit industry, such that the volume of skin exports in 1955 was less than 5% of the volume in 1948. The principal method of control has been by poisoning, primarily with arsenic or phosphorus, but since 1956 also with '1080' (sodium fluoroacetate). A large aerial agricultural top-dressing industry has developed in New Zealand and aerial baiting is extensively used, using the same aeroplanes. Between 1948 and 1964, after decommercialization, the number of Rabbit Boards increased from 100 to 200, the area covered from about eight to 18 million hectares and the annual cost of rabbit destruction stabilized at about $18 million (in 1983 $NZ). Financial benefits from improved farm production of up to $NZ33 million annually were claimed.

However, the success of the early years fuelled complacency, and the Rabbit Destruction Council's resolve to 'kill the last rabbit' weakened and lost farmer support. The eradication policy was abandoned in 1971 and gradually rabbit control was integrated into vertebrate pest management and special government support was phased out. To the surprise of most farmers, in most places the populations of rabbits did not explode when control was withdrawn in the mid-1980s, probably because farming practices had substantially changed the environment so that it was less suitable for rabbits. A few areas remain where rabbit control will always be needed, notably 'semi-arid' tussock grasslands of South Canterbury and Central Otago, both areas where physical conditions restrict land development but enhance rabbit survival. In May 1985 an Order in Council permitted the import and farming of ten specified breeds of domestic rabbits, provided that they were free of disease, securely housed and prevented from grazing pasture (Gibb and Williams, 1990).

Trials of myxomatosis

Following the spectacular spread of myxomatosis in Australia in 1950–51 (see p. 138), attempts were made to introduce myxomatosis into New Zealand in 1951–53 (Filmer, 1953). Several batches of captured rabbits were inoculated with the strain of virus used in Australia and either released or held in wire netting enclosures at each of 22 sites, at periods between November 1951 and February 1952 (Fig. 2.12). Some naturally infected rabbits were seen at some sites, but nowhere was there a significant decrease in rabbit numbers. Because the 1951–52 summer was unusually cold and wet, the trial was repeated next summer at the eight most promising sites, this time releasing all inoculated animals in heavily rabbit-infested areas. Although mosquitoes and sandflies were plentiful at some sites during the trials, the results were disappointing, and it was concluded that it was unlikely that myxomatosis would be of any use for rabbit control in New Zealand.

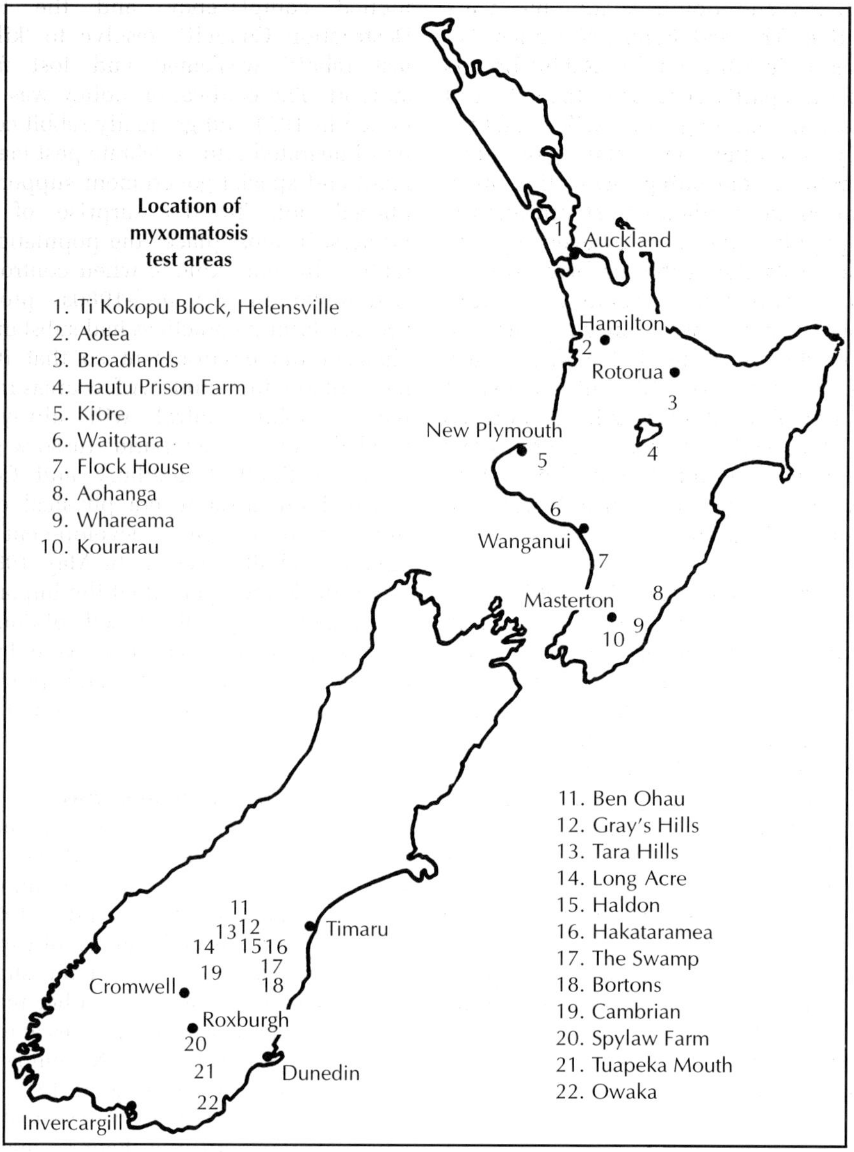

Fig. 2.12. Map of New Zealand showing sites where rabbits infected with myxoma virus were released in 1951–53. From Filmer (1953), with permission.

In the 1980s another attempt was made to introduce myxomatosis to control rabbits in the tussock grasslands of Central Otago[5], with support from some Australian scientists (Sobey, 1982). This was countered by New Zealand rabbit control experts (Gibb and Flux, 1983), who concluded that it was 'unwise to risk upsetting the present effective system for the dubious benefits of introducing myxomatosis', and in 1985, after receiving public comment, the Government refused to sanction importation of the virus[6]. In 1991 the New Zealand Federated Farmers again sought authority to introduce myxomatosis and the European rabbit flea into the tussock grasslands, but in 1993, after public comment on the proposal, permission was again refused. The general public distaste for myxomatosis was reinforced by the fear that the rabbit flea might spread to a New Zealand icon, the kiwi, a flightless bird[7].

Since 1989 the New Zealand Government has been collaborating with the Australian Government in trials of rabbit haemorrhagic disease virus as a possible means of biological control (see p. 246). Although the virus was officially released in Australia in 1996, in June 1997 the New Zealand government decided against its use. However, it was almost immediately introduced illegally by farmers and now occurs among wild rabbits in both South and North Islands.

Rabbit Control in South America

Separate introductions of European rabbits were made into central Chile, where initially they failed to thrive, and the Chilean part of the island of Tierra del Fuego, where they soon became a plague. A fox, *Dusicyon culpaeus,* had long been present there but made little impression on rabbit numbers. The sheep farmers attempted to control the rabbits by hunting and trapping and then by using cyanide gas. In 1951 twelve pairs of the fox *D. griseus* were introduced from the mainland. Jaksic and Yanez (1983) tried to evaluate the possible role of foxes in biological control by analysing their food. As in central Chile, *D. culpaeus* was a better hunter of rabbits than its congener, but neither made any impression on rabbit numbers.

Inoculation campaigns with myxoma virus, using a strain from Brazil, were initiated in Tierra del Fuego in 1954 (Sauer, 1954). The dense population, estimated at some 30 million animals in 1953 (Jaksic and Yanez, 1983), was decimated, but slowly increased again over the next 20 years, to an estimated 5% of the 1953 peak level. It was thought to be spread by direct contagion rather than by vectors (Sauer, 1954). It spread naturally to the Argentinian side of the island, where it was widely used by farmers. As a result, rabbits virtually disappeared from the treeless part of the north of the island, but persisted in the woodlands in the central and south[8]. In 1968 it was proposed to control rabbits in the Argentinian part of the island by the official use of myxomatosis, but this was forbidden by Argentinian national legislation, because of the extensive domestic rabbit trade in the Buenos Aires province[9].

Early Attempts to Control Rabbits in Australia

Legislative measures

Not surprisingly, the dramatic spread of rabbits in the 1870s caused great anxiety among pastoralists and they lobbied the colonial governments for help. With no experience and little biological understanding to guide them, the authorities tried to cope with the situation in various ways, primarily by requiring landholders to control rabbits on their properties (Rolls, 1984).

In 1869 the member for the Western District of Victoria (which consists of fine grazing land) tried to have a clause inserted into the Local Government Act to make the destruction of rabbits compulsory, but it was 1878 before a bill was introduced in Victoria, and this proposed introducing an inspection fee of twopence an acre to see whether land was infested. The landholders were outraged, and this provision was replaced by a bonus scheme. The first Rabbit Destruction Act was passed in South Australia in 1875, to ensure that landholders met their obligations to control rabbits, control being supervised by district councils or rabbit district boards. Rates were levied, and as rabbit numbers increased new legislation was introduced to enable the government to carry out rabbit control on the properties of uncooperative landholders on a cost-recovery basis. In 1880, the New South Wales government introduced a levy on landholders to pay scalp bonuses, but soon after amended the law to make it compulsory for landholders to control rabbits, with penalties for failures. In 1884 the farmers demanded that a bonus system should be re-introduced, but by 1887 newspaper editors and farmers alike called for its abolition, on the grounds that the $30 million (in $A at their 1990 value) spent that year 'might as well have been thrown into the sea'. Although rabbits did not become established in Western Australia until much later, the Western Australian government had legislated to make land managers responsible for rabbit control as early as 1883.

At the same time as these Acts were being introduced a thriving trade had developed in rabbit carcasses and skins, and a new occupation was born – the rabbiter, who trapped rabbits for the bonuses, and later for the meat and skins. Initially the rabbiters made fortunes from bonus payments (Rolls, 1984), and later planned their trapping so as not to destroy their resource, by siting their traps in such a way as to take bucks preferentially and moving on after they had harvested about 40% of the population. Rabbiting came into prominence again during the Great Depression of the 1930s (Fig. 2.13). A variety of other control methods were used, but none of them had much effect on the spreading of 'the grey blanket'. Some of these methods, updated by modern technology, remain useful today, others caused more harm than good.

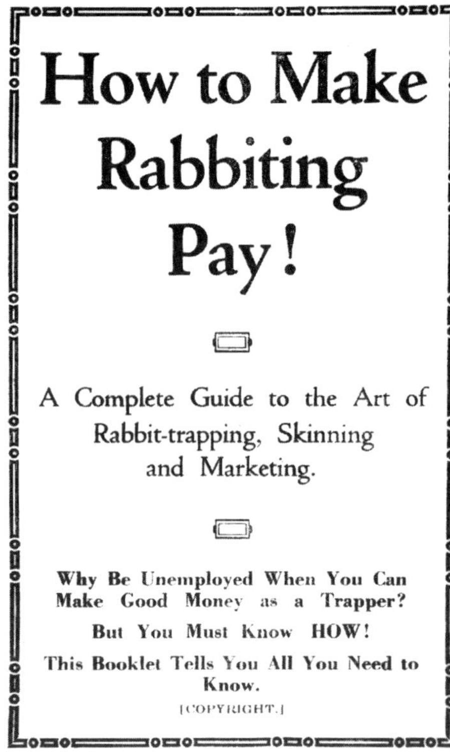
How to Make Rabbiting Pay!

A Complete Guide to the Art of Rabbit-trapping, Skinning and Marketing.

Why Be Unemployed When You Can Make Good Money as a Trapper?

But You Must Know HOW!

This Booklet Tells You All You Need to Know.

[COPYRIGHT.]

Fig. 2.13. The influence of the Great Depression of the early 1930s on the use of rabbits as a resource. Title page of booklet published by W.H. Downey in 1932, reduced to two-thirds natural size. Courtesy of the Mitchell Library, State Library of New South Wales.

Biological control: the release of predators

Since indigenous predators that could affect rabbit numbers were so rare in Australia, the earliest attempts at biological control of the rabbit involved the release of predators: stoats, weasels, mongooses and especially cats. In Australia, only the latter thrived. In the late 1880s they were bred for the purpose and released in thousands specifically to control rabbits, and in addition excess kittens bred on stations 'went wild'. In the late 1890s, before rabbits had reached that state in large numbers, the government of Western Australia released cats in order to kill rabbits as they invaded the state, but the numbers were too small to have any effect on the spreading rabbit plague. In New South Wales goannas (large lizards of the genus *Varanus*) were declared enemies of the rabbit and protected in the Rabbit Nuisance Act 1883, and a conference in Brisbane in 1888 resolved that because of their value for rabbit control, goannas, carpet snakes, native cats and feral cats should be protected. Perhaps the most outlandish predator was the South African carnivorous ant, proposed to the Institute for Science and Industry by the Farmers and Settlers' Association of Western Australia in 1919.

Ratcliffe (Fenner and Ratcliffe, 1965) likened predation of the rabbit in Australia to a poor handbrake on a car, which will hold the vehicle on a gentle slope, but becomes less and less effective as the car starts to move and gathers momentum. During their colonizing spread, feral cats and foxes were still comparatively rare and rabbits were like a car going downhill; a century later, in areas where relatively mild cases of myxomatosis occur, foxes, cats and raptors undoubtedly kill many rabbits that might otherwise have survived.

In New Zealand, where there were no natural predators, a more sustained attempt was made to introduce such animals for rabbit control (King, 1990). By the middle 1870s, when rabbits were becoming a serious agricultural pest, farmers demanded that the natural enemies of the rabbit should be introduced. Starting in 1882, thousands of ferrets were imported from Australia and Britain and thousands more bred locally by the Department of Agriculture and private landholders. They were liberated in rabbit-infested areas in inland Otago, and by the turn of the century were well established in the wild, but had themselves come to be regarded as a pest. Legal protection was removed in 1903 and control campaigns began in the 1930s. In 1883, despite the protests of ornithologists, the Chief Rabbit Inspector recommended that in addition to ferrets, stoats and weasels should be introduced, and importations began in 1885. Released on farmland where rabbits were a pest, they soon spread far beyond such sites. There were renewed protests from ornithologists, but it was not until 1936 that all legal protection was removed from the three mustelids. However, in objecting to the proposal to introduce myxoma virus in the 1980s, Flux (1986) noted that rabbit control had been very successful over large parts of New Zealand, and that 'predation by cats, ferrets and stoats keeps rabbits at these low levels over many areas of New Zealand even when Pest Board control is removed'.

The Rodier method

Among the most bizarre methods of biological control of rabbits proposed was that advocated with great persistence by William Rodier in New South Wales and Victoria between 1905 and 1925. In the press and by private publication, he promoted 'The Rodier Method', which consisted of catching as many rabbits as possible, killing all females and releasing the males[10]. The reasoning was:

> When the males exceed the females in numbers they will persecute them and prevent them from breeding, they will kill what young ones may be born and when they largely exceed the females in numbers, they will worry the remaining ones to death. By this means ALL the females are exterminated, and when this is done the males will die off by old age.

Needless to say, Rodier's method did not succeed, although it attracted enough attention to require testing by the Department of Agriculture of New South Wales (Stewart, 1906).

Trapping and shooting

Trapping is not a method of rabbit control (although purportedly introduced as such) but of harvesting a resource. At first rabbits were trapped for their scalps, for which bonus payments were made (although poisoning was less laborious and more effective), but usually for carcasses and sometimes skins. Rolls (1984) provides a vivid description of the manners and methods of the rabbiters. Commercial rabbit trappers no longer operate in Australia, and gin (leg-hold) traps are banned in many countries.

Besides being popular as a sport, initially practised by wealthy squatters, but later popular with farmers and their sons and employees, shooting can still be useful as a method of maintenance control. Head shooting, which is preferred by the rabbit meat industry, since it minimizes bruising of meat, is considered a humane method of killing rabbits.

Poisoning

When the rabbiters walked off because the bonus payments were withdrawn, baits containing strychnine, cyanide, arsenic and phosphorus were used for rabbit control; often the poisoned rabbits were skinned and the skins sold. All poisons are indiscriminate weapons; often causing deaths of non-target species such as birds and native mammals, and they are dangerous for humans as well. In 1950 a more effective poison, sodium fluoracetate ('1080'), an acute metabolic poison, was introduced (Lazarus, 1958; Poole, 1963). This poison is still widely used in rabbit management, primarily for surface-dwelling rabbits and to reduce rabbit populations before destroying their burrows by ripping (Williams *et al.*, 1995).

Stead (1935) was an enthusiastic proponent of fumigation with cyanogas, produced from calcium cyanide. Certain other poisons, such as chloropicrin (trichloronitromethane), which had been developed for gas warfare in the First World War, and phosphine (hydrogen phosphide) were used for fumigation, a procedure in which gas is introduced into a burrow in which most entrances have been blocked. Fumigation with chloropicrin is still used, primarily to kill rabbits that have been missed in ripping operations (Williams *et al.*, 1995). Before using fumigation, all surface-living rabbits are driven underground by a dog pack.

Ripping

Rabbits may live among surface shelter, such as rock piles, dense vegetation or fallen branches and logs, but they are unique among lagomorphs in that they usually live in underground burrows in large warrens. Breeding populations almost always live in warrens. Since they do not dig new warrens readily or regularly, the destruction of warrens greatly inhibits recolonization of an area. In the 19th century this was accomplished by digging; later it was accomplished much better by tractors with tynes at least 0.5 m deep (Williams *et al.*, 1995). Ripping with caterpillar tractors, supplemented by poisoning with 1080 and often followed by fumigation of recolonized burrows, today constitutes the most important methods of capitalizing on the destruction of rabbits caused by myxomatosis or rabbit haemorrhagic disease.

Rabbit-proof fencing

Wire-netting 'rabbit-proof' fencing has been used in two ways in Australia. It is valuable for keeping rabbits out of properties of high agricultural or sylvicultural value (Fig. 2.14), and was extensively used in such situations after the initial outbreaks of myxomatosis. Essentially, it consists of adding to a good cattle fence a strip of 3 cm hexagonal-mesh wire netting about 90 cm high, with 12–15 cm buried vertically and the same length extending horizontally beneath the ground (McKillop *et al.*, 1988). It is expensive (currently about $1.70 a metre), and to be effective, the entire length, including gates, must be maintained in good repair by inspections every one or two weeks. Early trials with electric fencing were disappointing, but more recently comparisons in the United Kingdom of wire-netting and portable electric fences suggested that the latter were more cost-effective in excluding rabbits from protected fields (McKillop and Wilson, 1987).

Nominally 'rabbit-proof' wire-netting fencing was also used on a very large scale in the panic that followed the spread of rabbits in Australia in the late 1880s, and vast 'barrier fences' were constructed in a dramatic but ineffectual bid to stem the relentless expansion of the rabbit (McKnight, 1969). Between 1880 and 1907 many thousands of kilometres of rabbit-proof fencing were hurriedly erected, initially along the boundaries between Victoria and South Australia and then in several places along the New South Wales–Queensland border (Fig. 2.15; Stead, 1928, 1935). The most famous was the 'No. 1 rabbit fence' in Western Australia, which ran for some 1830 km northward from near Esperance on the south coast to the Eighty Mile Beach between Port Hedland and Broome in the north (Anon., 1906). Spreading from South Australia, rabbits arrived at the border with Western Australia in 1894 and moved west at a rate of 180–200 km a year. In 1902 construction was started on the No. 1 rabbit fence. It

Fig. 2.14. Effectiveness of a well-built and well-maintained rabbit-proof fence in preserving the quality of pasture on a property in central New South Wales. From *Country Life*, Sydney.

took five years to build, the northern section passing through virtually unexplored country on the edge of the Great Sandy Desert. To provide water for the men and draught animals, bores had to be sunk every 25–30 km, and the average haulage distance from the railhead was 320 km, with a maximum distance of about 720 km. Everywhere, the barrier fences proved ineffective, not only because they were usually erected after there were rabbits on both sides, but also because it was clearly impossible to maintain them at anything like full efficiency, which was essential if they were to fulfil their expected role. However, they were of some use as dingo fences, and No. 3 fence, in Western Australia, is maintained as an emu barrier.

The Darling Downs-Moreton Rabbit Board (DDMRB) fence, in southern Queensland, near the northern limits of the rabbit range, has been maintained for over a hundred years (Pennycuick, 1994). David Berman describes the efficacy of this fence as follows[11]:

> The DDMRB is probably the only sizable area in Australia that is 'suitable' for rabbits that has never been overrun by rabbits. ... The Darling Downs black soil and the predominance of summer rainfall do not suit rabbits. However, not all the DDMRB is black soil and rabbits do survive and are a significant pest just outside the fence ... the wool clip of a shire [inside the fence] was from 10% to 17% higher than that of a shire just outside the fence. There was no difference between the shires in wool clip during a period when there was a concerted [rabbit] control effort outside the fence.

On high value land, rabbit-proof fencing is still useful in keeping rabbits out of properties from which they have been eradicated, usually by myxomatosis followed by poisoning and ripping. Special arrangements of rabbit-proof fencing are sometimes used in forest reserves, to

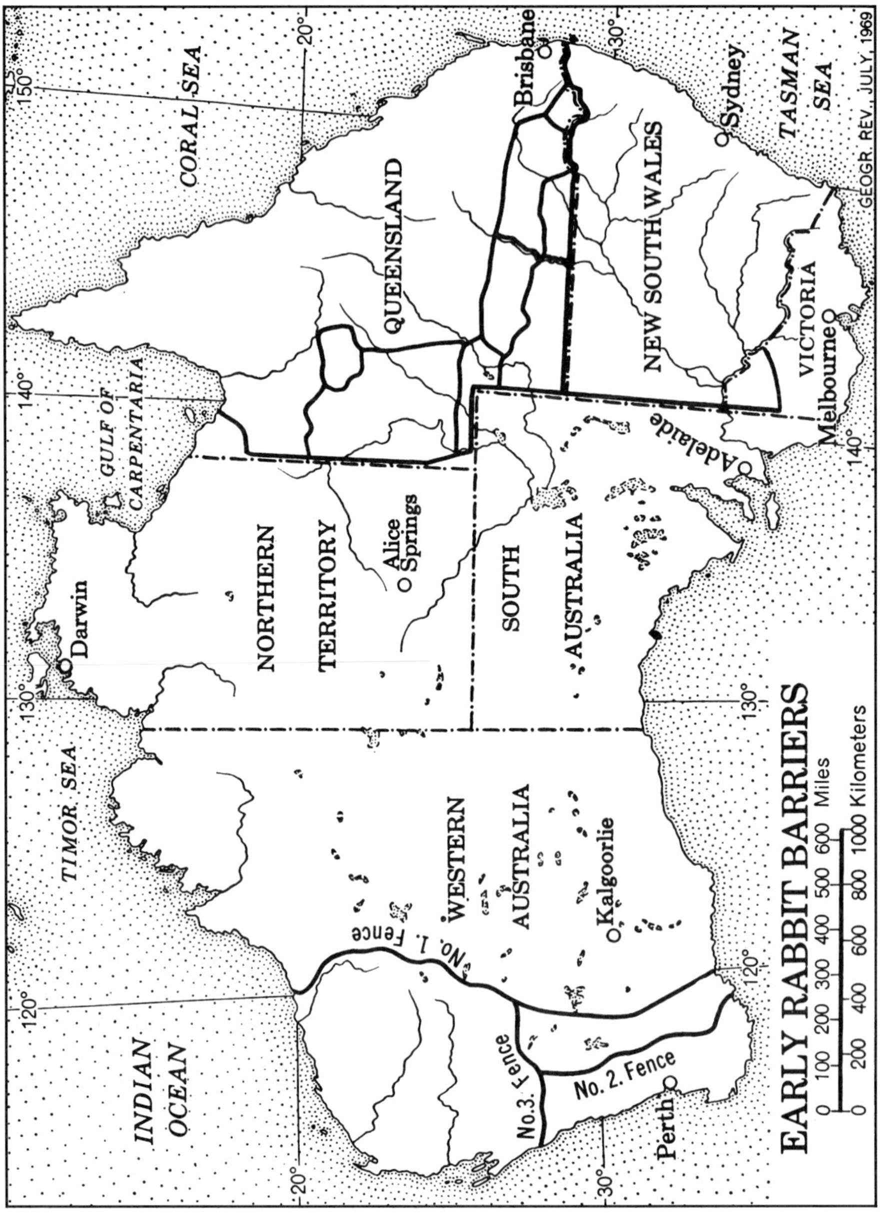

Fig. 2.15. Barrier fences in Australia, built between 1880 and 1910. From McKnight (1969), with permission.

protect young trees from rabbits (McKillop *et al.*, 1988).

The Economics of Rabbit Control in Australia

This subject is discussed at length by Williams *et al.* (1995) from which the following summary is largely drawn. Control of rabbits in agricultural areas lies in the hands of farmers and their local governing bodies. Control in national parks and in much of the arid rangelands is the responsibility of State pest control authorities, which will need to assess socially equitable means whereby governments can intervene to meet broad conservation benefits. Economic frameworks are needed setting out the economic problem posed by rabbits, data on the relative costs and benefits of various control strategies and an understanding why, in agricultural land, the actions of individual landholders do not necessarily lead to optimal rabbit management. However, the lack of reliable quantitative information about the relationship between rabbit density and the level of impact, and on the cost of control and its effectiveness in reducing damage, make economic cost–benefit modelling difficult. Since land managers have a legal obligation to control rabbits, they have to consider which impacts are most significant in their area of responsibility, estimate the costs of this damage in economic terms (including the value of maintaining biodiversity), and then assess the costs and benefits of rabbit control.

The costs of rabbit conventional control (i.e. excluding myxomatosis and rabbit haemorrhagic disease) depend on the type of country, how long the effects of various methods last and to what extent farmers use their own labour and equipment. Approximate costs per hectare of control in large-scale contracts are: poisoning (with 1080) $A6–8, warren ripping $A3–20, fumigation $10–20, and explosives, $A30–60. The best treatment combinations for long-term results include ripping with follow-on maintenance treatment, and since it is clearly impossible to eradicate rabbits, strategic, sustained management should be adopted wherever possible. It is important to note here that the scientists responsible for introducing myxomatosis and rabbit haemorrhagic disease always insisted that the extraordinarily high kills produced by the use of these biological control agents should always be followed up by conventional control, including especially the destruction of rabbits warrens.

Endnotes

[1]Readers seeking further details of the biology and history of the European rabbit are advised to read the book *The European Rabbit. The History and Biology of a Successful Colonizer* (Thompson, H.V. and King, C.M. eds) Oxford University Press (1994), from which much of the material of this chapter has been extracted.

[2]On 16 December 1997 the Anti-Rabbit Research Foundation of Australia launched a booklet *Rabbit Control and Rabbit Calicivirus Disease*, at the grand house at Barwon Park, which is now a Heritage Trust building (*The Age*, 17 December 1997).

[3]The Commonwealth Scientific and Industrial Research Organization (CSIRO) is the largest government research institution in Australia. In the immediate post-Second World War period it concentrated on research related to primary industry, and its officers were responsible for all Australian laboratory and field studies of myxomatosis between 1937 and 1950, and for most of the fieldwork after 1950.

[4]Rabbit Calicivirus Disease Project. *Report on the Concerns and Issues Raised by Aboriginal Communities in the Southern Central Land Council Region at Meetings Held in December 1995.* Report to the Central Land Council by Lynn Baker.

[5]John Bamford Associates (1985). *Environmental Impact Report on a Proposal to Introduce Myxomatosis as Another Means of Rabbit Control in New Zealand.* Commissioned by the Agricultural Pests Destruction Council.
[6]Agricultural Pests Destruction Council, Commission for the Environment, 24 October 1985. Decision not to proceed with the proposed introduction of myxomatosis and consequent suspension of the audit of the Environmental Impact Report.
[7]Announcement by New Zealand Minister for Agriculture, Mr J. Fallon, 3 June 1993: 'two main reasons why the Government has taken this decision [to decline the application to allow myxomatosis into New Zealand]: the great difficulty in establishing what possible effect the rabbit flea may have on New Zealand's national bird, the kiwi; and there is a potentially more humane biological control – RHD – now being researched in Australia. It could become available in a similar timeframe that it would take to introduce myxomatosis.'
[8]Basser Library Archives 143/25/5A. Letter from J. Amaya to Fenner, 23 August 1979.
[9]Basser Library Archives 143/25/5A. Letter from B.G. Cane to Fenner, 16 December 1993.
[10]Basser Library Archives 143/25/5A. *Rabbit and Rat Extermination by the Rodier Method.* Pamphlet published by W. Rodier, 1 March 1924.
[11]Basser Library Archives 143/25/5A. Letter from D. Berman to Fenner, 3 February 1998.
[12]Basser Library Archives MS143/4/T19. *Myxomatosis Investigations Carried Out in Central and South America.* Report to the Australian Wool Research Fund Committee by I.D. Marshall, 1961.
[13]Basser Library Archives, MS100, Adrien Loir archives.

References

Abbott, W.E. (1913) *The Rabbit-pest and the Balance of Nature.* 2nd edn. Angus & Robertson, Sydney, 20 pp.

Anon. (1906) The rabbit-proof fence (Report of findings by Mr Day). *Journal of the Department of Agriculture of Western Australia* 13, 157–160.

Arentsen, P. (1953) Plaga de conejos en Tierra del Fuego. *Boletin Ganadero* 3, 3–4.

Barrett-Hamilton, G.E.H. (1912) *A History of British Mammals.* Volume 2. Gurney and Jackson, London, 748 pp.

Bonino, N.A. and Amaya, J.N. (1984) Distribucion geografica, perjuicios y control del conejo silvestre europeo *Oryctolagus cuniculus* (L.) en la Republica Argentina. *IDIA, Instituto Nacional de Tecnologia agropecuria,* No. 429–432, 25–50.

Burdon, R.M. (1938) *High Country: the Evolution of a New Zealand Sheep Station.* Whitcomb and Tombs, Christchurch, 175 pp.

Corbet, G.B. (1983) A review of classification in the family Leporidae. *Acta Zoologica Fennica* 174, 11–15.

Corbet, G.B. (1994) Taxonomy and origins. In: Thompson, H.V. and King, C.M. (eds) *The European Rabbit. The History and Biology of a Successful Colonizer.* Oxford University Press, Oxford, pp. 1–7.

Crommelin, J.W.C. (1886) *Rabbits and How to Deal with Them.* George Robertson & Co., Sydney, 43 pp.

Downey, W.H. (1932) *How to Make Rabbiting Pay!* Shipping Newspapers, Sydney, 11 pp.

Feng-Yi, Z. (1990) The rabbit industry in China. *Journal of Applied Rabbit Research* 12, 278–279.

Fenner, F. and Ratcliffe, F.N. (1965) *Myxomatosis.* Cambridge University Press, Cambridge, p. 47.

Fennessy, B.V. (1958) Control of the European rabbit in New Zealand. *Wildlife Survey Section Technical Paper 1.* CSIRO, Melbourne.

Filmer, J.F. (1953) Disappointing tests of myxomatosis as rabbit control. *New Zealand Journal of Agriculture* 87, 402–404.

Flux, J.E.C. (1986) Ecological reasons for not introducing myxomatosis. *New Zealand Veterinary Journal* 34, 51–52.

Flux, J.E.C. (1994) World distribution. In: Thompson, H.V. and King, C.M. (eds) *The European Rabbit. The History and Biology of a Successful Colonizer.* Oxford University Press, Oxford, pp. 8–21.

Foster, M. and Telford, R. (1996) *Structure of the Australian Rabbit Industry: a Preliminary Analysis.* Australian Bureau of Agriculture and Resource Economics, Canberra, 33 pp.

Fox, R.R. (1974) Taxonomy and genetics. In: Weisbroth, S.H., Flatt, R.E. and Kraus, A.E. (eds) *The Biology of the Laboratory Rabbit.* Academic Press, New York, pp. 1–22.
Gibb, J.A. and Flux, J.E.C. (1983) Why New Zealand should not use myxomatosis in rabbit control operations. *Search* 14, 41–43.
Gibb, J.A. and Williams, J.M. (1990) European rabbit. In: King, C.M. (ed.) *The Handbook of New Zealand Mammals.* Oxford University Press, Oxford, pp. 138–160.
Gibb, J.A. and Williams, J.M. (1994) The rabbit in New Zealand. In: Thompson, H.V. and King, C.M. (eds) *The European Rabbit. The History and Biology of a Successful Colonizer.* Oxford University Press, Oxford, pp. 158–204.
Goodall, R.N.P. (1979) *Tierra del Fuego: Argentina, territoria nacionale de la Tierra del Fuego, Antarctica, e islas del Atlantico Sur.* Shanamaiim, Buenos Aires.
Greer, J.K. (1966) Mammals of Maleco province, Chile. *Publications of the Museum, Michigan State University Biological Series* 3, 49–152.
Hetzel, B.S. (1978) The changing nutrition of Aborigines in the ecosystem of Central Australia. In: Hetzel, B.S. and Frith, H.J. (eds) *The Nutrition of Aborigines in Relation to the Ecosystem of Central Australia.* CSIRO, Melbourne, pp. 39–47.
Housse, P.R. (1953) *Animales salvajes de Chile en su classificacion moderna.* Universidad de Chile, Santiago. Quoted by Jaksic and Yanez (1983).
Howard, W.E. and Amaya, J.N. (1975) European rabbit invades western Argentina. *Journal of Wildlife Management* 39, 757–761.
Jaksic, F.M. and Fuentes, E.R. (1991) Ecology of a succesful invader: the European rabbit in central Chile. In: Groves, R.H. and di Castri, F. (eds) *Biogeography of Mediterranean Invasions.* Cambridge University Press, Cambridge, pp. 273–283.
Jaksic, F.M. and Yanez, J.L. (1983) Rabbit and fox introductions in Tierra del Fuego: history and assessment of the attempts at biological control of the infestation. *Biological Conservation* 26, 367–374.
King, C.M. (ed.) (1990) *The Handbook of New Zealand Mammals.* Oxford University Press, Oxford, 600 pp.
Kirkpatrick, R.D. (1959) *San Juan Rabbit Investigation. Final Report.* Indiana Department of Conservation, Division of Fish and Game, Pittman-Robertson Project 2-R, 58 pp.
Kirkpatrick, R.D. (1960) The introduction of the San Juan rabbit (*Oryctolagus cuniculus*) in Indiana. *Proceedings of the Indiana Academy of Science* 69, 320–324.
Lazarus, M. (1958) Compound 1080. *CSIRO leaflet no. 22.*
McKillop, I.G. and Wilson, C.J. (1987) Effectiveness of fences to exclude European rabbits from crops. *Wildlife Society Bulletin* 15, 394–401.
McKillop, I.G., Pepper, H.W. and Wilson, C.J. (1988) Improved specifications for rabbit fencing for tree protection. *Forestry* 61, 359–368.
McKnight, T.L. (1969) Barrier fencing for vermin control in Australia. *Geographical Review* 59, 330–347.
Matthams, J. (1921) *The Rabbit Pest in Australia.* Specialty Press, Melbourne, 264 pp.
Morgan, C.L. (1898) *The Rabbit Question in Queensland.* Watson, Ferguson & Co., Brisbane, 132 pp.
Myers, K., Parer, I. and Richardson, B.J. (1989) Leporidae. In: Walton, D.W. and Richardson, B.J. (eds) *Fauna of Australia.* Australian Government Printing Service, Canberra, Volume 1B, pp. 917–931.
Pennycuick, R. (1994) *Keeping Rabbits Out. Darling Downs-Moreton Rabbit Board.* Darling Downs-Moreton Rabbit Board, Warwick, Queensland, 259 pp.
Poole, W.E. (1963) Field enclosure experiments on the technique of poisoning the rabbit. *CSIRO Wildlife Research* 8, 28–51.
Ramsay, B.J. (1994) *Commercial Use of Wild Animals in Australia.* Australian Government Printing Service, Canberra, pp. 120–138.
Richardson, B.J., Rogers, P.M. and Hewitt, G.M. (1980) Ecological genetics of the wild rabbit in Australia. II. Protein variations in British, French and Australian rabbits and the geographical distribution of the variation in Australia. *Australian Journal of Biological Science* 33, 371–383.
Rogers, P.M., Arthur, C.P. and Soriguer, R.C. (1994) The rabbit in continental Europe. In: Thompson, H.V. and King, C.M. (eds) *The European Rabbit. The History and Biology of a Successful Colonizer.* Oxford University Press, Oxford, pp. 22–63.
Rolls, E.C. (1984) (annotated edition) *They All Ran Wild.* Angus & Robertson, Sydney, 546 pp.
Sauer, P.A. (1954) Control biologico del conejo. *Boletin Ganadero* 43, 1–25.

Schwenk, S. (1986) The history and spread of the rabbit in Europe. In: *The Rabbit in Hunting, Agriculture and Forestry.* Conseil International de la Chasse, Paris, pp. 6–12.
Sobey, W.R. (1982) Should New Zealand use rabbit fleas and myxomatosis to assist with rabbit control? *Search* 13, 71–72.
Stead, D.G. (1928) *The Rabbit Menace in New South Wales: an Abridgement of the Report by D.G. Stead.* Government Printer, Sydney, 72 pp.
Stead, D.G. (1935) *The Rabbit Menace in Australia.* Winn and Co., Sydney, 108 pp.
Stewart, J.D. (1906) An experimental test of Rodier's method of rabbit extermination. *Agricultural Gazette of N. S. Wales, Miscellaneous Publication No. 1,002.*
Stodart, E. and Parer, I. (1988) *Colonization of Australia by the Rabbit* Oryctolagus cuniculus *(L.).* Project Report No. 6, CSIRO Division of Wildlife and Ecology, Canberra, 21 pp.
Thompson, H.V. (1955) The wild European rabbit and possible dangers of its introduction into the U.S.A. *Journal of Wildlife Management* 19, 8–13.
Thompson, H.V. (1994) The rabbit in Britain. In: Thompson, H.V. and King, C.M. (eds) *The European Rabbit. The History and Biology of a Successful Colonizer.* Oxford University Press, Oxford, pp. 63–107.
Thompson, H.V. and King, C.M. (eds) (1994) *The European Rabbit. The History and Biology of a Successful Colonizer.* Oxford University Press, Oxford, 245 pp.
Thomson, G.M. (1922) *The Naturalization of Animals and Plants in New Zealand.* Cambridge University Press, Cambridge, 607 pp.
Weisbroth, S.H., Flatt, R.E. and Kraus, A.E. (eds) (1974) *The Biology of the Laboratory Rabbit.* Academic Press, New York, 496 pp.
Williams, K., Parer, I., Coman, B., Burley, J. and Braysher, M. (1995) *Managing Vertebrate Pests: Rabbits.* Bureau of Resource Sciences/CSIRO Division of Wildlife and Ecology, Australian Government Publishing Service, Canberra, 284 pp.
Wilson, G., McNee, A. and Platts, P. (1992) *Wild Animal Resources: their Use by Aboriginal Communities.* Australian Government Printing Service, Canberra, pp. 120–138.
Wodzicki, K.A. (1950) *Introduced Mammals of New Zealand.* Bulletin No. 98, Department of Scientific and Industrial Research, Wellington, New Zealand, 250 pp.

3

Biological Control of Pests

Overview

The idea that the impact of animals, insects or plants that humans found troublesome (i.e. that were considered pests) could be lessened by exposing them to other organisms that fed on them or caused them serious diseases has been about for centuries. Described as 'biological control', it has been developed most rapidly and effectively during the last 150 years as a means of controlling insect pests of pastures and crops of agricultural importance and for controlling weeds.

The most effective means of biological control of insect pests and weeds have been the use of other insects, nematodes or fungi. Vaccination with attenuated strains or viral or bacterial products have proved a very cost-effective means of controlling infectious diseases of humans and livestock. Fungi have also provided an extremely effective means of controlling pathogenic bacteria by the production of poisons that are often highly specific; these are called antibiotics. Certain viruses are useful for the control of insect pests and viruses offer the best prospect of achieving biological control of vertebrate pests. The histories of the use of myxoma and rabbit haemorrhagic disease viruses for the control of the European rabbit in Australia form the main topic of this book. Other examples are the use of hog cholera virus to control pigs in Pakistan and of feline panleukopaenia virus to control cats on the sub-Antarctic Marion Island.

By the mid-1880s the rabbit had become such a serious pest in Australia and New Zealand that in 1887 the governments of five Australian colonies and New Zealand set up a Royal Commission and offered a reward of £25,000 for the discovery of a contagious disease or other successful method of controlling rabbits. Wide advertisement attracted over 1500 applications, among them one from Louis Pasteur. All proposals were carefully examined and detailed experiments were carried out by an Australian bacteriologist and Pasteur's assistants on the chicken cholera bacillus, which Pasteur had entered as an application for the reward. The Commission turned down all proposals, including Pasteur's application, because the disease he proposed, although often lethal, was not contagious and did not spread readily between sick and healthy rabbits. Experiments with another bacterial disease submitted by Dr Danysz some years later also gave negative results.

Pasteur's Germ Theory and the Idea of 'Life against Life'

The idea of using one living organism to control the numbers of another, which is the essence of biological control, has its historical roots in Pasteur's germ theory and the concomitant idea of 'life against life'. The germ theory, as Pasteur conceived it, was a general theory applicable to the whole of the living world, based on the idea that life is governed by general laws and that each living phenomenon is the individual manifestation of a specific form of life. The diversity and unique nature of the behaviour of each biological species was seen to explain the great diversity of life; notwithstanding that the basic chemistry was the same, specificity was determined by the specific 'plans' contained in each germ. The 'mystery of life' resides solely in the existence of the germ and its reproduction and development. 'Life is the germ with its becoming, and the germ is life' (Pasteur, 1870).

The relevance of the definition of life and life cycles is central to Pasteur's definition of the *Contagium vivum* as the result of a 'parasitic life cycle'; parasites were considered to be organisms with their own life cycle, which could involve major changes in body form and in metabolism. He visualized an infectious disease as the interaction between a parasite and its host. An infectious disease was thus seen as the result of competition between different forms of life, the expression within the body of the host of a foreign life, an 'alienum' in the classic sense. This led to the idea of using one form of life against another, in order to control its manifestations, in particular its reproduction.

The Concept of the Biological Control of Pests

The first explicit suggestion to use an infectious disease for the control of a pest can be traced to Pasteur (1880a). After commenting on the difficulty of controlling phylloxera infestation of vines by the pesticides then available, he went on to say:

> When life has a power equal to that manifest in the reproduction of *Phylloxera*, it is chiefly by life and by the power of superior reproduction that one can hope to triumph. Like all living species, *Phylloxera* must have its illnesses, its parasites, its natural causes of destruction. I will research these maladies and these parasites. From the latter I will study the properties, to understand if it is not possible to multiply them and oppose them to *Phylloxera*. In 1865, the silkworm species having been very nearly annihilated by the microscopic organism known as 'corpuscle de Cornalia' and there one did all possible to remove this enemy of the precious insect. Here, one must try to reverse the problem. We will find for the species *Phylloxera* a parasite, and far from combating the latter, we will make it multiply and attach it to *Phylloxera* until *Phylloxera* is destroyed, just as easily as it was to destroy the silkworm species by the parasite 'corpuscle de la pébrine.'

In 1882 (as related by Dubos, 1950) he reiterated this idea in a note to his assistant Adrien Loir, who was later to figure prominently in Pasteur's effort to control rabbits by means of a bacterial disease:

> To find a substance which could destroy phylloxera either at the egg, worm, or insect stage appears to me to be extremely difficult to achieve. One should look in the following direction. The insect which causes phylloxera must have some contagious disease of its own and it should not be impossible to isolate the causative microorganism of this disease. One should next study the techniques of cultivation of this microorganism, to produce artificial foci of infection in countries affected by phylloxera.

A few years later he was to put these ideas into a concrete form. Writing to *The Times* [London], on 27 November 1887, in response to the advertisement by the Intercolonial Commission for a method for the control of rabbits in Australia, he wrote (Pasteur, 1887):

> So far, one has employed chemical poisons to control this plague. ... Is it not preferable to use, in order to destroy living beings, a poison endowed with life and capable of multiplying at a great speed? ... I should like to see the agent of death carried into the burrows by communicating to rabbits a disease that might become epidemic.

However, it was to be many years before a successful method of controlling rabbits (or any other vertebrate pests) by disease was to be used in the field. In fact, in most parts of the world invertebrate pests of agricultural crops (especially insects and helminths) were economically much more important than vertebrate pests. Biological control, a term first used by Smith (1919), was developed primarily as a method of controlling such pests, usually by the use of other insects or entomophagous nematodes (Hagen and Franz, 1973). Several substantial books have been produced on this subject (DeBach, 1964; Huffaker and Messenger, 1976; Waterhouse and Norris, 1987; TeBeest, 1991; Bedding *et al.*, 1993). Since we are concerned primarily with the control of vertebrate pests by the use of pathogenic microbes, we shall not attempt to survey the large literature on the microbial control of insect pests and weeds (Burges and Hussey, 1971; Maramorosch and Sherman, 1985; Hoagland, 1990), but will briefly describe a few examples. In addition, because of its importance as a model of what some early investigators thought might be necessary if myxomatosis was to be used to control rabbits in Australia (see p. 47), we will also give a short account of a classical early example of the biological control of a weed, the use of the caterpillar *Cactoblastis* to control prickly pear in Queensland.

Biological control consists of using its natural enemies (or in rare cases biological enemies discovered by chance) to maintain a pest organism at a lower average abundance than it would reach in their absence. It applies principally to plants and animals that have been introduced into new environments where such enemies are absent, often as part of acclimatization movements. This has happened most often in Australia, New Zealand, the Americas and in Hawaii and smaller oceanic islands. Biological control has several advantages over other methods of pest control, some of which are:

1. If the control agent has been properly screened, it is highly specific for the pest organism and is unlikely to affect non-target organisms.
2. For widespread pests, biological control is often the only method practicable in national parks, rangelands and forests, since other methods may be uneconomic or environmentally damaging.
3. Once established, the control organisms are usually self-perpetuating.
4. Land tenure arrangements pose no problems and societal patterns are unaffected.
5. If successful, the benefit/cost ratio is very high compared with most other control methods.

Biological Control of Bacterial Diseases

Three methods of biological control have been proposed for the prevention or treatment of bacterial diseases of man and his domestic animals: vaccination, antibiotics and bacteriophages.

Vaccination

First systematically practised by Edward Jenner at the turn of the 18th century, as a way of preventing smallpox, vaccination is probably the most cost-effective way of preventing the many diseases, viral and bacterial, for which vaccines have now been developed (Fenner, 1983). Its history and practice have been comprehensively reviewed in several recent publications (Moulin, 1996; Plotkin and Fantini, 1996; Pastoret *et al.*, 1997). It is relevant to the diseases with which this book is concerned only as a means of protecting commercial, pet or sometimes wild rabbits against myxomatosis and rabbit haemorrhagic disease, or in some circumstances against both diseases. It will be discussed in this context in subsequent chapters.

Microbial antagonism: antibiosis

Although it is not usual to think of it this way, the most widely practised method of biological control of a pest in the modern world is the use of antibiotics to control bacteria that cause diseases in man and his domestic animals. Florey (1949) gives an excellent account of the history of antibiotics. He noted that apart from folk remedies utilizing moulds (fungi) for the treatment of superficial wounds, Pasteur was a pioneer in this field, as in so many others aspects of microbiology. After describing bacterial antagonism in experiments with anthrax bacilli (Pasteur and Joubert, 1877), Pasteur (1880b) described an experiment to demonstrate 'antibiosis' using the fowl cholera bacillus (which he was later to recommend for the control of rabbits, see below):

> Substances derived from a vital activity can act against a similar activity. In certain fermentations antiseptics are produced during and by the fermentation which bring the activity of the latter to an end long before the reaction is complete. It is possible that there could be formed in cultures of our microbe [the organism of fowl cholera] products whose presence would convincingly explain immunity and vaccination.
>
> Our laboratory cultures of the parasite will allow us to test this hypothesis. We take a laboratory culture of the microbe and evaporate it at low temperature *in vacuo*, then make it up to its original volume in broth. If this medium contained a microbial poison which was the cause of the non-growth of the microbe, a fresh inoculation would not grow, but this is not the case.
>
> (Translation from Florey, 1949.)

Other scientists in France, Germany and Italy took up the search and in 1889 Doehle produced an illustration (Fig. 3.1A) that anticipated Fleming's famous plate (Fig. 3.1B; Fleming, 1929) by nearly 40 years. The words 'antibiosis' and 'antibiotic', used in a general way by Vuillemin (1889), were applied specifically to microbial antagonism, as the converse of symbiosis, by Ward (1899).

During the intervening period a number of other examples of antibiosis were described, notably products of the bacterium *Pseudomonas pyocyanea*, which were used extensively in the early years of the 20th century for the treatment of patients with diphtheria and other diseases. Then in 1929 Fleming discovered penicillin, but its development to clinical use had to wait until the early 1940s, when Florey, Chain and their colleagues purified it sufficiently for clinical use (Abraham *et al.*, 1949). In the meantime Dubos (review: Hotchkiss, 1944) published his first papers on two polypeptide antibiotics, gramicidin and tyrocidine, obtained from a soil bacterium, *Bacillus brevis*. As Abraham and Florey (1949) remarked:

> This work and its continuation had the outstanding merit of considering the subject

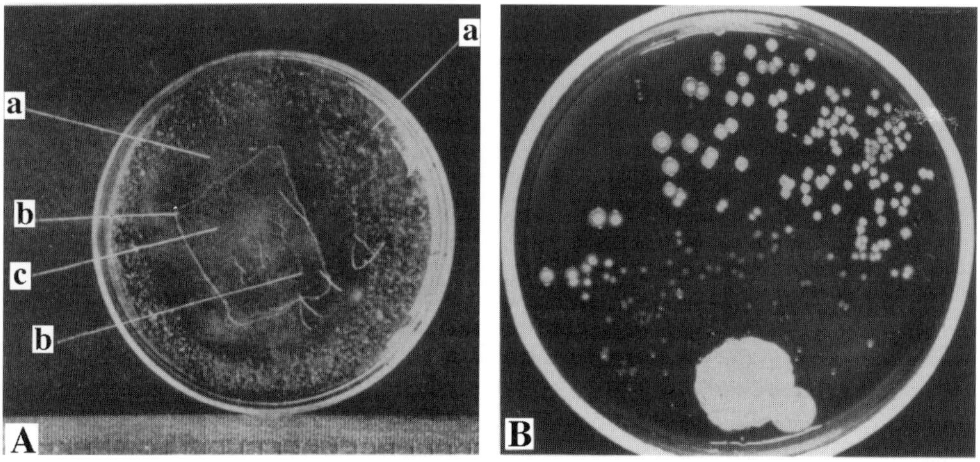

Fig. 3.1. Early demonstrations of antibiosis. (A) A plate was poured with gelatin containing anthrax bacilli (a). On the surface was planted a square of '*Micrococcus anthracotoxicus*' (b). Surrounding this square is a zone (c) in which no anthrax colonies have developed owing to the diffusion of an inhibitory substance from the micrococcus. From Florey (1949), reproduced from Doehle (1889). (B) Photograph of the original plate on which Fleming found a colony of *Penicillium* dissolving the surrounding colonies of staphylococci. From Fleming (1929), with permission.

> from many points of view – bacteriological, biochemical, biological and eventually clinical. This was in sharp contrast to all previous work, much of which was adequate on the bacteriological side but suffered from extreme limitation in chemical and other investigations, if indeed these were attempted.

Unfortunately, gramicidin and tyrocidine were too toxic for systemic use. It remained for Florey and his team to purify penicillin and produce it in large enough quantities to initiate the antibiotic era in which we now live, although many antibiotics are now synthesized and are thus 'chemical pesticides' rather than 'biological control agents'.

Control of bacterial diseases by viruses

The existence of bacterial viruses was discovered by Twort (1915, 1949), but his short paper was completely unnoticed until an analogous finding by d'Hérelle (1917, 1926), who was then apparently unaware of Twort's discovery. Some years later d'Hérelle (1949) gave a popular account of his discovery, parts of which are worth quoting at some length, because they reveal in addition to his discovery of a bacterial virus that as early as 1910 d'Hérelle was experimenting with microbial control of insect pests.

> In 1910, I was in Mexico, in the state of Yucatan, when an invasion of locusts occurred; the Indians reported to me that in a certain place the ground was strewn with the corpses of these insects. I went there and collected sick locusts, easily picked out since their principal symptom was an abundant blackish diarrhoea. This malady had not as yet been described, so I studied it. It was caused by bacteria, the locust coccobacillus, which was present in almost the pure state in the diarrhoeal liquid. I could start epidemics in columns of healthy insects by dusting cultures of the coccobacillus on plants in front of the advancing columns; the insects infected themselves as they devoured the plants.
>
> During the years that followed, I went from the Argentine to North Africa to spread this illness. In the course of these researches, at various times I noticed an anomaly shown by some cultures of the coccobacillus which intrigued me greatly ... [it] consisted of clear spots, quite circular, two or three millimeters in diameter, speckling the cultures grown on agar. I scratched the surface of the agar in these transparent patches, and made slides for the microscope; there was nothing to be seen. I concluded from this and other experiments that the something which caused the formation of the clear spots must be so small as to be filtrable, that is to say able to pass a porcelain filter of the Chamberland type, which will hold back all bacteria.

In 1915 d'Hérelle returned to the Pasteur Institute in Paris and was asked to investigate an epidemic of dysentery.

> I filtered emulsions of the faeces of the sick men, let the filtrates act on cultures of dysentery bacilli and spread them after incubation on nutritive agar in petri dishes; on various occasions I again found my clear spots. ... I resolved to follow one of these patients through from the moment of admission to the end of convalescence, to see at what time the *principle* causing the appearance of clear patches appeared. ... I made an emulsion with a few drops of the still bloody stools [of a case of Shiga dysentery], and filtered it through a Chamberland filter; to a broth culture of the dysentery bacillus isolated the first day, I added a drop of the filtrate; then I spread a drop of this mixture on agar. I placed a tube of the broth culture and the agar plate in an incubator at 37°.
>
> The next morning, on opening the refrigerator, I experienced one of those rare moments of intense emotion which reward the research worker for all his pains; at first glance I saw that the broth culture, which the night before had been very turbid, was perfectly clear: all the bacteria had vanished, they had dissolved away like sugar in water. As for the agar spread, it was devoid of all growth and what caused my emotion was that in a flash I had understood; what caused my clear spots was in fact an invisible microbe, a filtrable virus, but a virus parasitic on bacteria.
>
> Another thought came to me also: 'If this is true, the same thing has probably happened during the night in the sick man, who yesterday was in a serious condition. In his intestine, as in my test-tube, the dysentery bacilli will have dissolved away under the action of their parasite. He should now be cured.' I dashed to the hospital. In fact, during the night, his general condition had greatly improved and convalescence was beginning.

In contrast to Twort, d'Hérelle pursued his discovery with great vigour, believing that the bacteriophage, as he called it, was destined to be of immense practical significance as a method of treating human bacterial diseases, as he explained in his second note (d'Hérelle, 1918):

> the course of the illness [dysentery] is a result of the interaction of the Shiga bacillus and the bacteriophage, the disease and its aggravation corresponding to a deficiency of bacteriophage activity and convalescence to a restitution of the latter. Pathogenesis and pathology of dysentery are dominated by two opposing factors: the dysentery bacillus as pathogenic agent and the filtrable bacteriophage as agent of immunity.

It was a short step from this concept to the idea that large amounts of bacteriophage could be grown up in the laboratory and used for the treatment of the various bacterial diseases, and even for preventing such diseases in large populations in which they were endemic. In the context of this book, bacterial viruses were to be used for the biological control of bacterial diseases. Bacteriophages active against the causative bacteria of many diseases – anthrax, bronchitis, diarrhoea, scarlet fever, typhoid, paratyphoid, cholera, diphtheria, gonorrhoea, plague, osteomyelitis – were soon discovered and in the 1920s there was a flood of publications on bacteriophage therapy of a variety of bacterial diseases, especially dysentery and cholera. Even Burnet, a notable early contributor to studies of the fundamental nature of bacterial viruses (Stent, 1963), published several papers on their use in dysentery (Burnet, 1929; Burnet *et al.*, 1931). After a vast amount of work, but especially after the development of antibiotics, it was accepted that the biological control of bacterial diseases by bacteriophages would never be a practical method of therapy.

Biological Control of Insect Pests

The idea of suppressing insect pests by using their natural enemies arose from observations of one insect eating another, and as early as the 9th century in China predaceous ants were used against citrus pest insects. From late in the 19th century a great number of insect pests have been satisfactorily controlled by the use of other insects (e.g. cottony cushion scale of citrus in California by ladybird beetles from Australia) and by nematodes (e.g. Sirex wasps of *Pinus radiata* in Australia by *Deladenus siricidicola*), but there have been relatively few examples of microbial control of insect pests. Also, there are few examples of the use of vertebrates to control insect pests; one that merits mention here is the use of the fish *Gambusia* to control mosquitoes.

Examples of the control of insect pests by microbes

Not surprisingly, because of their economic importance from early times, microbial diseases of bees and silkworms were the first to be observed and recorded (Cameron, 1973). The first experimental demonstration that a microorganism could cause a disease of insects is attributed to Bassi (1835), who had recognized as early as 1816 that muscardine disease in silkworms (now known to be caused by the fungus *Beauveriana bassiana)* 'did not originate spontaneously in the insect, and that it needed an extraneous germ which entered the insect from the outside and caused the disease' (Steinhaus, 1956). *B. bassiana* is now recognized as one of the most widely occurring entomogenous fungi, and frequently causes epizootics among numerous insect species.

With the introduction of genetic engineering during the last two decades, the gene for the toxin produced by *Bacillus thuringiensis* is being introduced into the genomes of crop plants, e.g. cotton, to make them more resistant to insect pests such as the larval stage of *Heliothis*.

Control of the rhinoceros palm beetle with a baculovirus

It took many years for Pasteur's concept of microbial control of insect pests to be realized in practice, but several insect viruses have now been shown to be effective biocontrol agents, members of the family *Baculoviridae* being the most frequently used. A good example is the non-occluded baculovirus that is specific for the rhinoceros palm beetle *Oryctes rhinoceros* (reviews: Bedford, 1986; Waterhouse and Norris, 1987). This beetle, which is native to a region extending from India to Indonesia, was accidentally introduced to a number of South Pacific and Indian Ocean islands early in the 20th century (Bedford, 1981) and caused great damage in the coconut palm plantations. After unsuccessful attempts had been made to use arthropod and fungal natural enemies of the beetle, Huger (1966) discovered that *Oryctes* in Malaysia (and probably throughout its endemic range) were infected with a non-occluded baculovirus, which Marschall (1970) demonstrated could be used for its biocontrol in the Pacific. Free virus is unstable at environmental temperatures, but introduction of virus-infected beetles has been successful throughout the islands of the Pacific and Indian Oceans and has caused a very substantial reduction in the pest status of the beetle (Bedford, 1986).

Control of an insect pest by a vertebrate: Gambusia *for mosquitoes*

In special situations, such as shallow ponds, small streams, ornamental pools or cisterns, larvivorous fish may be useful for mosquito control (review: Gerberich and Laird, 1985). Criteria for selection include a preference for insect larvae over any alternative food, small size, rapid maturation and high fecundity, reasonable resistance to salinity or pollution and harmlessness to other valuable aquatic species. Among several possibilities, including certain 'annual fish', the most popular is *Gambusia affinis,* which were introduced from Florida into Spain in 1922, and from there to Italy (Grassi, 1923). It was subsequently distributed by the Malaria Experiment Station in Rome all over Europe, and to Australasia, Africa, India and the Far East. Hackett (1937) speaks glowingly of its efficacy in Europe, seeing it as 'a new and disturbing element introduced into a delicately balanced community in which an anopheline vector is fighting to maintain itself'. *Gambusia* are very small fish and are required in very large numbers; effective control requires 2–3 fish per square metre of water surface and sites must be restocked periodically. A disadvantage is that they may become predators of other ecologically important animals, such as larval stages of endemic fish.

The experience of Mosquito Abatement personnel in California suggests that *Gambusia* frequently drives mosquitoes to such a low level that they cannot be found by routine sampling. This usually occurs in temporary pools, which have to be restocked each season, but *Gambusia* are able to persist in some environments in southern California. In these habitats mosquitoes are driven to extinction but reinvade sporadically.

Biological Control of Weeds

Since many insects are pests of plants of agricultural value, it was clear that there should be opportunities to find insects which destroyed weeds but spared valuable plants. Many examples are given in books on the biological control of weeds (DeBach, 1964; TeBeest, 1991); only one will be described here, chosen because it was quoted by early proponents of myxomatosis as a possible model of the way that myxoma virus might have to be used.

A classical model:* Cactoblastis *for the control of prickly pear

The control of prickly pear was an early and outstanding example of successful biological control of a major pest plant, which although employing an insect, has relevance to the use of myxomatosis. Prickly pears are members of the genus *Opuntia* of the family *Cactaceae*, which is indigenous to North, Central and South America. *O. inermis* was probably introduced into New South Wales prior to 1839 and was spread into many properties in New South Wales and Queensland between 1850 and 1875, for use as hedges; *O. stricta* was introduced into central Queensland about 1860. By 1884 prickly pear was seen as a 'growing evil', by 1900 an area of four million hectares was affected, and by 1925, at its peak, over 26 million hectares of southern Queensland and northern New South Wales were rendered almost totally unproductive by the density of the plants (Fig. 3.2A). The idea of using its natural enemies as a means of control was broached as early as 1899, and in 1920 the Commonwealth Prickly Pear Board was established to investigate methods of biological control (Dodd, 1940, 1959). Some 150 cactus insect pests were investigated, the most promising being a moth from northern Argentina, *Cactoblastis cactorum,* the larvae of which tunnel in the stems and cladodes of the plant, after which they are rotted by fungi. After establishing its specificity and its effectiveness, large-scale rearing of the moth in cages was conducted at field breeding stations in Queensland until by the end of 1927 nine million eggs had been liberated at many centres. During 1928–30, pupae, and subsequently eggsticks, were collected in large numbers from the field for subsequent distribution, the total number of eggs used reaching the enormous number of three billion. *Cactoblastis* was extraordinarily effective; prickly pear has been controlled over 24 million hectares in Queensland and northern New South Wales (Fig. 3.2B). It was not so effective in cooler areas of New South Wales or Victoria. The control of prickly pear has been repeated in some 16 other countries (Julien, 1992).

Fig. 3.2. (A) Dense prickly pear (*Opuntia inermis*) in southern Queensland, in October 1926, prior to insect attack. (B) The same area of prickly pear infestation in October 1929, after the release of *Cactoblastis cactorum*. From Dodd (1940).

The 'prickly pear model', first suggested by Martin (1936), dominated the thoughts of those involved in the early days of myxomatosis investigations in Australia. Essentially, it consisted of the building up of a scientific organization comparable to the Commonwealth Prickly Pear Board, and a similar pattern of large-scale breeding (cultivation) and extensive and repeated release until the desired level of control had been achieved.

Control of weeds by microbes

One of the best examples of successful control of weeds by microbes is the control of the skeleton weed *Chondrilla juncea* by the rust fungus *Puccinia chondrillina*. Skeleton weed was accidentally introduced from Europe into western United States and south-eastern Australia about 1910 (Cullen and Groves, 1977) and soon became a major weed largely because it was freed from the pests and diseases that control populations of the plant in its homeland in the Mediterranean region (Wapshere, 1970). After extensive testing for host specificity (Hasan and Wapshere, 1973), two arthropods, the gall mite *Aceria chondrillae* and the gall midge *Cystophora schmidti*, and a rust fungus, *Puccinia chondrillina*, were released, and all have had significant effects on certain skeleton weed biotypes, the fungus being much the most important (Cullen *et al.*, 1973; Marsden *et al.*, 1980). Released in June 1971 from Wagga Wagga, a country town in the centre of its range in south-eastern Australia, the fungus was dispersed over the whole range of over 500,000 square kilometres in less than a year, with a dramatic effect. The density of the host plant, which was much greater than found in Europe, the virulence of the fungal pathogen, carefully selected from the many strains examined (Hasan, 1972) and the unusually wet summer together contributed to its success. Three genetically distinct biotypes of skeleton weed occur in Australia, characterized by narrow, intermediate and broad leaves (Hull and Groves, 1973). The strain of *Puccinia* that was released attacked only the then most common narrow-leaf strain of skeleton weed, and was extremely effective in reducing the numbers of this form (Cullen, 1978). However, the intermediate and broad-leaf forms then steadily increased their distribution and economic significance, and a new programme to discover strains of rust that could control these forms was initiated in the endemic home of skeleton weeds in Europe in 1983.

Heartened by the Australian success, in 1975 two strains of *Puccinia chondrillina* were introduced in California and spread rapidly, causing severe infections throughout several populations of skeleton weed in California and Oregon (Emge *et al.*, 1981). The fungus is now found in all four states of the Pacific Northwest and the weed is under control there (Lee, 1986).

Biological Control of Vertebrate Pests

Use of predatory animals

From early times small carnivores were used for the control of small vertebrate pests. In England, ferreting is a traditional way of capturing rabbits which is still widely used (Sheail, 1971). Careful investigations of the efficiency of ferreting in warrens on open chalk grassland showed that on average only 36% of the population was captured when each warren was ferreted once (Cowan, 1984). Although higher reductions would be achieved by repeated visits, the costs would probably be prohibitive.

After rabbits had been recognized as a serious pest in New Zealand in the 1870s, ferrets, stoats and weasels (family Mustelidae) were introduced over the period 1881 to 1897 in an attempt to control them, without success (Thomson, 1922). During the 18th century rabbits had been released on many islands throughout the world's oceans (see p. 18), to provide food for shipwrecked mariners. Later it became clear that they often caused severe ecological damage on these small islands, and attempts were made to eradicate them. As well as removal by trapping, poisoning, and gassing, many different predators were used: cats, ferrets, foxes, mongooses, stoats and weasels. Flux (1993) investigated the relative efficiency of different methods used for removing rabbits from 607 islands of known area. Rabbits had died out naturally on 11% of these islands; poisoning, shooting, etc. exterminated rabbits on 37 islands on which it was carried out systematically; cats killed out the rabbits on nine of the 80 islands on which they were released; and myxomatosis eliminated rabbits from 12 of the 119 islands into which it was introduced.

For over a century the mongoose (family Viverridae) has been released in many places to control rats. They failed to achieve any measure of pest control, but became a serious pest themselves, especially on islands, where the mongoose exterminated several endemic small animal species (Pimentel, 1955).

The fox occupies a special place in relation to rabbit predation in Australia (Saunders *et al.,* 1995). Like the rabbit, it was deliberately introduced by huntsmen, who had been hunting dingoes with horses and hounds since the 1820s (Rolls, 1984). Finding these animals unsatisfactory as quarry, they introduced foxes in the early 1870s. As with rabbits, foxes were deliberately introduced into new areas, and then spread over Australia in the wake of rabbits (compare Fig. 3.3 with Fig. 2.7, p. 22). They now occupy about the same parts of Australia as do rabbits, of which they are the principal predators, and in many areas rabbits are the principal prey of foxes. Foxes regulate rabbit populations when rabbit densities are low but not when they are high. When rabbit populations crash because of drought or myxomatosis, fox and feral cat populations also collapse,

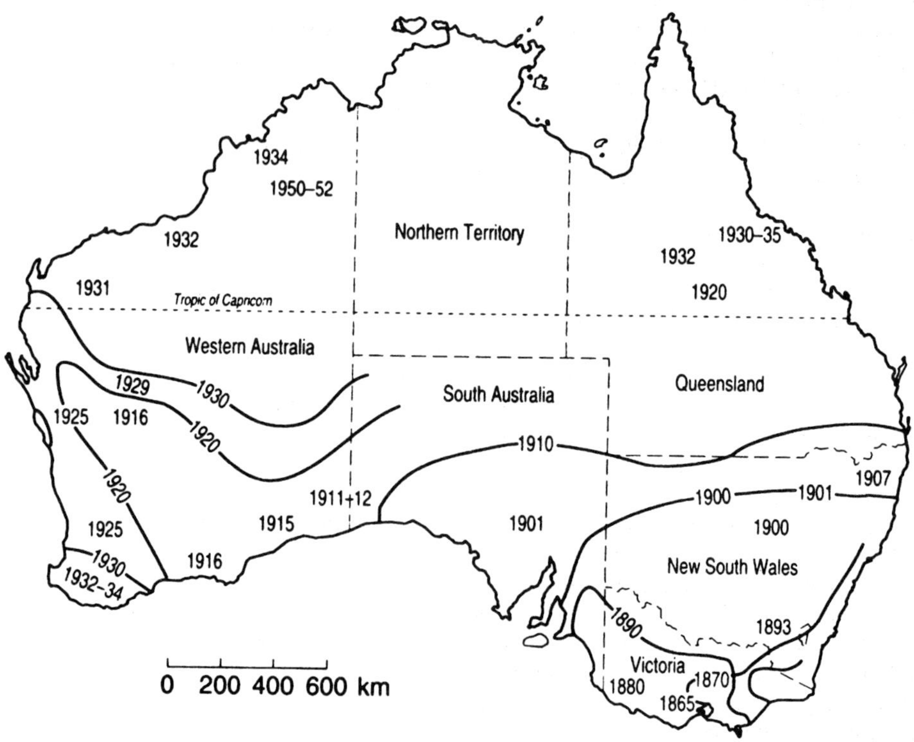

Fig. 3.3. Spread of the red fox in Australia. From Saunders *et al.* (1995), Commonwealth of Australia copyright reproduced by permission. This illustration is based on Jarman (1986).

after a lag period. There is concern from conservationists that during this lag period foxes prey heavily on native fauna; this was a matter of special concern in planning the introduction of the rabbit haemorrhagic disease virus in the 1990s (Newsome *et al.*, 1997; see Chapter 11).

Use of microbes

Microbes can control population numbers in vertebrate populations in three ways: as lethal agents, as agents that reduce fertility and as vectors for immunocontraception. Familiarity with infectious diseases of humans and domestic animals led naturally to the idea that some such diseases could be used to control animals that were regarded as pests. Two features were clearly of critical importance for their use as lethal agents: the microbe used had to be highly specific for the target pest animal, and the case-mortality rate needed to be high, since pest mammals have a high reproductive potential. Nevertheless, in the few cases in which they have been successful, virulent, species-specific microbes such as myxoma virus have so far proved to be the most effective of all methods of biological control of vertebrate pests.

Enzootic infection with certain non-lethal parasites may reduce the fertility of rodents. In south-eastern Australia the house mouse (*Mus domesticus*) causes plagues every few years (Redhead and Singleton, 1988). Infestation of mice with the hepatic nematode *Capillaria hepatica* reduced the number of litters produced and of young weaned over a period of three months (Singleton and Redhead, 1991). Foere *et al.* (1997) observed a similar effect in infections of bank voles (*Clethrionymus glareolus*) and wood mice (*Apodemus sylvaticus*) infected with cowpox virus.

The third method of using microbes for pest control is to use viruses like ectromelia virus or murine cytomegalovirus and myxoma virus as vectors for host genes such as those for the zona pellucida proteins to control mice and rabbits respectively by immunocontraception (see p. 202).

Salmonella *spp. for the control of rodents*

Having observed that *Salmonella typhimurium* wreaked havoc among his laboratory mice at Greifswald, Löffler (1892a,b) suggested that infection with this bacterium might be turned against wild rodents. He found that the field-mouse, *Arvicola arvalis*, was also susceptible, and successes were claimed after baiting both field and house mice. However, rats were not susceptible, and *S. typhimurium* was replaced as a potential biological control agent by a strain of *S. enteritidis* (var. *danysz*) that had been isolated from *Arvicola* (Danysz, 1900). Over the next fifty years *S. enteritidis* var. *danysz* and to a lesser extent *S. typhimurium* were used extensively for rodent control, for mice, rats and plagues of voles (*Microtus* spp.). Commercial baits entitled 'Ratin' and 'Liverpool virus' were sold on a large scale. However, rodent populations quickly developed a resistance to the salmonella serotype used in the baits; in addition the salmonellas caused disease in humans. Use of these baits has been repeatedly associated with outbreaks of food poisoning (Taylor, 1956), and bacterial rodenticides were banned in the United Kingdom in the early 1960s (Healing, 1991). Although the World Health Organization recommended that no further use should be made of them (WHO, 1967), as recently as 1995 it was reported that 50 tons of 'Biorat', a salmonella rodenticide prepared in Cuba, was exported to Nicaragua[1].

Viruses for the control of rabbits

Viruses have not been used for rodent control but two viruses, myxoma virus and rabbit haemorrhagic disease virus, have been used successfully for rabbit control. Both are highly host-specific, although some leporids other than the European rabbit can be infected with myxoma virus. The processes leading to the introduction of these agents and the results achieved with them will be described in Chapters 6–10 (myxoma virus) and Chapter 11 (rabbit haemorrhagic disease virus).

Viruses for the control of feral pigs

Feral pigs are a problem in many parts of the world, and of great concern in Australia because of their susceptibility to foot-and-mouth disease, should this virus ever be introduced into Australia. If it were, the outbreak might cover 10,000–30,000 square kilometres before being detected (Hone and Pech, 1990). Because of the importance of the domestic pig industry in Australia, control by swine-specific viruses such as hog cholera virus (genus *Pestivirus,* family *Flaviviridae*) has never been contemplated. This concern is not relevant in Pakistan, where the idea of using this virus for biological control was initiated by the Directorate of the Veterinary Research Institute, Lahore (Inayatullah, 1973). Following the release of a single inoculated boar and portions of the body of a pig dead of hog cholera in an irrigated forest plantation with a total area of 10,000 acres, 77 dead pigs were counted over two months, in an estimated population of 465 animals, and three months later the population was estimated at 87. These data were analysed by Hone *et al.* (1992), whose modelling indicated that hog cholera could be established in a small population of pigs and for a time exert substantial control, although would later disappear, a result observed in a naturally occurring outbreak in wild pigs in California (Nettles *et al.,* 1989).

Biological control of introduced mammals on islands

Rats, rabbits, goats, dogs and cats that have been introduced on oceanic islands present a major threat to the endemic fauna and flora of these islands, which harbour a high percentage of the endangered species of birds and mammals. Because the founding populations of the introduced mammals were small, in most cases they carried a limited subset of the parasites (bacteria, viruses and protozoa) found in mainland populations. Dobson (1988) suggested that it might be possible to introduce parasites absent from the island populations for biological control, since their low genetic variability and high population densities should favour transmission and increase the severity of disease. Freeland (1990) has pointed out that large feral animals in northern Australia (buffalo, horse, donkey, goat, banteng cattle) are in a somewhat similar situation. In all cases their maximum densities over minimum areas of 100 square kilometres are much higher than in their native habitats, and like island populations they have developed from small founder populations and lack many of the parasites found in their natural habitats. Conservation of the northern Australian savannas requires the control of feral ungulates, and host-specific microbial pathogens might well be the most cost-effective means of control.

We have already commented on the use of myxoma virus for rabbit control on oceanic islands. Another example of the successful use of a viral pathogen for the biological control of an introduced animal that had become a pest is the introduction of feline panleukopaenia virus (a member of the genus *Parvovirus*) for the control of cats on the Marion Island, a sub-Antarctic island administered by South Africa (van Rensburg *et al.,* 1987). Five cats introduced in 1949 had multiplied to over 2000 by 1975, and were severely affecting the indigenous bird populations. Feline panleukopaenia virus, a highly contagious, highly lethal and host-specific virus, was introduced in 1977 (Howell, 1984). By 1982 the cat population had decreased from over 3000 to just over 600, indicating an annual rate of decrease of about 30%, and after 1982 it continued to decrease at a rate of 8% a year.

The risks and benefits of microbial control of vertebrate pests

Just as there are risks in the use of chemical pesticides to control either plant or animal pests, the introduction of organisms for biological control carries some risks, primarily because they may not be specific and may, therefore, attack useful organisms. There are numerous examples of mistakes being made (Pimentel, 1995), such as the introduction of the Indian myna *(Acridotheres tristis)* for the control of army-worms in Hawaii, the Indian mongoose

(*Herpestes auropunctatus*) for the control of rats in Puerto Rico and Hawaii, and the giant toad (*Bufo marinus*) for the control of ground insects in sugarcane plantations in Queensland. The risks and benefits of various methods of biological control of insects and weeds have been reviewed by Hokkanen and Lynch (1995), but although their book has a general title, successful biological control of vertebrate pests is so unusual that myxomatosis is the only example of this category mentioned, and that very briefly. As Greathead (1995) noted, predators of vertebrate pests are usually opportunistic and their parasites are usually non-specific or have stages affecting non-target animals.

Further, the use of microbes for the control of vertebrate pests carries a greater risk than their use for control of weeds or insects, because of the potential risks to the health of humans, domestic and companion animals, and native animals. For this reason exhaustive studies of their species specificity should always be carried out before their introduction, as was the case with both myxoma virus (see p. 123) and rabbit haemorrhagic disease virus (see p. 249).

The benefits must be evaluated against the damage caused by the pest. In the case of the rabbit, Australian farmers had no doubt about the damage to agricultural production during the latter part of the 19th century, as evidenced by the many popular cartoons, a few of which are reproduced in Chapter 2, and by the setting up of the Intercolonial Rabbit Commission in 1888 (see p. 54). Later, when rabbits were more effectively controlled in agricultural country, studies in Australia's very extensive rangelands demonstrated that the rabbit also caused major ecological and environmental damage (see p. 276).

The economics of biological control

In a review of the management of vertebrate pests, Braysher (1993) points out that vertebrate pests such as the rabbit cost Australia many millions of dollars annually in lost agricultural production, as well as causing serious damage to native wildlife and the natural environment, whereas the harvesting of wild rabbits is worth only $10 million annually, without substantially reducing their density. Other benefits provided by rabbits are their use as a source of food in some remote Aboriginal communities, and their value as pets and laboratory animals.

In considering the economics of rabbit control it is useful to understand the varying relationships between rabbit density and the level of damage they inflict (Braysher, 1993; Fig. 3.4). Curve A represents a linear relationship between pest density and damage, and is often used to calculate the benefits of rabbit control to pastoralism. Assuming 16 rabbits are equivalent to one sheep (Short, 1985), reducing rabbit numbers by X should allow X/16 more sheep to be run (Sloane *et al.*, 1988). However, in some situations competition between sheep and rabbits only occurs at high rabbit densities or when pasture biomass is low, so that pest damage is low until a threshold density is reached, as represented in curve B. In other situations, such as where rabbits browse on regenerating native shrubs, rabbits at low densities (less than one per hectare) can prevent regeneration (Lange and Graham, 1983; curve C). If the cost of reducing pest density is known, the theoretical density at which it is most cost-effective to reduce pests is where the broken line (----) intersects the graph of damage versus density (curves A, B or C).

Williams *et al.* (1995) discuss the economic costs of rabbits in Australia in terms of public and private costs, as summarized below.

Public costs

The major public cost, and the most difficult to quantify in dollar terms, is the environmental impact, which includes land degradation, destruction of trees, shrubs, grasses and herbs, and loss and even extinction of native animals, especially in Australia's rangelands, which comprise 70% of the continent (see p. 274). Other public costs include the siltation of dams, forestry and tree plantation losses, the cost

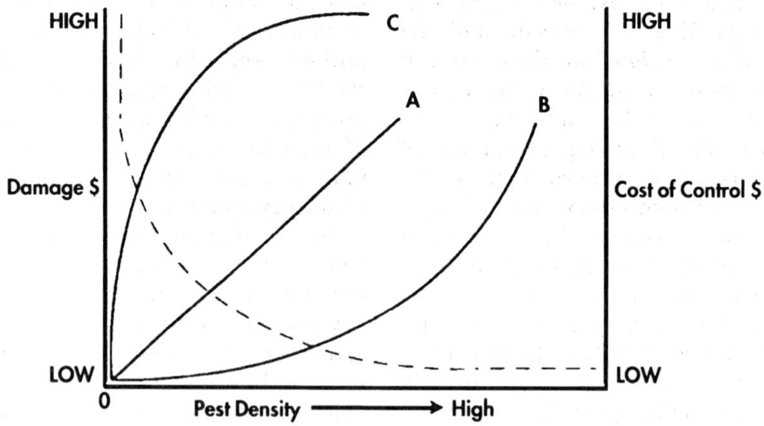

Fig. 3.4. The relationship between rabbit density and damage for three theoretical situations (A, B and C) and between cost of control and rabbit density (-----). Theoretically, the density at which control is most cost-effective is where the broken line intersects the curve of damage versus pest density. From Braysher (1993), Commonwealth of Australia copyright reproduced by permission.

of rabbit control on public lands, research, extension and administration costs associated with rabbit control and reduced tax revenue due to the reduced income of primary producers.

Private costs

Overall, the annual value of lost stock production due to rabbits is estimated to be $20 million for the pastoral districts of South Australia (Henzell, 1989) and $115 million for the whole of Australia (Sloane *et al.*, 1988). To these costs must be added the costs of reduced crop yields, estimated to be $6.5 million in South Australia (Henzell, 1989) and probably six or seven times that for the whole of Australia, damage to forestry and planted trees, degradation of agricultural land, higher costs during droughts and the direct costs of rabbit control to farmers.

Benefits of effective biological control

Except in the hot, dry parts of Australia, rabbits were effectively controlled by myxomatosis for at least 20 years after its introduction, and kept at a reduced level for much longer than that, thus reducing the many costs outlined above. Because myxoma virus was and has continued to be absolutely specific for rabbits, this was achieved without causing illness in humans or their domestic and companion animals, and at a relatively low financial cost in terms of research, implementation and administration. It was for these reasons that when another highly lethal and host-specific viral disease of rabbits, rabbit haemorrhagic disease, was observed in Europe in the late 1980s, moves were immediately made to investigate its potential for rabbit control in Australia (see p. 246). In addition, and this is a matter usually excluded from cost-benefit calculations, the scientific study of myxomatosis provided a unique set of observations of coevolution in action, which has been exploited by scientists working in many fields besides pest control (see p. 318).

Integrated Pest Management

The concept of integrated pest management, as distinct from pest control, envisages the use of all available pest control practices in an integrated way, so that different procedures are complementary (Smith and Pimentel, 1978). Sometimes, as with prickly pears by *Cactoblastis* in Queensland, biological control is so effective that no other methods are needed, but more frequently it is only one of several methods, including selection of plants resistant to the pest, modified farming procedures and the use of pesticides in a carefully controlled manner. The key words are integration, so that pesticides are applied when they will kill the pest rather than the biological control agent, and management, which implies that the aim is not eradication, but reduction of the pest below the level of economic injury in a way that keeps adverse effects on the environment at a minimal level.

From the foregoing considerations it is clear that integrated pest management is highly specific to the pest and, in agricultural systems, the crop under consideration. One of the early examples was malaria control. From the beginning of the 20th century, after the discoveries of Ross and Laveran, malaria control was slowly and painfully developed to become a relatively highly sophisticated science of integrated pest management (Boyd, 1949). However, as Jeffrey (1976) pointed out, with the advent of DDT and its use by the World Health Organization to drive the malaria eradication programme:

> this *science* was ... almost overnight converted to the rather simplistic *technology* of malaria eradication, which basically required that one know how to deliver 2 grams of something to every square meter of a sometimes elusive interior wall, and to manage a hopefully ever-diminishing Kardex file of cases.

As McGregor (1984) put it: 'Perhaps the most noteworthy casualty of the concept of malaria eradication was the experienced malariologist'.

Guidelines for integrated pest management strategies

In order to manage vertebrate pests it is necessary to develop guidelines for each pest animal (Braysher, 1993), based on:

1. defining the problem in terms of its impact on agricultural productivity and environmental damage;
2. determining objectives and performance indicators;
3. identifying and evaluating management options; and
4. implementing, monitoring and evaluating the management programme.

In relation to the rabbit as a pest in Australia, these guidelines were addressed comprehensively by Williams *et al.* (1995), within the framework of integrated pest management, which includes fencing, poisoning, fumigation, destruction of warrens by ripping and explosives, and biological control by myxomatosis and rabbit haemorrhagic disease. It was emphasized by the scientists during the early days of myxomatosis, but not always followed up on good agricultural land and ignored on rangelands, that myxoma virus was not a 'magic bullet', and that the kills obtained by its use should always be followed up by warren destruction. Similar warnings were issued after rabbit haemorrhagic disease virus was released in 1996, and continued to be emphasized.

The environmental impact statement produced to determine the potential value of introducing rabbit haemorrhagic disease virus (Coman, 1996) concluded that its strategic release would provide the basis of sustained and high-level control in a cost-effective manner, even in environmentally sensitive areas, such as National Parks, where chemical and physical control techniques were considered too intrusive. In commercial terms, the ratio of benefits to costs was thought to exceed 100:1 and would probably be higher, while the potential environmental benefits were 'enormous and would far outweigh any possible deleterious effects'.

Early Proposals for Biological Control of Rabbits in Australia

By the 1880s, rabbit damage to agricultural and pastoral land was so extensive in all colonies except Western Australia, and also in the neighbouring colony of New Zealand, that legislators recognized that this was a problem that needed a coordinated response. How might this be done? Prior to 1901 New Zealand and what are now the six states of Australia were separate colonies, together described as 'Australasia'. Occasionally, when faced with a serious problem, they acted together.

Establishment of the Intercolonial Rabbit Commission

In the 1880s rabbits were seen as a problem that needed such joint action, and in 1885 the Premier of Victoria wrote to the Colonial Secretary's office in Sydney proffering £10,000 towards a reward for a method of exterminating rabbits. After consultation with the other Australian states and New Zealand, on 16 April 1888 the New South Wales Government established a Royal Commission of Inquiry into Schemes for Extermination of Rabbits in Australasia (colloquially called the Intercolonial Rabbit Commission), which was enjoined 'to make a full and diligent inquiry as to whether or not the introduction of contagious diseases amongst Rabbits ... for promoting their destruction, will be accompanied by danger to human health ... or to animal life other than Rabbits'. To encourage international participation, it offered a reward of £25,000 (an amount, in present-day terms, of about $A2,000,000), in the following terms (Royal Commission, 1890):

> It is hereby notified that the Government of New South Wales will pay the sum of £25,000 to any person or persons who will make known and demonstrate at his or their own expense any method or process not previously known in the Colony for the effectual extermination of rabbits, subject to the following conditions, viz.:
>
> 1. That such method or process shall, after experiment for a period of twelve months, receive the approval of a Board appointed for that purpose by the Governor with the advice of the Executive Council.
> 2. That such method or process shall, in the opinion of the said Board, not be injurious, and shall not involve the use of any matter, animal or thing, which may be noxious to horses, cattle, sheep, camels, goats, swine, or dogs.
> 3. The Board shall be bound not to disclose the particulars of any method or process, unless such Board shall decide to give such method or process a trial.

The Commission consisted of 11 members:

Henry Norman MacClaurin, M.D.	New South Wales
William Camac Wilkinson, M.D.	New South Wales
Edward Quin	New South Wales
Harry Brookes Allen, M.D.	Victoria
Edward Harewood Lascelles	Victoria
Alfred Naylor Pearson, F.R.Met.Soc., F.C.S., A.I.C.	Victoria
Alfred Dillon Bell	New Zealand
Edward Charles Stirling, M.D.	South Australia
Alexander Stuart Paterson, M.D.	South Australia
Joseph Bancroft, M.D.	Queensland
Thomas Alfred Tabart	Tasmania

The large number of medical men on the Commission reflects the fact that there were few other biologists with expertise in infectious diseases in the colonies at that time; Allen and Stirling were university professors of pathology and physiology respectively and Bancroft a distinguished naturalist. Allen, who was appointed President of the Commission for its final meetings, was later to play a role in the consideration of Aragão's proposal, in 1919, to use myxoma virus for rabbit control. The Commission set up an Experiment Committee, under the chairmanship of Dr W.C. Wilkinson, and appointed Dr Oscar Katz, a bacteriologist who had studied under Robert Koch, as its chief expert officer. After some difficulty in finding a suitable place for work to proceed, a small laboratory was built on Rodd Island, in Sydney Harbour (Fig. 3.5). In addition, there was a dwelling house, an aviary, and an enclosure of a quarter of an acre for animals, with stalls, pens, and artificial burrows for rabbits, the whole area being protected with fly-proof wire gauze over a wide-meshed wire netting.

Fig. 3.5. Laboratory and other buildings on Rodd Island, Sydney Harbour, built for experiments by Loir and Katz for the Royal Commission and later used by the Pasteur Institute of Australia. From the *Sydney Illustrated News*, November, 1891.

Suggestions for the destruction of rabbits by disease were received from 115 correspondents and schemes for the destruction of rabbits otherwise than by disease from 1456 correspondents, from all parts of the world (Royal Commission, 1890).

Pasteur's interest in the Commission's reward

The offer of this substantial reward was published worldwide, and was advertised in *Le Temps* from 9 November to 2 December 1887. Before that, the rabbit problem in Australia and New Zealand had been brought to the attention of Louis Pasteur through letters from New Zealand politicians to the Colonial Secretary in Britain as early as 1885 (Chaussivert, 1988). Although at the time he was still deeply involved in research and with the establishment of the Pasteur Institute in Paris, Pasteur (Fig. 3.6) lost no time in responding to the advertisement. A letter written to Mrs Priestley (an English friend) in December 1888 makes it clear that his motive was primarily to raise money for the establishment of the Institute; both the £25,000 (625,000 francs) reward and the large subscription from grateful pastoralists which he expected would follow the success of the project (Chaussivert, 1988). On 27 November 1887 Pasteur wrote a letter to the Commission and to the editor of *Le Temps* (Pasteur, 1887), in which he proposed the use of the newly discovered chicken cholera bacillus (now known as *Pasteurella multocida*). Immediately after this letter was published, he carried out some laboratory experiments with his assistant Loir and then a small 'field trial' on the estate of Mme Pommery, of Reims, both of which he thought provided support for his proposal (Pasteur, 1888).

Pasteur dispatched a mission consisting of Adrien Loir (Fig. 3.7), his laboratory assistant and Mme Pasteur's nephew, and Dr Louis Germont, which left Naples for Australia on 27 February 1888, bringing their bacterial cultures with them (Rountree, 1983a). There were difficulties

Fig. 3.6. Louis Pasteur (1822–1895). While at the height of his powers and fame, when Pasteur was seeking to raise money to establish a 'Pasteur Institute' to carry on the Pasteur tradition, he became aware of the very substantial reward offered by the Intercolonial Commission for an effective method of controlling rabbits. Thinking that he had discovered such a method, using the virulent chicken cholera bacillus, he applied for the reward and sent two of his staff to Sydney to carry out experiments which would demonstrate the efficacy of the method. After considerable experimentation by his team, and by several Australian scientists, the method was never adopted because the bacillus was not host-specific and it was not contagious in rabbits. It took Pasteur some years to realize the validity of the Australian objections to the use of what was essentially a biocide for rabbits and could infect non-target species.

Fig. 3.7. Adrien Loir (1862–1947). A nephew of Pasteur's wife, Loir was born in Lyon and graduated in medicine from the University of Paris in 1886. As Pasteur's assistant he carried out anti-rabies treatments in various towns in France and then in Russia. In 1888 he was sent to Sydney to demonstrate Pasteur's proposal that the chicken bacillus should be used for the control of rabbits. The experimental work was carried out in the laboratory on Rodd Island, which he later used to prepare vaccines against anthrax and blackleg, which were important diseases of sheep in Australia at the time. He was registered as a medical practitioner in Sydney in 1892, but returned permanently to France in 1893, and subsequently worked in Tunisia and Canada before settling in Le Havre in 1908 as Director of the Bureau of Hygiene and Director of the Natural History Museum until his retirement in 1939.

in working so far from their base (it took a minimum of 40 days for letters to reach their destination), especially as Loir and Germont refused to undertake any experiments that had not been approved by Pasteur. Considerable experimentation was carried out by the Pasteur mission and by Katz, largely to determine the safety for birds and other animals of the chicken cholera bacillus and (primarily by Katz) to determine how contagious it was among rabbits.

In July 1888 Loir and Germont began work in the Rodd Island laboratory, under the supervision of the Expert Committee and Dr Katz. Initially they carried out the experiments designated by Pasteur, which showed that rabbits were killed by eating feed contaminated with the bacterium, and that a number of species of domestic animals were unaffected. No attempt was made to see whether the disease in rabbits was contagious, and Loir and Germont refused to carry out experiments to test this until Pasteur had agreed, which he was loth to do, regarding contagiousness as being unimportant[2]. Eventually, after the Commission had in effect issued Pasteur with an ultimatum, Loir carried out two experiments on contagion, with disappointing results. All except one of the deliberately infected rabbits died, but only five and seven respectively of the 20 contact animals in each experiment were infected.

Katz (1889) then carried out further experiments on contagion, which the Commission (in contrast to Pasteur) regarded as an essential prerequisite if the bacterium was to be used for the biological control of rabbits. It proved impossible to show effective transmission to other rabbits, because rabbits infected with the chicken cholera bacillus died of septicaemia, with little bacterial excretion, in contrast to fowls in which death was preceded by profuse diarrhoea, so that transmission via faeces readily occurred. Katz also showed that the organisms on dried threads did not survive for more than a few hours when exposed to sunlight, although they could survive in the carcasses of dead rabbits for almost three weeks. Further experiments by Katz showed that a large number of species of Australian birds died after feeding on the chicken cholera bacillus, including crows, which are carrion eaters.

On the basis of its danger to native birds and its lack of transmissibility from infected to healthy rabbits, on 3 April 1889, the Commission rejected Pasteur's proposal. As noted in its report (Royal Commission, 1890) and by local cartoonists (Fig. 3.8):

> the destruction of rabbits on a large scale can be effected only by feeding the rabbits with the microbes of the disease; and as other poisons such as arsenic and phosphorus, to the use of which no exception can be taken, will kill rabbits when they are administered, the Commission cannot recommend that permission be given to disseminate a disease which has not been shown to exist in these colonies, which in other countries prevails in disastrous epidemics among fowls, but which has never been known to prevail naturally among rabbits.

Pasteur was furious, and blamed the rejection partly on the employment of Katz, a 'German doctor' whom he regarded as an enemy, and partly on what he considered to be the corruption of all Australians, members of the Commission and politicians alike (Chaussivert, 1988). He could never understand why the Commission placed so much emphasis on the need for the infectious agent to be highly contagious; thus, in a letter to Mrs Priestley in June 1888 he said: 'Why put so much emphasis on contagion? All the Australian burrows are not linked and it will be necessary to proceed burrow after burrow, field after field'. In December 1889 the Commission issued a short final report in which all claims for the reward were rejected, and biological control of the rabbit had to wait for over 60 years before the efficacy of myxoma virus was demonstrated. The reward was withdrawn shortly after the Commission's report was published[3].

In 1896–97, after Pasteur's death, C.J. Pound, Director of the Queensland Stock Institute, showed that chicken cholera was already present in Australian poultry, and freed of concern about introducing a novel organism, carried out experiments with the bacillus in rabbits in a mile-square enclosure in Queensland. He reported substantial rabbit kills, but the disease did not spread effectively and soon died out (Pound, 1897). Not satisfied, another veterinary bacteriologist, J.A. Gilruth, carried out further experiments on a 35,000 acre property in New South Wales which had supplied trappers with almost half a million rabbits annually, using a pollard bait with added sugar. Many rabbits died, but the bait essentially acted as a direct poison, similar to but not as good as phosphorus (Gilruth, 1897).

In spite of his disappointment about the reward, Pasteur agreed to the request that Loir, who had returned to France in April

Fig. 3.8. The Pasteur Rabbit Process. A caricature of Pasteur's proposal to control rabbits in Australia by using the chicken cholera bacillus, which was not contagious for rabbits. From *The Tribune*, 12 April 1888.

1889, should return to Sydney in June 1890 and establish a 'Pasteur Institute of Australia', in the laboratories on Rodd Island, primarily for the production of anthrax vaccine (Rountree, 1983b,c; Todd, 1992). In 1893 Loir returned to France, and from 1894 Pasteur's two-dose vaccine was in competition with a single dose vaccine produced more cheaply by Australian competitors. After Loir's replacement in turn by Dr Louis Momont and then Dr Emile Rougier, the interests of the Pasteur Institute of Australia were disposed of in 1898.

The Visit to Australia of Dr Jean Danysz

Dr Jean Danysz (Fig. 3.9), a distinguished bacteriologist at the Pasteur Institute, had developed a deep interest in the biological

Fig. 3.9. Jean Danysz (1860–1928). Born in western Poland (then under German rule) in 1860, Danysz migrated to France in 1879 and was naturalized there. He graduated in physics and mathematics at the University of Sorbonne in 1882, and later worked under Professor Pouchet at the Museum of Natural History. Becoming interested in parasitism in agriculture and forestry, he discovered a new salmonella, *S. enteritidis* (var. *danysz*), which became known as the 'Danysz virus' and was sold by the Pasteur Insitute for rodent control, under the name of 'Ratin'. Metchnikov invited him to work in the Pasteur Institute in Paris and he was soon appointed to take charge of the Microbiological Department there. He was invited at various times to South Africa, Portugal and Russia to help control diseases of animals and plants, and later came to Australia to try to control the rabbit pest. He also made important discoveries about toxin–antitoxin reactions. During the First World War he supported Polish soldiers in France and was later honoured by the Polish Government.

control of vertebrate pests, principally rodents, in various countries around the world (Danysz, 1895). Early in the 20th century he recovered a virulent strain of *Pasteurella* sp. from an epidemic in domestic rabbits in France. In 1905 he wrote to the Premier of New South Wales to tell him that new microbes for rabbit destruction had become available, and offered to come out to Australia to demonstrate them. The Government was not interested, but a group of pastoralists set up a Rabbit Destruction Fund Committee, raised over £10,000 and brought Danysz to Australia in 1906 to test his ideas (Rountree, 1988). Before his arrival and after his experiments had started, some sections of the Australian press waged a hostile campaign against Danysz and the idea of using microbes for rabbit control (Paszkowski, 1969). Predictably, rabbit trappers[4], the trade unions[5] and rabbit processors[6] voiced their opposition, and there was public concern about possible effects of the microbes on native fauna.

The trials were carried out on Broughton Island, off the coast north of Newcastle, under the direction of an experienced Australian bacteriologist, Dr Frank Tidswell, and in housing, laboratory buildings and animal houses built for the purpose (Danysz, 1907). Although Danysz left in May 1907, Tidswell and Danysz' assistant Latapie continued with experiments on rabbits in pens and in the open country. Once again, the sticking points were that the organism was not specific and that it failed to spread between rabbits (Tidswell, 1907). The Commonwealth Government, on quarantine grounds, refused to allow the dissemination of the organism on the mainland, but later Tidswell showed that three strains of *Pasteurella* isolated from naturally infected rabbits in country areas of New South Wales were identical to Danysz' strain, showing that the latter strain had been in Australia for an indefinite period[7]. The visit ended, as had Pasteur's effort 20 years earlier, with disappointment and recriminations.

Endnotes

[1]*La Nacion*, Havana, 25 November 1995.
[2]See Vallery-Radot, P. (1940–51) *Correspondance de Pasteur, 1840–1895*. B. Grasset, Paris, pp. 241, 245, 249–250.
[3]*Riverine Grazier*, 2 January 1891.
[4]Australian Archives, Series A2, 1905/4877. Letter from J.E. Jarvis, of Young, NSW, to his Federal Member, J.C. Watson, MHR, objecting to the proposal to bring out Dr Danysz, as a member of the public worried about the infection of fowls and as a rabbit trapper. '... At the presant time there are some thousands of men in the Commonwealth who are making a desent (sic) living by trapping rabbits for export (from 3 to 5 pounds a week) ... It is not to exterminate rabbits that they want the desease (sic) as it is to exterminate the trade for export because now men are independent of employer ... One of the squatters said at their meeting the other day "if we can git (sic) a desease amongst the rabbits it will stop all exports and men won't be so independent." The letter was sent by Mr Watson to the Prime Minister, who referred it to the Premier of New South Wales as a State matter.
[5]The Melbourne Trades Hall Council and the Sydney Labour Council expressed their concern. Correspondence in *The Age*, Melbourne, 24 February 1906.
[6]After estimating the current annual value of the rabbit harvesting industry as £1 million, the proprietor of a rabbit freezing works wrote to say: 'If it is necessary, I will take action as a trader ... in getting an injunction to restrain the pastoralists from introducing a foreign pathogenic microbe of an unknown nature ... I have invested thousands of pounds in the development of a rabbit industry, and surely I have a right to protect my interests as a trader'. *The Argus*, 25 April 1906.
[7]Australian Archives, Series A2, 1908/998. Report of Conference of Chairmen of State Boards of Health on the Danysz bacillus, forwarded to Prime Minister on 25 February 1908, found that:
1. The Danysz bacillus was identical to bacteria that had been isolated from rabbits from Yalgoquin (in 1902), Gundagai and Picton.
2. Several domestic animals were susceptible by inoculation but were unlikely to be infected under natural conditions.
Conference recommended that prohibition of importation should be maintained.

References

Abraham, E.P. and Florey, H.W. (1949) Antibiotics from bacteria in the genus *Bacillus*. In: Florey, H.W., Chain, E., Heatley, N.G., Jennings, M.A., Sanders, A.G., Abraham, E.P. and Florey, M.E. (eds) *Antibiotics. A Survey of Penicillin, Streptomycin, and other Antimicrobial Substances from Fungi, Actinomycetes, Bacteria, and Plants*. Oxford University Press, Oxford, pp. 417–492.

Abraham, E.P., Chain, E., Florey, H.W., Florey, M.E., Heatley, N.G., Jennings, M.A. and Sanders, A.G. (1949) Penicillin: historical introduction. In: Florey, H.W., Chain, E., Heatley, N.G., Jennings, M.A., Sanders, A.G., Abraham, E.P. and Florey, M.E. (eds) *Antibiotics. A Survey of Penicillin, Streptomycin, and other Antimicrobial Substances from Fungi, Actinomycetes, Bacteria, and Plants*. Oxford University Press, Oxford, pp. 631–671.

Bassi, A. (1835) *Del mal del Segno, Calcinaccio o Moscardino, malattia che afflugge I bachi da Seta. Parte Prima*. Dalla Tipografia orcesi, Lodi. (Translated by P.J. Yarrow, University of Exeter, as *Phytopathological Classics No. 10*, American Phytopathological Society, 1958.)

Bedding, R., Akhurst, R. and Kaya, H. (1993) *Nematodes and the Biological Control of Insect Pests*. CSIRO Publications, East Melbourne, 178 pp.

Bedford, G.O. (1981) Control of the rhinoceros beetle by baculovirus. In: Burges, H.D. (ed.) *Microbial Control of Pests and Plant Diseases 1970–1980*. Academic Press, London, pp. 409–426.

Bedford, G.O. (1986) Biological control of the rhinoceros beetle (*Oryctes rhinoceros*) in the South Pacific by baculovirus. *Agriculture, Ecosystems and Environment* 15, 141–147.

Boyd, M.F. (ed.) (1949) *Malariology*, Volume II, pp. 1161–1477.

Braysher, M. (1993) *Managing Vertebrate Pests; Principles and Strategies*. Australian Government Publishing Service, Canberra, 58 pp.

Burges, H.D. and Hussey, N.W. (1971) *Microbial Control of Insects and Mites*. Academic Press, London, 861 pp.

Burnet, F.M. (1929) Bacteriophage in its clinical aspects. *Medical Journal of Australia* 1, 406–410.

Burnet, F.M., McKie, M. and Wood, I.J. (1931) A study of bacteriophage in relation to infantile dysentery. *Medical Journal of Australia* 2, 714–716.

Cameron, J.W.M. (1973) Insect pathology. In: Smith, R.K., Mittler, T.E. and Smith, C.N. (eds) *History of Entomology.* Annual Reviews Inc., Palo Alto, California, pp. 285–306.

Chaussivert, J.S. (1988) Letters regarding the "Pasteur Mission" in Australia. In: Chaussivert, J. and Blackman, M. (eds) *Louis Pasteur and the Pasteur Institute in Australia.* University of New South Wales Publication Section, Sydney, pp. 17–23.

Coman, B.J. (1996) *Environmental Impact Associated with the Proposed Use of Rabbit Calicivirus Disease for Integrated Rabbit Control in Australia.* Prepared for the Australia and New Zealand Rabbit Calicivirus Program, 96 pp.

Cowan, D.P. (1984) The use of ferrets (*Mustela furo*) for the study and management of the European wild rabbit (*Oryctolagus cuniculus*). *Journal of Zoology* 204, 570–574.

Cullen, J.M. (1978) Evaluating the success of the programme for the biological control of *Chondrilla juncea* L. In: *Proceedings of the IV International Symposium for the Biological Control of Weeds, 30 August–2 September 1976, Gainesville, Florida.* University of Florida, Gainesville, pp. 117–121.

Cullen, J.M. and Groves, R.H. (1977) The population biology of *Chondrilla juncea* L. in Australia. In: Anderson, D. (ed.) *Exotic Species in Australia – Their Establishment and Success.* Ecological Society of Australia, Canberra, pp. 120–134.

Cullen, J.M., Kable, P.F. and Catt, M. (1973) Epidemic spread of a rust imported for biological control. *Nature* 244, 462–464.

d'Hérelle, F. (1917) Sur un microbe invisible antagoniste des bacilles dysentériques. *Comptes rendus Hebdomadaires des Séances de l'Académie des Sciences. D: Sciences naturelles (Paris)* 165, 373–375.

d'Hérelle, F. (1918) Sur le rôle du microbe filtrant bacteriophage dans la dysenterie bacillaire. *Comptes rendus Hebdomadaires des Séances de l'Académie des Sciences. D: Sciences naturelles (Paris)* 167, 190–192.

d'Hérelle, F. (1926) *The Bacteriophage and its Behaviour.* (English translation). Williams and Wilkins, Baltimore, 629 pp.

d'Hérelle, F. (1949) The bacteriophage. *Science News, No. 14,* Penguin, Harmondsworth, pp. 44–68.

Danysz, J. (1895) Maladies contagieuses des animaux nuisibles: leurs applications en agriculture. *Annales de la Science Agronomique* 40, 1–85.

Danysz, J. (1900) Un microbe pathogene pour les rats (*Mus decumanus* et *Mus rattus*) et son application à la destruction de ces animaux. *Annales de l'Institut Pasteur* 14, 193–201.

Danysz, J. (1907) Rabbit destruction by a contagious disease. *Pastoralists' Review,* May 15, 273–275.

DeBach, P. (ed.) (1964) *Biological Control of Insect Pests and Weeds.* Chapman and Hall, London, 844 pp.

Dobson, A.P. (1988) Restoring island ecosystems: the potential of parasites to control introduced mammals. *Conservation Biology* 2, 31–39.

Dodd, A.P. (1940) *The Biological Campaign against Prickly-Pear.* Commonwealth Prickly Pear Board, Brisbane, 44 pp.

Dodd, A.P. (1959) The biological control of prickly pear in Australia. In: Keast, A., Crocker, R.L. and Christian, C.S. (eds) *Biogeography and Ecology in Australia.* Monographiae Biologicae, Junk, Den Haag, Vol. VIII, pp. 565–577.

Doehle, P. (1889) *Beobachtungen über einen Antagonisten des Milzbrandes.* Habilitatationsschrift, Kiel. (Quoted by Florey, 1949).

Dubos, R.J. (1950) *Louis Pasteur. Freelance of Science.* Little Brown, Boston, p. 310.

Emge, R.G., Melching, J.S. and Kingsolver, C.H. (1981) Epidemiology of *Puccinia chondrillina,* a rust pathogen for the biological control of rush skeleton weed in the United States. *Phytopathology* 71, 839–843.

Fenner, F. (1983) Biological control, as exemplified by smallpox eradication and myxomatosis. *Proceedings of the Royal Society of London* B218, 259–285.

Fleming, A. (1929) On the antibacterial action of cultures of a Penicillium, with special reference to their use in the isolation of *B. influenzae. British Journal of Experimental Pathology* 10, 226–236.

Florey, H.W. (1949) Historical introduction. In: Florey, H.W., Chain, E., Heatley, N.G., Jennings, M.A., Sanders, A.G., Abraham, E.P. and Florey, M.E. *Antibiotics. A Survey of Penicillin, Streptomycin, and other Antimicrobial Substances from Fungi, Actinomycetes, Bacteria, and Plants.* Oxford University Press, Oxford, pp. 1–73.

Flux, J.E.C. (1993) Relative effect of cats, myxomatosis, traditional control, or competitors in removing rabbits from islands. *New Zealand Journal of Zoology* 20, 13–18.

Foere, S.M., Bennett, M., Chantrey, J., Jones, T., Baxby, D. and Begon, M. (1997) The effect of cowpox virus infection on fecundity in bank voles and field mice. *Proceedings of the Royal Society of London* Series B 254, 1457–1461.

Freeland, W.J. (1990) Large herbivorous animals: exotic species in northern Australia. *Journal of Biogeography* 17, 445–449.

Gerberich, J.B. and Laird, M. (1985) Larvivorous fish in the biocontrol of mosquitoes, with a selected biography of recent literature. In: Laird, M. and Miles, J.W. (eds) *Integrated Mosquito Control Methodologies.* Academic Press, London, pp. 47–76.

Gilruth, J.A. (1897) Chicken cholera. *New Zealand Agriculture Department Report.*

Grassi, B. (1923) Acclimitazione della Gambusia in Italia. *Atti della Reale Accademia dei Lincei, Rend.* 32, 544–548.

Greathead, D.J. (1995) Benefits and risks of classical biological control. In: Hokkanen, H.M.T. and Lynch, J.M. (eds) *Biological Control: Benefits and Risks.* Cambridge University Press, Cambridge, pp. 53–63.

Hackett, L.W. (1937) *Malaria in Europe. An Ecological Study.* Oxford University Press, London, pp. 312–315.

Hagen, K.S. and Franz, J.M. (1973) A history of biological control. In: Smith, R.F., Mittler, T.E. and Smith, C.N. (eds) *History of Entomology.* Annual Reviews, Inc., Palo Alto, California, pp. 433–476.

Hasan, S. (1972) Specificity and host specialization of *Puccinia chondrillina. Annals of Applied Biology* 72, 257–263.

Hasan, S. and Wapshere, A.J. (1973) The biology of *Puccinia chondrillina*, a potential control agent of skeleton weed. *Annals of Applied Biology* 74, 325–332.

Healing, T.D. (1991) Salmonella in rodents: a risk to man? *Communicable Disease Reports* 1, R114–RR116.

Henzell, R. (1989) *Proclaimed Animal Research in South Australia – Cost-Benefits, Future Directions and Related Issues.* Animal and Plant Control Commission, Adelaide.

Hoagland, R.E. (1990) *Microbes and Microbial Products as Herbicides.* American Chemical Society, Washington, 340 pp.

Hokkanen, H.M.T. and Lynch, J.M. (1995) *Biological Control: Benefits and Risks.* Cambridge University Press, Cambridge, 304 pp.

Hone, J. and Pech, R. (1990) Disease surveillance in wildlife with emphasis on detecting foot-and-mouth disease in feral pigs. *Journal of Environmental Management* 31, 173–184.

Hone, J., Pech, R. and Yip, P. (1992) Estimation of the dynamics and rate of transmission of classical swine fever (hog cholera) in wild pigs. *Epidemiology and Infection* 108, 377–386.

Hotchkiss, R.D. (1944) Gramicidin, tyrocidine and tyrothricin. *Advances in Enzymology* 4, 153–199.

Howell, P.G. (1984) An evaluation of the biological control of feral cat *Felis catus* (Linnaeus 1758). *Acta Zoologica Fennica* 172, 111–113.

Huffaker, C.B. and Messenger, P.S. (eds) (1976) *Theory and Practice of Biological Control.* Academic Press, New York, 788 pp.

Huger, A.M. (1966) A virus disease of the Indian rhinoceros beetle, *Oryctes rhinoceros* (Linnaeus), caused by a new type of insect virus, *Rhabdionvirus oryctes* gen.n., sp.n. *Journal of Invertebrate Pathology* 8, 38–51.

Hull, V.J. and Groves, R.H. (1973) Variation in *Chondrilla juncea* L. in south-eastern Australia. *Australian Journal of Botany* 21, 113–115.

Inayatullah, C. (1973). Wild boar in West Pakistan. *Forest Research Institute, Peshawar* Bulletin No. 1, 1–17.

Jarman, P. (1986) The red fox – an exotic large predator. In: Kitching, R.L. (ed.) *The Ecology of Exotic Animals and Plants.* John Wiley and Sons, Brisbane, pp. 44–61.

Jeffrey, G.M. (1976) Malaria control in the twentieth century. *American Journal of Tropical Medicine and Hygiene* 25, 361–371.

Julien, M.H. (1992) *Biological Control of Weeds. A World Catalogue of Agents and their Target Weeds*, 3rd edn. CAB International, Wallingford, 186 pp.

Katz, O. (1889) Experimental researches with the microbes of chicken-cholera. *Proceedings of the Linnean Society of New South Wales*, Second series 4, 513–597.

Lange, R.T. and Graham, C.R. (1983) Rabbits and the failure of regeneration in Australian arid zone *Acacia. Australian Journal of Ecology* 8, 377–381.

Lee, G.A. (1986) Integrated control of rush skeleton weed (*Chondrilla juncea*) in the western United States. *Weed Science* 34, (Suppl. 1), 2–6.

Löffler, F. (1892a) Uber Epidemien unter den im hygienischen Institute Greifswald gechalten Mäusen und über der Feldmäuseplage. Bekampfung. *Zentralblatt für Bakteriologie* 11, 129–141.

Löffler, F. (1892b) Die Feldmäuseplage in Thessalien u. ihre erfolgegreiche Bekampfung mittels des "Bacillus typhi Murium". *Zentralblatt für Bakteriologie* 12, 1–17.

McGregor, I. (1984) Malaria – recollections and observations. *Transactions of the Royal Society of Tropical Medicine and Hygiene* 78, 1–8.

Maramorosch, K. and Sherman, K.E. (eds) (1985) *Viral Insecticides for Biological Control.* Academic Press, Orlando, 809 pp.

Marschall, K.J. (1970) Introduction of a new virus disease of the coconut rhinoceros beetle in Western Samoa. *Nature* 225, 288–289.

Marsden, J.S., Martin, G.E., Parham, D.J., Ridsdill Smith, T.J. and Johnston, B.G. (1980) *Returns on Australian Agricultural Research: The Joint Industries Commission – CSIRO Benefit–Cost Study of the CSIRO Division of Entomology.* CSIRO, Melbourne, 107 pp.

Martin, C.J. (1936) Observations on *myxomatosis cuniculi* (Sanarelli) made with a view to the use of the virus in the control of rabbit plagues. *Council for Scientific and Industrial Research Bulletin* No. 96, 28 pp.

Moulin, A.M. (ed.) (1996) *L'Aventure de la Vaccination.* Libraire Arthème Fayard, Paris, 498 pp.

Nettles, V.F., Corn, J.L., Erickson, G.A. and Jessup, G.A. (1989) A survey of wild swine in the United States for evidence of hog cholera. *Journal of Wildlife Diseases* 25, 61–65.

Newsome, A., Pech, R., Banks, P. and Dickman, C. (1997) *Potential Impacts on Australian Native Fauna of Rabbit Calicivirus Disease.* Biodiversity Group, Environment Australia, Canberra, 130 pp.

Pasteur, L. (1870) Sur la vie. In: Vallery-Radot, P. (ed.) (1922–39) *Oeuvres de Pasteur*, Masson, Paris, VII, p. 29.

Pasteur, L. (1880a) Observations sur les moyens propres a détruire le phylloxera. *Comptes Rendus Hebdominaires des Seances de l'Academie des Sciences* 90, 512–513; 514–515.

Pasteur, L. (1880b) Sur le choléra des poules; études des conditions de la non-récidive de la maladie et de quelques autres de ses caractères. *Comptes Rendus Hebdominaires des Seances de l'Académie des Sciences* 90, 952–958.

Pasteur, L. (1881) Le vaccin du charbon. *Comptes Rendus Hebdominaires des Seances de l'Académie des Sciences* 92, 666–668.

Pasteur, L. (1887) Letter to *The Times*, 27 November 1887. In: Vallery-Radot, P. (ed.) (1922–39) *Oeuvres de Pasteur*, Masson, Paris, VII, pp. 88–89.

Pasteur, L. (1888) Sur la destruction des lapins en Australie et dans la Nouvelle-Zélande. *Annales de l'Institut Pasteur* 2, 1–8.

Pasteur, L. and Joubert, M.J. (1877) Charbon et septicemie. *Comptes Rendus Hebdominaires des Seances de l'Academie des Sciences* 85, 101–115.

Pastoret, P.-P., Blancou, J., Vannier, P. and Verschueren, C. (eds) (1997) *Veterinary Vaccinology.* Elsevier, Amsterdam, 853 pp.

Paszkowski, L. (1969) Dr Jan Danysz and the rabbits of Australia. *The Australian Zoologist* 15, 109–120.

Pimentel, D. (1955) Biology of the Indian mongoose in Puerto Rico. *Journal of Mammalogy* 36, 62–68.

Pimentel, D. (1995) Biotechnology: environmental impacts of introducing crops and biocontrol agents in North American agriculture. In: Hokkanen, H.M.T. and Lynch, J.M. (eds) *Biological Control: Benefits and Risks*, pp. 13–29.

Plotkin, S.A. and Fantini, B. (eds) (1996) *Vaccinia, Vaccination, Vaccinology. Jenner, Pasteur and Their Successors.* Elsevier, Amsterdam, 379 pp.

Pound, C.J. (1897) The destruction of rabbits by means of the microbes of chicken-cholera. *Agricultural Gazette, New South Wales* 8, 538–573.

Redhead, T. and Singleton, G. (1988) The PICA strategy for the prevention of losses caused by plagues of *Mus domesticus* in rural Australia. *EPPO Bulletin* 18, 237–248.

Rolls, E.C. (1984) (annotated edition) *They All Ran Wild.* Angus & Robertson, Sydney, 546 pp.
Rountree, P.M. (1983a) Pasteur in Australia. Part I. The Rabbit Commission. *Australian Microbiologist* 4(3), 5–9.
Rountree, P.M. (1983b) Pasteur in Australia. Part II. Vaccination against anthrax. *Australian Microbiologist* 4(4), 9–11.
Rountree, P.M. (1983c) Pasteur in Australia. Part III. The Pasteur Institute of Australia. *Australian Microbiologist* 4(5), 9–14.
Rountree, P.M. (1988) Jean Danysz – a forgotten visitor to Australia. *Australian Microbiologist* 9(1), 35–42.
Royal Commission (1890) *Royal Commission of Inquiry into Schemes for Extermination of Rabbits in Australasia.* Progress Report, Minutes of Proceedings, Minutes of Evidence, and Appendices. Government Printer, Sydney, 216 pp.
Saunders, G., Coman, B., Kinnear, J. and Braysher, M. (1995) *Managing Vertebrate Pests: Foxes.* Australian Government Publishing Service, Canberra, 140 pp.
Sheail, J. (1971) *Rabbits and their History.* David and Charles, Newton Abbot, 226 pp.
Short, J. (1985) The functional response of kangaroos, sheep and rabbits in an arid grazing system. *Journal of Applied Ecology* 22, 435–437.
Singleton, G, and Redhead, T. (1991) Future prospects for biological control of rodents using micro- and macro-parasites. In: Quick, G.R. (ed.) *Rodents and Rice.* International Rice Research Institute, Los Banos, Philippines, pp. 75–82.
Sloane, Cook and King Pty Ltd (1988) Other pests. In: *The Economic Impact of Pasture Weeds, Pests and Disease on the Australian Wool Industry.* Report prepared for the Australian Wool Corporation, pp. 68–77.
Smith, E.H. and Pimentel, D. (1978) *Pest Control Strategies.* Academic Press, New York, 334 pp.
Smith, H.S. (1919) On some phases of insect control by the biological method. *Journal of Economic Entomology* 12, 288–292.
Steinhaus, E.A. (1956) Microbial control – the emergence of an idea. *Hilgardia* 26(2), 107–157.
Stent, G.S. (1963) *Molecular Biology of Bacterial Viruses.* W.H. Freeman, San Francisco, pp. 15–16.
Taylor, J. (1956) Bacterial rodenticides and infection with *Salmonella enteritidis. Lancet* 1, 630–633.
TeBeest, D.O. (ed.) (1991) *Microbial Control of Weeds.* Chapman and Hall, New York, 284 pp.
Thomson, G.M. (1922) *The Naturalization of Animals and Plants in New Zealand.* Cambridge University Press, Cambridge, 607 pp.
Tidswell, F. (1907) *Rabbit Destruction, Broughton Island Experiments: Report upon a Virus Proposed by Dr. Jean Danysz for Destruction of Rabbits.* Government Printer, Melbourne, 55 pp.
Todd, J.H. (1992) Adaptation to environment – the Pasteur anthrax vaccine in Australia. *Australian Veterinary Journal* 69, 318–321.
Twort, F.W. (1915) An investigation on the nature of the ultra-microscopic viruses. *Lancet* 2, 1241–1243.
Twort, F.W. (1949) The discovery of the bacteriophage. *Science News* No. 14, pp. 33–43. Penguin, Harmondsworth.
van Rensburg, P.J.J., Skinner, J.D. and van Aarde, R.J. (1987) Effects of feline panleucopaenia on the population of feral cats on Marion Island. *Journal of Applied Ecology* 24, 63–73.
Vuillemin, P. (1889) Antibiosis. *Association francaise pour l'Avancement des Sciences,* Part 2, p. 525.
Wapshere, A.J. (1970) Assessment of the biological control potential of the organisms attacking *Chondrilla juncea* L. *Proceedings of the First International Symposium on Biological Control of Weeds, Delemont 1969.* Miscellaneous Publication No. 1, CIBC, pp. 81-89.
Ward, H.M. (1899) Symbiosis. *Annals of Botany* 13, 540–562.
Waterhouse, D.F. and Norris, K.R. (1987) *Biological Control: Pacific Prospects.* Inkata Press, Melbourne, 454 pp.
WHO (1967) *Joint FAO/WHO Expert Committee on Zoonoses. Third Report.* WHO Technical Report Series, No. 370. World Health Organizatin, Geneva, pp. 38–39.
Williams, K,. Parer, I., Coman., Burley, J. and Braysher, M. (1995) *Managing Vertebrate Pests: Rabbits.* Australian Government Publishing Service, Canberra, 284 pp.

4
The Discovery of Myxoma Virus

Overview

Myxoma virus was among the first viruses to be discovered, after a new, extremely severe disease had affected European rabbits in Sanarelli's laboratory in Montevideo, Uruguay, in 1896. In 1927 myxoma virus was correctly grouped by Aragão with the viruses that caused smallpox, fowlpox and molluscum contagiosum, now classified as the family *Poxviridae*.

Brazilian scientists showed that myxoma virus was transmitted mechanically by fleas and mosquitoes, and that the reservoir host in Brazil was *Sylvilagus brasiliensis*, in which it produced a localized fibroma. In the early 1930s outbreaks of myxomatosis were observed among domestic rabbits in southern California. In 1959 it was shown that the natural host of myxomatosis in California was *Sylvilagus bachmani*, in which the virus caused a localized fibroma from which it could be transmitted mechanically by mosquitoes. The strain of myxoma virus found in *S. bachmani* produces a lethal disease in European rabbits which differs in some respects from that produced by South American strains.

Myxomatosis can be transmitted from one infected *Oryctolagus* to another by contact or by the respiratory route, but the most common mode of transmission is mechanical transfer by insect bite, which is the only mode from fibromas in *Sylvilagus* rabbits. Investigations in Australia confirmed that myxoma virus could be transmitted by any insect that probed through a skin lesion and then probed or fed on a normal rabbit. Since transmission is mechanical there is no extrinsic incubation period of the kind found with arboviruses.

The dominant vector(s) in the field varies; mosquitoes are the most important in the Americas and in Australia (until fleas were introduced in 1966), the European rabbit flea in Britain, and mosquitoes and fleas in France. In European rabbits highly attenuated strains of myxoma virus do not reach high enough titres in the skin overlying the lesions to be efficiently transmitted, whereas highly and moderately virulent strains reach high titres in the skin and readily contaminate the mouthparts of probing mosquitoes. Because rabbits infected with highly virulent strains die so soon after their lesions become infectious, moderately virulent strains are more likely to survive through the Australian winter, when mosquitoes are scarce.

Fleas also transmit mechanically. Studies of the breeding habits of European fleas (*Spilopsyllus cuniculi*) showed that egg maturation is dependent on hormones produced by pregnant rabbits; on the other hand the Spanish rabbit flea (*Xenopsylla cunicularis*), also an efficient vector, shows no such hormone dependence.

The Development of the Concept of 'Virus'

The word 'virus' had been long used to denote a 'morbid poison'. The history of development of the concept of 'virus' as now understood has been discussed at length by Hughes (1977) and Wilkinson (1979), and more recently by van Helvoort (1994). In relation to the discovery of myxoma virus by Sanarelli (1898) it is useful to summarize the position as it was seen at the end of the 19th century.

From the 1860s French workers named Jenner's vaccine 'le virus vaccin'. By the early 1880s, Pasteur, although he usually used the words 'germ' and 'microbe' to describe infectious agents, referred without ambiguity to 'the anthrax (or chicken cholera) virus' and 'the anthrax (or chicken cholera) bacterium' in the same paper, although he always used 'virus' when writing about rabies. A decade later Ivanovski (1894) and Beijerinck (1898) realized that the causative agent of tobacco mosaic disease, which Beijerinck called a '*contagium vivum fluidum*', differed from bacteria because it could pass through a filter that held back all bacteria. They later used the word 'virus' as a term to denote this filterable infectious agent, as did Loeffler and Frosch (1898) in their classical paper on foot-and-mouth disease virus. Likewise, in his detailed report of myxomatosis of rabbits, which was published in June 1898, just after that of Loeffler and Frosch (March 1898) but before that of Beijerinck (November 1898), Sanarelli described the causative agent as an 'invisible virus'. Using a Chamberland filter, he tried but could not demonstrate that it was filterable; this was achieved later by Moses (1911), using the slightly coarser Berkefeld filter instead of a Chamberland filter. Sanarelli summarized his views about the infectious agent thus: myxoma virus was not related to the pathogenic microorganisms then familiar to microbiologists (bacteria, protozoa and fungi). He thought it unlikely that unorganized (non-cellular) infectious agents existed, because they could reproduce themselves, which 'chemical ferments' could not.

By the early years of the 20th century most microbiologists had accepted the notion that there was a class of very small infectious agents with the following properties: they would pass through filters that retained bacteria, they were very difficult or impossible to visualize by light microscopy, and they could not be cultured on bacteriological media. To distinguish these agents from more familiar microorganisms such as bacteria and protozoa, the term 'invisible microbe' was suggested by Roux (1903) and 'filterable viruses' by Remlinger (1906); soon the latter term was widely accepted. In a review of filterable viruses in the mid-1920s, Rivers (1927a) summed up the position as he and many colleagues saw it: 'filterable viruses appear to be obligate parasites in the sense that their reproduction is dependent upon living cells. Whether this reproduction occurs intra- or extra-cellularly is a debated question'. Gradually the adjective 'filterable' was dropped altogether and by about 1940 scientists spoke of these invisible, ultrafilterable, infectious agents as 'viruses', and it was universally accepted that their reproduction occurred only within susceptible cells. Initially the virus particles that could be visualized with the light microscope were called 'elementary bodies', then 'elementary particles'; later the particulate forms of all viruses were called 'virus particles'. Eventually virologists accepted Lwoff's proposal, supported by an international group of virologists (Caspar *et al.*, 1962), to use the term 'virion' to describe the mature virus particle, using the word 'virus' to embrace all phases of the viral life cycle. The concept of viruses as a unique kind of infectious agent was brilliantly presented by Lwoff (1957) in the Third Majory Stephenson Memorial Lecture, epitomized by his phrase: 'viruses should be considered as viruses because viruses are viruses'.

The Discovery of Myxomatosis in Montevideo, Uruguay

In 1896, to further his experimental studies, Sanarelli (see Fig. 4.1), who had set up a laboratory in Montevideo, Uruguay, imported some laboratory rabbits from Brazil. Some of them became sick with a novel, highly specific disease characterized by swelling of the ears and eyes ('leonine facies' – see Fig. 4.2). He named the disease infectious myxomatosis (from the Greek *muxa*, mucus; *oma*, tumour) (Sanarelli, 1898), to describe the syndrome it produced in European rabbits – an infectious, usually lethal disease characterized by numerous lumps on the skin which exuded mucus when sectioned. In addition, such animals had swollen heads, their eyes were closed, the perineal region was swollen, and they almost invariably died within 14 days of infection.

Fig. 4.1. Guiseppe Sanarelli (1864–1940). Born in Monte San Savino, in Italy, Sanarelli graduated in medicine at the University of Siena in 1889. He then worked at the Pasteur Institute for two years, publishing work on the pathogenesis of typhoid fever. In 1893 he returned to Siena as Professor of Hygiene, but later that year went to the University of Montevideo, in Uruguay, to establish an Institute of Experimental Hygiene. Like so many other scientists in the Americas at that time, at first he worked on yellow fever, and like others, thought that he had discovered a bacterium that caused the disease. Having observed a strange, new lethal disease in his laboratory rabbits in 1896, he published the first description of myxomatosis in 1898. That year he returned to Italy, where he became involved in politics and did not return to scientific work until 1912. In 1914 he was appointed Professor of Hygiene at the University of Rome, where he carried out important work on tuberculosis and cholera.

The Classification of Myxoma Virus

In 1927 Aragão, who had previously worked with variola (smallpox) virus, published excellent illustrations of the 'elementary bodies' of myxoma virus as seen in stained smears by high power microscopy (Fig. 4.3A), which he noted were very similar to those of smallpox, molluscum contagiosum and epithelioma of fowls (fowlpox) viruses. This similarity has been confirmed by modern electron microscopy (see Fig. 4.3B,C). However, myxoma virus was not included by Goodpasture (1933) in the first description of what became the 'poxvirus group', although, surprisingly, the apparently extinct virus of 'horsepox' was included.

In 1941, the distinguished plant virologist F.C. Bawden suggested that viral classification should be based on the properties of the virus particle. Starting at the Fifth International Congress of Microbiology in Rio de Janeiro in 1950, efforts were made by international committees to devise schemes for viral classification that placed major emphasis on the size, shape and chemistry of the virus particle. At the next International Congress, a Subcommittee on the Nomenclature of Viruses was established by the Judicial Commission of the Committee on Bacteriological Nomenclature, and in 1955 this subcommittee commissioned selected virologists to produce formal descriptions and classifications of five groups of viruses.

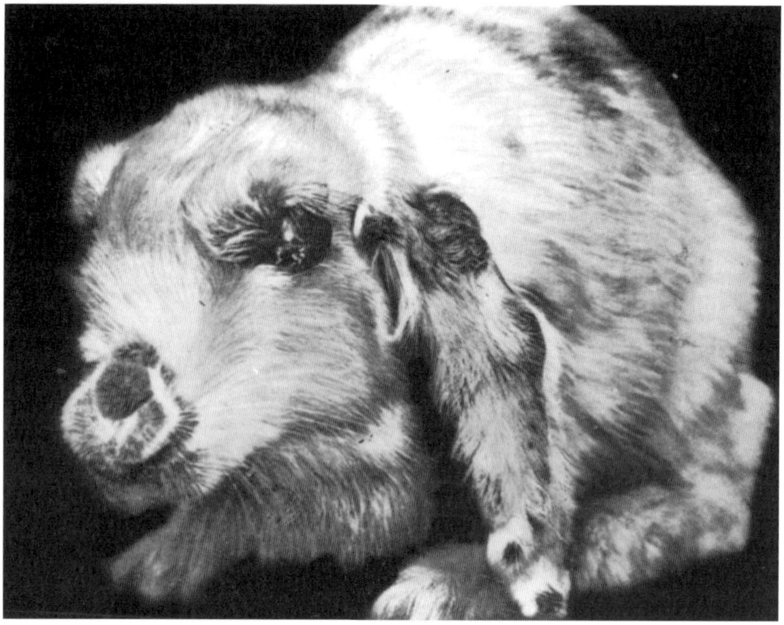

Fig. 4.2. Myxomatosis in a spontaneously infected laboratory rabbit, at an advanced stage of the disease, showing the 'leonine facies'. From Aragão (1927), with permission.

Fenner and Burnet (1957), writing on behalf of the Poxvirus Subcommittee, grouped myxoma virus with rabbit fibroma virus and squirrel fibroma virus as a subgroup within

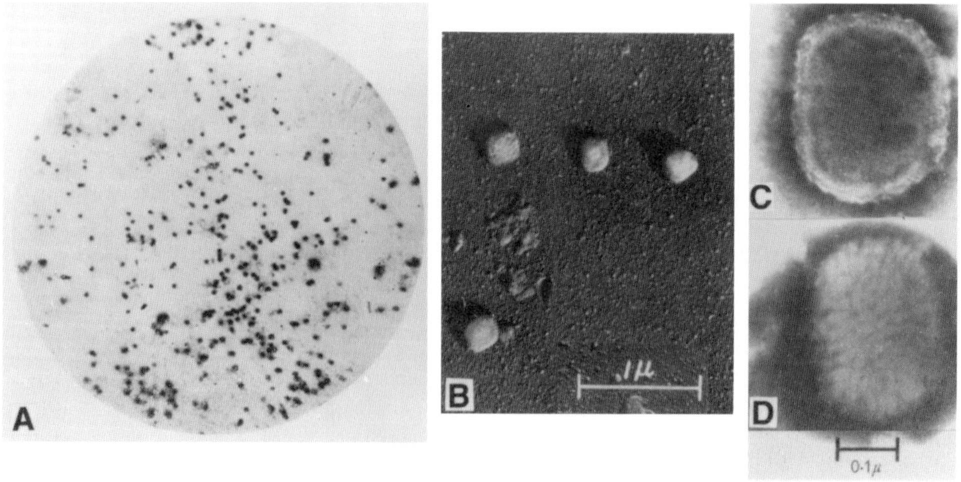

Fig. 4.3. Virions of myxoma virus. (A) Enlarged 2500 times, as demonstrated by light microscopy. (B) Fixed with osmium tetroxide and shadowed with uranium, electron-micrograph. (C) Negatively stained with phosphotungstic acid, electron-micrograph, enveloped form. (D) Negatively stained with phosphotungstic acid, electron-micrograph, non-enveloped form. (A) From Aragão (1927), with permission. (B) From Farrant and Fenner (1953), with permission. (C, D) From Padgett *et al.* (1964), with permission.

the poxvirus group. Eventually, with the development of a coherent system of virus classification by the International Committee on the Taxonomy of Viruses, myxoma virus was definitively classified as a member of the genus *Leporipoxvirus* in the family *Poxviridae* (Fenner, 1976). By this time Bawden's suggestion that classification should be based on the properties of the virion (virus particle) was universally accepted. The family *Poxviridae* was based on the possession by all members of a large brick-shaped virion with a genome consisting of a single, long molecule of double-stranded DNA. Within this family, genera were distinguished by cross-protection in experimental animals; later it was found that all members of each genus defined in this way had a distinctive genome, as defined by restriction mapping and sequencing. Further studies over the last 20 years have confirmed the validity of the family *Poxviridae* and the genus *Leporipoxvirus*, to which six distinctive viruses have now been allotted. Four of these occur in lagomorphs (rabbits and hares); the other two in squirrels (Table 4.1).

Shope's fibroma virus

The first leporipoxvirus to be studied extensively outside of Brazil was rabbit fibroma virus, discovered among eastern cottontails in New Jersey (*Sylvilagus floridanus*) by R.E. Shope, of the Rockefeller Institute for Medical Research (Shope, 1932). It produces localized fibromas in both cottontail and domestic rabbits (Fig. 4.4), which remain localized except when newborn or immunosuppressed rabbits are infected, when generalized fibromatosis may occur (Duran-Reynals, 1940). Cases have been observed in cottontail rabbits in the United States and Canada from the east coast to Wisconsin in the west and Texas in the south (see Fig. 4.7). It is transmitted mechanically by biting insects, especially mosquitoes, and cottontails infected as juveniles may serve as long-term reservoirs of infection (Kilham and Dalmat, 1955).

Table 4.1. Types of clinical disease produced by viruses of the genus *Leporipoxvirus* (family *Poxviridae*) in their natural hosts and in the European rabbit (*Oryctolagus cuniculus*).

Virus	Natural host	Endemic area	Clinical signs in *Oryctolagus cuniculus*	Eponym[a]
Rabbit fibroma virus	*Sylvilagus floridanus*	Eastern United States	Localized benign fibroma	Shope's fibroma
Brazilian myxoma virus	*Sylvilagus brasiliensis*	South and Central America	Generalized, lethal disease, gross external signs	Aragão's fibroma
Californian myxoma virus	*Sylvilagus bachmani*	Western United States, Baja California	Generalized, lethal disease, few external signs	Marshall–Regnery fibroma
Hare fibroma virus	*Lepus europaeus*	Europe	Localized benign fibroma	
Squirrel fibroma virus[b]	*Sciurus carolinensis*	Eastern United States	Localized benign fibroma (non-transmissible)	
Western grey squirrel fibroma virus[c]	*Sciurus griseus griseus*	Western United States	Not tested	

[a]Since the disease in *S. floridanus* was called 'Shope's fibroma', it is not unreasonable to give the other two American leporipoxviruses that produce fibromas in *Sylvilagus* rabbits the eponyms of 'Aragão's fibroma' and 'Marshall–Regnery fibroma', to emphasize that the type of lesion that they produce in their natural hosts is a fibroma.
[b]Kilham (1955).
[c]Regnery (1975).

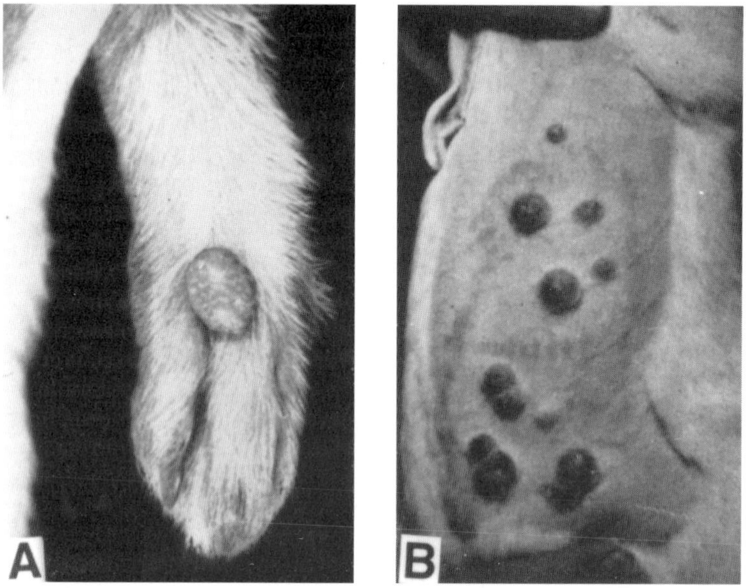

Fig. 4.4. Shope fibromas. (A) Fibroma produced in an eastern cottontail (*Sylvilagaus floridanus*) two months after being bitten by a mosquito that had fed on another cottontail fibroma a week earlier. (B) Some of the fibromas produced on the back of a domestic rabbit by some 40 mosquitoes eight days after they had fed on a cottontail fibroma. From Kilham and Dalmat (1955), with permission.

Occasionally, in places where cottontails and mosquitoes are common, cases of fibromatosis may occur in rabbits housed in unscreened commercial rabbitries (Joiner *et al.*, 1971; Raflo *et al.*, 1973). The fibromas produced by fibroma virus in *Oryctolagus* regress within three weeks of inoculation, compared with months for fibromas in *S. floridanus*. Mosquito transmission is difficult to demonstrate with fibromas of *Oryctolagus* but easy in those of *S. floridanus* (Day *et al.*, 1956; Dalmat, 1959; Dalmat and Stanton, 1959).

As would be expected for poxviruses belonging to the same genus, infection with fibroma virus provides protection against myxomatosis (Shope, 1932). The Boerlage strain of virus was found to provide serviceable protection for 12 months (Fenner and Woodroofe, 1954), and in France some ten million doses a year were used for the first few years of the epidemic of myxomatosis there, mainly in domestic rabbits but also among wild rabbits on some hunting estates.

Hare fibroma virus

Hare fibroma (Fig. 4.5) is something of an enigma. It was reported in European hares (*Lepus europaeus*) during the 1950s and is the only leporipoxvirus not native to the Americas. Some years ago the disease was investigated by Leinati *et al.* (1961) and the virus by Fenner (1965), but extensive correspondence in 1993 with wildlife microbiologists in Austria, Belgium, France, Germany, Italy and the United Kingdom showed that it had not been recognized in those countries since 1964 (Fenner, 1994). Since hare fibroma and the two squirrel fibromas (diseases native to North America) are peripheral to the topic of this book they will not be further discussed.

Further Studies of South American Strains of Myxoma Virus

Ten years elapsed after Sanarelli's discovery before another paper on myxomatosis was published, although the disease was

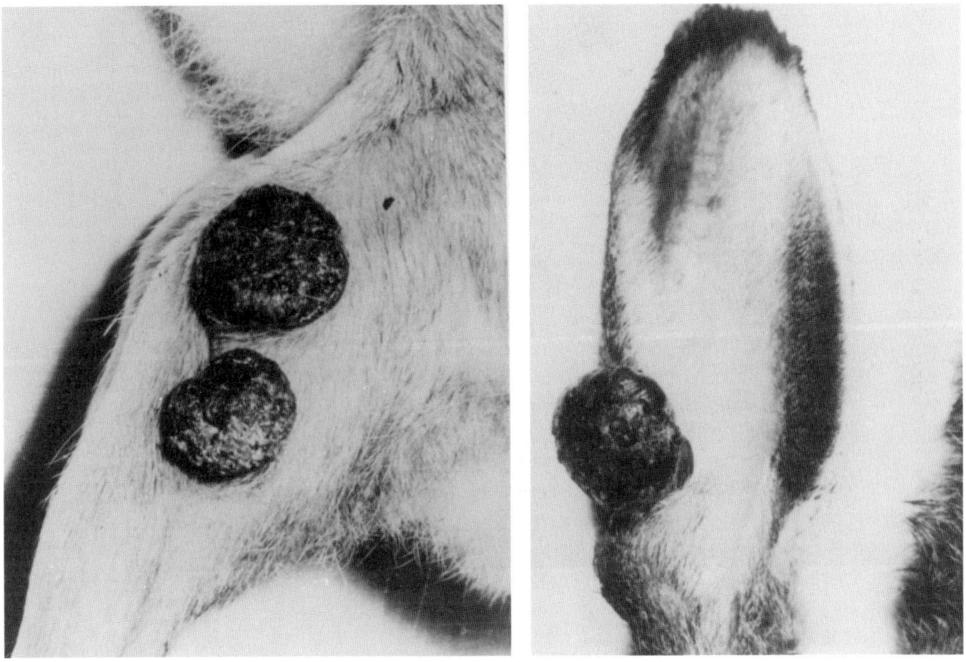

Fig. 4.5. Naturally occurring hare fibromas caused by hare fibroma virus. Courtesy Dr A. Leinati.

mentioned by Roux (1903) in an early review on filterable viruses. The next publication came from the bacteriological laboratory of the Portuguese Hospital in São Paulo, Brazil (Splendore, 1909). Splendore recognized that myxomatosis had been the cause of deaths among European rabbits coming to his laboratory from the town market in São Paulo. Like Sanarelli, he was unable to demonstrate that the infectivity was filterable. Noting that 'myxoma cells' stained with Giemsa contained inclusions similar to those found in trachoma, he speculated about the possible protozoal nature of the infection, an idea that was to persist for some years.

Three years later Moses (1911), working at the Oswaldo Cruz Institute in Rio de Janeiro, isolated myxoma virus from a locally infected rabbit and showed that it could be passed through Berkefeld but not through Chamberland filters, suggesting that it was one of the larger filterable viruses. In 1926 Moses responded to a request from scientists in the United States for a strain of myxoma virus by sending one isolated at the Instituto Oswaldo Cruz (possibly his 1911 strain, possibly a later isolate) to E.H. Simon at the Johns Hopkins University. Subsequently this strain was sent to A. Carrel at the Rockefeller Institute for Medical Research, where it was used in turn by Rivers (1927b) for laboratory studies of myxoma virus and by Shope (1932) in his characterization of rabbit fibroma virus. In 1934 Dr L.B. Bull, of the [Australian] Council for Scientific and Industrial Research, took this 'Moses' strain from Shope's laboratory to C.J. Martin in Cambridge, for his experimental investigation of its suitability for rabbit control in Australia (Martin, 1936). It was subsequently brought to Australia and used (as the 'Standard Laboratory Strain') to introduce myxomatosis into the Australian wild rabbit population.

In 1911 Aragão (Fig. 4.6), who was to become a major figure in studies of the

Fig. 4.6. Henrique de Beaurepaire Rohan Aragão (1879–1956) began laboratory work at the Manguinhos Institute while studying at the Rio de Janeiro Faculty of Medicine, from which he graduated with distinction in 1905. He then joined the Institute as assistant to Dr Oswaldo Cruz and spent his life working there. His work covered a very wide spectrum of infectious diseases, in some 25 different areas of bacteriology, protozoology, acarology and virology, and he is regarded as the founding father of virological and protozoological research at the Oswaldo Cruz Institute. In virology, he carried out distinguished work on smallpox and yellow fever, and contributed major insights into the natural history of myxomatosis, identifying its reservoir host in South America and demonstrating the importance of mechanical transmission of the virus by arthropods. In 1918 he suggested to the governments of Australia and Argentina that myxomatosis could be used for the biological control of the rabbit pest in those countries. Further biographical data in Coura (1994).

natural history of myxoma virus, published a short note in which he proposed that the causative agent could be visualized as small granules in the nucleus, a suggestion that he subsequently withdrew. He did not publish again until 1920, and then only a short paper on the transmission of myxomatosis by cat fleas (Aragão, 1920), an observation followed up by the demonstration by Torres (1936) that mosquitoes could act as vectors. However, as early as 1918 Aragão was sufficiently impressed with the lethality of myxoma virus for European rabbits, and its host specificity, that he wrote to the Australian government suggesting that myxoma virus should be used for rabbit control (see p. 117).

Some years later Aragão (1927) summarized his studies. He described the symptomatology of the disease in European rabbits and the way it spread; by contact between infected and susceptible rabbits or by introducing susceptible rabbits into hutches that had previously been occupied by diseased rabbits, and also by the bites of fleas, for at least 24 hours after they had fed on an infected rabbit. Having worked previously with variola virus, he noted the resemblance between the particles seen in smears of pus from cases of smallpox and from mucous material from sections of lesions of myxomatosis (Fig. 4.3A). He confirmed Sanarelli's observations concerning the high species specificity of the virus, as judged by its lack of infectivity for domestic animals and all laboratory animals except the rabbit. Reiterating his written suggestion of 1918, he concluded the paper with the proposal that it could be used for rabbit control in the Argentine and Australia, noting that trials arranged by the New South Wales Department of Agriculture were then in progress (White, 1929).

The reservoir host

For many years the origin of myxomatosis in Sanarelli's laboratory rabbits in 1896 was a complete mystery, as were subsequent outbreaks in hutch and laboratory rabbits in Rio de Janeiro, São Paulo and elsewhere. Bearing in mind the high host specificity of myxoma virus, it seemed to Aragão (1942, 1943) that it would be reasonable to test Brazilian wild rabbits, the tapeti or tropical forest rabbit (then called *Sylvilagus minensis* – now *S. brasiliensis*) for its sensitivity to infection. He found that among 42 such rabbits bought in the local market, about 40% were susceptible to infection. In contrast to the lethal disease found in laboratory rabbits, characterized by numerous swellings in the skin of many parts of the body, infection of *Sylvilagus* rabbits by mosquito bite or scarification of the skin produced only a small, localized, raised tumour, or sometimes a larger, flat tumour, and the rabbit showed few signs of illness. He also showed that the disease could be transferred from one tapeti to another, or from the tumour on a tapeti to a laboratory rabbit, by *Aedes aegypti* and *Ae. scutellaris* mosquitoes. Mosquitoes were infectious for several successive bites for up to 17 days after an infectious meal on a tumour, and subinoculation of the probosis, thorax and abdomen showed that virus was found only on the proboscis. On this evidence, and because there was no extrinsic incubation period such as was found with yellow fever virus, with which he had had considerable experience, he concluded that transmission was mechanical.

The suggestion that the tapetis that were not susceptible were immune because of prior infection was supported by the capture of a naturally infected animal with a lesion in the State of Rio, and the occurrence of transmission by mosquitoes explained the summer incidence of outbreaks of myxomatosis in commercial European rabbit breeding establishments. Such outbreaks continued to occur, and in a summary of diseases of domestic rabbits in the State of São Paulo between 1963 and 1967, Giorgi (1968) listed 31 diagnoses of myxomatosis among 1006 examinations of material from sick hutch rabbits.

S. brasiliensis is widely distributed in South and Central America (Fig. 4.7). Little has been published on the occurrence of myxomatosis in other countries of South America, but investigations during visits to several South American countries by Marshall[1] showed that since 1949 outbreaks that presumably originated from *S. brasiliensis* have been recorded among European rabbits in Argentina, Brazil, Colombia, Panama, Uruguay and Venezuela (Fig. 4.7). Other countries in which myxomatosis has been reported to occur are Costa Rica, Ecuador, Mexico and Panama (P. Arambulo, personal communication, 1993).

In addition to myxomatosis resulting from transfer from *S. brasiliensis* to hutch or laboratory rabbits, myxoma virus has been used for the biological control of pest European rabbits in the island of Tierra del Fuego in Chile, where there are no *Sylvilagus* rabbits (see p. 29).

Characteristics of infection in Sylvilagus brasiliensis

The only description of lesions due to myxoma virus in *S. brasiliensis* is that provided by Aragão (1943). Small localized tumours (Fig. 4.8), which are fibromas by histological criteria, appear 5–7 days after probing by infective mosquitoes and develop slowly, to reach a diameter of about 1 cm before regressing some 10–40 days later. Secondary lesions do not occur and there are no generalized signs of infection. Transfer by mosquitoes from these lesions produces a very severe disease with gross lesions in laboratory rabbits (Fig. 4.8C).

Host specificity

Although Sanarelli claimed that one of the dogs he inoculated was infected, Aragão (1927) could not confirm this, nor were any of the other animals tested (horses, cattle, fowls, ducks, pigeons, goats, sheep, monkeys, guinea-pigs, mice, rats, ferrets, or hamsters) susceptible to infection. Later Bull and Dickinson (1937) again tested these animals and eight native Australian mammals, three native lizards and five native birds with negative results. The only introduced wild animal that they tested was the European hare (*Lepus europaeus*); no lesions developed in any of the nine animals tested.

Several years later several workers reported that cases of myxomatosis had occurred in hares during the explosive spread of myxomatosis in Europe in the early 1950s (Magallon *et al.*, 1953; Lucas *et al.*, 1953; Jacotot *et al.*, 1954a; Collins,

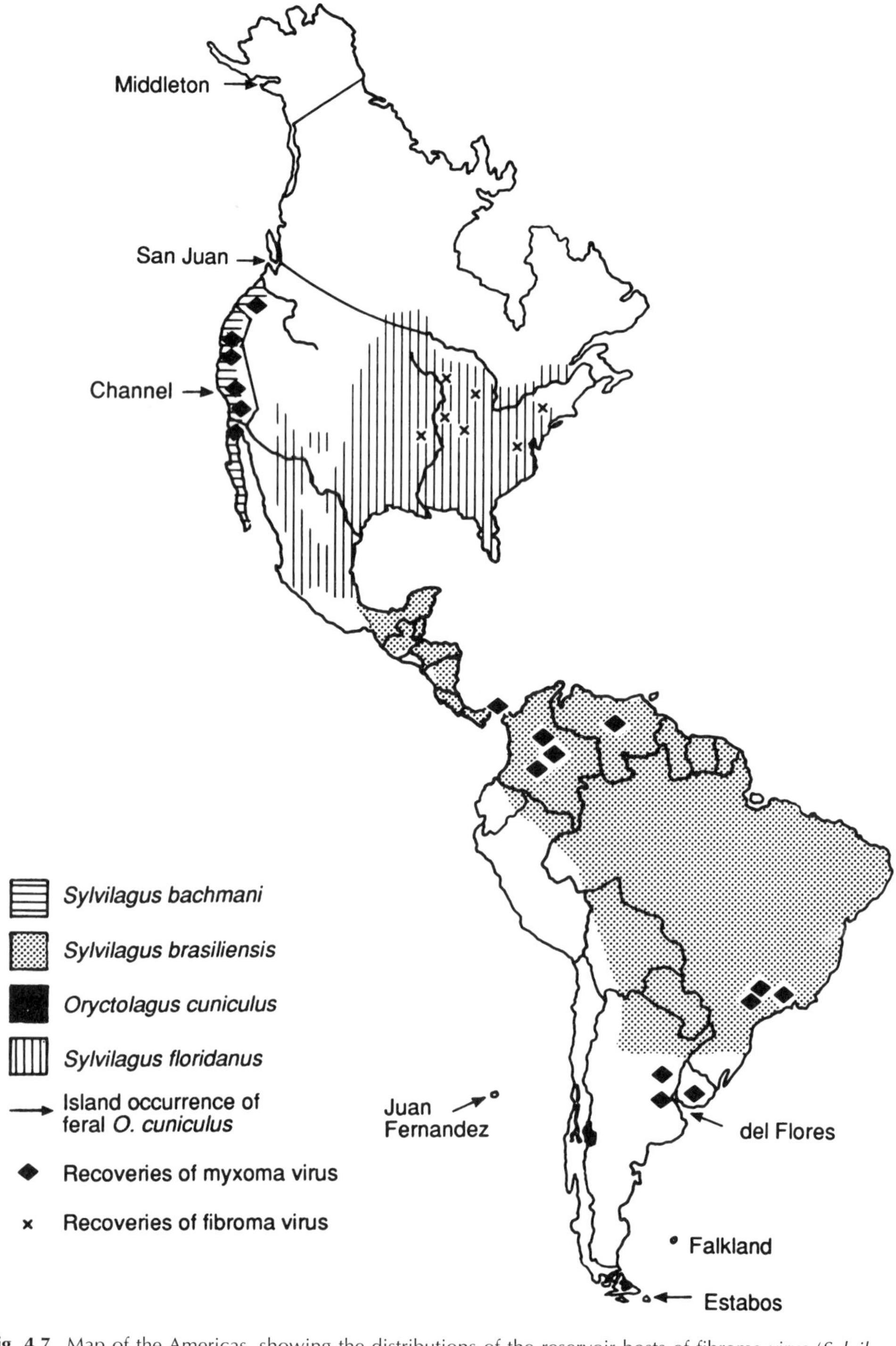

Fig. 4.7. Map of the Americas, showing the distributions of the reservoir hosts of fibroma virus (*Sylvilagus floridanus*), South American myxoma viruses (*Sylvilagus brasiliensis*) and Californian myxoma virus (*Sylvilagus bachmani*), and of feral *Oryctolagus cuniculus* in the Americas. Symbols indicate some of the sites from which the different leporipoxviruses have been isolated from their natural hosts.

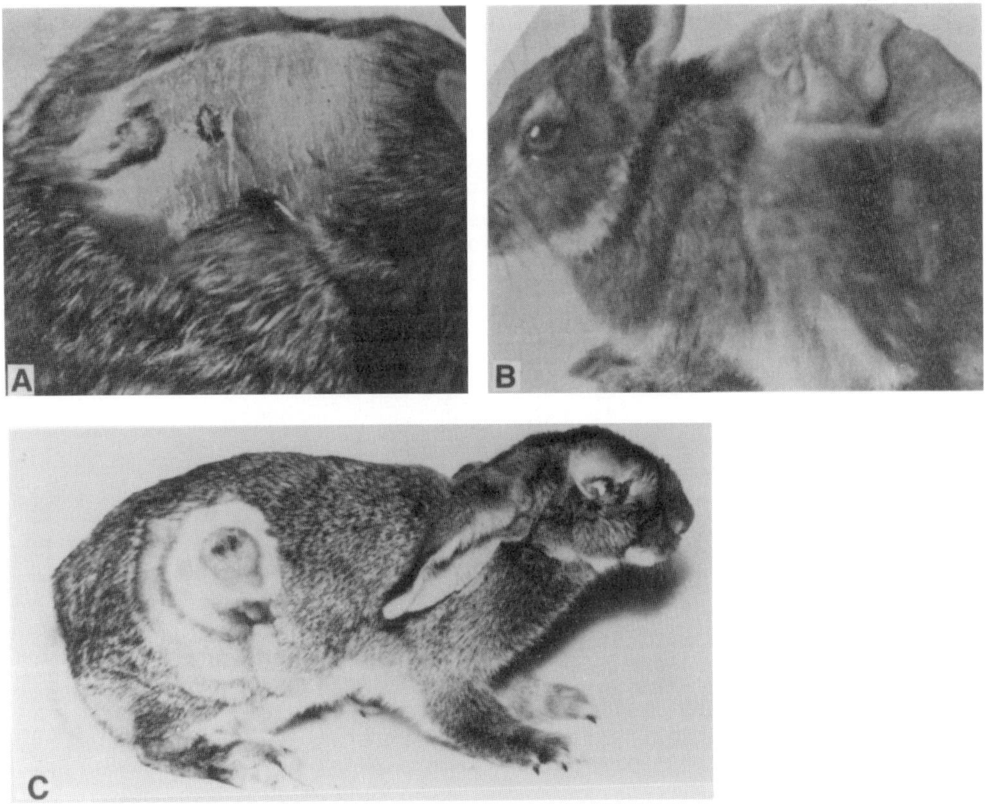

Fig. 4.8. Lesions produced by the Brazilian strain of myxoma virus in the reservoir host in South America, the tapeti, *Sylvilagus brasiliensis* and on transfer to *Oryctolagus cuniculus.* (A) Small fibroma produced in *Sylvilagus brasiliensis* by the bite of a mosquito which had probed a lesion in a domestic rabbit. (B) Large flat lesion produced in *Sylvilagus brasiliensis* by inoculation of a suspension of myxoma virus. (C) Domestic rabbit (*Oryctolagus cuniculus*) 10 days after inoculation with a small dose of myxoma virus. (A, B) from Aragão (1943), with permission.

1955; Kejdana, 1955; Whitty, 1955). Subsequently Jacotot *et al.* (1955) inoculated 13 hares from different parts of Europe with large doses of virus obtained from a fatal case of myxomatosis in either a rabbit or a hare. In one of the inoculated hares a small lump developed which was shown to contain myxoma virus when it was removed 15 days later; virus was also recovered from the testes of three rabbits nine, 12 and 15 days after intratesticular inoculation. Subsequent observations have confirmed that natural infection of hares is rare, and that very rarely a hare may suffer from severe generalized myxomatosis (Fig. 4.9).

Regnery (1971) tested the susceptibility of three species of North American cottontails (*S. audubonii, S. floridanus* and *S. nuttallii*) to infection with a Brazilian strain of myxoma virus. Prominent tumours developed on all three, and in *S. nuttallii* the Brazilian strain of virus produced severe disease with extensive secondary tumours (similar to the disease in *Oryctolagus cuniculus*), which could be passed by mosquito bite. He suggested that if it were introduced into wild populations, the Brazilian strain of myxoma virus could become established in *S. audubonii* and *S. nuttallii* populations and might reduce the population density of *S. nuttallii.*

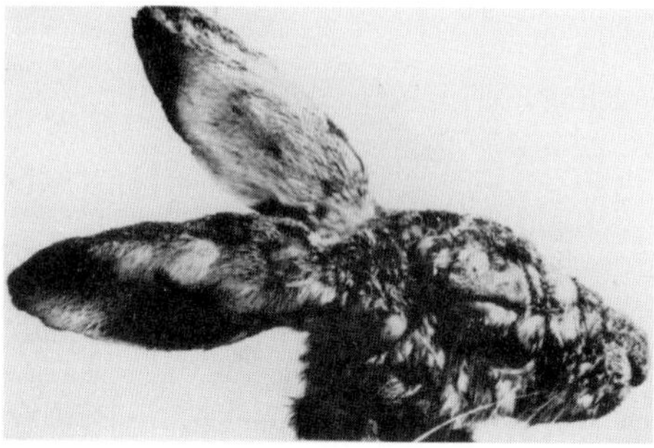

Fig. 4.9. Head of a European hare (*Lepus europaeus*) naturally infected with myxoma virus. Such cases were extremely rare. From Jacotot *et al.* (1954a), with permission.

Characteristics of infection in Oryctolagus cuniculus

In addition to the Brazilian strain used to initiate infection of wild rabbits in Europe in 1952 (see p. 213), we were able to examine additional recently isolated strains of myxoma virus from Brazil and strains from Uruguay, Argentina, Colombia and Panama (Fenner and Marshall, 1957; Fenner, 1965; Woodroofe and Fenner, 1965). All these strains produced a syndrome in laboratory rabbits similar to the severe, lethal disease described by Sanarelli and Aragão (Fenner and Marshall, 1957). A skin lesion appeared 4–5 days after the bite of an infective mosquito and enlarged to become a hard, hemispherical, purple tumour about 3 cm in diameter by the ninth or tenth day. The eyelids became thickened on the sixth day and the eyes were usually completely closed by the ninth day (Fig. 4.8C), and there was a semipurulent ocular discharge. Secondary lesions were widely distributed over the body from the sixth or seventh day and there was an oedematous swelling of the head, base of the ears and genitalia. Death was almost invariable, 8–15 days after infection. The strain used to initiate the disease in wild rabbits in Australia also derived from Brazil, but it had been passaged for many years in laboratory rabbits and produced less protuberant tumours than recently isolated strains.

Myxomatosis in Western United States

Vail and McKenney (1943) reported the occurrence of myxomatosis among domestic rabbits in commercial rabbitries in San Diego County in the summer of 1928. Within four years cases were seen in the Los Angeles area and in 1937 near Corvallis, Oregon. They noted that myxomatosis was called 'mosquito disease' by some rabbit breeders, because it was frequently found when mosquitoes were numerous around rabbitries. Kessel *et al.* (1931) reported 12 outbreaks in rabbitries in the regions of Santa Barbara, Ventura and San Diego in the summer of 1930; outbreaks also occurred in early summer in 1931, 1932 and 1933 (Kessel *et al.*, 1934). In comparisons with the Moses strain, provided to them by Dr Rivers, they noted that the Californian strain was somewhat less virulent. Subsequently, in a letter to J.S. Simmons (quoted in Fenner and Ratcliffe, 1965), Kessel commented: 'In the rabbitries in which the epidemic was encountered, the incidence was about 60% while the mortality of those that were infected was 100%. The outbreaks were usually sporadic in some ten rabbitries each season, and we usually heard nothing from breeders except during the months of May, June and July'. Vail and McKenney (1943) implied that the introduction of myxomatosis into California was associated with an importation of infected domestic rabbits to be used for laboratory purposes from Baja California, Mexico, to San Diego, California, in the late summer of 1927, but Kessel *et al.* (1934) suggested that indigenous wild rabbits might harbour the virus. Subsequently D.C. Regnery (Fig. 4.10) and his colleagues have shown that myxoma virus is enzootic in western North America in the local species of rabbit, *Sylvilagus bachmani*, so that the outbreaks in southern California probably had a local origin. Recent studies show that outbreaks in Baja California, Mexico, also derive from *S. bachmani*[2].

The reservoir host

In 1959–60 I.D. Marshall (see Fig. 5.1, p. 94), who had been a major contributor to

Fig. 4.10. David C. Regnery (1918–). After graduating AB from Stanford University in 1941, Regnery obtained a PhD degree from California Institute of Technology in 1947, on *Neurospora* genetics. The whole of his subsequent career was at Stanford University, where he became a Professor in the Department of Biological Sciences in 1953. He worked on *Chlamydomonas* genetics and histocompatibility of scale grafts in fish before becoming involved in research on myxomatosis, which arose after a lecture at Stanford University by Fenner on myxomatosis in 1957. In collaboration with Marshall (see Fig. 5.1), he carried out studies on myxomatosis in California in August 1959 which elucidated the natural history of the Californian strain of myxoma virus.

studies on myxomatosis in the period after its successful release in Australia, spent two years with Dr W.C. Reeves at the University of California, Berkeley, working on arbovirus diseases. Armed with letters of introduction to several scientists in the Bay Area, he established contact with Regnery, who worked at Stanford University, a biologist with an interest in the local mammals.

In August 1959 Regnery learnt of two outbreaks of myxomatosis in California, one among pet rabbits in Palo Alto and the other in four rabbitries near San Diego. During his weekends Marshall teamed up with Regnery to investigate these outbreaks. Noting that outbreaks in European rabbits had been interspersed with long quiescent periods, and knowing that a *Sylvilagus* rabbit was the natural host in South America, Marshall and Regnery (1960) decided to see whether the wild rabbits in the vicinity of the outbreak in Palo Alto, which were brush rabbits (*Sylvilagus bachmani*), were carriers of the virus. They trapped three brush rabbits and were able to recover myxoma virus from a small tumour on a leg of one of them; sera from all three contained antibodies to myxoma virus. In a study of the outbreaks near San Diego, myxoma virus was recovered from six diseased domestic rabbits (Marshall *et al.*, 1963). The isolates from *S. bachmani* and the domestic rabbits were tested in laboratory rabbits under standardized conditions in Canberra and found to resemble prototype Californian strain of myxoma virus (MSW, Fenner and Marshall, 1957) (see below).

Virus was isolated from two out of 71 pools of blood sucking diptera caught near the Palo Alto site; both were pools of *Anopheles freeborni*, which was much the most common insect. Using laboratory-bred and wild-caught mosquitoes, Grodhaus *et al.* (1963) showed that myxoma virus could be serially transmitted in brush rabbits if individual *A. freeborni* mosquitoes probed through the skin over tumours on donor *Sylvilagus* rabbits and then fed on marked sites on recipient rabbits (either *Sylvilagus* or *Oryctolagus*). Positive results were obtained with five other species of mosquito, and from tumours on brush rabbits that had just appeared (7 days after intradermal inoculation) up to the time that the localized tumour had become encrusted with a scab, usually 30–40 days later, and on one occasion almost 90 days later.

The epidemiology of myxoma virus infection in *S. bachmani* was further elucidated in an epizootic that occurred in an isolated population of brush rabbits in Almeda County (Regnery and Miller, 1972). Over the spring and summer of 1964 over 95% of a population of several hundred rabbits were infected. Complement-fixing antibody declined to low levels by the third month after infection, but animals were still immune to reinfection, although by 18 months they were often susceptible again.

As with *S. brasiliensis*, the mechanism by which myxoma virus survives in *S.*

bachmani throughout the winter has not been elucidated. The explanation may be that proposed by Kilham and Dalmat (1955) for the related Shope fibroma virus in cottontail rabbits (*S. floridanus*). They found that fibromas that developed in baby cottontails did not regress for almost a year and served as an effective source for mosquito transmission for at least 10 months. A wild-caught cottontail that had fibromas of natural origin remained infective for mosquitoes from December to May.

Characteristics of infection in Sylvilagus bachmani

Like the Brazilian strain of myxoma virus in *S. brasiliensis*, the Californian strain produced only a benign fibroma in *S. bachmani* (Fig. 4.11A). The earliest lesions, seen 7 days after mosquito probing, consisted of slightly thickened areas of skin, which became sharply delineated tumours about 1 cm in diameter. Scab formation and regression occurred between two and eight weeks later. At no time were there secondary lesions or signs of illness, in either mature, immature or pregnant animals, although occasionally there were multiple primary lesions, presumably because of multiple infective bites.

Comparisons of the behaviour of the Californian and Brazilian strains of myxoma virus in several Californian leporids suggested that the Californian strain is exquisitely adapted to survive in *S. bachmani* (Regnery and Marshall, 1971).

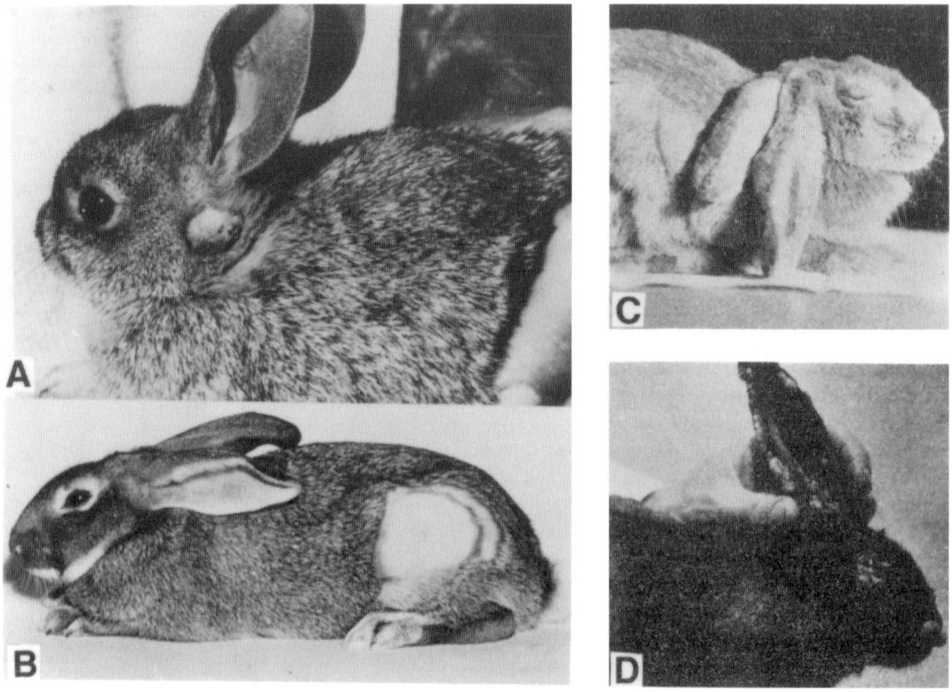

Fig. 4.11. Lesions produced by the Californian strain of myxoma virus in its reservoir host, *Sylvilagus bachmani*, and in *Oryctolagus cuniculus*. (A) Fibroma produced in *Sylvilagus bachmani* by the bite of a mosquito which had probed a similar lesion in another *Sylvilagus bachmani*. (B) Myxomatosis in *Oryctolagus cuniculus* due to infection with a Californian strain of myxoma virus, seven days after the intradermal inoculation of a small dose of virus. Although the external lesions were minimal, the rabbit died next day. (C) Acute disease, and (D), advanced disease in *Oryctolagus cuniculus* after inoculation with a strain of virus recovered in 1930; both cases were fatal. (A) From Regnery and Miller (1972), with permission. (B) From Fenner and Marshall (1957), with permission. (C, D) From Kessel *et al.* (1934).

Using fibromas produced in *S. bachmani* by Californian myxoma virus as sites for mosquito probing, infectivity was tested in five species of *Sylvilagus* and in *Oryctolagus cuniculus*. If a local lesion was produced in the recipient rabbit, mosquitoes were induced to probe on this and then tested for infectivity on the most susceptible host, which was *Oryctolagus cuniculus*, and on another animal of the species on which the mosquitoes had originally probed. Tumours developed in each of five species of *Sylvilagus* (*S. audubonii, S. bachmani, S. floridanus, S. idahoensis* and *S. nuttallii*), and probes from each produced lesions in *Oryctolagus*, but serial transfer by mosquito bite was successful only from lesions in *S. bachmani*. *S. brasiliensis* could not be infected with the Californian strain of myxoma virus but were susceptible to infection with a South American strain. On the other hand, although the South American strain of virus produced fibromas in *S. bachmani*, these did not contain sufficient virus for serial mosquito transmission (Marshall and Regnery, 1963). These results suggest a high degree of coevolution between the Californian strain of myxoma virus and the brush rabbit, *S. bachmani*, a matter that is further discussed in Chapter 14 (see p. 313).

Characteristics of infection in Oryctolagus cuniculus

In their description of infection of domestic rabbits with Californian strains of myxoma virus, Kessel *et al.* (1931, 1934) noted that although invariably lethal, the disease appeared to be less virulent than that caused by the Moses strain, in that the incubation period was longer and the average length of life after infection longer. Their illustrations (Fig. 4.11C, D) depict a disease that looks very similar to myxomatosis caused by South American strains of virus. This contrasts with later studies, in which it was observed that the signs of disease (Fig. 4.11B) were 'muted' compared with those of myxomatosis due to South American strains.

With all strains available to us (Fenner and Marshall, 1957; Marshall *et al.*, 1963) the primary lesion appeared on the third day and slowly increased in size, but never became prominent and the edges merged into normal skin. Swelling of the eyelids was first seen on the seventh day but the eyes were never closed and the conjunctival discharge was slight in amount and thin. Secondary tumours and swelling of the anogenital region appeared about the ninth day. Signs of nervous system disease, consisting of very rapid tremor or convulsions, were not uncommon, and rabbits often died on the seventh or eighth day, before most of the signs of 'classical' myxomatosis appeared. On the other hand, the illustrations in a paper by Patton and Holmes (1977), reporting outbreaks of myxomatosis among European rabbits in 26 rabbitries in western Oregon (a region in which *S. bachmani* is endemic), show rabbits with advanced lesions of the kind commonly seen in infections with slightly attenuated South American strains (see Fig. 4.11C, D). R. Maria Licon (personal communication, 1996) found that a naturally infected domestic rabbit from northern Baja California (where 16% of *S. bachmani* were serologically positive)[2] showed similar signs. Not surprisingly, strains with different virulence for European rabbits may occur in different parts of the range of *S. bachmani*, and perhaps also within the same geographic area.

Other Comparisons of Myxoma Viruses from the Americas

Only a few tests other than symptomology of the disease in laboratory rabbits have been made comparing strains of myxoma virus from different parts of the Americas. Woodroofe and Fenner (1965) found that all the Californian isolates produced small clear plaques on rabbit embryo fibroblast and rabbit kidney cell monolayers, whereas most South American strains produced larger plaques. In gel diffusion tests, most strains from South America could be readily distinguished from the Californian strains, but some strains from Colombia and Panama resembled the Californian strains in gel diffusion tests (see p. 101) but produced typical 'Brazilian-type' lesions in European rabbits.

Mechanisms of Transmission of Myxomatosis

The investigations just described had demonstrated that myxoma virus can be transmitted mechanically by mosquitoes between reservoir hosts in South America (*S. brasiliensis*) and California (*S. bachmani*), and from these animals to European rabbits. However, early observations in laboratory and hutch rabbits had shown that myxomatosis could also be transmitted by other mechanisms.

Mechanisms other than arthropod bite

Sanarelli (1898) had noted the infectious nature of the new lethal disease that affected his laboratory rabbits in 1896, and Aragão (1927) showed that transmission could occur by contact between diseased and healthy animals, or by contact of a healthy rabbit with hutches that had housed diseased animals. These findings are consistent with the high virus content of the early ocular discharge and of fluid oozing from abraded skin lesions. All investigators agree that infection by the oral route is of negligible importance. Infection by the respiratory route may occur, but is rare unless rabbits are exposed to artificially produced aerosols containing virus (Martin, 1936; Mykytowycz, 1958), although some strains may be more infectious than others by the respiratory route. For example, Sobey and Chapple and Boulter (quoted by Fenner and Ratcliffe, 1965) found that the Glenfield strain (see p. 160) spread within an animal house much more readily than other strains, an observation confirmed in experiments with the Henderson apparatus (P.J. Chapple and E.A. Boulter, personal communication 1963). However, by far the most important mechanism of transmission in populations of wild *Sylvilagus* and European rabbits is via arthropod vectors (Fenner and Ratcliffe, 1965). Indeed, since such common ectoparasites of rabbits as *Haemodipsus ventricosus* (a blood-sucking louse) and *Cheyletiella parasitovorax* (a mite) can transmit myxomatosis (Mykytowycz, 1958), and in Europe, *Spilopsyllus cuniculi,* even 'contact' transmission may be due to insect bites.

In the laboratory myxomatosis can be transmitted by any route of inoculation, an observation that was extended when Australian scientists were asked to consider the possibility of using myxomatosis for rabbit control. Realizing that this would require that many wild rabbits should be infected, a modified rabbit trap was developed that would inoculate the rabbit with virus but not catch the animal (Anon., 1942). In 1937 they undertook a variety of preliminary studies on myxomatosis in European rabbits, including transmissibility, which were summarized by Bull and Mules (1944). By chance, the rabbit flea (*Spilopsyllus cuniculi*) had not been brought to Australia when rabbits were introduced. However, noting that mosquitoes and the stick-fast flea (*Echidnophaga myrmecobii*), a parasite of macropods, were part of the environment of wild rabbits during the warm months in Australia, they undertook experiments to determine whether these insects could act as vectors. These showed that *E. myrmecobii* and several species of mosquitoes were effective vectors, mosquitoes transmitting infection immediately after feeding on infected rabbits and for up to 14 days afterwards, hence they considered that transmission was mechanical.

Mosquito transmission

No further experiments on transmission were carried out until myxomatosis had spread through much of the rabbit population of southeastern Australia in the summer of 1950–51. As related in Chapter 6 (p. 141), Fenner, newly appointed as Professor of Microbiology in the Australian National University, decided in February 1951 to make virological studies on myxomatosis his major research project. Located temporarily in Melbourne until November 1952, he travelled periodically to Canberra and early in 1951 commenced studies on mosquito transmission of myxomatosis in collaboration with M.F.

Day (Fig. 4.12), a Canberra entomologist interested in insect transmission of plant viruses. In a series of papers (Fenner *et al.*, 1952, 1956; Day *et al.*, 1956; summarized in detail in Fenner and Ratcliffe, 1965) they demonstrated several aspects of insect transmission which were crucial for the understanding of the epidemiology and evolution of myxomatosis in wild European rabbits in Australia.

Transmission by mosquitoes is mechanical

The initial experiments (Fenner *et al.*, 1952) utilized female *Aedes aegypti* mosquitoes and sharp pins to determine the source of virus and the period over which transmission could occur. Mosquitoes were fed (or pinpricks made) on the rabbit's ear, in places where there were no skin lesions, and through the skin over a tumour. Six days after inoculation the rabbit was viraemic and a local lesion had developed at the inoculation site. From that time on, titration of the heads (including proboscis) and the abdomens of the mosquitoes showed that immediately after a blood feed on normal skin the abdomens always yielded virus, but the heads were negative. However, after feeding through the skin over a tumour both head and abdomen were positive, although due to the amount of blood it contained the titre of virus was much higher in the abdomen. Mosquitoes that had obtained a blood feed through normal skin never transmitted myxomatosis, even after four weeks had elapsed to allow for replication of virus in the mosquito (if it occurred), whereas many of those that had fed through the tumour transmitted immediately after feeding and at intervals over the next three weeks. Pinpricks gave similar results. Electron micrographs of a mosquito's mouthparts showed that virions were attached to the maxilla of a mosquito that had probed through a tumour (Fig. 4.13). Thus transmission depended not on contamination of the mouthparts with viraemic blood, but with virus from epidermal cells over the skin lesion (Fig. 4.14).

Fig. 4.12. Maxwell Frank Cooper Day (1915–). After graduating in biology at the University of Sydney in 1937, Day joined the CSIR Division of Economic Entomology. Almost immediately he went to Harvard University to study for a PhD degree, which he was awarded in 1941. After working at Washington University in St Louis in 1941–42, he undertook wartime jobs at the Australian Embassy in Washington from 1942 until 1947, when he rejoined the CSIRO Division of Entomology, where his special interest was in plant viruses and their transmission by insects. In the period 1951 to 1955 he collaborated with Fenner in studies of the transmission of myxomatosis by mosquitoes. After becoming Assistant Chief of the Division of Entomology in 1963, he moved to the CSIRO Executive from 1966–76 and then became Chief of the Division of Forest Research until his retirement in 1980. Day was appointed a Fellow of the Australian Academy of Science in 1956 and in 1977 he received the award of Officer of the Order of Australia (OA).

The infectivity for probing mosquitoes of various accessible parts of the body of advanced cases of myxomatosis showed that 100% of mosquitoes became infective after probing through the skin over the primary lesion, 78% after probing the swollen eyelids, 58% by probing the swollen base of

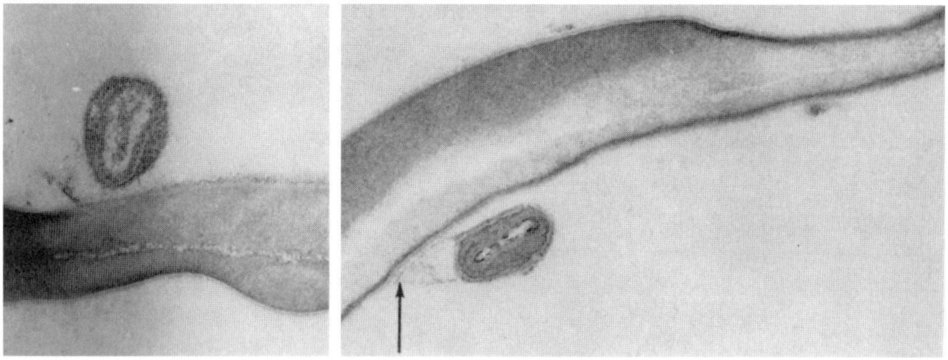

Fig. 4.13. Electron micrograph of myxoma virions on the maxilla of an *Aedes aegypti* mosquito after probing through a skin tumour. From Filshie (1964), with permission.

the ears and 15% after probing secondary skin lesions. Probing in unaffected skin gave negative results. When mosquitoes were allowed to probe through tumours at times ranging from 5 to 10 days after inoculation (the last in a recently dead rabbit), and then tested by probing on susceptible rabbits, positives rose from 38% on day 5 to 71% on day 9, and to 92% on the dead rabbit (probably because they probed repeatedly in an attempt to obtain blood). When individual infective mosquitoes fed on test rabbits on successive days the results were irregular, but in general were most often positive on the first day and then fell off.

These results confirmed the findings of Aragão (1943) and Bull and Mules (1944), namely that transmission of myxomatosis

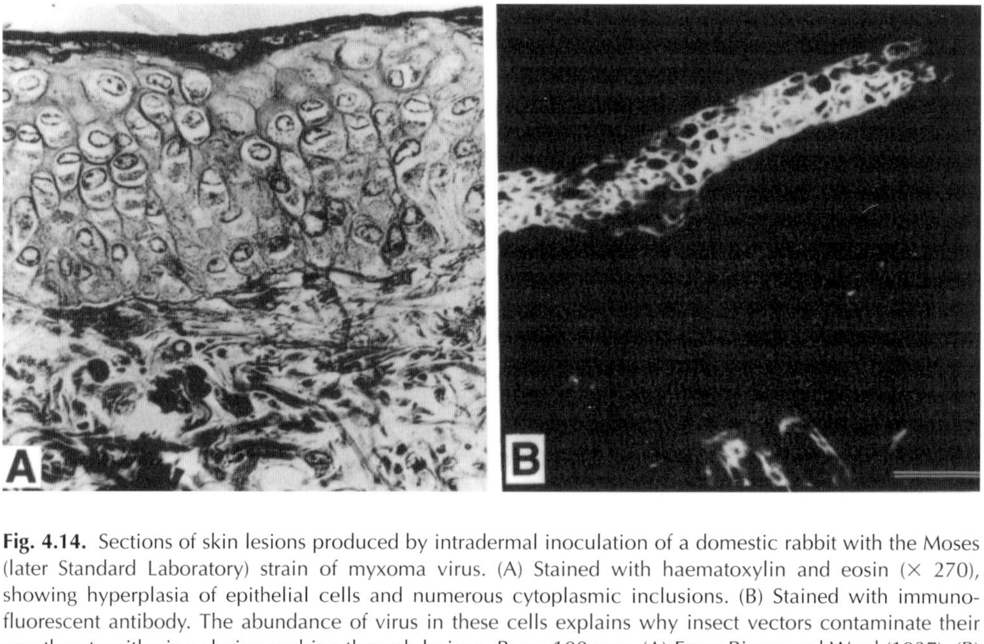

Fig. 4.14. Sections of skin lesions produced by intradermal inoculation of a domestic rabbit with the Moses (later Standard Laboratory) strain of myxoma virus. (A) Stained with haematoxylin and eosin (× 270), showing hyperplasia of epithelial cells and numerous cytoplasmic inclusions. (B) Stained with immunofluorescent antibody. The abundance of virus in these cells explains why insect vectors contaminate their mouthparts with virus during probing through lesions. Bar = 100 mm. (A) From Rivers and Ward (1937). (B) Courtesy Sandra Best.

by mosquitoes was mechanical. However, at about this time Kilham and coworkers (Kilham and Woke, 1953; Kilham and Dalmat, 1955), working with Shope fibroma virus in *Sylvilagus floridanus*, and Jacotot *et al.* (1954b), working with the strain of myxoma virus that occurred in Europe, suggested that although mechanical transmission occurred, these viruses might multiply in the mosquito. The problem was therefore reexamined, paying particular attention to the possibility of replication of the virus in the mosquito (Day *et al.*, 1956). The results were conclusive; viral multiplication in the mosquito did not occur. For example, groups of individually housed mosquitoes were allowed to probe repeatedly on susceptible rabbits at various times from 2 to 18 days after probing through a tumour, whereas the 18-day controls had not probed on the infected rabbit until that day. The steady fall in positive results in the test mosquitoes suggested that virus on the proboscis was eventually 'wiped-off', and that there was no enhancement of infectivity by replication. In another experiment large numbers of *Aedes aegypti* and *Anopheles annulipes* mosquitoes were fed through skin lesions containing high titres of virus and individually tested for virus at intervals after the acquisition feed. There was a progressive diminution in the virus titre of the mosquito suspensions and no evidence whatever of a rise after a latent interval.

Efficiency of transmission of strains of low virulence

The final series of experiments (Fenner *et al.*, 1956) was designed to examine the epidemiological significance of mosquito transmission of myxoma virus in relation to the attenuation of field strains of virus that had been observed by this time (see p. 172). The experiments just described had established that the important virus from the point of view of mosquito transmission was that in the superficial layers of skin through which the mosquito probed in search of a blood meal. The concentrations of virus in skin slices taken from the skin over tumours at different times from laboratory rabbits that had been infected with different strains of myxoma virus were therefore titrated. In parallel, on each day mosquitoes were induced to probe through these lesions and the numbers of infections resulting from 20 successive probes on marked sites on the back of susceptible rabbits determined. The results of the titrations of skin slices are shown in Fig. 4.15.

More recent work with a Grade V field strain (not available in the 1950s) has shown that as with all other strains (except neuromyxoma virus, see below), the titre rises to a level infectious for mosquitoes by the fourth day after infection but in contrast to Grade III and Grade IV strains, falls below that level by the twelfth day (S.M. Best and P.J. Kerr, unpublished observations). Highly virulent (Grade I) strains produced high enough titres of virus in the skin for effective transmission from about the fourth day after infection until the rabbit died on the tenth or eleventh day. Even the highest skin titre of a highly attenuated laboratory variant (the neuromyxoma strain of Hurst, 1937) was almost two logs lower than that of highly virulent strains, and mosquito transmission was correspondingly poor – only 12 out of 136 positive probes compared with 53 out of 88 for a virulent strain. The most important result, epidemiologically, was that skin titres in rabbits infected with somewhat attenuated strains, with longer survival times (Grade III and IV strains), remained high for over 20 days after infection and sometimes longer. These results proved of great value in interpreting the evolution of virus in the field in Australia. Little work was done on mosquito transmission in England, where fleas were much more important vectors, but Andrewes *et al.* (1956) showed that mosquitoes might be important for overwintering of the virus, since at hibernating temperatures (0–8°C) some mosquitoes remained infective for 220 days.

Flea transmission

Aragão (1920) had demonstrated that cat fleas (*Ctenocephalides felis*) could transmit

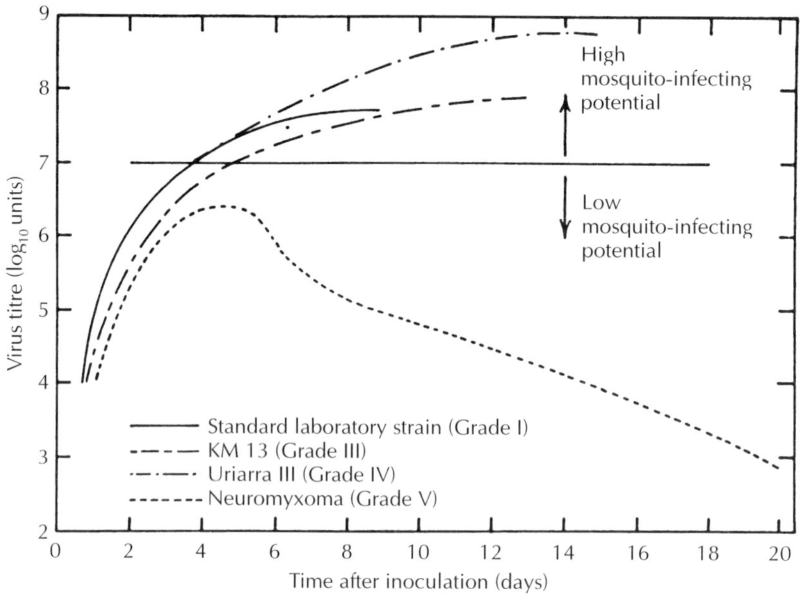

Fig. 4.15. Changes with time of the titre of virus in skin slices taken from the surface of lesions produced by the intradermal inoculation of rabbits with large doses of various strains of myxoma virus. Standard laboratory strain = highly virulent strain used to initiate myxomatosis in Australian wild rabbits; neuromyxoma = laboratory strain of very low virulence produced by Hurst (1937); KM 13 and Uriarra III are Australian field strains of moderate virulence (see Chapter 7). Lesions produced by highly attenuated strains rarely reach concentrations high enough to contaminate the proboscis of a probing mosquito; lesions produced by moderately or highly virulent strains reach high skin titres, but rabbits infected with highly virulent strains die soon after. Lesions caused by moderately virulent viruses may have highly infectious skin lesions for many days after infection, especially in the 10% or more animals that survive. From Fenner and Ratcliffe (1965), with permission.

myxomatosis and in early trials in Australia Bull and Mules (1944) found that the stick-fast flea (*Echidnophaga myrmicobii*) was an efficient vector. However, the European rabbit flea (*Spilopsyllus cuniculi*) did not occur in Australia, and it was not until myxomatosis spread in Britain that detailed research was carried out on the role of this flea in transmission of myxomatosis, by scientists of the Ministry of Agriculture, Fisheries and Food (review, Mead-Briggs, 1977).

Transmission by fleas is mechanical

As with mosquitoes, transmission by fleas is mechanical, positive results being obtained immediately after removal from a diseased rabbit and after several days of starvation (Muirhead-Thompson, 1956). The piercing mouthparts of a flea are about 300 mm long, about one-tenth as long as the proboscis of a mosquito, so fleas are capable only of relatively shallow probing. However, the highly developed cutting plates of the laciniae of fleas are better adapted for retaining large numbers of virus particles than the scanty teeth on the maxillae of mosquitoes, and as Fig. 4.14 indicates, there are numerous infected epithelial cells in the epidermis over a myxomatous skin lesion.

The importance of the rabbit flea as a vector

The year-round occurrence of cases of myxomatosis focussed attention on the European rabbit flea (*Spilopsyllus cuniculi*) as the important vector in Britain. During their experiments on the introduction of myxomatosis into the Heisker Islands (Shanks *et al.*, 1955), Allan and Shanks had shown that these fleas would transmit the infection, an observation confirmed by Lockley (1954). Rothschild (Fig. 4.16A), a world authority on fleas, emphasized the ability of the rabbit flea to maintain infectivity through the winter (Rothschild, 1953). This was confirmed by demonstrations that rabbits released in deserted burrows 50 days after the inhabitants had died of myxomatosis became infested with fleas and died of myxomatosis (Brown *et al.*, 1956) suggesting that fleas could act as a reservoir of infection for several months after rabbits had deserted a burrow. The potential prolonged infectivity of fleas was confirmed by the observation that some fleas that had fed through lesions of a rabbit with myxomatosis and were then buried in the ground in glass tubes were infective for as long as 112 days (Chapple and Lewis, 1965). In France, quiescent rabbit fleas were recovered from soil scrapings from deep burrows that had been abandoned by rabbits ten weeks earlier following autumn epizootics of myxomatosis, and myxoma virus was recovered from these fleas (Joubert *et al.*, 1969) These observations led to intensive investigations into the biology of the rabbit flea, notably into its breeding cycle, which have been admirably reviewed by Mead-Briggs (1977) (Fig. 4.16B).

Flea biology relevant to myxomatosis

When fleas are placed on a rabbit, most of them migrate to the head and many become firmly attached to the ears. *S. cuniculi* had long been thought to be a stationary species, rarely leaving their host unless the rabbit died (Rothschild, 1915). This belief prompted the suggestion that as vectors they would favour the persistence in Britain of highly virulent strains of myxoma virus (Fenner and Marshall, 1957; Andrewes *et al.*, 1959; Rothschild, 1960). However, experiments with marked fleas showed that movement between rabbits, and from the burrow floor on to rabbits, was much greater than had been thought (Mead-Briggs, 1964a). In addition, as explained below, rabbits infected with strains of moderate virulence carried a higher percentage of infectious fleas than rabbits infected with strains of very high or very low virulence.

Efficiency of transmission of strains of diminished virulence

Strains of diminished virulence had been found in Britain as early as 1955 (Hudson and Mansi, 1955), and by 1962 strains of moderate virulence (Grade III) constituted 63% of over 200 samples of myxoma virus recovered from rabbits in 80 counties in Britain (Fenner and Chapple, 1965). The influence on transmissibility by fleas of the type of disease produced by viruses of varying grades of virulence was investigated by Mead-Briggs and Vaughan (1975), whose results are summarized in Fig. 4.17. Few fleas (12%) from rabbits infected with fully virulent strains were infective, and few (8%) from individual rabbits that recovered from infection with attenuated strains. Rabbits which died within 44 days of infection with moderately virulent strains had, on average, the highest proportion of infective fleas (30–50%). At the time the wild rabbits used in this experiment were captured (before 1970) there had not been any substantial increase in genetic resistance in Britain; as in transmission by mosquitoes, the balance between host, virus and vector could change if host resistance increased.

Reproductive biology of Spilopsyllus cuniculi

Although other fleas parasitize rabbits (see p. 321), *Spilopsyllus cuniculi* is unusual in that egg maturation in the female is dependent on hormones found only in female rabbits at the late stages of pregnancy (Mead-Briggs and Rudge, 1960; Mead-Briggs, 1964b; Rothschild and Ford, 1964); for successful reproduction male fleas also need to have probing contact with a rabbit in the final stages of pregnancy or with a newborn nestling (Mead-Briggs and Vaughan, 1969). The life cycle of the European rabbit flea, worked out by Mead-Briggs and Rothschild and Ford (1964, 1972), is illustrated in Fig. 4.18 (see also

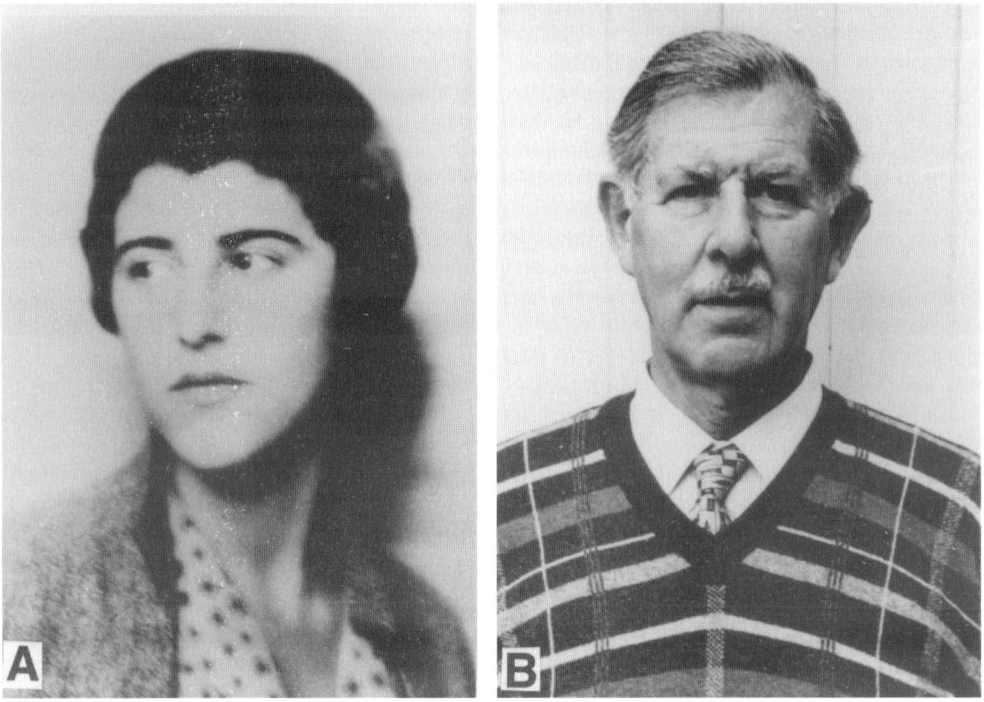

Fig. 4.16. (A) Miriam Louisa Rothschild (1908–). Daughter of distinguished entomologist N.C. Rothschild, Miriam Rothschild was educated at home and followed her father's interests in fleas, on which she is the recognized world authority. This expertise was of signal value to the Advisory Committee on Myxomatosis in the United Kingdom when it became clear that European rabbit fleas were the principal vectors in that country. She played a major role in elucidating their peculiar biology; their breeding cycle is controlled by the hormones of pregnant rabbits. Author of over 300 scientific papers and several books, she was made a Commander of the Order of the British Empire (CBE) in 1982 and was appointed a Fellow of the Royal Society in 1985.
(B) Anthony Mead-Briggs (1929–). After graduating with a BSc in Zoology at the University of Birmingham in 1953, Mead-Briggs obtained a PhD degree in the same university with a thesis on insect physiology. He then joined the Pest Infestation Control Laboratories of the Ministry of Agriculture, Fisheries and Food as a research scientist. His principal research in relation to myxomatosis, carried out between 1956 and 1978, related to the biology of the European rabbit flea and its importance as a vector of myxomatosis in Britain. As well as developing a method of culture of fleas, based on their unusual reproductive biology, he made important observations on the ability of fleas to locate rabbits and their ability to transmit strains of myxoma virus of differing virulence with differential effectiveness.

Fig. 14.1, p. 322). During the last few days of pregnancy, corticosteroid hormone levels in the blood of the doe increase, making her more attractive to fleas, which accumulate and become firmly attached, especially to the ears. The defaecation rate of the fleas increases from the normal rate of once every 20 minutes to one each minute at the time of birth of the rabbit's litter, so that the fleas are pumping blood into the nest, where the droplets dry and provide food for the larvae. Stimulated by the increased hormone levels before birth, the eggs in the female fleas mature, and a few hours after the rabbit drops its litter the fleas move on to the newborn kittens, feed avidly, copulate and lay eggs in the nest, maximally during the first 24 hours. A further specialization is that copulation will not occur unless a kitten is present in the nest, although the fleas do not necessarily have to feed on the kitten.

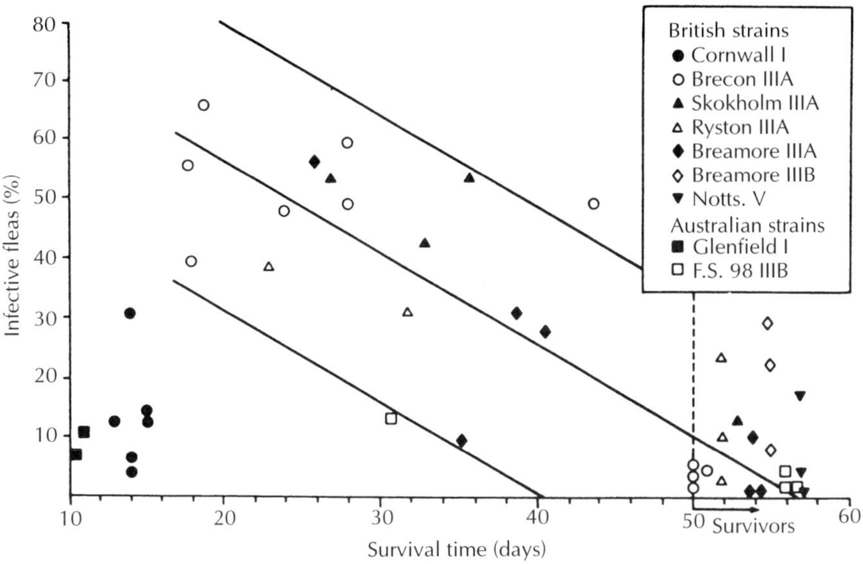

Fig. 4.17. The relationship between the survival times of wild rabbits infected with myxoma virus and the percentage of infective fleas obtained at standardized times during the infection. The regression line (shown with its 95% confidence limits) excludes the results for rabbits infected with the Grade I Cornwall and Glenfield strains, which killed the rabbits within 14 days of infection. From Mead-Briggs and Vaughan (1975), with permission.

Ten to twenty days after the birth of the kittens the fleas leave the nest and return to the doe. The eggs in the nest hatch and the larvae feed on the dried blood deposited earlier, turn into pupae and emerge as adult fleas, in two episodes. The first emergence

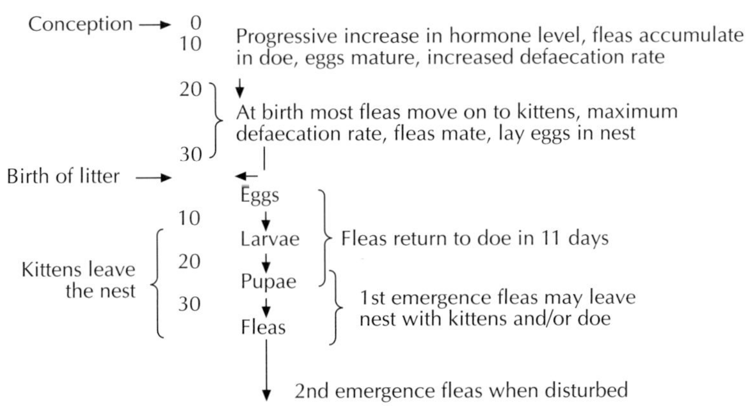

Fig. 4.18. Diagram illustrating the synchronization between the breeding cycle of the rabbit and the European rabbit flea (*Spilopsyllus cuniculi*). From Sobey (1977), with permission.

occurs 15–30 days after birth of the litter; these fleas leave the nest on the doe or the weaned kittens. If the nest is disturbed, a second emergence occurs at 30 days, or this emergence may be deferred for months, for example when a doe returns to clean out the nest for a second litter.

This complicated life cycle is found in *S. cuniculi* and *Cediopsylla simplex* (see p. 321), but not in *Xenopsylla cunicularis* (Cooke, 1990); its elucidation was crucial to the success of the importation and distribution of *S. cuniculi* in Australia (see Chapter 8).

The Spanish flea

Although *Spilopsyllus cuniculi* spread widely in temperate Australia after its introduction into the wild rabbit population in 1968 (Sobey and Conolly, 1971), it cannot cope with arid conditions (Cooke, 1984), and was therefore useless in what had by the 1980s become the problem areas for rabbit control in Australia. There are several species of fleas in Spain, and in an effort to enhance the efficacy of myxomatosis in arid parts of Australia the Spanish rabbit flea (*Xenopsylla cunicularis*), which is a vector of myxomatosis in drier parts of Spain and France, was investigated (Cooke, 1990). Proving relatively easy to breed, without the need to feed on pregnant rabbits, it was imported in 1990 (Bartholomaeus, 1991) and introduced in many sites in inland South Australia, New South Wales, Queensland and Northern Territory (Cooke, 1995). Its spread and effectiveness in Australia are described on p. 189.

Transmission by other arthropods

Since transmission is mechanical, any arthropod that probes through a skin tumour of an infected rabbit and then bites another rabbit is potentially a vector of myxomatosis. Data collected over the period 1944 to 1958, summarized in Fenner and Ratcliffe (1965), and later information, support this view. The critical features are the extent to which the arthropod feeds on rabbits and moves from one rabbit to another.

Endnotes

[1]Basser Library Archives, 143/25/5A. Marshall, I.D. (1961) Myxomatosis investigations carried out in Central and South America, 31st January to 21st March, 1961. Report to the Australian Wool Research Fund Committee.

[2]Basser Library Archives, 143/25/5A. Letter from R. Maria Licon to Fenner, 5 November 1994.

References

Andrewes, C.H., Muirhead-Thompson, R.C. and Stevenson, J.P. (1956) Laboratory studies of *Anopheles atroparvus* in relation to myxomatosis. *Journal of Hygiene* 54, 478–486.

Andrewes, C.H., Thompson, H.V. and Mansi, W. (1959) Myxomatosis: Present position and future prospects in Great Britain. *Nature* 184, 1179–1180.

Anon. (1942) A mechanical device for the spread of disease agents amongst rabbits. *Journal of the Council for Scientific and Industrial Research* 15, 83–84.

Aragão, H.B. (1911) Sobre o microbio do myxoma dos coelhos. *Brasil Medico* 25, 471–473.

Aragão, H.B. (1920) Transmissão do virus do myxoma dos coelhos pelas pulgas. *Brasil Medico* 34, 753–754.

Aragão, H.B. (1927) Myxoma of rabbits. *Memorias do Instituto Oswaldo Cruz* 20, 237–247.

Aragão, H.B. (1942) Sensibilidade do coelho do mato ao virus do mixoma; transmissão pelo *Aedes scapularis* e pelo *Stegomyia. Brasil Medico* 56, 204–220.

Aragão, H.B. (1943) O virus do mixoma no coelho do mato (*Sylvilagus minensis*), sua transmissão pelos *Aedes scapularis* e *aegypti. Memorias do Instituto Oswaldo Cruz* 38, 93–99.

Bartholomaeus, F.W. (1991) Rabbit fleas and myxomatosis: update on the Spanish connection.

Unpublished data. In: *Working Papers; 9th Australian Vertebrate Pest Control Conference, Adelaide 1991,* pp. 101–105. Quoted with the author's permission.
Beijerinck, M.W. (1898) Ueber ein *Contagium vivum fluidum* als Ursache der Fleckenkrankheit der Tabaksblätter. *Centralblatt für Bakteriologie und Parasitenkunde* Abteilung II, 5, 27–33.
Brown, P.W., Allan, R.M. and Shanks, P.L. (1956) Rabbits and myxomatosis in the N.E. of Scotland. *Scottish Agriculture* 35, 204–207.
Bull, L.B. and Dickinson, C.G. (1937) The specificity of the virus of rabbit myxomatosis. *Journal of the Council for Scientific and Industrial Research* 10(4), 291–294.
Bull, L.B. and Mules, M.W. (1944) An investigation of *myxomatosis cuniculi* with special reference to the possible use of the disease to control rabbit populations in Australia. *Journal of the Council for Scientific and Industrial Research* 17(2), 1–15.
Caspar, D.L.D., Dulbecco, R., Klug, A., Lwoff, A., Stoker, M.P.G., Tournier, P. and Wildy, P. (1962) Proposals. *Cold Spring Harbor Symposia on Quantitative Biology* 27, 49.
Chapple, P.L. and Lewis, N.D. (1965) Myxomatosis and the rabbit flea. *Nature* 207, 388–389.
Collins, J.J. (1955) Myxomatosis in the common hare – *Lepus europaeus. Irish Veterinary Journal* 9, 268–269.
Cooke, B.D. (1984) Factors limiting the distribution of the European rabbit flea, *Spilopsyllus cuniculi* (Dale) (Siphonaptera), in inland South Australia. *Australian Journal of Zoology* 32, 493–506.
Cooke, B.D. (1990) Notes on the comparative reproductive biology and the laboratory breeding of the rabbit flea *Xenopsylla cunicularis* Smit (Siphonaptera: Pulicidae). *Australian Journal of Zoology* 38, 527–534.
Cooke, B.D. (1995) Spanish rabbit fleas, *Xenopsylla cunicularis,* in arid Australia: a progress report. In: *Proceedings of the 10th Australian Vertebrate Control Conference, Hobart, 1995,* pp. 399–401.
Coura, J.R. (1994) Great lives at Manguinhos. Henrique de Beaurepaire Rohan Aragão. *Memorias do Instituto Oswaldo Cruz* 89(3), I–III.
Dalmat, H.T. (1959) Arthropod transmission of rabbit fibromatosis (Shope). *Journal of Hygiene* 57, 1–30.
Dalmat, H.T. and Stanton, M.F. (1959) A comparative study of the Shope fibroma in rabbits in relation to transmissibility by mosquitoes. *Journal of the National Cancer Institute* 22, 595–615.
Day, M.F., Fenner, F., Woodroofe, G.M. and McIntyre, G.A. (1956) Further studies on the mechanism of mosquito transmission of myxomatosis in the European rabbit. *Journal of Hygiene* 54, 258–283.
Duran-Reynals, F. (1940) Production of degenerative inflammatory or neoplastic effects in the newborn rabbit by the Shope fibroma virus. *The Yale Journal of Biology and Medicine* 13, 99–110.
Farrant, J.L. and Fenner, F. (1953) A comparison of the morphology of vaccinia and myxoma viruses. *Australian Journal of Experimental Biology and Medical Science* 31, 121–125.
Fenner, F. (1965) Viruses of the myxoma-fibroma subgroup of the poxviruses. II. Comparison of soluble antigens by gel diffusion tests, and a general discussion of the subgroup. *Australian Journal of Experimental Biology and Medical Science* 43, 143–156.
Fenner, F. (1976) Classification and Nomenclature of Viruses. Second Report of the International Committee on Taxonomy of Viruses. *Intervirology* 7, 1–116.
Fenner, F. (1994) Hare fibroma virus. In: Osterhaus, A.D.M.E. (ed.) *Virus Infections of Rodents and Lagomorphs.* Elsevier Science, Amsterdam, pp. 77–79.
Fenner, F. and Burnet, F.M. (1957) A short description of the poxvirus group (vaccinia and related viruses). *Virology* 4, 305–314.
Fenner, F. and Chapple, P.L. (1965) Evolutionary changes in myxoma virus in Britain. *Journal of Hygiene* 63, 175–185.
Fenner, F. and Marshall, I.D. (1957) A comparison of the virulence for European rabbits (*Oryctolagus cuniculus*) of strains of myxoma virus recovered in the field in Australia, Europe and America. *Journal of Hygiene* 55, 149–191.
Fenner, F. and Ratcliffe, F.N. (1965) *Myxomatosis.* Cambridge University Press, Cambridge, 379 pp.
Fenner, F. and Woodroofe, G.M. (1954) Protection of laboratory rabbits against myxomatosis by vaccination with fibroma virus. *Australian Journal of Experimental Biology and Medical Science* 32, 653–668.
Fenner, F., Day, M.F. and Woodroofe, G.M. (1952) The mechanism of transmission of myxomatosis in the European rabbit (*Oryctolagus cuniculus*) by the mosquito *Aedes aegypti. Australian Journal of Experimental Biology and Medical Science* 30, 139–152.
Fenner, F., Day, M.F. and Woodroofe, G.M. (1956) Epidemiological consequences of the mechanical transmission of myxomatosis by mosquitoes. *Journal of Hygiene* 54, 284–303.

Filshie, B.K. (1964) Observations with the electron microscope of myxoma virus on mosquito mouthparts. *Australian Journal of Biological Science* 17, 903–906.
Giorgi, W. (1968) Diseases of domestic rabbits observed in the State of São Paulo between 1963 and 1967. *Biológico* 34, 71–82.
Goodpasture, E.W. (1933) Borreliotoses: fowl-pox, molluscum contagiosum, variola-vaccinia. *Science* 77, 119–121.
Grodhaus, G., Regnery, D.C. and Marshall, I.D. (1963) Studies in the epidemiology of myxomatosis in California. II. The experimental transmission of myxomatosis in brush rabbits (*Sylvilagus bachmani*) by several species of mosquitoes. *American Journal of Hygiene* 77, 205–212.
Hudson, J.R. and Mansi, W, (1955) Attenuated strains of myxoma virus in England. *Veterinary Record* 67, 746.
Hughes, S.S. (1977) *The Virus. A History of the Concept.* Heinemann Educational Books, London, 140 pp.
Hurst, E.W. (1937) Myxoma and the Shope fibroma. II. The effect of intracerebral passage on the myxoma virus. *British Journal of Experimental Pathology* 18, 15–22.
Ivanovski, D.I. (1894) Über die Mosaickrankheit der Tabakspflanze. *Bulletin de l'Académie Impériale des Sciences de St. Petersbourg* 3, 67–70.
Jacotot, H., Vallée, A. and Virat, B. (1954a) Sur un cas de myxomatose chez le lièvre. *Annales de l'Institut Pasteur* 86, 105–107.
Jacotot, H., Toumanoff, C., Vallée, A. and Virat, B. (1954b) Transmission expérimentale de la myxomatose au lapin par *Anopheles maculipennis atroparvus* et *A. stepheni*. *Annales de l'Institut Pasteur* 87, 477–485.
Jacotot, H., Vallée, A. and Virat, B. (1955) Étude sur la transmission experimentale de la myxomatose au lièvre. *Annales de l'Institut Pasteur* 88, 1–10.
Joiner, G.N., Jardine, J.H. and Gleiser, C.A. (1971) An epizootic of Shope fibromatosis in a commercial rabbitry. *Journal of the American Veterinary Medical Association* 159, 1583–1587.
Joubert, L., Chippaux, A., Mouchet, J. and Oudar, J. (1969) Entretien hiverno-vernal du virus myxomateux dans les terriers. Myxomatose d'inoculation par la puce du lapin et myxomatose du fouissement. *Bulletin de l'Académie véterinaire de France* 42, 93–101.
Kejdana, S. (1955) Myxomatosis in hares. *Médecine Veterinaire, Varsovie* 11, 136.
Kessel, J.F., Prouty, C.C. and Meyer, J.W. (1931) Occurrence of infectious myxomatosis in southern California. *Proceedings of the Society for Experimental Biology and Medicine* 28, 413–414.
Kessel, J.F., Fisk, R.T. and Prouty, C.C. (1934) Studies with the Californian strain of the virus of infectious myxomatosis. *Proceedings of the Fifth Pacific Science Congress*, Volume IV, pp. 2927–2939.
Kilham, L. (1955) Metastastizing viral fibromas of gray squirrels: pathogenesis and mosquito transmission. *American Journal of Hygiene* 61, 55–63.
Kilham, L. and Dalmat, H.T. (1955) Host-virus-mosquito relations of Shope fibromas in cottontail rabbits. *American Journal of Hygiene* 61, 45–54.
Kilham, L. and Woke, P.A. (1953) Laboratory transmission of fibromas (Shope) in cottontail rabbits by means of fleas and mosquitoes. *Proceedings of the Society for Experimental Biology and Medicine* 83, 296–301.
Leinati, L., Mandelli, G., Carrara, O., Cilli, V., Castrucci, G. and Scatozza, F. (1961) Richerche anatomo-istopathologiche e virologiche sulla malattia cutanea nodulare delle lepri padane. *Bollettino Istituto Sierter. Milan* 40, 295–328.
Lockley, R.M. (1954) The European rabbit flea, *Spilopsyllus cuniculi*, as a vector of myxomatosis in Britain. *Veterinary Record* 66, 434.
Loeffler, F and Frosch, F. (1898) Berichte der Kommission zur Erforschung der Maul-und Klauenseuche bei dem Institut für Infektionskrankheiten in Berlin. *Zentralblatt für Bakteriologie, Parasitenkunde und Infektionskrankheiten* Abteilung I, 23, 371–391.
Lucas, A., Bouley, D., Quinchon, C. and Toucas, L. (1953) La myxomatose du lièvre. *Bulletin Office internationale des Epizooties* 39, 770–776.
Lwoff, A. (1957) The concept of virus. *Journal of General Microbiology* 17, 239–253.
Magallon, P., Bazin, O. and Bazin, J. (1953) La myxomatose du lièvre. *Bulletin Office internationale des Epizooties* 39, 765–769.
Marshall, I.D. and Regnery, D.C. (1960) Myxomatosis in a Californian brush rabbit (*Sylvilagus bachmani*). *Nature* 188, 73–74.

Marshall, I.D. and Regnery, D.C. (1963) Studies in the epidemiology of myxomatosis in California. III. The response of brush rabbits (*Sylvilagus bachmani*) to infection with exotic and enzootic strains of myxoma virus and the relative infectivity of the tumours for mosquitoes. *American Journal of Hygiene* 77, 213–219.

Marshall, I.D., Regnery, D.C. and Grodhaus, G. (1963) Studies in the epidemiology of myxomatosis in California. I. Observations on two outbreaks of myxomatosis in coastal California and the recovery of myxoma virus from a brush rabbit (*Sylvilagus bachmani*). *American Journal of Hygiene* 77, 195–204.

Martin, C.J. (1936) Observations on Myxomatosis cuniculi (Sanarelli) made with a view to the use of the virus in the control of rabbit plagues. *Bulletin of the Council for Scientific and Industrial Research, Australia*, No. 96, 28 pp.

Mead-Briggs, A.R. (1964a) Some experiments concerning the interchange of rabbit fleas, *Spilopsyllus cuniculi* (Dale), between living rabbits. *Journal of Animal Ecology* 33, 13–26.

Mead-Briggs, A.R. (1964b) The reproductive biology of the rabbit flea, *Spilopsyllus cuniculi* (Dale), and the dependence of this species upon the breeding of its host. *Journal of Experimental Biology* 41, 371–402.

Mead-Briggs, A.R. (1977) The European rabbit, the European rabbit flea and myxomatosis. *Applied Biology* 2, 183–261.

Mead-Briggs, A.R. and Rudge, A.J.B. (1960) Breeding of the rabbit flea, *Spilopsyllus cuniculi* (Dale); requirement of a "factor" from a pregnant rabbit for ovarian maturation. *Nature* 187, 1136–1137.

Mead-Briggs, A.R. and Vaughan, J.A. (1969) Some requirements for mating in the rabbit flea, *Spilopsyllus cuniculi* (Dale). *Journal of Experimental Biology* 51, 495–511.

Mead-Briggs, A.R. and Vaughan, J.A. (1975) The differential transmissibility of myxoma virus strains of differing virulence grades by the rabbit flea *Spilopsyllus cuniculi* (Dale). *Journal of Hygiene* 75, 237–247.

Moses, A. (1911) O virus do mixoma dos coelhos. *Memorias do Instituto Oswaldo Cruz* 3, 46–53.

Muirhead-Thompson, R.C. (1956) Observations on the European rabbit flea (*Spilopsyllus cuniculi*) in relation to myxomatosis in England. *Report to the Scientific Subcommittee of the Myxomatosis Advisory Committee*, Ministry of Agriculture, Fisheries and Food, London.

Mykytowycz, R. (1958) Contact transmission of infectious myxomatosis of the rabbit, *Oryctolagus cuniculus* (L.). *C.S.I.R.O. Wildlife Research* 3, 1–6.

Padgett, B.L., Wright, M.J., Jayne A. and Walker, D.L. (1964) Electron microscopic structure of myxoma virus and some reactivable derivatives. *Journal of Bacteriology* 87, 454–460.

Patton, N.M. and Holmes, H.T. (1977) Myxomatosis in domestic rabbits in Oregon. *Journal of the American Veterinary Medical Association* 171, 560–562.

Raflo, C.P., Olsen, R.G., Pakes, S.P. and Webster, W.S. (1973) Characterization of a fibroma virus isolated from naturally-occurring skin tumours in domestic rabbits. *Laboratory Animal Science* 23, 525–532.

Regnery, D.C. (1971) The epidemic potential of Brazilian myxoma virus (Lausanne strain) for three species of North American cottontails. *American Journal of Epidemiology* 94, 508–513.

Regnery, D.C. and Marshall, I.D. (1971) Studies in the epidemiology of myxomatosis in California. IV. The susceptibility of six leporid species to Californian myxoma virus and the relative infectivity of their tumours for mosquitoes. *American Journal of Epidemiology* 94, 508–513.

Regnery, D.C. and Miller, J.H. (1972) A myxoma virus epizootic in a brush rabbit population. *Journal of Wildlife Diseases* 8, 327–331.

Regnery, R.L. (1975) Preliminary studies on an unusual poxvirus of the western grey squirrel (*Sciurus griseus griseus*) of North America. *Intervirology* 5, 364–366.

Remlinger, P. (1906) Les microbes filtrants. *Bulletin de l' Institut Pasteur, Paris* 4, 337–345; 385–392.

Rivers, T.M. (1927a) Filterable viruses. A critical review. *Journal of Bacteriology* 14, 217–257.

Rivers, T.M. (1927b) Changes observed in epidermal cells covering myxomatous masses induced by virus myxomatosum (Sanarelli). *Proceedings of the Society for Experimental Biology and Medicine* 24, 435–437.

Rivers, T.M. and Ward, S.M. (1937) Infectious myxomatosis of rabbits. Preparation of elementary bodies and studies of serologically active materials associated with the disease. *Journal of Experimental Medicine* 66, 1–14.

Rothschild, M. (1953) Notes on the European rabbit Flea. *Report to the Myxomatosis Advisory Committee*, Ministry of Agriculture, Fisheries and Food, 6 December 1953.

Rothschild, M. (1960) Myxomatosis in Britain. *Nature* 185, 257.

Rothschild, M. and Ford, B. (1964) Breeding of the rabbit flea (*Spilopsyllus cuniculi* (Dale)) controlled by the reproductive hormones of the host. *Nature* 201, 103–104.
Rothschild, M. and Ford, B. (1972) Factors influencing the breeding of the rabbit flea (*Spilopsyllus cuniculi*): A spring-time accelerator and a kairomone in nestling rabbit urine. *Journal of Zoology* 170, 87–137.
Rothschild, N.C. (1915) A synopsis of British Siphonaptera. *Entomology Monthly Magazine* 51, 49–112.
Roux, E. (1903) Sur les microbes dits 'invisible'. *Bulletin de l'Institut Pasteur, Paris* 1, 7–12; 49–56.
Sanarelli, G. (1898) Das myxomatogene Virus. Beitrag zum Studium der Krankheitserreger ausserhalb des Sichtbaren. *Zentralblatt für Bakteriologie, Parasitenkunde, Infektionskrankheiten und Hygiene* (Abteilung I) 23, 865–873.
Shanks, P.L., Sharman, G.A.M., Allan, R., Donald, L.G., Young, S. and Mar, T.G. (1955) Experiments with myxomatosis in the Hebrides. *British Veterinary Journal* 111, 25–30.
Shope, R.E. (1932) A filtrable virus causing a tumor-like condition in rabbits and its relationship to virus myxomatosum. *Journal of Experimental Medicine* 56, 803–822.
Sobey, W.R. (1977) Rabbit fleas. *Wool Technology and Sheep Breeding*, September/ October, 1977.
Sobey, W.R. and Conolly, D. (1971) Myxomatosis: the introduction of the rabbit flea *Spilopsyllus cuniculi* (Dale) into wild rabbit populations in Australia. *Journal of Hygiene* 69, 331–346.
Splendore, A. (1909) Ueber das Virus myxomatosum der Kanichen. *Zentralblatt für Bakteriologie, Parasitenkunde, Infektionskrankheiten und Hygiene* (Abteilung I) 48, 300–301.
Torres, S. (1936) Transmissão da mixomatose dos coelhos pelo *Culex quinquefasciatus. Boletim da Sociedade Brasileira de Medicina Veterinária* 6, 4–6.
Vail, E.L. and McKenney, F.D. (1943) *Diseases of Domestic Rabbits.* Conservation Bulletin No. 31, Fish and Wildlife Service, U.S. Department of the Interior.
van Helvoort, T. (1994) History of virus research in the twentieth century: the problem of conceptual continuity. *History of Science* 32, 185–235.
White, H.C. (1929) Observations on rabbit myxoma. *New South Wales Department of Agriculture Veterinary Research Report* No. 5, 1927–28, pp. 45–47.
Whitty, B.T. (1955) Myxomatosis in the common hare – *Lepus europaeus. Irish Veterinary Journal* 9, 267.
Wilkinson, L. (1979) The development of the virus concept as reflected in corpora of studies on individual pathogens. 5. Smallpox and the evolution of ideas on acute (viral) infections. *Medical History* 23, 1–28.
Woodroofe, G.M. and Fenner, F. (1965) Viruses of the myxoma-fibroma subgroup of the poxviruses. I. Plaque production in cultured cells, plaque-reduction tests, and cross-protection in rabbits. *Australian Journal of Experimental Biology and Medical Science* 43, 123–142.

5

The Disease Myxomatosis in the European Rabbit

Overview

To understand the changes that occurred in myxoma virus and its host since its introduction in the 1950s, we need to have a general knowledge of the disease in the European rabbit: the clinical signs, the pathogenesis of the disease, the immune response, methods of assay of virus and antibodies, ways of comparing different strains and ways of immunizing rabbits against infection. This chapter provides some basic information on these matters; it does not attempt to cover the subject comprehensively[1].

When it was first introduced myxoma virus produced a very lethal disease in Australian wild rabbits, but as time progressed less severe cases were seen; the clinical features of the disease caused by the original virus and some less virulent strains that occurred in Australian wild rabbits are described.

To follow evolutionary changes in myxomatosis, it was necessary to devise a way of measuring changes in the virulence of large numbers of field strains of myxoma virus, using this term as a synonym for lethality. Because it was impractical to measure case-fatality rates directly, a statistical measure was developed by which they could be inferred by calculating the mean survival times in groups of five or six rabbits. To handle data obtained on tests with hundreds of strains of virus, five (later six) 'virulence grades' were identified in terms of mean survival times after the intradermal inoculation of small doses of virus. Until plaque assays became available, viral infectivity was titrated by counting the pocks produced on the chorioallantoic membrane of the developing egg.

The pathogenesis of myxomatosis was studied by tracing the spread of virus through the body. After the injection of a small dose intradermally, virus replicated in the skin, then in the draining lymph nodes, and then entered the bloodstream, whence it was taken to many organs but localized and multiplied to highest titres in other parts of the skin, producing mucinous tumours (hence the name 'myxoma'), and in the testes of male rabbits, producing temporary sterility.

Immunity persisted for life in animals that recovered; several methods were developed for measuring the antibody response. Some strains of virus temporarily suppressed the immune response to secondary infections, allowing bacteria on the respiratory mucosa and conjunctiva to produce 'snuffles' and conjunctivitis. Maternal antibodies provided a high level of protection to baby rabbits, which otherwise died quickly, even after infection with attenuated strains. High environmental temperatures reduced the severity and lethality of the disease, low temperatures had the opposite effect.

The only successful vaccines were live virus vaccines. The first to be used was Shope's fibroma virus, another leporipoxvirus. Later attenuated strains of myxoma virus were developed by serial passage of selected strains in cultured cells. Recently, one such vaccine strain has been genetically engineered so as to immunize rabbits against rabbit haemorrhagic disease as well as against myxomatosis.

Fig. 5.1. Ian David Marshall (1922–). After service in the Royal Australian Navy during the Second World War, Marshall graduated with a BAgrSc from the University of Melbourne in 1951 and immediately joined the Department of Microbiology, John Curtin School of Medical Research, as a Research Assistant, working with Fenner on myxomatosis. He was then awarded his doctorate in 1956, later becoming a Research Fellow, Fellow and Senior Fellow before his formal retirement in 1987. He played a major role in virological investigations of myxomatosis between 1951 and 1959, when he went to work on arboviruses with W.C. Reeves at the University of California at Berkeley for two years. While there he also collaborated with D.C. Regnery of Stanford University on classical studies of the epidemiology of myxomatosis in California. On his return he established an arbovirus laboratory in the John Curtin School which became one of the major Australian centres of arbovirus research.

Clinical Signs

In its natural host in South America, *Sylvilagus brasiliensis,* infection with myxoma virus produces a small localized fibroma, without generalized signs, whereas in laboratory rabbits it causes a very severe, lethal disease (see p. 78). The strain of virus used to introduce the disease into Australia (the 'Standard Laboratory Strain') had been passaged for many years in laboratory rabbits before it was released (see p. 71), and it produced flatter lesions than the more protuberant tumours produced by strains recently derived from *S. brasiliensis* (Fenner and Marshall, 1957).

After myxomatosis had been spreading in Australia and Europe for a few years, less virulent strains became common in both continents. In general, the lesions produced by these variants maintained the characteristics of the initiating virus strains; in Australia the skin lesions were relatively flat, whereas in Europe they were usually but not always protuberant. In genetically unselected laboratory rabbits the survival times of fatal cases infected with these less virulent strains were usually prolonged and some rabbits recovered.

Designations of 'virulence grades'

Monitoring the changes of virulence from year to year and over geographically wide areas became a major research activity, especially in Australia. It was important to devise an economical way of measuring changes in virulence, using this term as a synonym for lethality. To measure changes from 99.5% to 95% case-fatality rates at the 5% significance level would have required some 45 rabbits per test, numbers that were beyond the capacity of the laboratory except for a few selected strains. It was impossible to carry out tests on a less expensive laboratory animal such as the mouse and there were no *in vitro* tests of virulence. Some surrogate for case-fatality rates was needed, and a statistical measure was developed by which lethality could be inferred by calculating the mean survival times in groups of five or six rabbits inoculated in a standard way, using certain conventions for including recovered rabbits for all except the lowest virulence grade (Fenner and Marshall, 1957). In

1956, when the grading system was developed, there was a dearth of information about what the range of strains with differing levels of virulence might be. At that time no strains had been recovered which lay between Grade I (>99% case-fatality rate) and what was designated as Grade III (70–95% case-fatality rate), so an intermediate Grade II strain was designated as being characterized by a case-fatality rate of 95–99% and a mean survival time of 14–16 days. Later, over the period 1955 to 1980, it emerged that judging by mean survival times in groups of 5–6 laboratory rabbits, strains of intermediate virulence were by far the most common to be recovered, and Grade III, which covered a broad spectrum, was subdivided into IIIA and IIIB. Grade V was set up to accommodate highly attenuated strains, which could be recognized on the basis of survival rates, even in such small groups of rabbits. At the time the only strain with a case-fatality rate of <50% was the highly attenuated 'neuromyxoma' strain of Hurst (1937b), but no strain as attenuated as this has ever been recovered from wild rabbits, although a few strains with case-fatality rates of <50% have been found. The ultimate criteria adopted for grading virulence, for both Australian and European strains, are given in Table 5.1.

The reliability of three of these virulence grades was assessed by testing large numbers of rabbits with 'prototype strains' allocated to virulence grades with particular case-fatality rates (CFR) on the basis of mean survival times (MST) (Fenner and Marshall, 1957), with the results given in Table 5.2.

Features of myxomatosis caused by selected Australian strains

The appearance of laboratory rabbits infected with some of the prototype strains at the height of the disease are described briefly and illustrated in Fig. 5.2, infection being produced by the intradermal inoculation of very small doses of virus, to simulate infection by mosquito bite (Fenner and Marshall, 1957). The lesions caused by selected European strains are described and illustrated in Chapter 9.

Standard Laboratory Strain, prototype Grade I strain (Fig. 5.2A)

This virus was derived from the Moses strain, which was isolated in the Oswaldo Cruz Institute in Rio de Janeiro in 1911, sent to the Rockefeller Institute of Medical Research in 1926, used there by Shope and later by Martin in trials of its effectiveness for rabbit control in Cambridge in 1936, brought to Australia in 1937 and eventually successfully released in Australia in 1950. During this period of nearly 40 years it was used for research by all the early workers in the United States (Berry, Hyde, Parker, Rivers, Shope, Thompson) and England (Andrewes, Hurst, Lush), and was passaged in rabbits many times.

The lump produced at the inoculation site was first seen on the third day and later became large, hard and convex. The roughly circular margin merged gradually into the surrounding skin, with no sharp demarcation. Secondary skin lesions were seen by the sixth or seventh day, and by the ninth day were widely distributed over the body and ears. Lumps were found on the limbs only if these were the site of the

Table 5.1. Criteria for grading virulence of strains.

Virulence grade	Mean survival time	Inferred case-fatality rate (%)
I	≤13 days	>99
II	14–16 days	95–99
III	17–28 days	70–95
IIIA	17–22 days	90–95
IIIB	23–28 days	70–90
IV	29–50 days	50–70
V	not applicable	<50

Table 5.2. Results of testing rabbits to assess reliability of the virulence grades.

Grade I	Standard Laboratory Strain	43 rabbits, MST 10.8 days, CFR 100%
Grade III	KM13 strain	77 rabbits, MST 21.5 days, CFR 88%
Grade IV	Uriarra strain	45 rabbits, MST 26.2 days, CFR 58%

primary lesion, or occasionally in animals that survived longer than usual. Thickening of the eyelids was first seen on the sixth or seventh day and by the ninth day the eyes were usually completely closed. There was an opalescent discharge from the eyes which became more copious and turbid during the last few days of life. Oedematous swelling of the head, the base of the ears and the perineum became pronounced in the later stages, and the respiration then slowed, often accompanied by a semipurulent nasal secretion. The animal continued to eat and drink well until just before death and the body fat remained abundant. In the many tests conducted in laboratory rabbits early death was almost invariable. The virus content of the epidermal cells over the skin lesions was very high, providing abundant virus for probing mosquitoes.

KM13 strain, prototype Grade III strain (Fig. 5.2B)

This strain was isolated from a pool of *Anopheles annulipes* collected in Corowa in December 1952 (Myers *et al.*, 1954). Signs appeared at the same times as with the Standard Laboratory Strain but their development was more gradual. The eyes were rarely closed before the 14th or 15th day and the eyelids were irregularly distorted rather than generally thickened. The swelling of the head and perineum

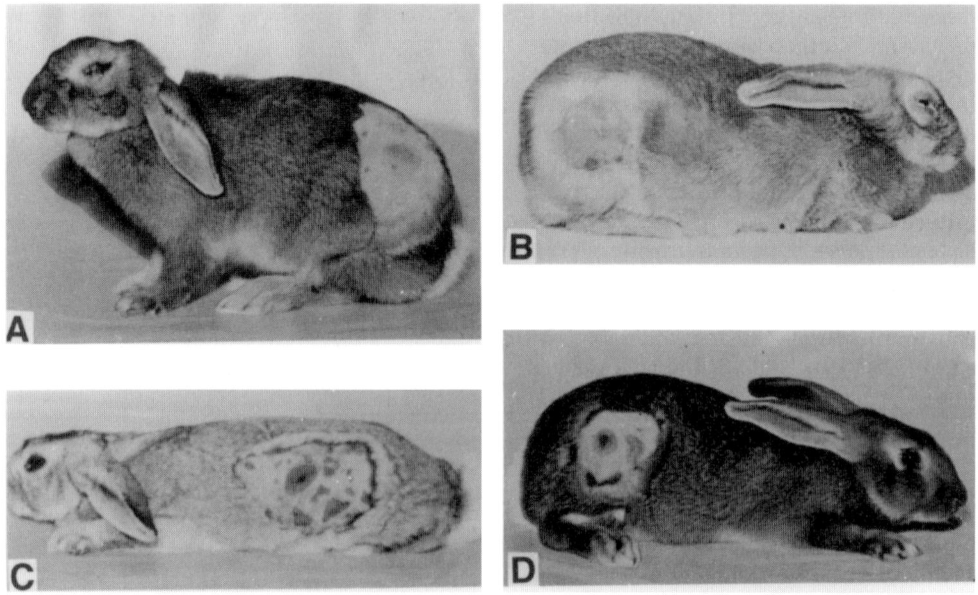

Fig. 5.2. Appearance of laboratory rabbits infected by the intradermal inoculation of small doses of the various strains of myxoma virus. (A) Standard Laboratory Strain, 10 days after inoculation. (B) KM13 strain, Grade III virulence, 21 days after inoculation. (C) Uriarra III strain, Grade IV virulence, 24 days after inoculation. (D) Neuromyxoma virus, at the height of the disease, 10 days after inoculation. From Fenner and Marshall (1957), with permission.

was less pronounced, but severe cases that survived for more than 20 days presented a wretched picture. Breathing was laboured, the eyes were closed and bulbous, there was a purulent nasal discharge, the animal was emaciated, and strings of mucus were passed with the faeces. The virus content of the epidermal cells over the skin lesions was very high, for as long as 40 days in some survivors.

Uriarra strain, prototype Grade IV strain (Fig. 5.2C)

This strain was derived from a case at Uriarra, near Canberra, in February 1953 (Mykytowycz, 1953). The local lesion appeared on the third or fourth day and soon became hard, red and slightly convex. Except in acutely fatal cases it became clearly demarcated from the surrounding skin by the 12th day. The centre of the flat primary lesion became necrotic and scabbed during the third or fourth week. Secondary skin lesions were numerous, flat and red, with clearly demarcated margins. The virus content of the epidermal cells over the skin lesions was always high between the eighth and 14th days; in some animals it then fell rapidly, but in others it remained high for another two weeks.

Neuromyxoma, an artificially attenuated strain (Fig. 5.2D)

This strain was developed in England by serial intracerebral passage of the Moses strain in rabbits (Hurst, 1937b). It was highly attenuated when inoculated intradermally; Fenner and Marshall (1957) recorded no deaths among 22 cases. The local lesion at the inoculation site was hard, red and convex, but became depressed and purple very early and the periphery was well demarcated as early as the fifth day, regression and scabbing following soon afterwards. Secondary skin lesions were scanty and generalized oedema of the head and perineum were never seen. The general health of the rabbit was little affected. The virus content of the epidermal cells over the primary lesion reached a peak on the sixth day but was always much lower than with other strains of myxoma virus. Experimentally, Fenner *et al.* (1956) rarely obtained positive results after mosquito probing through lesions caused by neuromyxoma (12 positives out of 169 attempts). No isolates as attenuated as neuromyxoma virus have ever been recovered from field cases, although a few have been found which have case-fatality rates of less than 50% in laboratory rabbits; these have been classified as Grade V strains.

Pneumonic or non-myxomatous myxomatosis

In 1980 a new clinical syndrome caused by infection with myxoma virus was observed in French rabbitries operated under intensive husbandry, in wire mesh cages containing large numbers of rabbits, in buildings from which insects were excluded (Brun *et al.*, 1981a; Joubert *et al.*, 1982; Arthur and Louzis, 1988). The disease was manifested by eye and genital involvement, the absence of nodular skin lesions, and pronounced pulmonary signs, with lacrimation and a mucopurulent nasal discharge. This 'pneumonic' syndrome was initially seen in intensive husbandry rabbitries following the administration of SG 33 vaccine (Brun *et al.*, 1981b), but it also occurred in non-vaccinated rabbits. It is transmitted by the airborne route. It was found that survivors were often sterile and farmed does abandoned their newborn litters.

There was little seasonal variation in its incidence, the morbidity rate varied between 5% and 40%, the mortality rate between 30% and 35%, and the incubation period and evolution of the disease were longer than in the classical disease (17–28 days). Under conditions of traditional husbandry (small farms and backyard hutches) most infections with myxomatosis are caused by viruses of the classical nodular type, introduced by winged vectors which had acquired virus from wild rabbits (Arthur and Louzis, 1988). The pneumonic form also occurred, along with the classical nodular form of the disease, in wild rabbits, although the nodular form was more common in major epizootics. The pathogenesis of the two forms are compared on p. 104, their epidemiology in Chapter 9.

Assay Methods for Virus

Pockcounting on the chorioallantoic membrane

From the outset of their studies of myxomatosis in 1951, it was clear to Fenner and his colleagues that there was an immediate requirement for effective methods of titrating virus and antibody. Lush (1937) had demonstrated that myxoma virus would produce pocks on the chorioallantoic membrane of the developing egg. A few years later Parker (1940) showed that the infectivity could be assayed by intradermal inoculations at many sites on the backs of rabbits, and concluded that a single virus particle could initiate infection. Having been trained by Macfarlane Burnet, the scientist who pioneered the pock counting method (Burnet, 1936; Beveridge and Burnet, 1946), and since eggs were much cheaper and more convenient to use than rabbits, Fenner adopted pock counting as the standard method of assay of myxoma virus for all work in his group between 1951 and 1965. Using six strains of myxoma virus of widely differing virulence, Fenner and McIntyre (1956) found that there was no difference in the infectivity of different strains for either the rabbit or the egg, although the severity of the signs in rabbits varied greatly. The results were consistent with the hypothesis that a pock or a skin lesion was initiated by one infective dose of virus, but the rabbit was about two-and-a-half times more sensitive than the chorioallantoic membrane.

Comparison of pocks on the chorioallantoic membrane

All strains of myxoma virus produce pocks on the chorioallantoic membrane of the developing egg; Shope's rabbit fibroma virus replicates in the chorioallantoic membrane (Smith, 1948) but produces no pathognomic lesions, gross or microscopic. The pocks produced by Californian strains of myxoma virus were smaller than those produced by Brazilian strains; the neuromyxoma strain of Hurst also produced a smaller pock than its virulent parent (Fenner and Marshall, 1957; Fig. 5.3).

Identification of mixed infection; Nottingham 55 strain

In the early days of myxomatosis, Ratcliffe (quoted by Mykytowycz, 1953) and Jacotot *et al.* (1955a,b) suggested that the first attenuated strains recovered from wild rabbits (Uriarra in Australia and Loiret 55 in Europe) were mixtures of a highly virulent and greatly attenuated components. Recovery and testing of several single pocks from membranes exhibiting only one pock showed that the strains studied by these authors were not mixed (Fenner and Marshall, 1957). Hudson and Mansi (1955) had reported that virus recovered from a naturally infected rabbit in Nottinghamshire, England, in April 1955 was highly virulent if obtained early in the course of disease but of reduced virulence if obtained later in the course of infection. When the original material was tested in 10 rabbits in Canberra, three died of classical 'Lausanne-type' disease, and seven recovered. Passage from lesions of each of these subgroups bred true, i.e. the original material was a mixture. This was confirmed by testing of three single pock isolates, and it was shown that the attenuated substrain (subsequently designated 'Nottingham attenuated') produced pocks with a mean diameter of 0.26 mm compared with 0.40 mm for the virulent substrain (see Fig. 5.3).

Plaque counting on cultured cells

Plaque production by animal viruses in cultured cells was first developed as an assay method by Dulbecco and Vogt (1954). A practical method of plaque assay of myxoma virus was first reported by Schwerdt and Schwerdt (1962) and Padgett *et al.* (1962), using monolayers of rabbit kidney cells, work which was followed up by Woodroofe and Fenner (1965). After Padgett *et al.* (1962) had demonstrated that plaque assay in rabbit kidney cells was as sensitive as assay in rabbits and that fibroma virus (which did not produce countable pocks on the chorioallantoic membrane) could also be assayed in this way, plaque assays became the standard method of titration of infectivity and neutralizing antibody. Woodroofe and

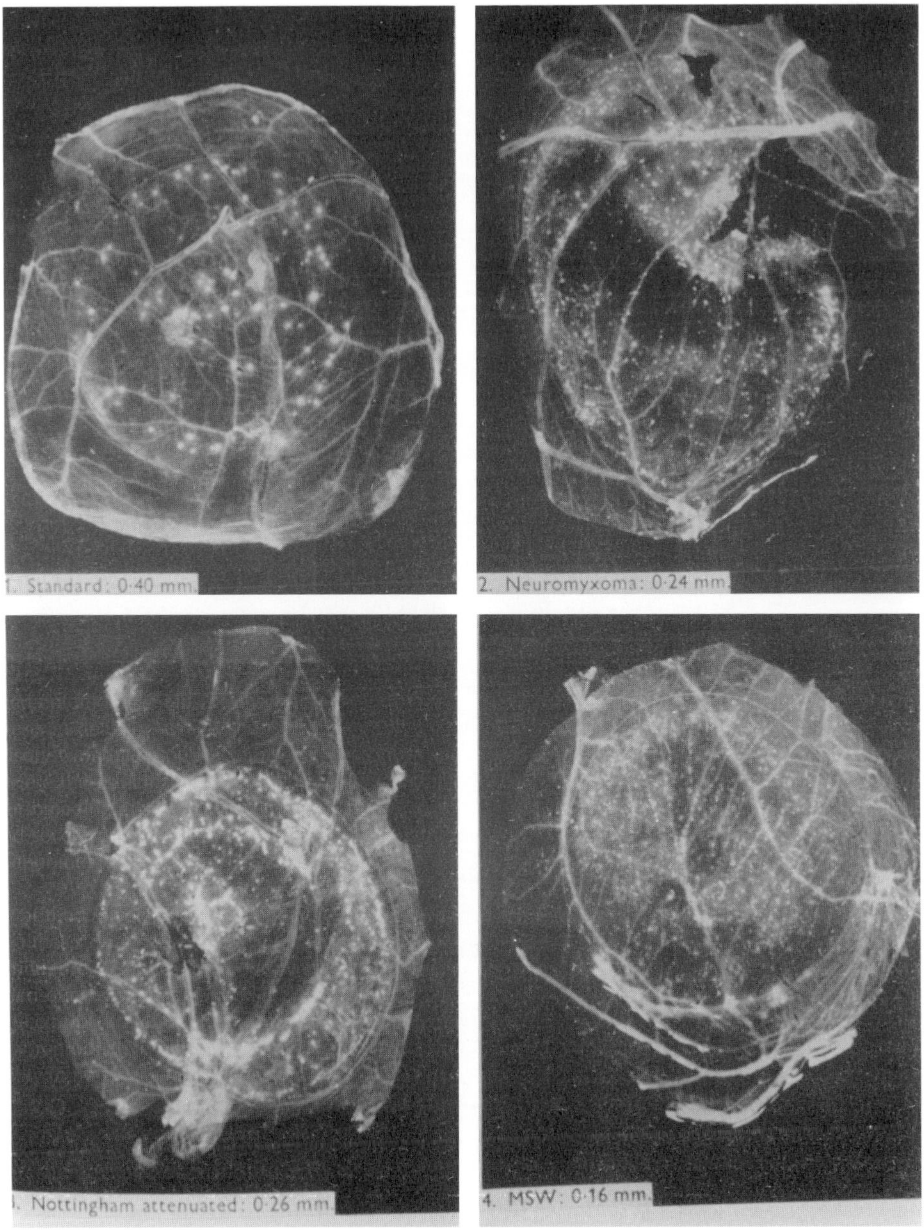

Fig. 5.3. Appearance of pocks produced by different strains of myxoma virus on the chorioallantoic membrane of the developing egg. Twelve-day-old eggs were incubated at 35°C for three days after inoculation. Except for the Standard Laboratory Strain, the numbers of pocks on the membranes illustrated are considerably greater than used for pock counting or for the measurement of pock diameter. (1) Standard Laboratory Strain; average diameter 0.40 mm. (2) Neuromyxoma virus; average diameter 0.24 mm. (3) Nottingham attenuated strain; average diameter 0.26 mm. (4) Californian strain MSW; average diameter 0.16 mm. From Fenner and Marshall (1957), with permission.

Fenner (1965) showed that all leporipoxviruses produced plaques in rabbit embryo fibroblasts and that all strains of myxoma virus, but not other leporipoxviruses, produced plaques in chick embryo fibroblasts. From time to time several different continuous cell lines have been used for titrations of myxoma virus. SIRC cells (derived from *Oryctolagus cuniculus* cornea), RK13 cells (rabbit kidney-derived), and Vero and BGMK cells (both derived from African green monkey kidney) are now the most commonly used.

Comparison of plaques on cultured cells

All strains of myxoma virus, and Shope's fibroma virus, produce plaques on monolayers of rabbit embryo fibroblasts and rabbit kidney cells (Padgett *et al.* 1962; Woodroofe and Fenner, 1965). The plaques produced on different types of cell differed substantially in size, and isolates of myxoma virus from California (which were highly virulent viruses) produced smaller plaques than most strains from South America (Fig. 5.4). A small plaque mutant of a Brazilian strain was found to be less virulent than the parent strain.

Methods of Assaying Antibodies

Virtually all the standard methods of assaying sera for the presence of antibodies (White and Fenner, 1995) have been used at one time or another. Those commonly used have been neutralization of infectivity by pock reduction (Lush, 1937; Fenner *et al.*, 1953) or plaque reduction tests (Woodroofe and Fenner, 1965), complement fixation (Fenner *et al.*, 1953), immunodiffusion (Mansi and Thomas, 1958; Sobey *et al.*, 1970) and most recently enzyme-linked immunosorbent assay (ELISA) (Wetherall *et al.*, 1983; Kerr, 1997).

Antibody assays are used primarily for epidemiological studies, for which the prime requirements are reliability (in terms of absence of false positives), durability (in terms of false negatives, i.e. negative results in rabbits that are known to have recovered from infection) and technical simplicity. Rabbits that have recovered from myxomatosis usually retain high titre neutralizing antibodies for the life of the rabbit, although reinoculation of recovered rabbits many months after infection sometimes causes an increase in titre. However, they are relatively

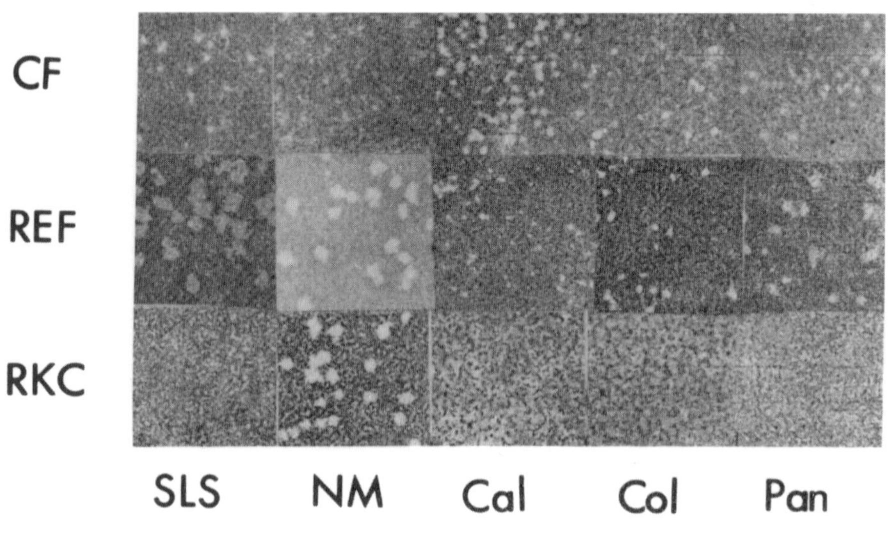

Fig. 5.4. Appearance of plaques produced in monolayers of chick fibroblasts (CF), rabbit embryo fibroblasts (REF) and rabbit kidney cells (RKC) by several strains of myxoma virus. SLS = Standard Laboratory Strain, NM = neuromyxoma virus, Cal = MSW strain of Californian myxoma virus, Col = strain obtained from Colombia, Pan = strain obtained from Panama. From Woodroofe and Fenner (1965), with permission.

slow and difficult to perform. Complement-fixation and immunodiffusion tests are simple to perform, but titres fall relatively quickly. In addition, the immunodiffusion test may give fluctuating titres (alternately positive and negative) over short periods of time (Williams *et al.*, 1973), making it unsatisfactory for epidemiological studies. These problems did not occur with ELISA, which was simple, sensitive and accurate. Using ELISA, IgG antibodies could be distinguished from IgM antibodies. IgG antibodies, assayed by ELISA or plaque reduction tests, were measurable for many months after recovery and did not fluctuate in titre. Since IgM could be detected at about the time of onset of clinical signs but declined within about five weeks, shortly after persistent IgG antibodies had peaked, calculation of IgM:IgG ratios could be used to distinguish recent infections (Kerr, 1997). For all these reasons, ELISA is the preferred test for epidemiological studies.

Comparisons of Other Characteristics of Leporipoxviruses

Investigations of different strains of myxoma virus and of Shope's fibroma virus revealed a number of strain and species differences additional to the clinical features in domestic rabbits, pock size and plaque size, namely gel diffusion antigen patterns and restriction maps.

Antigenic structure

Being large and complex viruses, leporipoxviruses produce a number of antigens and these show differences between strains and species. Using gel diffusion precipitin tests, Reisner *et al.* (1963) showed that viruses of Brazilian and Californian origin had at least one common and one specific antigen. Fenner (1965) followed this up by testing soluble antigens extracted from tumours produced by several species of leporipoxviruses, and from isolates of myxoma virus from different parts of the Americas. Nine isolates of Californian myxoma virus (natural host, *Sylvilagus bachmani*) gave identical patterns, which were readily distinguishable from those of Brazilian strains (natural host, *S. brasiliensis*) (Fig. 5.5, Panel A). Three isolates from Brazil, two from Colombia, one from Uruguay and one from Argentina were antigenically indistinguishable from each other (Fig. 5.5, Panel B), whereas two strains from Colombia and two from Panama (both places where the natural host was probably *S. brasiliensis*) were antigenically closer to the Californian than the Brazilian viruses (Fig. 5.5, Panel C).

Restriction fragment length patterns

Short of sequencing, which is not feasible for viruses with such a large genome, the best way of comparing the genomes of poxviruses is by restriction fragment cleavage patterns. Russell and Robbins (1989) found that the patterns of two Australian strains (the virulent Glenfield strain and the attenuated Uriarra strain) and the Lausanne strain, which had been used to introduce myxomatosis into Europe, gave identical patterns with seven out of eight restriction enzymes. One enzyme, *Kpn*I, produced a site with Lausanne DNA not found with the DNAs of the two derivatives of the Standard Laboratory strain. Subsequently P.J. Kerr (personal communication, 1997) used restriction patterns to study the epidemiology of myxomatosis in several locations near Canberra.

Figure 5.6 illustrates the differences between the restriction fragment length patterns of six different leporipoxviruses and the orthopoxvirus, vaccinia virus, after cleavage with two restriction enzymes, *Hin*dIII and *Sal*I. The restriction patterns given after cleavage of DNA from three strains of Shope fibroma virus (2, 3, 4 in Fig. 5.6) with each enzyme were very similar to each other, but distinctly different from those of the myxoma virus strains. Myxoma virus from Venezuala (6) and Brazil (Lausanne strain, 7) showed similar patterns after cleavage with *Hin*dIII but several different bands in *Sal*I digests, whereas a strain of myxoma virus from Panama (5) showed distinctive patterns with both enzymes. None of the leporipoxviruses had any resemblance to the patterns of vaccinia virus (8).

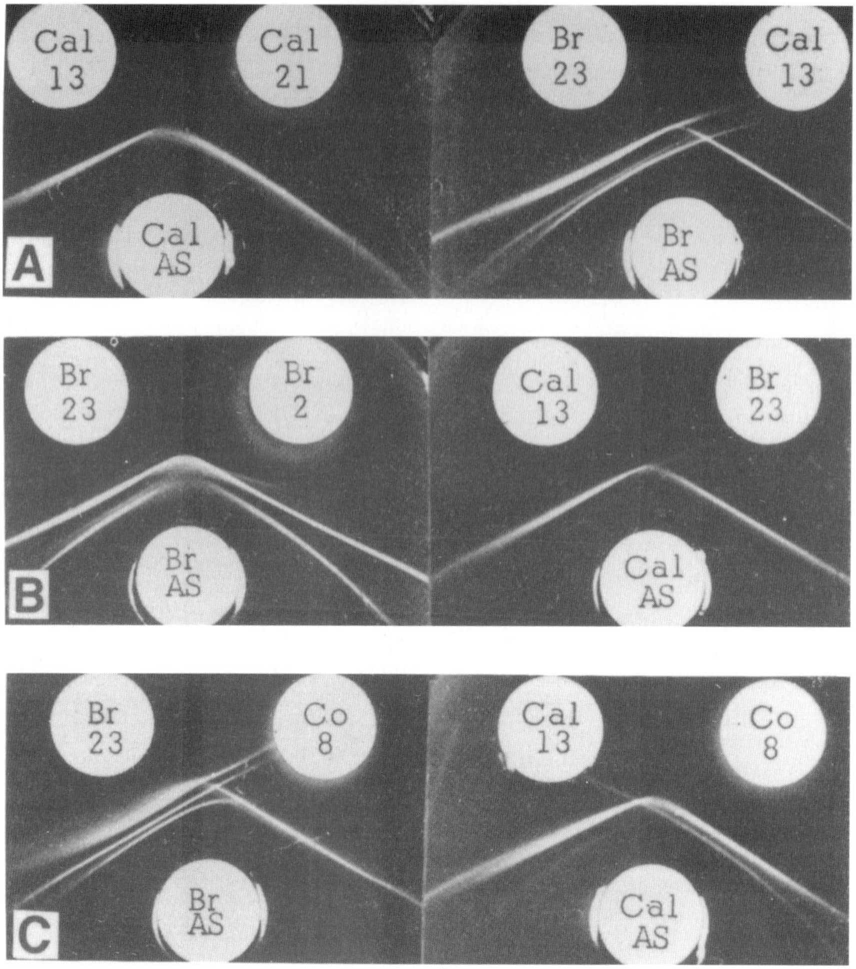

Fig. 5.5. Differences in soluble antigens of strains of myxoma virus derived from different countries in the Americas as determined by gel diffusion tests. The lower wells contain antibodies to a Brazilian (Br) or a Californian (Cal) strain of myxoma virus; the upper wells contain antigens from tumours produced by strains of myxoma virus from different countries. Each panel is representative of the results obtained with antigens derived from several different strains of virus, as indicated in the legends. From Fenner (1965), with permission. *Panel A.* The identity of the soluble antigens of strains of eleven strains of myxoma virus from California and their differences from soluble antigens of the Standard Laboratory Strain (derived from Brazil). *Panel B.* The identity of the soluble antigens of the Standard Laboratory Strain and three other strains of myxoma virus from Brazil, two from Colombia, one from Argentina and one from Uruguay and their difference from antigens of the Californian strains. *Panel C.* The lack of identity of soluble antigens of two strains of myxoma virus from Colombia and two from Panama with those of Brazilian myxoma virus, and their close resemblance to, but not identity with, those from a Californian virus.

Pathogenesis of Myxomatosis

Spread of virus around the body

Early in his scientific career Fenner (1948) had studied the way virus spread around the body in mousepox, a poxvirus disease of mice with generalized skin lesions. He applied the same approach to the pathogenesis of myxomatosis (Fenner and Woodroofe, 1953), with essentially similar

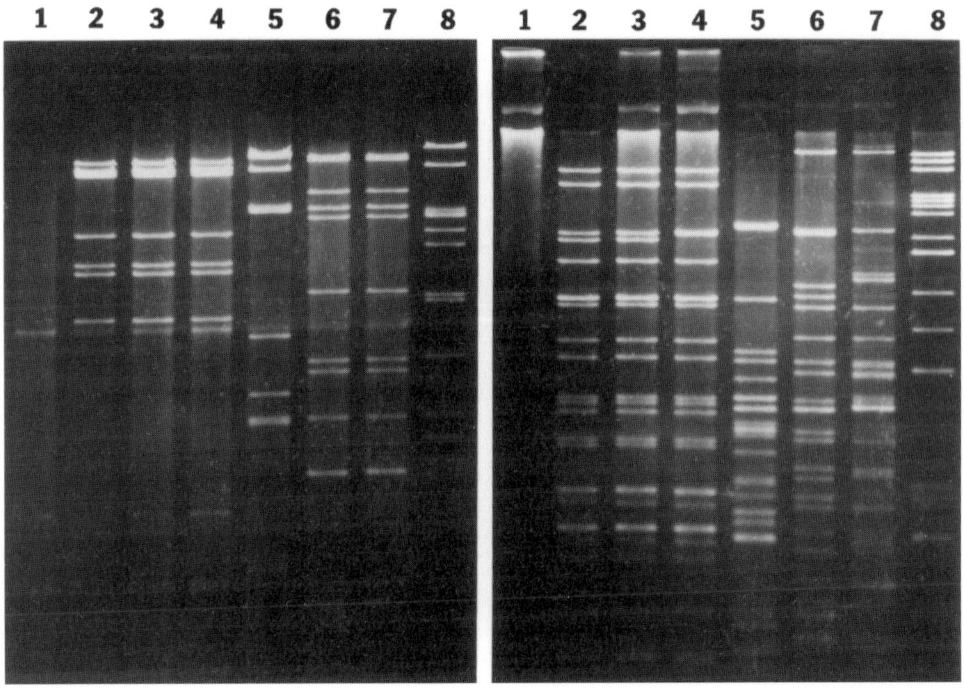

Fig. 5.6. Restriction patterns of several poxviruses after cleavage with *Hin*dIII (on left) and *Sal*I (on right). Lane 1, mitochondrial DNA; 2, Patuxent fibroma virus; 3, Kasya fibroma virus; 4, Shope OA fibroma virus; 5, Panama myxoma virus; 6, Venezuala myxoma virus; 7, Lausanne myxoma virus; 8, Lister vaccinia virus. Courtesy Dr J.J. Esposito.

results (Fig. 5.8). After the intradermal inoculation of a very small dose of virulent myxoma virus, there was a stepwise infection of the skin, the local lymph node, and the small arterioles. Release from some of the lymph node cells produced viraemia, which peaked on the fourth day. Generalized infection then occurred, with rising titres of virus in localized areas of skin distant from the inoculation site and to a high titre in the testis. Fenner and Woodroofe did not do more extensive titrations, but judging from histological findings, virus replication occurred throughout the lymphoid system, bone marrow, and spleen, the lungs, surface of the peritoneum and omentum. The viraemia was cell-associated, in lymphocytes.

Recent studies with genetically resistant wild rabbits and non-resistant domestic rabbits, using both a highly virulent and an attenuated strain of myxoma virus, showed that the viral titres in the draining and in distal lymph nodes were significantly reduced in the wild rabbits compared with the domestic rabbits, although the titres at the injection sites in the skin were similar (S.M. Best and P.J. Kerr, unpublished observations, 1998).

Shope's fibroma virus produces only a localized skin infection in normal European rabbits (see p. 69). Unlike myxoma virus, it cannot replicate in rabbit lymphocytes *in vitro,* instead the infected lymphocytes undergo apoptosis. The failure to replicate in lymphocytes is probably a key event in preventing the development of systemic disease in infections of rabbits with Shope fibroma virus; deletion of genes that affect replication in lymphocytes greatly reduces the virulence of myxoma virus (McFadden and Barry, 1998).

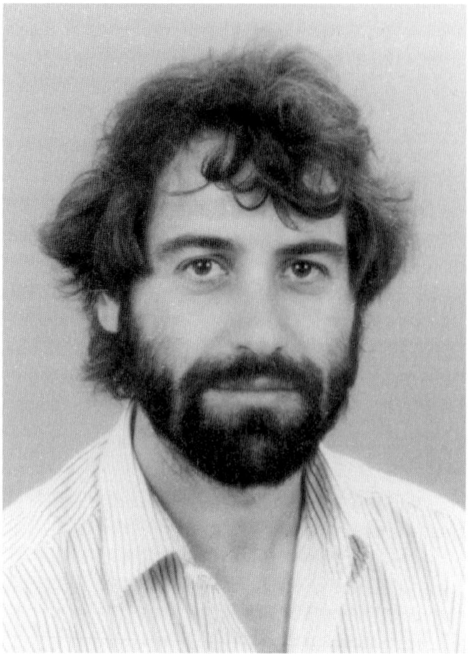

Fig. 5.7. Peter John Kerr (1956–). After graduating with a BAgrSc from the University of Melbourne in 1978, Kerr worked in a rural veterinary practice and in agribusiness before undertaking graduate studies of the molecular virology of arboviruses at the Australian National University, where he obtained the degree of PhD in 1990. That year he joined the CSIRO Division of Wildlife and Ecology, to work on the epidemiology and pathogenesis of myxomatosis. He developed molecular methods of characterizing field isolates of myxoma virus that have proved valuable in elucidating the epidemiology of myxomatosis. His studies on pathogenesis are beginning to provide an understanding of the basis of genetic resistance to myxomatosis.

Replication in the testes

However infection was induced, the testes of male rabbits were always swollen, contained virus to a high titre (Fenner and Woodroofe, 1953) (Fig. 5.9) and showed necrosis of tubular cells and epididymitis (Hurst, 1937a). Male domestic rabbits that had recovered from myxomatosis were infertile for up to 12 months (Sobey and Turnbull, 1956). As judged by the absence of spermatozoa, male wild rabbits were also infertile after epidemics of myxomatosis (Poole, 1960), although this seemed to have minimal impact on female fecundity in the wild (Marshall *et al.*, 1955). Fountain *et al.* (1997) reinvestigated the phenomenon, using domestic rabbits infected with a highly attenuated strain of virus. They confirmed the occurrence of interstitial orchitis and epididymitis, associated with decreased testosterone and increased luteinizing hormone concentrations. Infectious virus was cleared within 20–30 days and concurrently testosterone and luteinizing hormone concentrations returned to normal. The animals were sterile for 60 days after infection, but normal fertility then returned.

Comparison of pathogenesis of pneumonic and nodular forms of myxomatosis

Earlier in this chapter we noted that in the 1980s French workers recognized the emergence of a novel 'pneumonic' form of myxomatosis, occurring mainly in commercial rabbitries in which intensive husbandry was practised. Duclos *et al.* (1983) have compared the pathohistology of the pneumonic and the nodular forms of myxomatosis. In the classical nodular form the virus is introduced into the skin by an insect vector, replicates there, producing a primary skin lesion, and proceeds via the local lymph nodes to cause a generalized viraemia and secondary lesions in the skin, the testes and elsewhere, as illustrated in Fig. 5.8. In the pneumonic form, airborne virus enters the alveoli of the lung and replicates locally, followed by invasion of the local lymph nodes and viraemia. Virus spreads through the lungs and to the eyes, facilitating secondary bacterial infection, bronchopneumonia and severe conjunctivitis. There are few secondary skin lesions, which appear as congestive macules or small nodules.

The immunological response

Highly virulent virus killed rabbits too quickly for free antibody to be titratable by the methods available in the mid-1950s, but on the sixth and seventh day the serum showed anticomplementary activity and a day later complement-fixing antigen was present in the serum, to such an extent that

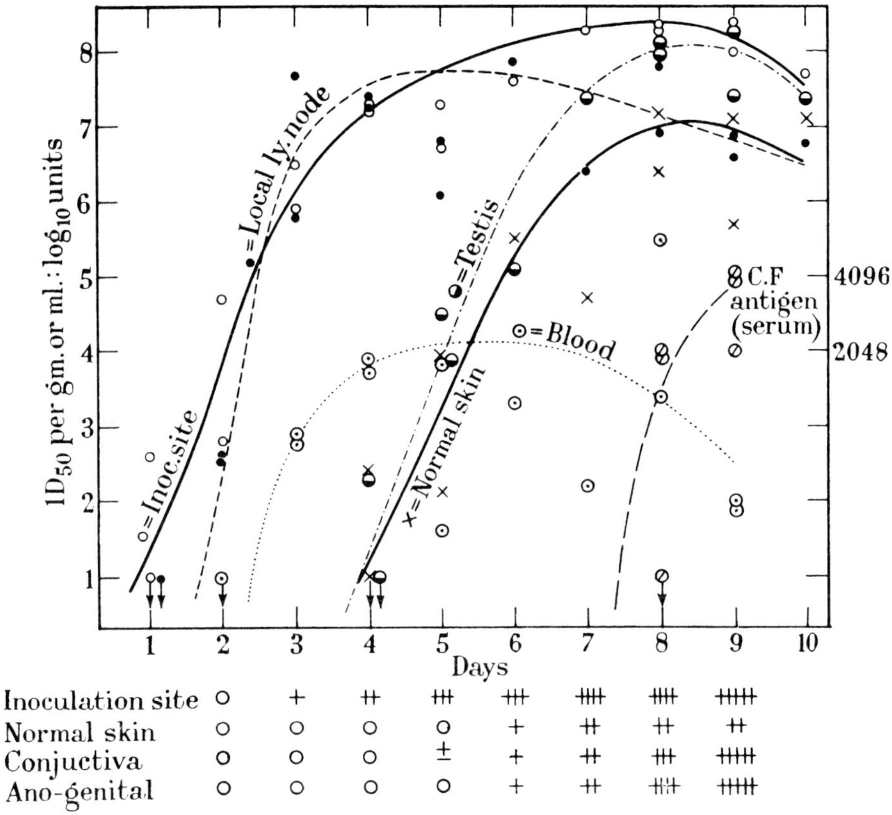

Fig. 5.8. The spread of myxoma virus through the body after the intradermal inoculation of a small dose of the Standard Laboratory Strain, as shown by infectivity titrations of several organs and the blood, and by the appearance of complement-fixing antigen in the serum. The symbols at the bottom indicate the day of appearance and the severity of physical signs in the sites indicated. From Fenner and Woodroofe (1953), with permission.

it was more viscous than normal serum. The antibody response was therefore studied in rabbits infected with the less virulent neuromyxoma virus. Free complement-fixing antigen could not be detected, but anticomplementary activity was present from the 7th to the 28th day and complement-fixing antibody was first detected on the 11th day, rising to a very high titre by the 28th day. Neutralizing antibody was detected on the 10th day, rising to a peak four days later. The persistence of antibodies over a period of 19 months was studied in two wild rabbits that had survived an early epidemic of myxomatosis (Fenner *et al.*, 1953). Within six months of capture the initial very high titre of complement-fixing antibody fell to a low but stable level, but the level of neutralizing antibody remained high. Nineteen months later there was no clinical or serological response to challenge infection with a small dose of virus, but after a large dose one rabbit developed a skin nodule about 25 mm in diameter and the level of complement-fixing antibody of this rabbit increased fourfold; the other rabbit failed to respond. At the time this work was done, methods of assaying the cellular immune response had not been developed.

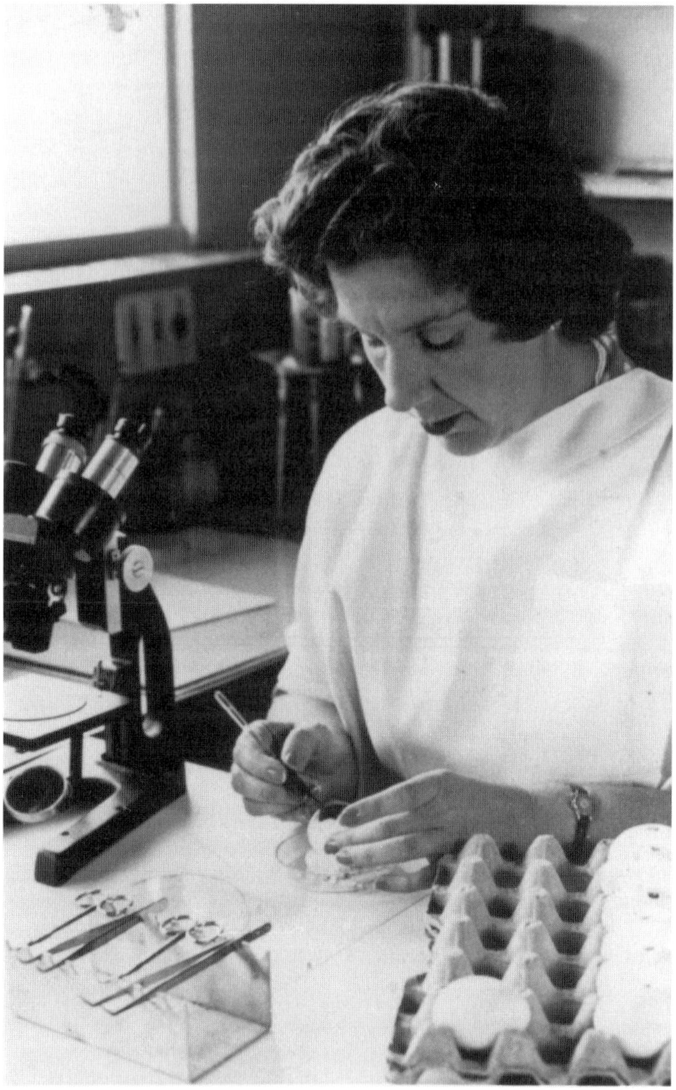

Fig. 5.9. Gwendolyn Marion Woodroofe (1918–). After graduating with a BSc(Hons) in botany from the University of Adelaide in 1940, Woodroofe gained an MSc degree in bacteriology for work on salmonellosis before joining the Department of Microbiology, John Curtin School of Medical Research, in 1951 as a research assistant, working on myxomatosis. She later became a Research Fellow, gaining a PhD degree in 1962, and then a Fellow until her retirement in 1978. She played a major role in virological investigations of myxomatosis between 1951 and 1966, when she went to work with I.D. Marshall on arboviruses. The illustration shows her reaping chorioallantoic membranes infected with myxoma virus for titration by pock-counting. As was characteristic at the time (the 1950s), the work was done on an open bench, without gloves. After her retirement she supported the work of UNICEF in Canberra, for which she was awarded a Medal of the Order of Australia in 1997.

Immunosuppressive effects

Strayer and his colleagues have reported extensively on the immunosuppression produced by infection of rabbits with what they call 'malignant rabbit fibroma virus' (review: Strayer, 1992). This virus is a recombinant between myxoma virus and Shope's fibroma virus that occurred inadvertedly (Strayer *et al.* 1983); about 90% of its unique sequences are derived from myxoma virus, but 10 kb of Shope fibroma-derived sequences have substituted for a similar number of myxoma virus sequences (Block *et al.*, 1985). The histological features of malignant rabbit fibroma are somewhat different from those of myxomatosis (Strayer and Sell, 1983; Strayer *et al.*, 1983), notably in the absence of virus in the skin overlying tumours and in epithelial cells in the liver, kidney and lung. However, since the involvement of the reticuloendothelial cells and lymphocytes of the spleen, lymphoid system and bone marrow was similar in both infections, many of their findings about immunosuppression are probably applicable to myxomatosis. Clinically, 'snuffles' (due to *Pasteurella multocida* or other bacterial infection) is a common feature of infections with malignant rabbit fibroma virus. Snuffles was almost always present in laboratory rabbits that did not die of acute myxomatosis, probably reflecting the endemic bacterial infections present in the close confines of the animal house. According to workers with extensive field experience, snuffles was rarely seen in myxomatous wild rabbits in the field, in either Australia or Britain, although it was universal in hutch rabbits exhibiting the pneumonic form of myxomatosis (Duclos *et al.*, 1983).

Effects of maternal antibodies

Young rabbits could be protected against myxomatosis for about three weeks by the intraperitoneal inoculation of myxoma-immune serum. Some resisted infection completely, whether attempted by intradermal inoculation or by mosquito bites; others were infected but survived substantially longer than controls (Fenner and Marshall, 1954). Animals not infected were completely susceptible when challenged some months later. Similar results occurred with young borne by mothers that had recovered from myxomatosis. However, Sobey and Conolly (1975) found that although kittens (<4 weeks of age) were protected, there was no evidence of protection among older rabbits borne by recovered mothers when exposed to infection by contact or flea transmission, compared with adult non-immune animals similarly infected.

Effects of age on the response to infection

If they lack maternal antibody, very young rabbits die of myxomatosis much more rapidly than adult animals (Fenner and Marshall, 1954). Thus groups of six kittens aged 9–14 days infected with a virus of Grade I virulence had a mean survival time of 5.4 days, with minimal lesions. Resistance increased slowly with advancing age; the mean survival time of animals 21–27 days old at the time of infection was 6.0 days, whereas adult rabbits had a mean survival time of 10.8 days. The greater susceptibility of young animals was more obvious when a strain of Grade IV virulence was used for challenge; all animals in the younger two age groups died after mean survival times of 11.7 or 17.0 days respectively, whereas there were 40% survivors and a mean survival time of 26.2 days in adult rabbits. The susceptibility of young rabbits is probably responsible for large numbers of unrecognized deaths in field situations where myxomatosis occurs in spring, especially after an absence of myxomatosis for several years. Uninfected kittens less than three weeks old are likely to die if their mothers die of myxomatosis.

In experiments on the genetic resistance of rabbits (see p. 174), the risk of immaturity or passive immunity complicating interpretation of the results was avoided by holding captured young wild rabbits for 16 weeks before carrying out inoculations with myxoma virus.

Effects of environmental temperature

Thompson (1938) and Parker and Thompson (1942) found that maintenance of infected rabbits at temperatures higher than 36°C profoundly reduced the severity of infections with both fibroma and myxoma viruses, the ameliorating effect increasing with increasing temperature. Realizing that ambient temperature might influence mortality in outbreaks of myxomatosis in Australia, Marshall (1959) investigated the effects of holding rabbits inoculated with myxoma virus at temperatures that might occur in the field, namely 37–39°C, 20–22°C (animal house) or 4°C for 16 of each 24 hours, and at 20–22°C for the other eight. Viruses of Grade I and Grade III virulence were used for challenge, and control rabbits were held at animal house temperatures throughout. The results are summarized in Table 5.3 and illustrated in Figs 5.10 and 5.11. Especially with the Grade III virus, it is clear that maintenance of rabbits at high temperatures greatly reduced, and at low temperatures greatly increased the severity of myxomatosis. Not only were the signs much less (at high temperatures) or more severe (at low temperatures) than at animal house temperature, but so was the extent of the viraemia and the level of antigen in the serum; antibody was found only in the serum of the rabbits that had been exposed to hot room conditions.

Sobey *et al.* (1968) confirmed the ameliorating effect of high environmental temperatures, and found that exposure to a temperature of 85°F (29.4°C) for 24 hours 3–4 days after infection with virulent virus of rabbits with some genetic resistance was useful in breeding for genetic resistance to myxomatosis (Sobey, 1969).

Immunization against Myxomatosis

Soon after myxomatosis spread in Australia there were outbreaks in laboratory rabbits in medical establishments in Melbourne and Sydney, so it was necessary to provide a vaccine to protect such animals. The need was even more urgent when myxomatosis spread in Europe after mid-1952, because there was a very large domestic rabbit industry in many countries. Initial efforts were based on the immunological cross-reactivity between myxoma and Shope fibroma viruses; later, in California and France, attenuated strains of myxoma virus were used as vaccines.

Shope fibroma virus

In his original paper on fibroma in cottontails, Shope (1932) commented on the immunological relationship between

Table 5.3. Myxomatosis in rabbits held at animal house temperature (20–22°C) for 8 hours each day, and for the other 16 hours at low (0–4°C), moderate (20–22°C) or high (37–39°C) temperatures, after inoculation with a virus of grade III virulence[a].

Inoculum	Result	Cold room (0–4°C)	Mild room (20–22°C)	Hot room (37–39°C)
Myxoma virus Grade III	Mortality	36/39	23/35	2/23[b]
	Clinical signs	Severe, progressive generalized disease	Variable	Mild, with early demarcation of lesions
	Viraemia	High and prolonged	Moderate and prolonged	Low and transient
	Antigenaemia	In 4/5 from day 10 to day 15	In 2/4 on day 12	Absent
	Serum antibody	Absent except in one survivor (from day 12)	Present in 3/4 from day 13	Present in 3/5 from day 11
Myxoma virus Grade I	Mortality		6/6	6/6
	Survival time (days)	—	10, 11, 11, 11, 11, 13	9, 9, 10, 12, 12, 12,
Nil (controls)	Mortality	0/36	—	0/31[b]

[a]Data from Marshall (1959).
[b]Seven deaths not due to myxomatosis omitted (five deaths in control group).

Fig. 5.10. The appearance of rabbits 20 days after being inoculated intradermally with a strain of myxoma virus of Grade III virulence and held for 16 of every 24 hours either in a cold room (4°C) (A), or a hot room (37–39°C) (B), and at ambient temperature (20–22°C) for the other 8 hours of each day. From Marshall (1959), with permission.

myxoma virus and the fibroma virus of eastern cottontail rabbits (*Sylvilagus floridanus*). Subsequently he and other workers (Hurst, 1938; Shope, 1938; Hyde, 1939) demonstrated that fibroma virus protected European rabbits against myxomatosis for at least a few weeks after vaccination. Protection developed rapidly, some modification of myxomatosis being evident when fibroma virus had been inoculated as recently as two days before challenge, but there was no information on the duration of protection. In reviewing the situation in 1939, Moses concluded that there was no satisfactory method of vaccination (Moses, 1939).

Fenner and Woodroofe (1954) reinvestigated the problem, using several strains of fibroma virus and testing rabbits for protection for as long as 12 months after vaccination. They found that the vaccination with commonly used OA laboratory strain of fibroma virus was protective against generalized myxomatosis for less than two months. However, the Boerlage strain, isolated from a cottontail rabbit in New Jersey in 1947 (Smith, 1952), gave substantial protection against generalized disease for over 12

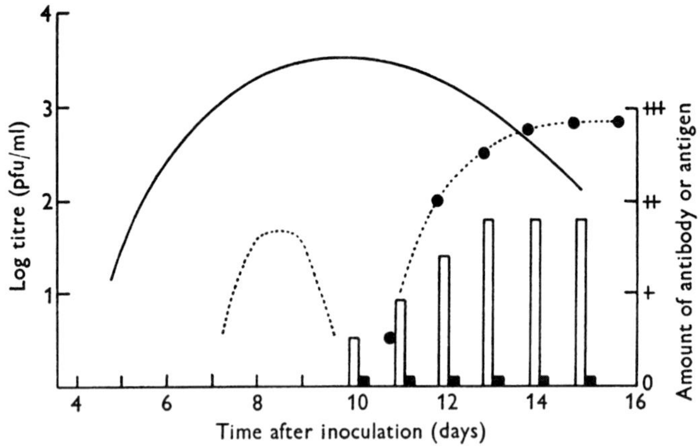

Fig. 5.11. Level of viraemia and concentrations of serum antigen and antibody at various times after infection in rabbits held in a cold room (4°C) or a hot room (37–39°C) for 16 hours out of each 24 hours, and at animal house temperature (20–22°C) for the other 8 hours. Curves on left and centre indicate viraemia in cold room rabbits (solid curve) and hot room rabbits (dotted curve). Dotted line through solid circles indicates antibody in hot room rabbits – antibody was not detected in cold room rabbits. Columns indicate antigen in the serum of cold room rabbits; solid blocks indicate that no antigen was seen in the serum of hot room rabbits. From Fenner and Ratcliffe (1965), with permission.

months, and this virus was developed for use as a vaccine for laboratory rabbits in Australia. Duran-Reynals (1945) had found that newborn rabbits suffered from generalized fibromatosis after inoculation with fibroma virus, and experiments with the Boerlage strain had shown that lesions were severe in rabbits less than two weeks old. Further, while lesions produced by fibroma virus in adult rabbits were not readily transmitted by mosquitoes, the lesions produced in infant rabbits were readily transmissible (Dalmat and Stanton, 1959). It was therefore suggested that laboratory rabbits in danger of being naturally infected with myxomatosis should be vaccinated with the Boerlage strain of fibroma virus, except for rabbits less than 14 days old, and in Australia freeze-dried virus for this purpose was made available to hospital and research laboratories by the Commonwealth Serum Laboratories[2].

While the difficulty of mosquito transmission from adult rabbits vaccinated with fibroma virus was clearly an advantage in Australia, where it was essential that fibroma virus should not become established in wild rabbits, in France the possibility that it might be spread by mosquitoes was deemed an advantage because of the widespread wish by hunters to control myxomatosis in wild rabbits. Although there are several ways in which more persistent tumours, which became infectious for mosquitoes, could be produced, such as inoculating by baby rabbits or by treating adult rabbits with X-rays or carcinogens (Dalmat and Stanton, 1959), these were impractical for use in the field.

As soon as myxomatosis broke out in France there was an immediate demand for a vaccine, for both domestic rabbits, and among chasseurs, for wild rabbits. Jacotot *et al.* (1955c, 1958) noted the variable effects of vaccination with fibroma virus, but it was used on a large scale, initially at the rate of about ten million doses a year. In Britain the demand was less, because of the rarity of mosquito transmission and therefore the lower risk of infection of domestic rabbits. Rowe *et al.* (1956) recommended vaccination of domestic or pet rabbits with the Boerlage strain, with revaccination six months later.

Attenuated myxoma viruses as vaccines

As with other poxviruses, killed vaccines are useless, so steps were taken in California and in France to produce attenuated live-virus vaccines.

Attenuated Californian strain

The first attenuated myxoma virus vaccine was developed by Saito *et al.* (1964), in response to a severe outbreak of myxomatosis in the main rabbit-raising area in California in 1949 and dissatisfaction with the efficacy of fibroma virus as a vaccine. The MSD (Californian) strain of myxoma virus was still fully virulent after 161 serial transfers on the chorioallantoic membrane, but was somewhat attenuated after 32 passages in rabbit kidney cell cultures and highly attenuated after 40 passages. Rabbits vaccinated with this strain were refractory to experimental challenge for at least nine months after vaccination.

The vaccine was produced commercially by the Poultry Health Laboratories (PHL) and small lots were sold from time to time. In 1978 the PHL supplied a request for 1500 doses for vaccination of a herd of Rex rabbits. About a week after supplying the vaccine PHL was notified by the attendant veterinarian that many of the vaccinated rabbits had died, and a few months later a law suit was launched by the owner (Hulme), naming PHL and the University of California at Davis as defendants and asking for several million dollars in damages. The trial was held in 1984; the University was dropped from the defendants, and the case became Hulme vs. PHL. It transpired that prior to vaccination each rabbit had been carefully examined and any showing signs of pasteurellosis placed in a separate building (infirmary) and not vaccinated. The ranch owner then left the property in care of a high school student who could not cope with the work. Large numbers of deaths occurred in all buildings including the infirmary, *Pasteurella* but not myxoma virus was isolated from all fatal cases examined by a diagnostic laboratory, and after a short deliberation the jury decided that the vaccine was not to blame and dismissed the case[3].

Attenuated Lausanne strain (France)

There was a demand in France for vaccination of laboratory and domestic hutch rabbits against myxomatosis, and also a desire by chasseurs to vaccinate wild rabbits in hunting reserves. Initially Shope's fibroma virus was used, and then the Saito vaccine, but Jacotot *et al.* (1967) found that this vaccine recovered some of its pathogenicity on passage. It was therefore replaced by the attenuated Lausanne strain SG33, developed by Saurat *et al.* (1978) from a strain isolated from a wild rabbit near Toulouse in 1973 by serial passage in rabbit kidney cells and chick embryo fibroblasts at 33°C. High titre virus could be obtained from cultured cells, and this appeared to be of very low virulence for rabbits, by any route of inoculation, and it was not transmissible from inoculated to susceptible rabbits. However, with wide scale use the SG33 vaccine was found to have an initial immunosuppressive effect which, in commercial rabbit farms with poor standards of hygiene, led to secondary bacterial infections (Brun *et al.*, 1981b). It is now recommended that fibroma vaccine should be given first, followed by vaccination with SG33 six to eight weeks later (Joubert *et al.*, 1982).

Another serious viral disease, rabbit haemorrhagic disease (RHD), spread through Europe between 1987 and 1989 (see Chapter 11) and a recombinant RHDV capsid-baculovirus vaccine was developed which induced good protection against RHD (Laurent *et al.*, 1994). Using SG33, modified by inactivation of its thymidine kinase or certain 'pathogenicity' genes, Bertagnoli *et al.* (1996) then produced a myxoma virus–RHDV capsid recombinant vaccine. After intradermal inoculation, this gave good protection against both rabbit haemorrhagic disease and myxomatosis, but no information is yet available on its efficacy in the field.

Endnotes

[1]A more comprehensive coverage of scientific studies on the virus and the disease up to the early 1960s is provided in Fenner and Ratcliffe (1965).
[2]CSL Archives series C47. File 58/2419. Fibroma virus vaccine.
[3]Basser Library Archives 143/25/5A. Letter, R.W. Wichmann to Fenner, 3 March 1997.

References

Arthur, C.P. and Louzis, C. (1988) A review of myxomatosis among rabbits in France. *Revue scientifique et technique de l'Office International des Epizooties* 7, 937–957, 959–976.

Bertagnoli, S., Gelfi, J., Le Gall, G., Boilletot, E., Vautherot, J.-F., Rasschaert, D., Laurent, S., Petot, F., Boucraut-Baralon, C. and Milon, A. (1996) Protection against myxomatosis and rabbit viral hemorrhagic disease with recombinant myxoma viruses expressing rabbit hemorrhagic disease virus capsid protein. *Journal of Virology* 70, 5061–5066.

Beveridge, W.I.B. and Burnet, F.M. (1946) *The Cultivation of Viruses and Rickettsiae in the Chick embryo.* Special Report No. 256. Medical Research Council, London, 92 pp.

Block, W., Upton, C. and McFadden, G. (1985) Tumorigenic poxviruses: Genome organization of malignant rabbit fibroma virus, a recombinant between Shope fibroma virus and myxoma virus. *Virology* 140, 115–124.

Brun, A., Saurat, P., Gilbert, Y., Godard, A. and Bouquet, J.F. (1981a) Données actuelle sur l'épidémiologie, la pathogénie et la symptomatologie de la myxomatose. *Revue de Médecine Vétérinaire* 132, 585–590.

Brun, A., Godard, A. and Moreau, Y. (1981b) La vaccination contre le myxomatose, vaccins hétérologues et homologues. *Bulletin de la Societé de Science Vetérinaire Medécine Compareé* 83, 251–254.

Burnet, F.M. (1936) *The Use of the Developing Egg in Virus Research.* Special Report Series Medical Research Council, London, No. 220, 58 pp.

Dalmat, H.T. and Stanton, M.F. (1959) A comparative study of the Shope fibroma in rabbits in relation to transmissibility by mosquitoes. *Journal of the National Cancer Institute* 22, 593–615.

Duclos, Ph., Tuaillon, P. and Joubert, L. (1983) Histopathologie de l'atteinte cutanéo-muqueuse et pulmonaire de la myxomatose. *Bulletin Académie Veterinaire de France* 56, 95–104.

Dulbecco, R. and Vogt, M. (1954) One-step growth curves of Western equine encephalomyelitis virus on chick embryo cells grown in vitro and analysis of viral yields from single cells. *Journal of Experimental Medicine* 99, 185–199.

Duran-Reynals, F. (1945) Immunological factors that influence the neoplastic effects of the rabbit fibroma virus. *Cancer Research* 5, 25–39.

Fenner, F. (1948) The pathogenesis of the acute exanthems. *Lancet* 2, 915–920.

Fenner, F. (1965) Viruses of the myxoma-fibroma subgroup of the poxviruses. II. Comparison of soluble antigens by gel diffusion tests, and a general discussion of the subgroup. *Australian Journal of Experimental Biology and Medical Science* 43, 145–156.

Fenner, F. and McIntyre, G.A. (1956) Infectivity titrations of myxoma virus in the rabbit and the developing chick embryo. *Journal of Hygiene* 54, 246–257.

Fenner, F. and Marshall, I.D. (1954) Passive immunity in myxomatosis of the European rabbit (*Oryctolagus cuniculus*): the protection conferred on kittens born by immune does. *Journal of Hygiene* 52, 321–336.

Fenner, F. and Marshall, I.D. (1957) A comparison of the virulence for European rabbits (*Oryctolagus cuniculus*) of strains of myxoma virus recovered in the field in Australia, Europe and America. *Journal of Hygiene* 55, 149–191.

Fenner, F. and Ratcliffe, F.N. (1965) *Myxomatosis.* Cambridge University Press, Cambridge, 379 pp.

Fenner, F. and Woodroofe, G.M. (1953) The pathogenesis of infectious myxomatosis: the mechanism of infection and the immunological response of the European rabbit (*Oryctolagus cuniculus*). *British Journal of Experimental Pathology* 34, 400–411.

Fenner, F. and Woodroofe, G.M. (1954) Protection of laboratory rabbits against myxomatosis by vaccination with fibroma virus. *Australian Journal of Experimental Biology and Medical Science* 32, 655–668.

Fenner, F., Marshall, I.D. and Woodroofe, G.M. (1953) Studies in the epidemiology of infectious myxomatosis of rabbits, I. Recovery of Australian wild rabbits (*Oryctolagus cuniculus*) from myxomatosis under field conditions. *Journal of Hygiene* 51, 225–244.

Fenner, F., Day, M.F. and Woodroofe, G.M. (1956) Epidemiological consequences of the mechanical transmission of myxomatosis by mosquitoes. *Journal of Hygiene* 54, 284–303.

Fountain, S., Holland, M.K., Hinds, L.A., Janssens, P.A. and Kerr, P.J. (1997) Interstitial orchitis with impaired steroidogenesis and spermatogenesis in the testes of rabbits infected with an attenuated strain of myxoma virus. *Journal of Reproduction and Fertility* 110, 161–169.

Hudson, J.R. and Mansi, W. (1955) Attenuated strains of myxomatosis in England. *Veterinary Record* 67, 746–747.

Hurst, E.W. (1937a) Myxoma and the Shope fibroma. I. The histology of myxoma. *British Journal of Experimental Pathology* 18, 1–15.

Hurst, E.W. (1937b) Myxoma and the Shope fibroma. II. The effect of intracerebral passage on the myxoma virus. *British Journal of Experimental Pathology* 18, 15–22.

Hurst, E.W. (1938) Myxoma and the Shope fibroma. V. Myxoma in the fibroma-immune rabbit, with a summary of present knowledge of the relationship between myxoma and fibroma viruses. *Australian Journal of Experimental Biology and Medical Science* 16, 205–208.

Hyde, R.R. (1939) Infectious myxomatosis of rabbits (Sanarelli) versus the fibroma virus (Shope) with especial reference to the time interval in the establishment of concominant immunity. *American Journal of Hygiene* 30, 47–55.

Jacotot, H., Vallée, A. and Virat, B. (1955a) Étude d'une souche attenuée de virus du myxome (Uriarra III d'Australie). *Annales de l'Institut Pasteur* 89, 8–15.

Jacotot, H., Vallée, A. and Virat, B. (1955b) Apparition en France d'un mutant naturellement attenué du virus de Sanarelli. *Annales de l'Institut Pasteur* 89, 361–364.

Jacotot, H., Vallée, A. and Virat, B. (1955c) Considerations sur la durée et le mecanisme de l'immunité engendrée par le virus du fibrome de Shope contre le virus du myxome de Sanarelli. *Annales de l'Institut Pasteur* 89, 381–384.

Jacotot, H., Vallée, A. and Virat, B. (1958) Sur l'immunisation contre le virus du myxome infectieux par inoculation du fibrome de Shope. *Annales de l'Institut Pasteur* 94, 282–293.

Jacotot, H., Virat, B., Reculard, P. and Vallée, A. (1967) Étude d'une souche attenuée du virus du myxome infectieux obtenue par passages en culture cellulaires (McKercher et Saito, 1964). *Annales de l'Institut Pasteur* 113, 221–237.

Joubert, L., Duclos, Ph. and Tuaillon, P. (1982) La myxomatose des garennes dans le Sud-Est. La myxomatose amyxomateuse. *Revue de Médecine Vétérinaire* 133, 739–753.

Kerr, P.J. (1997) An ELISA for epidemiological studies of myxomatosis: persistence of antibodies to myxoma virus in European rabbits (*Oryctolagus cuniculus*). *Wildlife Research* 24, 53–65.

Laurent, S., Vautherot, J.-F., Madelaine, M.-F., Le Gall, G. and Rasschaert, D. (1994) Recombinant rabbit hemorrhagic disease virus capsid protein expressed in baculovirus self-assembles into virus-like particles and induces protection. *Journal of Virology* 68, 6794–6798.

Lush, D. (1937) The virus of infectious myxomatosis of rabbits on the chorioallantoic membrane of the developing egg. *Australian Journal of Experimental Biology and Medical Science* 15, 131–139.

McFadden, G. and Barry, M. (1998) How poxviruses oppose apoptosis. *Seminars in Virology* 8, 429–442.

Mansi, W. and Thomas, V. (1958) Serological investigation in the study of myxoma and fibroma viruses. II. The gel diffusion test. *Journal of Comparative Pathology and Therapeutics* 68, 188–200.

Marshall, I.D. (1959) The influence of ambient temperature on the course of myxomatosis in rabbits. *Journal of Hygiene* 57, 484–497.

Marshall, I.D., Dyce, A.L., Poole, W.E. and Fenner, F. (1955) Studies on the epidemiology of infectious myxomatosis of rabbits. IV. Observations on disease behaviour in two localities near the northern limit of rabbit infestation in Australia, May 1952 to April 1953. *Journal of Hygiene* 53, 12–25.

Moses, A. (1939) Infectious rabbit myxomatosis. A review of recent observations. *Report of Proceedings, Third International Congress for Microbiology*, p. 342.

Myers, K., Marshall, I.D. and Fenner, F. (1954) Studies in the epidemiology of infectious myxomatosis of rabbits. III. Observations on two succeeding epizootics in Australian wild rabbits on the riverine plain of southeastern Australia 1951–1953. *Journal of Hygiene* 52, 337–360.

Mykytowycz, R. (1953) An attenuated strain of myxomatosis virus recovered from the field. *Nature* 172, 448–449.

Padgett, B.L., Moore, M.S. and Walker, D.L. (1962) Plaque assays for myxoma and fibroma viruses and differentiation of the viruses by plaque form. *Virology* 17, 462–469.

Parker, R.F. (1940) Studies on the infectious unit of myxoma. *Journal of Experimental Medicine* 71, 439–444.

Parker, R.F. and Thompson, R.L. (1942) The effect of external temperature on the course of infectious myxomatosis of rabbits. *Journal of Experimental Medicine* 75, 567–673.

Poole, W.E. (1960) Breeding of the wild rabbit, *Oryctolagus cuniculus* (L.), in relation to the environment. *Australian Wildlife Research* 5, 21–43.

Reisner, A.H., Sobey, W.R. and Conolly, D. (1963) Differences among the soluble antigens of myxoma viruses originating in Brazil and in California. *Virology* 20, 539–541.

Rowe, B., Mansi, M. and Hudson, J.R. (1956) The use of fibroma virus (Shope) for the protection of rabbits against myxomatosis. *Journal of Comparative Pathology and Therapeutics* 66, 290–298.

Russell, R.J. and Robbins, S.J. (1989) Cloning and molecular characterization of the myxoma virus genome. *Virology* 170, 147–159.

Saito, J.K., McKercher, D.G. and Castrucci, G. (1964) Attenuation of the myxoma virus and the use of the living attenuated virus as an immunizing agent for myxomatosis. *Journal of Infectious Diseases* 114, 417–428.

Saurat, P., Gilbert, Y. and Ganière, J.P. (1978). Étude d'une souche de virus myomateux modifié. *Revue de Medécine Vetérinaire* 129, 415–451.

Schwerdt, P.R. and Schwerdt, C.E. (1962) A plaque assay for myxoma virus infectivity. *Proceedings of the Society for Experimental Biology and Medicine* 109, 717–721.

Shope, R.E. (1932) A transmissible tumor-like condition in rabbits. A filtrable virus causing a tumor-like condition in rabbits and its relationship to virus myxomatosum. *Journal of Experimental Medicine* 56, 795–802.

Shope, R.E. (1938) Protection of rabbits against naturally acquired infectious myxomatosis by previous infection with fibroma virus. *Proceedings of the Society for Experimental Biology and Medicine* 38, 86–89.

Smith, M.H.D. (1948) Propagation of rabbit fibroma virus in the embryonated egg. *Proceedings of the Society for Experimental Biology and Medicine* 69, 136–140.

Smith, M.H.D. (1952) The Berry-Dedrick transformation of fibroma into myxoma in the rabbit. *Annals of the New York Academy of Science* 54, 1141–1152.

Sobey, W.R. (1969) Selection for resistance to myxomatosis in domestic rabbits (*Oryctolagus cuniculus*). *Journal of Hygiene* 67, 745–754.

Sobey, W.R. and Conolly, D. (1975) Myxomatosis: passive immunity in the offspring of immune rabbits (*Oryctolagus cuniculus*) infected with fleas (*Spilopsyllus cuniculi* Dale) and exposed to myxoma virus. *Journal of Hygiene* 74, 45–55.

Sobey, W.R. and Turnbull, K. (1956) Fertility in rabbits recovering from myxomatosis. *Australian Journal of Biological Sciences* 9, 455–461.

Sobey, W.R., Menzies, W., Conolly, D. and Adams, K.M. (1968) Myxomatosis: the effect of raised ambient temperature on survival time. *Australian Journal of Science* 30, 322–323.

Sobey, W.R., Conolly, D., Haycock, P. and Edmonds, J.W. (1970) Myxomatosis. The effect of age upon survival of wild and domestic rabbits (*Oryctolagus cuniculus*) with a degree of genetic resistance and unselected rabbits infected with myxoma virus. *Journal of Hygiene* 68, 137–149.

Strayer, D.S. (1992) Determinants of virus-related suppression of immune responses as observed during infection with an oncogenic poxvirus. *Progress in Medical Virology* 39, 228–255.

Strayer, D.S. and Sell, S. (1983) Immunohistology of malignant rabbit fibroma virus – a comparative study with rabbit myxoma virus. *Journal of the National Cancer Institute* 71, 105–116.

Strayer, D.S., Cabirac, G., Sell, S. and Leibowitz, J.L. (1983) Malignant rabbit fibroma virus: observations on the culture and histopathologic characteristics of a new virus-induced tumor. *Journal of the National Cancer Institute* 71, 91–104.

Thompson, R.L. (1938) The influence of temperature on the proliferation of infectious fibroma and infectious myxomatosis *in vivo*. *Journal of Infectious Diseases* 62, 307–312.

Wetherall, J.D., Clay, D.E. and King, D.R. (1983) Humoral immunity to myxoma virus in wild rabbits. *Australian Wildlife Research* 10, 277–285.

White, D.O. and Fenner, F. (1995) *Medical Virology*, 4th edn. Academic Press, San Diego, pp. 210–216.

Williams, R.T., Dunsmore, J.D. and Sobey, W.R. (1973) Fluctuations in the titre of antibody to a soluble antigen of myxoma virus in field populations of rabbits. *Journal of Hygiene* 71, 487–500.

Woodroofe, G.M. and Fenner, F. (1965) Viruses of the myxoma-fibroma subgroup of the poxviruses. I. Plaque production in cultured cells, plaque-reduction tests, and cross-protection tests in rabbits. *Australian Journal of Experimental Biology and Medical Science* 43, 125–142.

6

The Introduction of Myxomatosis into Australia

Overview

The use of myxoma virus for the biological control of rabbits was first proposed by the Brazilian scientist H.B. Aragão in 1919, but the Australian government rejected his scheme, primarily on the grounds that it might constitute a health hazard and that it 'wouldn't work'. Fourteen years later the prominent Melbourne paediatrician Jean Macnamara was impressed with what she saw of rabbits with myxomatosis in the laboratory of R.E. Shope in Princeton, New Jersey. She persuaded the Australian government to commission the eminent British scientist C.J. Martin to make a preliminary study of the feasibility of using this disease for biological control of the rabbit pest in Australia. Since Martin reported that myxomatosis was promising as a method of rabbit control, field trials were made on islands in Britain and Denmark and on an estate in southern Sweden, but the disease did not become established.

In Australia, L.B. Bull and M.W. Mules, of the Council for Scientific and Industrial Research, undertook further experiments in Melbourne which confirmed its high species specificity and demonstrated that it could be mechanically transmitted by mosquitoes and fleas. The Chief Quarantine Officer, H.J.L. Cumpston, insisted that any field trials would have to be conducted in remote areas, and since such places were usually very arid, insect transmisson was not very effective and it seemed as though myxomatosis was not a practicable method of rabbit control. Since Australia was then (1943) in the middle of a major war, the proposal was shelved.

After the War, rabbits were particularly numerous in most areas of southern Australia. In 1948 a Wildlife Survey Section was set up under the leadership of zoologist F.N. Ratcliffe, and was given the task of thoroughly testing myxomatosis in the field. There were acrimonious disputes in the media between Bull and Ratcliffe on the one side and Macnamara on the other. Partly because of Macnamara's insistence that myxomatosis should be trialled in well-watered country, experiments were conducted in five field sites in the Murray Valley in the winter of 1950. Although it appeared to die out in most sites, in January 1951 it spread explosively throughout the areas near the waterways of the Murray-Darling Basin. After dying down through the winter of 1951, it again spread explosively throughout the Basin in the spring and summer of 1952.

A complication in 1951 was a concurrent epidemic of encephalitis among people living in the Murray Valley. A classical encephalitis virus was recovered from a fatal human case, but public anxiety was not assuaged until a Minister reported to the Federal Government that three eminent scientists had been deliberately inoculated with myxoma virus, with no ill-effects.

The chapter concludes with an analysis of the reasons – medical, economic, political and psychological – that accounted for the delay of nearly 30 years between the suggestion of Aragão in 1919 that myxomatosis could be used for rabbit control and its successful introduction in late 1950.

Aragão's Proposal to Use Myxoma Virus for Rabbit Control

Dr H. de Beaurepaire Aragão, of the Instituto Oswaldo Cruz in Rio de Janeiro, made important contributions to the understanding of the nature of myxoma virus and its natural history in Brazil (see p. 72). Having observed the lethality of myxomatosis for European rabbits, its specificity for the rabbit, and the fact that among laboratory rabbits it was infectious by contact, Aragão conceived the idea that it might be useful for the control of pest wild rabbits in Argentina and Australia. Presumably because he knew him and because of their mutual interest in tropical medicine, Aragão wrote to Dr A. Breinl, who in 1910 had been appointed Director of the Australian Institute of Tropical Medicine, in Townsville. In a letter dated 15 January 1919,[1] he enclosed a report on myxomatosis (dated December 1918) which contained the suggestion that: 'the virus of this disease could be employed with great advantage and without any danger to kill the rabbits in the countries where they have become a plague as in Australia, New Zealand, R. Argentina and other countries'. The report then went on to describe experiments carried out in cages and on open ground 'of 40 square meters' which showed that the virus spread by contact from infected to uninfected rabbits. He wrote a similar letter to Mr F.H. Taylor, who was at the time entomologist to the Institute of Tropical Medicine, to which reference was made in the official journal of the Commonwealth Institute of Science and Industry (Anon., 1919).

Breinl and Taylor sent their letters to the Commonwealth Institute of Science and Industry[a], which first discussed the idea at its meeting in July 1919. The proposal was submitted to Sir Harry Allen, the Professor of Pathology at the University of Melbourne, who had been a member of the Intercolonial Rabbit Commission 20 years earlier. Allen wrote to the Secretary of the Commonwealth Institute of Science and Industry[2] expressing doubt about: 'any scheme for destroying rabbits by Infectious Diseases', and concluding that:

> In my opinion the introduction of Rabbit Myxoma virus should be forbidden, except for purely laboratory purposes under approved control, unless it has previously been shown that a large continuing epidemic mortality is caused by Rabbit Myxoma among rabbits on a large scale, under conditions approaching natural conditions, and unless further information of a reassuring character is given as to the relationships of the disease, I have reason to believe that importation would be forbidden unless these conditions are complied with.

The secretary of the Commonwealth Institute of Science and Industry wrote to Aragão on 19 December 1919 in a somewhat different vein[3]. After thanking him for bringing the matter to their attention, he said that:

> Since the experiments of Dr. Danysz were carried out on an island off the coast of New South Wales, the trade in rabbits both fresh and frozen, either for local food or for export, has grown to be one of great importance and popular sentiment here is opposed to the extermination of the rabbit by the use of some virulent organism.

The letter concluded with the statement that the cultures sent by Aragão had been kept in the care of the Quarantine Department and that they were sent to 'the fine new Commonwealth Serum Laboratories', and that 'there is no possibility of making experiments with it at the present time'.

[a]The Commonwealth Institute of Science and Industry was the precursor of the Council for Scientific and Industrial Research (CSIR) (see Currie and Graham, 1966). The history of the CSIR from its establishment in 1926 until its transformation into the Commonwealth Scientific and Industrial Research Organization (CSIRO) in 1949 is described in Schedvin (1987).

The letter written by the Secretary of the Commonwealth Institute of Science and Industry did not reflect the view of Australian pastoralists. Correspondence between the President of the Stockowners' Association of New South Wales and Aragão in 1919 and 1920 was followed in 1924 by a request for virus to be sent by Aragão to the Director of Veterinary Research at the Glenfield Veterinary Research Station, Dr H.R. Seddon. Seddon had taken the precaution of obtaining official permission for the importation, and after receipt of two batches that proved to be non-viable on arrival, he received a viable sample in November 1926. During 1927 a number of experiments on its specificity and transmissibility were carried out (White, 1929), and Seddon summed up his views in a letter to the Chief Veterinary Surgeon of New South Wales as follows[4]:

> The conditions necessary for its spread from rabbit to rabbit are such that it is extremely doubtful if it would even lead to death of the whole of the rabbits in a colony. And it is very doubtful if it would spread regularly from colony to colony. ... The two points which are in favour of Rabbit Myxoma being used for the destruction of rabbits are (1) That the disease is apparently invariably fatal, and (2) That it is not contracted by animals of other species. Regarding the first – other diseases have eventually led to a state of balance between virulence and susceptibility of animals and, in fact, all epizootics tend to arrive at this stage. Regarding the second point – we have not investigated this side of the matter thoroughly but have been unable to infect other small laboratory animals. ... It would be absolutely necessary, of course, to test the virus on all species of native fauna.

Early Field Trials in Europe: 1936–1938

The first field trials of myxomatosis as a method of rabbit control were made on islands in Britain and Denmark and on an estate in southern Sweden, but the disease did not become established. Following Martin's experiments in Cambridge (see p. 121, below), several attempts were made between 1936 and 1938 to introduce myxomatosis for rabbit control in parts of the United Kingdom, Denmark and Sweden (Table 6.1). The first of these was initiated by Martin himself (Lockley, 1940):

> In 1936 Sir Charles [Martin] approached me [R.M. Lockley] with the proposal that I allowed him to make a field test on a larger scale, for which he required a small island well isolated from the mainland. I was then living on Skokholm Island, Pembrokeshire, which carried at that time a dense population each autumn of up to 10,000 rabbits on its 240 acres. ... In the autumn of 1936 Sir Charles stayed with us on Skokholm. Between us we caught, marked and inoculated eighty-three rabbits with strain B of the virus, and released them at points scattered over the whole island. Of these marked rabbits, twelve were found dead in the open during the next fortnight, after which the island was left unvisited for the winter. ... in the spring ... no trace of myxomatosis ... rabbits as numerous as ever.

Further introductions were made on Skokholm Island in 1937 and 1938, with similar results (Lockley, 1940). The reason for this disappointing result became apparent many years later (Lockley, 1955)[5]:

> There are no rabbit fleas on Skokholm but why there are none on that island, and plenty on the island of Skomer, two miles distant, is not known. Myxomatosis has since ravaged Skomer, but never Skokholm.

Agriculturists in Scandinavia were also interested in exploring the possibility of using myxomatosis for rabbit control, using Martin's strain B virus[6]. On the Danish island of Vejro, groups of about 150 rabbits were caught, inoculated with myxoma virus, and released during one day of each of three successive years, 1936, 1937 and 1938 (Schmit-Jensen, 1939). About twice as many uninoculated as inoculated rabbits were found dead, but the infection appeared to be confined to families and myxomatosis did not persist for more than two or three months. Experiments on the Dufeke Estate in southern Sweden were more promising. Virus introduced by the

Table 6.1. Early attempts to introduce myxomatosis into Europe.

	Skokholm Island (UK)	Vejro Island (Denmark)	Dufeke Estate (Sweden)
Time of introduction	1936, late autumn; 1937, summer 1938, spring	1936, autumn 1937, late autumn 1938, early spring	1938, spring
Type of rabbits	Wild	Domestic rabbits run wild	Wild
Habitat	Warrens infested also with sea-birds	Underground warrens	In stone fences
Vectors	Rabbit fleas absent[a]	Not recorded	Rabbit fleas present[b]
Method of introduction	Local rabbits inoculated and released; 83, 55, and 7 in successive years	Local rabbits inoculated and released; 148, 156 and 163 in successive years	18 local rabbits inoculated
Result	Intrafamilial spread; died out	Intrafamilial spread; died out	Widespread infection over whole estate, over-wintered, then died out
Reference	Lockley (1940)	Schmit-Jensen (1939)	Hvass and Schmit-Jensen (1939)

[a]Lockley (1955).
[b]Common in southern Sweden (Borg, 1962).

release of 18 inoculated rabbits in May 1938 caused an extensive epizootic and persisted until the following spring, but then died out (Hvass and Schmit-Jensen, 1939). Subsequently, in a report on the rabbit problem on the island of Gotland which was published in May 1952, before the introduction of myxomatosis into France by Delille but after the great Australian epizootics, Notini (1952) considered that:

> Myxomatosis, a disease specific for the rabbit, is proposed as the best conceivable agent [for rabbit control] ... Probably it would by no means bring about total *extermination* other than in circumscribed areas; but such heavy decimation of the wild rabbits might be anticipated that the remainder could be dealt with by ... conventional devices.

Australian Investigations of Myxomatosis: 1934–1943

The repeated efforts of Aragão to interest the Australian government in the possible value of myxomatosis as a method for the biological control of the rabbit pest came to nothing because myxomatosis did not seem to be highly infectious by contact, and Australian authorities believed that it would need to be shown to be contagious if it were to be used for the control of wild rabbits. This view recalls the rejection of Pasteur's proposal 30 years earlier (see p. 57).

Dr Jean Macnamara's proposal

The proposal to test myxomatosis for its usefulness for the biological control of Australian rabbits was revived by a scientifically able and politically astute paediatrician, Dr Jean Macnamara, who knew nothing of Aragão's earlier initiative or of White's trial in New South Wales. The story is told in detail, from Macnamara's point of view, in her biography (Zwar, 1984). As will become apparent, Bull and Ratcliffe, the CSIR scientists responsible for field testing the virus in Australia, saw it somewhat differently. Macnamara (Fig. 6.1A) had graduated in medicine in 1922, being in the same class at the University of Melbourne as Frank Macfarlane Burnet. When Burnet returned to Melbourne in

Fig. 6.1. (A) Annie Jean Macnamara (1899–1968). After graduating in medicine at the University of Melbourne in 1922 she became a resident at the Melbourne Hospital and then at the Children's Hospital, where she graduated MD in 1925 and moved into private practice, with a particular interest in poliomyelitis. On an overseas trip in 1932–1933 she visited R.E. Shope at the Princeton laboratories of the Rockefeller Institute and first saw rabbits with myxomatosis, and interested the Chairman of CSIR in the possibility of using myxomatosis for controlling rabbits. In 1935 she was created a Dame of the British Empire for her work for victims of poliomyelitis. After the 1939–1945 war she conducted a spirited campaign to convince CSIRO scientists to go on experimenting with myxomatosis, and when it spread in 1951 she took a major role in promoting its use by Victorian Government departments. Further details in Smith (1968) and Zwar (1984). (B) Richard Edwin Shope (1901–1966). Soon after graduating in medicine at the University of Iowa in 1924, Shope joined the Rockefeller Institute for Medical Research, and worked in their plant and animal diseases laboratories in Princeton, New Jersey. Although initially he worked on tuberculosis, Shope was more interested in veterinary medicine. As well as making the first isolation of an influenza virus (from swine, in the late 1920s), he discovered two viruses of cottontail rabbits that caused tumours of the skin: rabbit papilloma virus and rabbit fibroma virus. He demonstrated the relationship between rabbit fibroma virus and myxoma virus, and it was the sight of his rabbits with myxomatosis that awakened Macnamara's interest in that disease. Shope was elected to the US National Academy of Sciences in 1940. In 1962 he came to Australia to advise the Victorian government on the use of fibroma virus as a vaccine to protect domestic rabbits against myxomatosis (see p. 176). Further details in Andrewes (1981).

1928, after three years at the Lister Institute in London, he set up a virus group in the Walter and Eliza Hall Institute, and Macnamara joined him as a part time field epidemiologist, working first on poliomyelitis and later, after Burnet's second trip to London, on psittacosis (Burnet, 1971).

In 1931 she was awarded a Rockefeller Foundation Travelling Scholarship to further her knowledge of paediatrics and especially poliomyelitis. Her work with Burnet influenced her choice of places to visit. Thus while in New York early in 1933 she visited the laboratories of the Rockefeller Institute for Medical Research; first those in New York City and then the Institute's laboratories at Princeton, New Jersey, where she met Dr Richard E. Shope (Fig. 6.1B). Shortly before this, Shope had discovered the viral causation of the fibromas (non-cancerous lumps on the skin) found in cottontail rabbits (*Sylvilagus floridanus*). Further, he demonstrated that although rabbit fibroma virus (later called Shope's fibroma virus) produced only a benign fibroma in domestic rabbits, it was immunologically related to myxoma virus (Shope, 1932). He told Macnamara that he was testing fibroma virus for use as a vaccine to protect hutch rabbits in California, which were suffering from outbreaks of myxomatosis at the time (see p. 76). Seeing rabbits with myxomatosis in Shope's laboratory, learning that it was almost invariably lethal, and being aware of the rabbit problem in Australia, she conceived the idea (she thought for the first time) that myxoma virus might be used for the biological control of the rabbit. In a letter to her family she wrote[7]:

> I had a lovely day out at Princeton, the branch of animal and plant pathology [of the Rockefeller Institute for Medical Research] ... There is a man there I would love to take home to work on our animal diseases. Shope is his name and he has something which kills rabbits though he has not tried ours. I am going to send some to Ivan[b], to give him the chance to become famous by killing off the rabbits.

Under instructions from Dr H.J.L. Cumpston, the Director of Quarantine, the virus samples sent to Connor by Macnamara were destroyed by quarantine officers on arrival. However, Macnamara was persistent. Remembering that she had treated a niece of the Australian High Commissioner in London, former Australian Prime Minister S.M. Bruce, during a poliomyelitis outbreak in 1926, she wrote to him expressing her disappointment over the destruction of the virus that she had sent to Australia. Bruce replied by return mail, and at his request, she prepared a 'preliminary memorandum'[8], in which she said:

> In view of the species specificity so far recorded, the extreme infectiousness, the high mortality, it would appear that further experimental research is justified, with a view to the possible contribution to the problem presented by the rabbits in Australia.

[b]Dr Ivan Connor, Acting Director of the Walter and Eliza Hall Institute, whom she later married.

She suggested further topics of research, and asked Bruce to see whether the British government would permit importation of the virus.

On 6 September 1933 F.L. McDougall, the CSIR representative at the Australian High Commission in London, sent the memorandum to A.C.D. Rivett, the Chief Executive Officer of the Council for Scientific and Industrial Research (CSIR). Rivett thought that it was 'quite an interesting document' and justified the 'very closest attention'. Following the conventional wisdom of the time, he doubted whether much could be expected from the use of an infectious disease for rabbit control, as he explained to McDougall[9]:

> The difficulty here is, of course, that unless you get a complete kill, the capacity of a quite small surviving fraction to multiply enormously is such that actually very little advantage is likely to be gained. Drought probably does far more in reducing the rabbit than any disease could ever accomplish in the time available before an immunity was set up.

However, he agreed that something should be done.

Investigations in Cambridge by Sir Charles Martin

Sir Charles Martin (Fig. 6.2) was Chief of the CSIR Division of Nutrition and Rivett's principal adviser on the Macnamara memorandum. He was about to retire and return to Cambridge, and since Rivett had been assured by Bruce that experiments with myxoma virus would be permitted in the United Kingdom, he agreed to carry these out. Macnamara was determined to talk with Martin about the experiments, so she booked a trip from Canada to England, at her own expense, and went to Cambridge to discuss his plans.

Martin collected all available information on myxomatosis, and by the end of September 1933 he wrote to Rivett expressing cautious optimism, since his survey of

Fig. 6.2. Charles James Martin (1866–1955). Born in England, Martin worked for substantial periods of time in the Universities of Sydney and Melbourne and served with distinction in the medical services of the Australian Imperial Forces in the First World War. After his retirement from the directorship of the Lister Institute in London in 1930, he came to Australia for a few years as Professor of Physiology and Biochemistry at the University of Adelaide and the first Chief of the CSIR Division of Animal Nutrition. He retired in Cambridge in 1933, but in 1934–1945 carried out investigations on myxomatosis which led to the virus being trialled in Australia. Further details in Chick (1956) and Morison (1986).

the literature had confirmed the host specificity and extraordinary virulence of myxoma virus. However, he was concerned as to whether the virulence might decline after release in the field, so that 'You mustn't imagine that the virus has only to be let loose in Australia to slowly exterminate the rabbits', nevertheless the virus 'has possibilities and is the most promising disease which has been brought to my notice'[10]. He established contact with Aragão, who sent him a specimen of myxoma virus (later called 'strain A'). Aragão also made further enquiries into the epidemiology of the disease in hutch rabbits in America, and suggested to Martin that myxoma virus might have to be used in a manner similar to that used for *Cactoblastis cactorum* for control of *Opuntia* (see p. 45). This was the first mention of the 'prickly pear model', later invoked by both Martin and Bull. By the end of 1933 Martin was sufficiently interested to agree to carry out trials in colonies of rabbits in Cambridge if CSIR would provide the necessary finance. Throughout the next six years he kept Rivett and his colleagues in Australia (Clunies Ross, Gilruth and Bull) informed of progress by regular letters[11].

Before starting his experiments, he set out his criteria for the control of vermin by means of a pathogenic microbe (Martin, 1936):

1. It produces a disease which is sufficiently infectious to spread throughout the population.
2. It has such virulence that few animals recover to build up gradually a resistant race.
3. It must maintain its virulence when passing from one animal to another by natural means, or the epizootics will 'peter out'. Should any animals recover and become immune to further infection, their progeny must not inherit sufficient resistance to escape infection.
4. It is specifically dangerous to the animal it is desired to exterminate, and harmless for domestic and other useful animals.
5. It is not too troublesome to propagate, keep in an active condition, and apply in practice.

The experiments were not part of a formal CSIR programme and Martin did not receive a salary, although in 1936 he was given an honorarium of 150 guineas. With the help of the Department of Comparative Pathology in Cambridge and a few hundred pounds made available by the Rabbit Destruction Committee of New South Wales, he carried out a series of experiments in colonies of rabbits living 'under as natural conditions as I could contrive'. The results were summarized in a report sent to Rivett and Bull in December 1935 and published by CSIR in

1936 (Martin, 1936). A netted compound of about 46 square metres was constructed to house the rabbits, and two strains of myxoma virus were used: strain A, supplied by Aragão, and strain B, which had been brought from Shope's laboratory to Martin by Bull. Strain B, later to become the 'Standard Laboratory Strain' of Australian scientists, was the more virulent. Martin concluded that myxoma virus satisfied 'the essential requirements ... for the control of a population of rabbits in a circumscribed area, and this without danger to domestic animals'. The virus appeared to be of stable high virulence during natural rabbit-to-rabbit passage over a period of 15 months, and the progeny of a few survivors of the slightly less virulent strain A were fully susceptible to challenge with the fully virulent strain B. Like all other investigators, Martin found the virus to be highly host-specific. The main problem, as he saw it and as Seddon had seen it before him, was that the disease was not highly contagious. Martin suggested that flies, feeding on the exudate on lesions, especially those on the eyelids, might be effective in dispersing the virus, and he was aware that Aragão had demonstrated transmission by cat fleas. (In retrospect, it seems highly likely that much of the transmission that occurred in Martin's colony experiments was due not to contact spread, but to the activity of rabbit fleas (*Spilopsyllus cuniculi*) with which his wild rabbits would have been infested.) He concluded his report as follows:

> Although, with proper management, individual colonies of rabbits may be exterminated, I cannot predict to what extent the infection will spread to colonies in the neighbourhood. This does not depend upon the virus but upon the amount of intercourse between the animals of separated colonies. It can best be ascertained by field tests on an insulated area in Australia.
>
> If such field experiments afford promising results, the application of the method on a large scale will necessitate the gradual building up of an appropriate scientific organization, such as was found necessary for the control of prickly-pear in Australia.

Tests of myxoma virus for host specificity

Martin's experiments were followed up promptly in both Europe (see p. 118) and in Australia. Martin had warned Rivett that 'the susceptibility of other animals would have to be enquired into very much more thoroughly and with greater numbers than has hitherto been done'[10]; if these tests were satisfactory, field trials should be carried out in Australia[12,13]. Responsibility for work on myxomatosis in Australia devolved upon Dr L.B. Bull (Fig. 6.3A), an experienced veterinarian and bacteriologist who had been appointed Deputy Chief of the CSIR Division of Animal Health early in 1934. Before replacing Gilruth as Chief of the Division in June 1935, he undertook an extensive overseas trip, during which he visited Shope, from whom he obtained the 'strain B virus'[c], which he personally delivered to Martin when he visited him in Cambridge. In February 1936, shortly after his return to Australia, Bull summarized his expectations of the value of myxomatosis as follows[14]:

> We do not expect that the liberation of *Virus myxomatosum* could of itself control rabbit populations in Australia. There has been no suggestion that any such views are held by those of us who have been interested in the problem. We do suggest that the virus may give us *another* weapon with which to attack the pest. We do hold that the problem is very difficult and intricate and that much study and investigation is needed before the problem can be fully defined.

There was still another hurdle to be surmounted, namely Australia's strict quarantine restrictions, which had been the reason for asking Martin to carry out investigations in England, where quarantine was much less restrictive. The Director-General of Health and Chief Quarantine Officer, Dr H.J.L. Cumpston (Fig. 6.3B), who had been involved in earlier discussions about myxoma virus in

[c]'Strain B virus' was eventually used to introduce myxomatosis into Australia, and was designated as the 'Standard Laboratory Strain' (for full history, see p. 71).

Fig. 6.3. (A) Lionel Bately Bull (1889–1978). After graduating in veterinary science in Melbourne in 1911, Bull worked from 1912 to 1934 as bacteriologist in the South Australian Laboratory of Bacteriology and Pathology. He was appointed Deputy Chief of the Division of Animal Health of the Council of Scientific and Industrial Research early in 1934 and undertook an extensive overseas trip prior to taking up appointment as Chief of the Division in June 1935. During this trip he visited Shope at the Rockefeller Institute for Medical Research at Princeton, discussed the potential of myxomatosis with him, and took a strain of myxoma virus to Martin in Cambridge. Back in Australia, he took a major role in testing the species specificity of myxoma virus, especially in relation to native animals, and led the work on field testing of myxoma virus in South Australia. Further details in French (1993) and French and Stewart (1983).
(B) John Howard Lidgett Cumpston (1880–1954). Soon after graduating in medicine at the University of Melbourne in 1902, Cumpston developed a deep interest in public health and preventive medicine. After postgraduate studies in England, in 1907 he was appointed medical officer to the Central Board of Health in Western Australia. He joined the Federal quarantine service when it was established in 1908, being appointed acting Director of Quarantine in 1913. Soon after the post-war outbreak of influenza, a Federal Department of Health was established and Cumpston became Director-General of Health and Chief Quarantine Officer. He first entered the debate about myxoma virus in 1933, and later insisted that field experiments should be carried out in remote areas. His role in the quarantine of myxoma virus was a minor part of a distinguished career, which saw the establishment of the National Health and Medical Research Council in 1937 and major contributions to child health and the control of human infectious diseases. He retired in 1945. Further details in Roe (1981).

1919 and 1924 and had authorized the destruction of the sample sent by Macnamara in 1933, was sceptical about the value of any infectious disease as a method of rabbit control, and expressed his concerns to Rivett[15]:

> (1) That vague apprehension that, in endeavouring to use Nature without sufficient knowledge, we may find Nature too strong for us. That is a valid fear but not valid enough to justify stopping any of your own work.
> (2) That the introduction of a new virus will be associated with dangers at present unimagined.
> (3) That the introduction of a new virus will not achieve the results anticipated.

Nevertheless, he gave permission for further tests to be carried out in Australia under strict quarantine, in 'enclosed' colonies of rabbits, and stipulated that further tests of the host specificity of the virus should be carried out on Australian native animals.

Tests carried out at the CSIR Animal Health Laboratory in Parkville, in central Melbourne, showed that all of a wide range of native, domestic and laboratory animals were totally resistant to the virus (Bull and Dickinson, 1937)[d]. Four experiments carried out on the spread of myxomatosis in small 'enclosed' colonies of rabbits showed that Martin's 'strain B' killed all the rabbits in such colonies in about 40 days (Bull and Mules, 1944)[e], a result which led Cumpston to authorize a limited field experiment.

Field experiments in South Australia

Cumpston's conditions stipulated that the field experiment was to be conducted in a remote part of Australia under quarantine conditions and in a compound of about 20 hectares enclosed within a double-netted wire fence.

There were some difficulties in obtaining a suitable site. The first proposal, which had been made by the Tasmanian Department of Agriculture in 1934, on the strength of Martin's results at that stage, was to use Clarke Island in the Furneaux Group in Bass Strait[16]. However, when Bull visited Clarke Island and developed a detailed proposal in 1937, it had to be abandoned at a late stage in the planning because the Tasmanian government refused permission[17]. Eventually Wardang Island, a dry and barren island of some 300 km^2 in Spencer Gulf in South Australia, was chosen. From the quarantine point of view it was excellent – remote, difficult of access, and an Aboriginal reserve and therefore unlikely to be casually visited by white Australians[f]. There were some 14 warrens on the island, and no foxes, dogs or other predators. Experiments commenced in November 1937 in an area of 36 hectares at the south of the island and were conducted by M.W. Mules.

The initial experiment was designed to test whether the introduction of inoculated rabbits 'foreign' to a colony would result in extensive spread of myxomatosis through the colony. Twenty foreign rabbits inoculated with virus into the conjunctival sac were introduced into an enclosure containing some 400–500 rabbits. Most did not join the established colonies, but a contact rabbit was found infected 18 days after the introduction and by the 100th day, when the disease appeared to have died out, 238 rabbits, belonging to five colonies, had been found dead above ground. About half of the enclosed colony appeared to have been killed. In the second experiment, late in June 1938, infected rabbits were introduced into the enclosure, which then contained about 1000 adult rabbits and an unknown number of kittens, in 33 warrens. Since the disease appeared to die out, further inoculations were made in August, November and December. By January 1939, when the experiment was terminated, 337 rabbits had been inoculated and died, 934 rabbits had died of myxomatosis, 633 rabbits had died of heat and starvation, and 1078 rabbits remained alive and were killed by poisoning as a control operation. Thus in spite of the high mortality rate in infected rabbits, the population had succeeded in replacing itself before the poisoning began.

Rumours about the experiments had spread rapidly among the rural community in the late 1930s, especially in Western

[d]There is no evidence that the possibility was ever considered that myxoma virus might mutate so that it could infect animals thought to be insusceptible, but almost nothing was known about mutation in viruses at that time.

[e]This paper, originally submitted as a preliminary report from Bull to Rivett in December 1943, contains brief accounts of all the studies carried out by Bull and his colleagues between 1938 and 1943. No more detailed account of this work was ever produced.

[f]This lack of concern for Aboriginal Australians reflects the view at the time that they were a vanishing race.

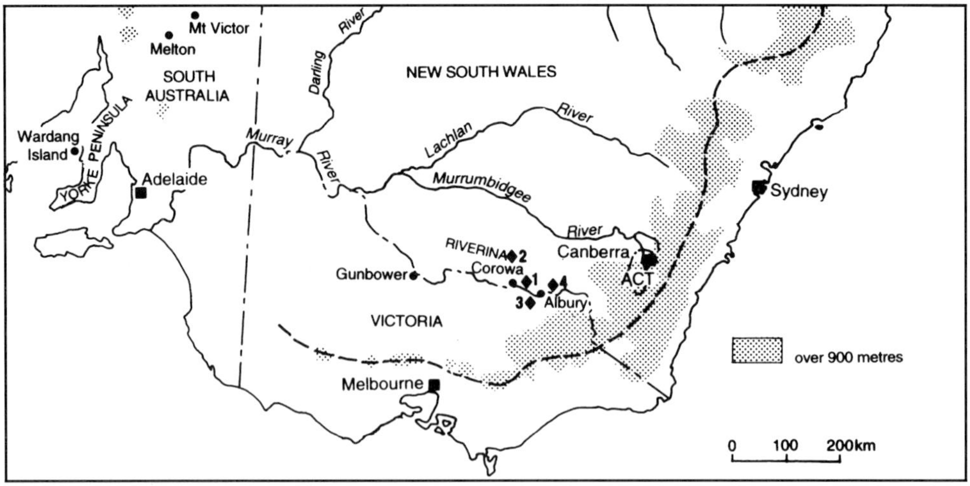

Fig. 6.4. Map showing the sites used between 1937 and 1950, during studies on the introduction of myxomatosis into Australia. The sites in South Australia were used by Bull and Mules in 1937–1939 (Wardang Island) and in 1942–1943 (Mount Victor and Melton). The sites in New South Wales and Victoria were used in 1950 by Ratcliffe and his colleagues; Gunbower in May and the other sites – Balldale (1), Coreen (2), Rutherglen (3) and Wymah (4) in September–December. The broken line indicates the divide, between streams flowing south and east and those flowing north and west.

Australia, suggesting that deliverance from the rabbit pest might be at hand, and Casey, as the Minister in Charge of CSIR, complained about being bombarded with requests from members [of Parliament] from Western Australia. This led to statements by the Director-General of Health that he was not prepared to relax quarantine restrictions on the use of myxoma virus[18], and CSIR issued a series of public statements emphasizing that the results were preliminary, and that myxomatosis was unlikely to produce an epizootic in the field. In a note prepared for Rivett in August 1938, Bull commented[19]:

> Attempts are being made in two or three places in Europe to use the virus for the control of rabbit populations. Three times, at intervals of months, infected rabbits were released on Skokholm Island, off the coast of Wales[g]. Observations have failed to show any evidence that the disease has appreciably reduced the population of rabbits on the island. ... Our own experience on Wardang Island ... has given no more promise of success. The spread of the disease is too limited. ... It is worse than useless to make any premature attempt to use the virus in the field, for failure is certain and future attempts under more favourable conditions would be prejudiced.

Since Bull and Mules (1944) had demonstrated that both fleas and mosquitoes could transmit the virus mechanically, and there were no fleas of any kind on Wardang Island, and few mosquitoes, permission was sought from the Director-General of Health to continue the experiments on the mainland. Cumpston was not enthusiastic[20], reiterating his earlier opinion that:

> at the outset of these experiments, I expressed my opinion that no disease under natural conditions could be expected continuously to eliminate any considerable proportion of an animal population. Nothing that has happened or been recorded in the literature since modifies that opinion which I at present firmly hold.

Nevertheless, he concurred with the proposal for experimentation on the mainland, with the proviso that the precautions

[g]It was later shown that Skokholm Island was devoid of rabbit fleas (see p. 118).

previously stipulated were maintained. A site in northeastern South Australia where stickfast fleas (*Echidnophaga myrmecobii*) were abundant[h] could not be used because of drought, so an experiment was conducted early in 1941 in an enclosure of 36 hectares at Point Pearce, which is on Yorke Peninsula not far from Wardang Island (see Fig. 11.7, p. 255). The result was much more impressive than the earlier experiments. The enclosed 36 hectares contained about 500 rabbits infested with stickfast fleas, and more stickfast fleas bred in the laboratory by Mules (1940) were added. Twenty-eight rabbits from about half the warrens were inoculated. The disease spread much more rapidly than on Wardang Island, probably because of the presence of the fleas. Over a period of 60 days all except 17 of the 500 rabbits, in 13 warrens, had died from myxomatosis; all survivors were susceptible when challenged. This result heartened Bull; he saw that there might be a role for myxomatosis if the spread of the disease could be assisted by an insect vector[21]. However, he was emphatic that the virus would not:

> act continuously to eliminate populations and prevent their development to plague proportions ... Such a suggestion may have been advanced by over-enthusiastic and over-anxious pastoralists, but as scientific workers we have never shared these views, nor have we advocated the use of the virus of myxomatosis for the control of the rabbit population.

Because of the views of Bull and Cumpston, further investigations were being pushed in one direction by two untested assumptions: firstly, that myxomatosis would not act 'continuously' to eradicate rabbit populations, and secondly, that if it did have a role, this would be limited to the dry interior, where, admittedly, there was a great need for a cheap method of rabbit control.

Permission was sought from the Director-General of Health for further trials, in marginal pastoral land in the interior of the continent. In May 1941 approval was given for trials on three pastoral properties, Melton, Koonamore and Mount Victor, but there were a number of frustrating delays. First, some method of introducing the virus other than by capture, inoculation and release, which was feasible in enclosures, was needed for these more extensive experiments. This problem was solved by Mules, who developed a modified rabbit trap – a steel gin trap with attached inoculating pins and only one active jaw, so that the rabbit would be inoculated with myxoma virus but could escape without physical damage (Anon., 1942). Secondly, the exceptionally dry seasons of 1940–41 and 1941–42 in the area of South Australia selected for the trials had greatly reduced the local rabbit populations. A proposal to carry out the experiment in Cunnamulla in southern Queensland was refused by the State Department of Agriculture because it was: 'politically ... very undesirable at present, in those parts, to be associated with any move to reduce the highly payable rabbit population'. And thirdly, by this time Australia was deeply involved in the Second World War, and long-term biological research in CSIR was downgraded, Bull's assistant Mules being transferred to another Division. Bull was tempted to give up the project at this stage[22], but was advised against this by Rivett[23]:

> I have felt a bit uneasy ever since we decided that the myxomatosis work should be indefinitely postponed ... Sooner or later we are going to get some fierce cry for finality on the virus question and, at the risk of being rather a nuisance to you, I would like to ask whether you yourself feel quite happy about ruling out this work on the ground that its immediate war interest is low.

It was agreed that field trials should be recommenced in the spring of 1942, and in November 1942 Bull and Mules (seconded for the project) proceeded to carry out field trials in a paddock of 142 hectares at Mt Victor, a station in the semi-arid bush country in the north-east of South Australia,

[h]Until their introduction in 1966 (see p. 181) there were no rabbits fleas (*Spilopsyllus cuniculi*) in Australia. Stickfast fleas (*Echidnophaga myrmecobii*), sometimes called 'red fleas', were parasites of native marsupials that readily parasitized rabbits.

and subsequently at an adjoining sheep station, Melton (Bull and Mules, 1944; see Fig. 6.4, p. 126). The investigations at Mt Victor were rendered difficult by predation of sick and dead rabbits by foxes and eagles, and in spite of attempts to reduce the numbers of these predators, the general conclusion was that the death rate from the disease failed to counter the natural increase in population. At Melton, the Mules trap was successful in introducing the disease into each of 12 warrens in which traps were set, but untreated warrens were not affected. Overall, the results were more promising, the population being reduced from about 250 to 12 within 18 days, but the disease then died out. Similar results were observed in another trial at Melton begun in January 1943, i.e. good results were obtained in trapped warrens but there was no extension beyond them. In a final trial, commenced in June 1943, some 1550 traps were sprung at 503 warrens, but deaths from myxomatosis failed to counter the natural increase in the rabbit population.

Although they were restricted by the quarantine requirement that experiments should be carried out in remote and relatively inaccessible places, and inevitably such places were arid and dry (and therefore not favoured by mosquitoes), Bull and Mules (1944) reached a pessimistic conclusion:

> The general results of the study, and especially those of the field experiments and field trials, show that myxomatosis cannot be used to control rabbit populations under most natural conditions in Australia with any promise of success. Nevertheless, it seems possible that in some parts of Australia under special conditions, including the presence of insect vectors in abundance and the absence of predatory animals, the disease could be used with some hope of temporary control of a rabbit population. In any case, to be of any real value the disease would have to be used when the population density was moderate, not high, and a re-introduction would be necessary from time to time, probably annually, in order to keep the population density as low as possible.

Bull's extreme pessimism is difficult to understand. Independently of Aragão (1943), he had demonstrated that mosquitoes as well as stickfast fleas were capable of transmitting myxomatosis (Bull and Mules, 1944), and under favourable conditions mosquito carriage could be expected to provide the mechanism for the transfer between colonies, the absence of which was the principal reason for his caution. However, this pessimistic view was to colour his own and subsequently Ratcliffe's attitude to the introduction of myxomatosis, although as Assistant Director of Entomology for the Australian Army in 1943–45, Ratcliffe was well aware of the effectiveness of the mosquito transmission of diseases like malaria and dengue.

By October 1943 the field experiments were completed. Australia was in the middle of a desperate world war that had for the first time reached the Australian mainland. In spite of the importance of the rabbit problem, research on myxomatosis was dropped, not to be resumed until the Wildlife Survey Section was set up in 1948, within the Commonwealth Scientific and Industrial Research Organization (CSIRO), the successor to CSIR.

Francis Ratcliffe, a 'biological scout'

Understandably for a country with the population of Australia in the 1930s (6.6 million), the work of CSIR was focused on problems of primary production, although several Divisions had programmes that touched on environmental protection (Soils: salinization of irrigated land; Animal Health: introduced diseases of livestock; Entomology: the sheep blowfly; Plant Industry: introduced weeds). However, in 1929 CSIR attempted to get an insight into a few specific 'environmental' problems by attaching a young Oxford graduate in zoology, Francis Ratcliffe (Fig. 6.5A) to CSIR Headquarters, as what Rivett was to call CSIR's 'biological scout'. His first problem, in 1929–31, was to study flying foxes (*Pteropus* spp.), which were thought to be causing considerable damage to orchards in Queensland and northern New South Wales. He wrote numerous long

Fig. 6.5. (A) Francis Noble Ratcliffe (1904–1970). Born in Calcutta, Ratcliffe studied zoology at the University of Oxford. In 1928 he came to the notice of F.L. McDougall, then correspondent in England for the CSIR. This led to his invitation to Australia to work on flying foxes and erosion in arid lands. He returned to Britain in 1932 as Lecturer in Zoology in Aberdeen, but was invited back to Australia as a scientific adviser to the CSIR Executive in 1935. In 1937 he was transferred to the Division of Economic Entomology to work on termites. In 1942 he joined the Australian Army and served with distinction as Assistant Director of Entomology. After demobilization he served briefly as assistant to the Chief of the Division of Entomology, but in 1948 he was appointed Officer-in-Charge of the Wildlife Survey Section of CSIRO. Initially he had to work on rabbit control, and after some disappointments succeeded in introducing myxomatosis. Study of this disease preoccupied the Section for several years, but then he was able to broaden studies of the biology of the rabbit and introduce biological studies of native animals as an important part of the work of the Section (now the Division of Wildlife and Ecology). In 1961 he returned to the Division of Entomology as Assistant Chief, and with Fenner produced a summary of the rabbit work of the Section in a book *Myxomatosis* that became a classic. He retired from CSIRO in 1969. He played a major role in the setting up in 1964 of the Australian Conservation Foundation, to which he devoted a great deal of time until forced in 1970 to relinquish this work, for health reasons. Further details in Mackerras (1971), Tyrrell *et al.* (1971) and Coman (1998).

(B) William Ian Clunies Ross (1899–1959). After graduating in veterinary science at the University of Sydney in 1921, Clunies Ross spent some years in veterinary practice and undertook research in parasitology at the University of Sydney. In 1926 he was appointed as a parasitologist within the newly formed Council for Scientific and Industrial Research (CSIR), and in 1931 was appointed officer-in-charge of its Animal Health Laboratory. After service on the International Wool Secretariat in London he returned to Australia in 1939 as Professor of Veterinary Science in the University of Sydney, but in 1941 was seconded to the Commonwealth Manpower Directorate. After the war he was appointed a full-time member of the Executive Committee of CSIR, with responsibility for agricultural and biological sciences. He immediately became involved with the rabbit problem, which had assumed great economic and political importance. In 1949 CSIR became CSIRO, Rivett retired as Chairman, and Clunies Ross replaced him. He recruited Ratcliffe to head the Wildlife Survey Section, and provided continued support for him throughout the difficult period before myxomatosis was established as an effective method of biological control. Clunies Ross was a visionary leader who played a key role in the rise of CSIRO to become the premier scientific organization in Australia. Further details in Clunies Ross (1977), Schedvin (1993) and O'Dea (1997).

letters to his family in Britain as well as to CSIR, as well as official reports, all of which were to prove valuable when he wrote a book that became a classic, *Flying Fox and Drifting Sand* (Ratcliffe, 1938). Having concluded his study of *Pteropus* (Ratcliffe, 1931), he returned to the United Kingdom to an academic position, but in 1935 willingly accepted another project for CSIR, namely to study and advise on measures to control soil erosion and sand drift in the arid and semi-arid grazing country in northern South Australia and south-western Queensland. During this investigation he had his first encounter with the rabbit problem in the arid pastoral zone (Ratcliffe, 1936). He thought that the destruction caused by rabbits had been exaggerated, but nevertheless acknowledged their capacity for survival under difficult conditions and their great reproductive potential. At that time Martin was investigating the potential of myxomatosis for rabbit control, and had concluded that:

> although myxomatosis may provide a means of destroying large numbers of rabbits, and in areas where this would not be practicable by any other known method, it cannot be expected to solve the problem of their permanent and general control. The very fatality of the disease will probably ensure that the infection will die out in the wild population when an epizootic has completed its course, and re-infection will be necessary whenever the animals breed up to a dangerous extent.

Sixty years later, with hindsight, those concerned with rabbit control might well wish that the last speculation had been correct; it would be splendid to have such a lethal biocide periodically spread through rabbit populations which would not have developed genetic resistance.

After the arid land study, Ratcliffe was appointed to the Division of Entomology, where he worked on termites, and in 1942, after Japan had entered the Second World War, he joined the Australian Army Medical Corps and was appointed Assistant Director of Entomology. Here he played an important role in investigations into the control of mosquitoes and scrub typhus mites in the New Guinea campaigns. After the war he returned to the Division of Economic Entomology in CSIR as assistant to the Chief, A.J. Nicholson. However, Ratcliffe found that he was fully engaged in administration, and he hated being chained to the desk, unable to return to the field or his own research. He poured out his frustrations in long letters to Clunies Ross, then an executive officer of CSIR[24]. In his replies, Clunies Ross told Ratcliffe that he had been granted the year's leave of absence in England that he had asked for, and asked him:

> to see something of Elton [a famous British animal ecologist], and also of the Wildlife Preservation Service in the United States, since I am convinced that we must get into work in this field as soon as possible.

Ratcliffe spent 1948 in England, working for most of that time in the Bureau of Animal Population in Oxford, with C.S. Elton.

The Establishment of the Wildlife Survey Section of CSIRO

During the war a combination of good seasons and the abandonment of rabbit control because of shortages of materials (wire netting, poisons and fumigants), and of labour, because most of the younger men in the country areas had joined the army, had led to a tremendous build-up of rabbit numbers (Fig. 6.6). Rabbits were in plague numbers in most districts, erosion had intensified, and with the end of the war the cost of materials and of labour were inflated. CSIR had been asked to help, but could only respond by saying that conventional methods of control were the most effective and that myxomatosis would not help[25]. Then early in 1948 The Graziers' Federal Council of Australia requested that a comprehensive programme should be undertaken on the general biology and ecology of the rabbit, together with further study of poisons[26]. Some of this information was passed on to Ratcliffe in England, and in a letter to Clunies Ross[27], Ratcliffe commented, *vis-à-vis* the 'rabbit work': 'If

Fig. 6.6. Photographs illustrating the density of rabbit populations just before the spread of myxomatosis in 1951. (A) Scene in the Victorian Mallee, near Ouyen. (B) Rabbit warrens in the Victoria Alpine country. Photographs by G.W. Douglas.

the CSIR is going to branch out in this field, I want to be in the work – but if I am in it, I want to be *right* in it, and from the start'.

Along with the concern of rural landowners about the rabbit problem, there was increasing pressure from zoologists and botanists for a general survey of the biological resources of Australia. This idea was first launched by the zoologist–anthropologist Baldwin Spencer when he was President of the Australasian Association for the Advancement of Science in 1921 (AAAS, 1921), and revived at the meetings of the Association in Adelaide in 1946 and in Perth in 1947. At the request of the CSIR Executive, Ratcliffe provided a personal assessment of the feasibility and value of a 'Commonwealth Biological Survey'[28]. CSIR was cautious about becoming involved in such an ambitious undertaking in what might be seen as 'pure science', but saw that a study of animals which had become economic pests was feasible[i].

After an enquiry into the organization of government science in 1948, CSIR was transformed into CSIRO in May 1949, the new Chairman of CSIRO being Ian Clunies Ross (Fig. 6.5B), a veterinarian by training who since 1945 had been one of two executive officers in CSIR. He was a man of great charisma and considerable administrative skill (Clunies Ross, 1977; O'Dea, 1997). He saw the potential for combining the request for scientific work on the rabbit pest with a general biological survey by setting up a small group to undertake research on the biology and ecology of the rabbit, and later go on to work on Australian native animals. From his personal knowledge of Ratcliffe, he knew that he had the man to head this new development, which was inaugurated as the Wildlife Survey Section[j] when Ratcliffe returned to Australia in December 1948[29]. As Officer-in-Charge, Ratcliffe reported directly to the CSIRO Executive, rather than through one of the biological Divisions.

[i]It was not until 1973 that the Commonwealth Government established the Australian Biological Resources Study (later named the Bureau of Flora and Fauna) to undertake a general survey of the biological resources of Australia. See Fenner (1995).

[j]The Wildlife Survey Section underwent several changes of name as time passed; rabbit control has remained an important part of its programme since its inception. It had grown substantially in size and scope by the time Ratcliffe transferred to the Division of Entomology in 1961, and in 1962 it was elevated to divisional status, with a name change to the CSIRO Division of Wildlife Research. In 1982 its name was changed to the Division of Wildlife and Rangelands Research, and in 1988 to the Division of Wildlife and Ecology.

Preliminary Discussions about the Work of the Wildlife Survey Section

The rabbit was to be the first and most important subject for investigation, the field being broadened as opportunities arose. Influenced by his recent experience with Elton, Ratcliffe wished to undertake studies of rabbit behaviour and reproduction, food preferences and feeding range, and effects on the environment – research that could not be expected to yield useful results for years[30]. However, good seasons and continuing shortages of labour and materials led to repeated requests to CSIRO to look again at myxomatosis as a method of rabbit control as a matter of urgency. Repetition of the official response, that the Director-General of Health was unlikely to release the virus from quarantine, that it might be difficult to supply enough virus, and that in any case such attempts were likely to be futile[31], was now beginning to be seen as obstructionist. Since Bull, who had been responsible for so much of the early work, was now Chief of the CSIRO Division of Animal Health and a powerful figure in CSIRO affairs, he and Ratcliffe consulted closely on what to do.

Ratcliffe accepted Bull's view that myxomatosis had little potential, but he thought it necessary to exhaust all possibilities. The idea of trying myxomatosis once more was enhanced by a trip to western New South Wales, where the rabbit density was extremely high and the rabbits, living in surface cover, were highly mobile: circumstances that should make contact infection more effective[32]. However, he then found that conditions at the time of the second Wardang Island experiment were very similar, and myxomatosis had not spread well. This fortified his pessimism[33]:

> the close and prolonged contacts necessary for the successful transmission of myxomatosis are not provided by mere crowding and mixing and jostling. So I learned that myxomatosis was not a weapon that had any value against a local rabbit plague.

The controversy between Macnamara and Bull/Ratcliffe

Another element now entered the scene. As outlined earlier, Macnamara, by now a Dame of the British Empire, regarded myxomatosis as her 'discovery' - it was her personal hobby, in which she believed passionately. She was deeply concerned about the future of rural Australia, and knew a good deal about the rabbit plague from her large circle of friends from the prosperous Western District of Victoria. In February 1949, she wrote to Shope, raising the possibility that the Food and Agriculture Organization might take up the use of myxomatosis. Shope contacted Dr K.B.L. Kesteven, Australian representative on FAO[34], who raised the matter informally with the Director-General of FAO and forwarded the letter to Clunies Ross[35]. Not surprisingly, Ross was not impressed with the idea that 'Australia was in line to be treated as a developing country in need of international interference in its domestic pest control. ... as though CSIR did not exist' (O'Dea, 1997).

On 4 April 1949 Ratcliffe gave an address on the national radio station on 'The Rabbit in Australia' in which he made the comment:

> I doubt if more than a handful of people realise how thoroughly it [myxomatosis] was tested. The men who did the job were very keen to demonstrate the practical value of the disease; and I may say they were quite optimistic about their chances of doing so. But they were disappointed.

Macnamara obtained a transcript of Ratcliffe's address, and two weeks later a statement, prepared by Macnamara, was released by the Liberal Party claiming that Australian research should have been pursued much more vigorously. Since the statement was published in a minor newspaper, she thought that it might be overlooked or ignored, so she wrote a long article in the Melbourne *Herald* on 11 May, 1949, demanding further trials, noting, among other things, that:

> Few advances would have been made in medical research if work had been

> abandoned after such a pathetically limited inquiry ... It is difficult to understand why the work was given up, why promising possibilities were not explored, why the virus was not tried on areas more favourable.

As a medical practitioner with more than the usual acquaintance with research, Macnamara knew of many examples in medicine in which experiments had to be carried out, sometimes over decades, and in a wide variety of conditions, before success was achieved. She kept on pointing out that all the unsuccessful trials had been carried out in dry, arid country, and to insist that trials should be carried out in good country. She was to continue in this vein until myxomatosis finally spread over south-eastern Australia in 1951.

Reassured by endorsement of his attitude as 'quite sound' by Macfarlane Burnet, and the latter's further comment that Macnamara was 'inclined to consider herself something of an authority on subjects about which others might perhaps be better qualified to speak'[36], Bull responded vigorously in the *Herald*, 19 May 1949. He objected that seven years' work (1937–1943) should not be described as 'pathetically limited', although in fact only 14 months had been spent on the critical field trials, and these had all been conducted (because of quarantine restrictions)[k] in dry, arid country. Bull concluded with a dogmatic and hostile statement:

> It would be unfortunate if the outburst of wishful thinking by Dame Jean led our distressed and overburdened primary producers to believe that hope for the control of rabbit populations lay in the use of any disease-producing virus at present known to man.

A week later Ratcliffe came to Bull's defence in a statement published in the rural newspaper *Stock and Land*, 25 May 1949, which restated and perhaps overstated his opposition to further work on myxomatosis:

> The conclusion that I came to was that the experiments showed clearly that myxomatosis would not and could not do what was hoped and expected of it; that though the disease could be used to kill rabbits, the process would be laborious and uncertain at best, rarely if ever showing any advantage over killing by other methods; that further trials would be a waste of time and that Dr Bull was right and wise in his decision to discontinue the work.

Macnamara responded in the same newspaper on 8 June 1949, by referring to what she had learned during her overseas trip during the early 1930s:

> I had the opportunity of learning how much patient effort and persistence, how many trials under differing conditions have been necessary before knowledge acquired by laboratory research can be turned into useful practical applications ... not by wondering or by assumption or by generalization drawn from a few experiments under certain conditions, but by work and trial – more work and more trials under different conditions and by using modifications of methods.

Especially to the point, she reiterated her proposition that there was need for research in 'good' country (i.e. in Australia, well-watered country):

> Work to determine whether the virus unaided is or is not of economic value in good country; and, if the infectivity of the virus, when adequately tested in good country, is less than required for an agent to be of economic value.

In all this controversy there is no evidence that Macnamara had considered the possibility of mosquito transmission; what she seems to have had in mind was that the disease might have spread more

[k]The numerous letters from Cumpston to Rivett and Bull (in Australian Archives, Series A 1928/1, files 225/9, Sections 1 and 2) show how insistent the Director-General of Health was that he and he alone had responsibility for quarantine restrictions and how reluctant he was to relax quarantine on myxoma virus (partly because of his conviction that it would not be useful for rabbit control), a fact that may help to explain Bull's irritation with Macnamara's criticisms of his work. See also Basser Library Archives, MS143/1/D5. Letter from Bull to Fenner, 3 August 1955.

effectively in cooler and more humid conditions than in the dry interior, where the virus had been tested previously, possibly because the virus would survive better. It is more surprising that neither Bull nor Ratcliffe took the possibility of mosquito transmission more seriously, since Bull had demonstrated mechanical transmission by mosquitoes long before, and Ratcliffe had been deeply involved with two mosquito-borne diseases, malaria and dengue, during the war. Yet in a letter to Fenner years later, Bull commented[37]:

> One point that has received scant or no attention in recent publications but in which I take pride is the fact that I was the first to conceive the idea that a winged insect vector might make the spread of myxomatosis sufficiently wide and sufficiently fast to prove of practical value in the control of rabbit populations in the field. I mentioned this to C.J. Martin when we were observing his experiments in Cambridge but was rather surprised that he did not spark as I expected of him, especially as he played a leading part in the elucidation of the importance of the rat flea in the spread of bubonic plague.

The public controversy went on for years, long after myxomatosis had spread across Australia, as Macnamara sought to see Australia capitalize to the full the opportunity for rabbit control that was afforded by the spread of myxomatosis.

Field Trials by the Wildlife Survey Section, 1950

However, annoyed though he was, Ratcliffe was influenced by the immediate controversy to look more closely at his strategy[38]: 'The dissipation of my initial anger at Dr Macnamara's tactics and techniques in pushing myxomatosis has enabled me to clear my mind, and assess the position and prospects better than before'. He went on to say: '... to give Macnamara her due, we have not squeezed the last drop out of the myxomatosis lemon, and Bull fully appreciated this when he left the door slightly ajar in his report'. He discussed the need for 'significant help from an insect vector', but noted that:

> We cannot do anything useful with local species of ectoparasites, nor with free-ranging blood-suckers like mosquitoes – they either occur or they don't and it is not practicable policy to try and alter their distribution or artificially bump up their local densities.

He then went on to say:

> The European rabbit flea *Spilopsyllus cuniculi* has not been recorded for Australia. It may actually not be here, or more likely, I think, it is here but has not been picked up simply because no study of the rabbit's ectoparasites has been made. ... If we cannot find it we should introduce it.

Ratcliffe remained ambivalent about myxomatosis, but continued to conduct field surveys which led to a long report to the CSIRO Advisory Council on the rabbit problem[39] and prepared a summary of information about myxomatosis that was made available to enquirers[40].

Over the next few months, while Ratcliffe was building up field staff, he came across a few situations where rabbits were so numerous that he thought that perhaps contact transmission might prove effective. In October 1949 he visited a small property, Gunbower Estate[I], adjacent to the floodplains of the Murray Valley and about 30 km north-west of Echuca (see Fig. 6.4, p. 126). This property was so heavily infested that the rabbits had dug for themselves an almost continuous warren, where he thought the use of myxomatosis might be seriously considered[41], as an alternative to fumigation[42]:

> Here are the essential features. It is an area of small holdings ... The rabbit population is (apart from periodic migrations from the Mallee, which occur at intervals of years) stable, and warren-tied. Surface living rabbits are practically unknown; the litters are reared in the main warren; stick-fast fleas are abundant at suitable situations in most years, and the district is heavily mosquito-infested. There are a few netting fences, and

[I]The owner of Gunbower Estate, C.R.G. Reid, had been corresponding with Rivett and Ratcliffe since December 1948, with suggestions about ways of controlling rabbits, and offering his property for experimental work by CSIRO.

landowners do not, as a rule, have much energy and time left for rabbit control after they have finished their routine jobs of irrigating and milking ...

The particular property in which I am interested ... carries a quite stupendous rabbit population. The owner and I estimate that there are between 150 and 200 well-established warrens on the place. They form a pattern that would enable us to gain a lot of useful data from a single well-planned liberation of myxomatosis....

Ratcliffe saw other possibilities at Albury 'where the rabbits find impregnable harbour in a series of rocky outcrops' and at Balldale near Corowa where a small patch of native pines (*Callitris*) harboured an almost continuous warren subject to constant reinfestation, all circumstances that produced propinquity. He was now anxious to start a trial, but it was several months before this could be commenced.

Up to now Ratcliffe had been alone, but with the end of the 1949 academic year he was able to appoint a few new staff members from the ranks of new university graduates. The first recruits were appointed in January 1950: B.V. Fennessy (Fig. 6.7A), was an agricultural scientist, while J.H. Calaby, a chemist (Fig. 6.7B) and M. Lazarus, a biochemist, were employed to work on poisoning techniques. Somewhat later J. le Gay Brereton and K. Myers (Fig. 6.8), both zoologists with an ecological training who had just graduated from the University of Sydney, joined the staff.

Frustrations lay ahead. Even before he took up his appointment on 2 February, Fennessy was detailed to prepare a stock of virus. Rabbits were inoculated with reconstituted freeze-dried virus, but had to be housed under the floor of the Veterinary Research Institute at Parkville and covered with fly-proof drapes. After Fennessy had gone to Gunbower to prepare for the trial, an inoculated rabbit died, probably of excess heat. This set back the preparation of material for inoculation, and it was not possible to get virus ready for the field until May 1950, thus missing the 1949–50 mosquito season[43].

However, in May 1950 the Gunbower trial started, with the aim of answering the question[44]:

If myxomatosis is introduced into a warren colony, what are the chances of (a) complete extermination, (b) a substantial reduction in numbers, and (c) comparative failure necessitating re-infection or re-treatment with some other agent.

Infection was successfully introduced into a number of warren colonies in different parts of the experimental area, using the modified gin trap devised by Mules. During the weeks that followed, a total of 77 diseased rabbits were seen (in a population of over 4000) well distributed over the area, but by mid-June the sighting of sick animals became less and less frequent, and by the end of July the disease appeared to have died out. Ratcliffe terminated the experiment and prepared a draft press statement on 31 July 1950 which concluded with a promise to test the virus under other conditions in the near future[45].

At this time Ratcliffe was convinced that the best chance of producing an epidemic was by contact infection[46], and the new sites were chosen with this in mind. In September a new series of trials was begun at four sites (see Fig. 6.4, p. 126), all in country that was representative of the worst rabbit-infested environments in the eastern Riverina – at three sites in New South Wales: Wymah, 30 km north-east of Albury, in rocky hills; Coreen, about 30 km north of Corowa, also in rocky hills, and Balldale, about 20 km north-east of Corowa, in stands of native pine (*Callitris*) on sand; and at Rutherglen, in north-eastern Victoria, close to Corowa, amid mine dumps. None of the trial sites had a river frontage. At each site rabbits were caught by digging out or netting at night, or with the use of ferrets, inoculated with myxoma virus, and released.

At Wymah, Fennessy observed that after the release of the inoculated rabbits there was low-grade disease activity for some weeks that noticeably reduced the rabbit density in one part of the property before the infection appeared to die out. There

Fig. 6.7. (A) Bernard Vincent Fennessy (1923–). After graduating in Agricultural Science at the University of Melbourne in 1946, Fennessy worked in the School of Agriculture on a survey of the management of sheep properties in the Western District of Victoria. In 1950 he was recruited by Ratcliffe to join the newly formed Wildlife Survey Section of CSIRO. He was closely involved in myxomatosis field trials in 1950, and then in monitoring the spread and effectiveness of myxomatosis as a method of rabbit control. Subsequently he became interested in other methods of rabbit control and especially in ensuring that the research findings of the CSIRO Division of Wildlife Research were implemented. He was involved in the integration of research findings into strategies and policies for rabbit control at State and regional levels throughout Australia. From 1972 until his retirement in 1988 he was scientific assistant to the Chief of the Division. In 1982 he was awarded a Medal of the Order of Australia (OAM).
(B) John Henry Calaby (1922–1998). Educated at the Ballarat School of Mines, Calaby was an Experimental Officer in the CSIRO Division of Entomology from 1945 to 1950, and was then recruited by Ratcliffe to join the newly formed Wildlife Survey Section of CSIRO. He was closely involved in initial field trials of myxomatosis in 1950 and in the early followup of this work, including extensive surveys of the performance of the disease in southern Victoria and western New South Wales in 1951. From 1952 to 1956 he investigated the disease in southwestern Western Australia. As the interests of the Section (later Division) broadened, he became deeply involved in the study of Australian mammals. He has also published on the history of the discovery of Australian mammals and birds and on Aboriginal history. He served as Assistant Chief of the Division of Wildlife and Ecology from 1985 until his retirement in 1987, and was awarded an honorary DSc of the Australian National University in 1977 and made an Officer of the Order of Australia (AO) in 1994.

were many aedine mosquitoes at Balldale (but no *Anopheles annulipes*), and in the trials at Balldale and Coreen there were few signs of the disease except in the inoculated rabbits; by the beginning of December it appeared to have died out.

The most detailed observations were those carried out at the Rutherglen site, which was a fenced area of 5.2 hectares surrounding some mullock heaps near an abandoned gold mine (Myers, 1954). Careful attention was paid to all aspects of possible epidemiological importance, including the presence of mosquitoes, counts of diseased rabbits were made daily and carcasses were counted and removed,

Fig. 6.8. Ken Myers (1923–). After service in the Royal Australian Air Force during the Second World War, Myers graduated in Zoology in 1950 and joined the Wildlife Survey Section of CSIRO. He immediately became involved in field studies of myxomatosis. During the next 5 years he elucidated the roles of various insect vectors and cooperated with Fenner and Sobey in studying the coevolution of myxoma virus and the rabbit. In 1960 he became leader of the rabbit programme of the Section and initiated a broad study of the biology of the rabbit in different climatic regions of Australia. In 1971 he moved to Canada to become Professor and Chairman of the Department of Zoology at the University of Guelph. From here he sent students to study rabbits in Spain and France. In 1976 he returned to Australia for a year to evaluate the CSIRO rabbit programme, and during that year he was awarded the degree of DSc by the University of Sydney. Returning to Guelph, in 1979 he organized the First World Conference on Lagomorphs. From 1980 to 1986 he worked in the CSIRO Division of Water and Land Resources, initiating studies integrating fauna into land use surveys. He retired in 1986, but spent 2 years with the CSIRO Divisions of Water and Land Resources and Wildlife Research before leaving Canberra.

and at intervals population counts were made, which although incomplete were valuable in assessing changes in population density.

On 7 September 1950 the population count was about 700. Twenty-seven rabbits were inoculated but myxomatosis did not appear to become established, so a further 66 rabbits were inoculated on 27 October. Between then and 16 December, when observations ceased, four generations of natural infection could be distinguished, each smaller than the preceding one (Fig. 6.9). From carcass counts, myxomatosis was known to have killed some 70 rabbits and coccidiosis and known predation, by a cat and a pair of eagles, accounted for about 400 more, mostly kittens. In spite of this, the population count rose to about 1000. A few mosquitoes, including *Anopheles annulipes,* were seen, and the conditions for contact infection were as good as could be imagined: very high rabbit density, no selective predation of diseased animals, and a high proportion of diseased rabbits remaining within their warren colonies until they died. Yet the rate of infectivity was only 0.6, i.e. the infection rate decreased with each generation of disease and looked as though the disease would certainly peter out.

With these five trials myxomatosis appeared to have been given every chance to prove itself as a method of rabbit control, but had failed in each site. The 1950 trials appeared to provide a clear confirmation of the views of Bull and Ratcliffe, as most recently expressed in the controversy with Macnamara, namely that myxomatosis was not an effective method for the biological control of rabbits. On 30 October 1950 Ratcliffe wrote to Clunies Ross[47]:

> ... Our four myxomatosis studies down here seem to be failing nicely in their different ways. The best crop of cases is being produced by Fennessy in his experiment in the hill country [the Wymah site] ... The rabbits in Myers' three trials are relatively undisturbed, and spend a good deal of their time above ground.

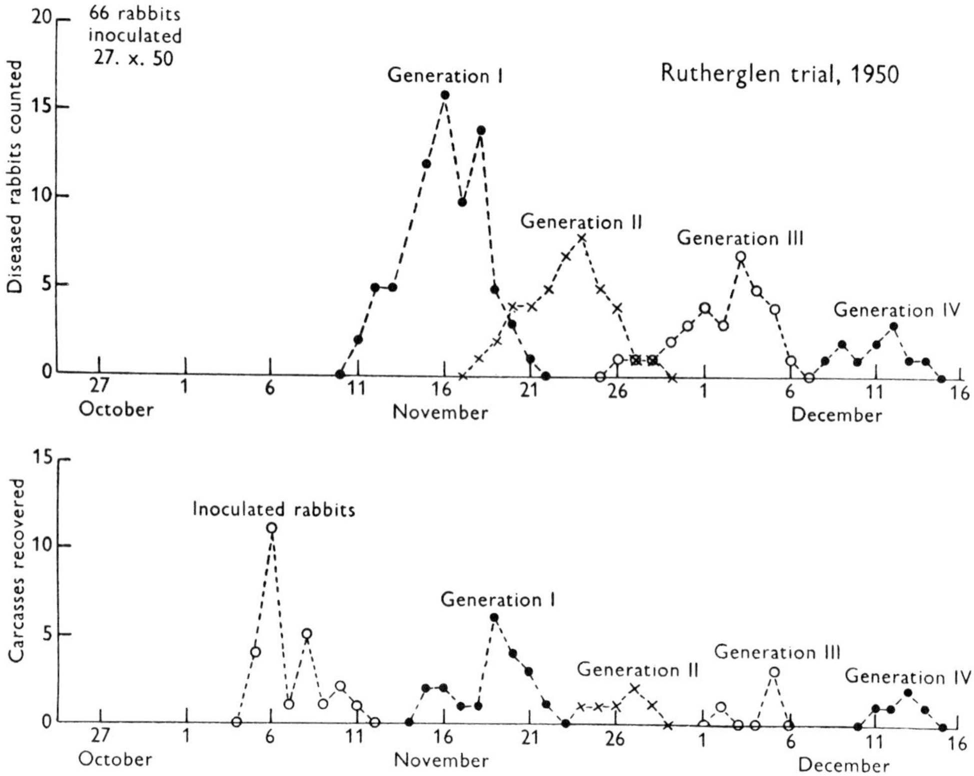

Fig. 6.9. Trial of myxomatosis at Rutherglen in northern Victoria in late 1950. As judged by counts of diseased rabbits and carcasses, there were four 'generations' of myxomatosis and then the disease appeared to have petered out. From Myers (1954), with permission.

The Escape: Spread through South-Eastern Australia, 1951

Then came one of the most remarkable events in the history of infectious diseases, the development of an epizootic that for scale and speed of spread is probably without parallel. Before any decision could be made about what should be done to follow up the apparently abortive Murray Valley releases, in late December a telephone call was received from Balldale to say that sick rabbits had been seen in large numbers at the experimental site. Almost immediately after this, sick rabbits were seen on the Corowa Common, a site on the River Murray some 15 km from Rutherglen. From there reports of sick rabbits came from other river frontages, first from the Murray about 16 km south of the Balldale site, and then, during January 1951, from further downstream on the Murray and from various places along its northern tributaries – the Murrumbidgee, the Lachlan and the Darling Rivers[48].

From the nature of its spread Ratcliffe quickly deduced that a winged vector must have been responsible, and mosquitoes were much the most likely. In January a start was made to map the spread, co-opting to the Wildlife Survey Section team colleagues from other divisions of CSIRO and obtaining the cooperation of numerous field officers of State Departments of Lands and Agriculture. By mid-February, myxomatosis had appeared in practically all parts of a huge area that was to be affected before winter set in, and the peak of activity had passed in most districts. High mortalities, about 90% or more of the rabbits in an area, were often achieved within a month of the appearance of the first sick rabbits. With some reluctance, the Director-General of Health cancelled the quarantine restrictions under which use of the virus was supposed to be contained within places remote from humans, and in February CSIRO circulated to State departments information on methods of spreading the disease in rabbit populations[49]. Macnamara took advantage of the success to emphasize the importance of her role in getting the trials undertaken[50]:

> Mr Ratcliffe in *Stock and Land* 8 June 1949 writes 'It was she who apparently suggested etc.' It involved a good deal more than a casual verbal suggestion, several months of part-time study – a trip to Baltimore from New York – and a change of plans for my trip home from New York from the comparative comfort of the *Mariposa* to a trip across the Atlantic in a cattle ship back to England and a wait of several weeks at my own expense in order that Sir Charles Martin should return to England, that I should discuss the matter with him and hand over all the data I had collected ready for him to start work. ...

There were a few outbreaks as late as April and May, but by June the disease appeared to have died out. The extent of the spread before June, as far as could be determined in the brief time available to make observations, is shown in Fig. 6.10. The disease spread naturally but patchily along the river systems of south-eastern Australia over an area measuring more than 1600 km from north to south and 1760 km from east to west. The few outbreaks along the coast of Victoria and New South Wales

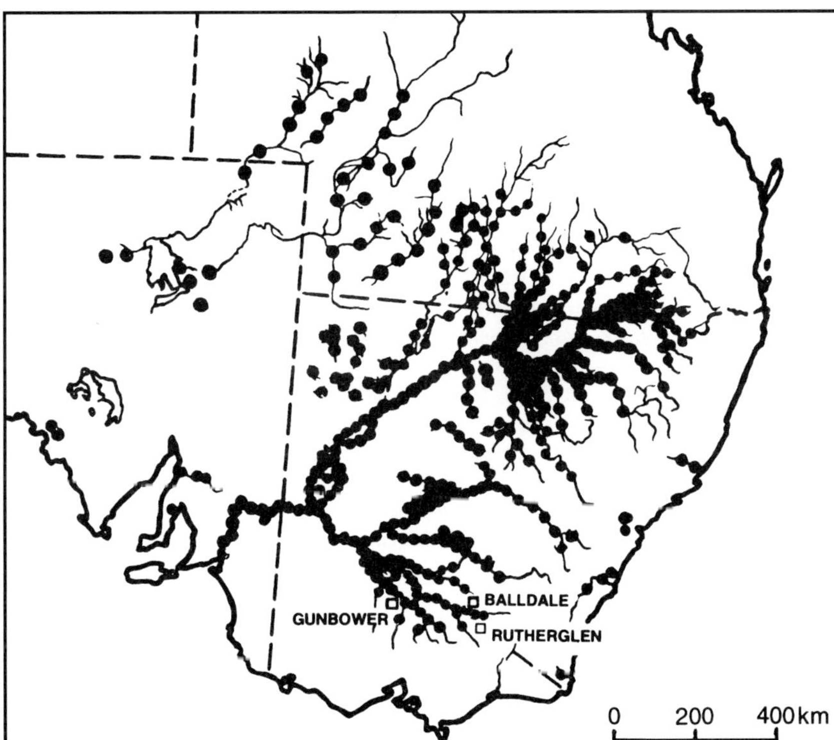

Fig. 6.10. The spread of myxomatosis in south-eastern Australia between December 1950 and May 1951. Compiled by the Wildlife Survey Section of CSIRO from data summarized in Progress Reports 1 (26 February 1951) to 5 (25 May 1951).

were the result of local inoculations organized by State authorities; in other regions such as the northern tableland of New South Wales inoculations were unsuccessful. Apart from these outliers the map is a record of natural spread.

Clearly, the Murray–Darling river system dominated the distribution of cases, which is explicable from what was later learned about the insect vectors of myxomatosis (see p. 166). Anticipating this information, the natural spread can be interpreted as follows. The epizootic distribution depended on mosquitoes that bred in summer, in persistent water, notably *Culex annulirostris*. The rainfall during 1950 had an unusual distribution. In Victoria and southern New South Wales 1950 was a year of average or below-average rainfall and by mid-summer the countryside was very dry, conditions that did not favour a spring emergence of one of the principal vectors of myxomatosis, *Anopheles annulipes*, and the conditions restricted summer breeding of the other major vector, *Culex annulirostris,* to areas close to the rivers and large swamps. In contrast, the northern parts of New South Wales and southern Queensland received record-breaking rains, with flooding on an unprecedented scale. Although the last rains occurred in November, abundant surface water lingered for at least two months. Corresponding to this distribution of breeding places for *Culex annulirostris,* the epizootic along the Murray and Murrumbidgee Rivers was restricted to a very narrow strip, often only a few hundred metres wide, along the flood plains and river banks, except where irrigation channels complicated the picture. Further north, along the Lachlan, epizootics of high intensity occurred in a belt up to 5 km back from the river and its associated lagoons. Further north still, in the headwaters and tributaries of the River Darling, where there had been extensive flooding, the disease lost its association with the rivers and was widespread and general. Nevertheless, even in southern Queensland the occurrence of severe outbreaks was variable[51].

Away from the Murray–Darling river system, scattered localized outbreaks occurred, some associated with inoculation campaigns operated by State rabbit control authorities, as in coastal Victoria and New South Wales, some, as in central Victoria and in Eyre Peninsula in South Australia, with the introduction of sick rabbits by local farmers. The question in everybody's mind was: would the disease survive through the winter and recur naturally in spring? Before we answer this question, the immediate reorganization of responsibilities for future work and two other matters that relate directly to the outbreak of summer 1951 need to be mentioned, namely, the initiation of virological investigations and the encephalitis scare.

Reorganization of responsibilities

In February 1951 a conference of all interested persons and groups was held in Melbourne to discuss future activities[52], the first of many such conferences (see p. 158). A number of initiatives were agreed, not the least important being that every effort should be made to follow up the success of myxomatosis by redoubled efforts at getting rid of as many remaining rabbits as possible, as outlined in a press release. The best response came from New South Wales, where the Minister for Agriculture announced that £80,000 would be spent on equipment to destroy warrens in badly infested country[53].

Virus for the field trials in 1950 had been produced by the Veterinary Research Institute of the University of Melbourne. With the likely introduction of widespread inoculation campaigns in spring, arrangements needed to be made for the production of virus on a larger scale. Initially H.E. Albiston of the Veterinary Research Institute, which was located adjacent to Bull's CSIRO Division of Animal Health in Parkville, accepted this responsibility, and in New South Wales the Glenfield Veterinary Research Station produced freeze-dried virus. From October 1952 the Commonwealth Serum Laboratories produced freeze-dried virus for inoculation (see p. 159). The New South Wales Department of Agriculture and the Victorian Department of Lands accepted responsibility for carrying out inoculations in their States, and agreed to assist the Wildlife Survey Section with field observations.

Ratcliffe took stock of what his Section should concentrate on in the forthcoming spring and summer, and outlined his ideas in a staff circular[54]:

> ... Our main task this year is to try and get a reliable assessment of the potentialities and limitations of myxomatosis as a weapon of practical control, which means getting to understand where and when the infection can be relied on to give a useful ... kill, and conversely where and when it cannot – and find out the true 'why' in both cases.

Initiation of virological studies of myxomatosis

Although in the early 1940s Bull had undertaken some laboratory studies of myxoma virus, Ratcliffe's team, recruited after 1949, consisted of zoologists, and there were at that time no virologists in any of the Divisions of CSIRO. Indeed, in 1950 the only substantial centre of virological studies in Australia was the Walter and Eliza Hall Institute in Melbourne, where Macfarlane Burnet was Director. As the only Australian centre of virological expertise, he saw it as his responsibility to undertake investigations on any virological problem of great public importance. In his diary dated 31 January 1951 he noted[55]:

> rang Clunies Ross re possibility of being associated with myxomatosis work of F. Ratcliffe. ... Bull is of the opinion that some pathological and virological work should be done in association with the zoological work of the section.

To explain the next entry in Burnet's diary, another piece of history, which concerns one of the authors of this book (Fenner), needs to be recorded. Fenner had trained in virology with Burnet, working with mousepox virus, over the period February 1946 to August 1948. He then went to work for a year at the Rockefeller Institute for Medical Research in New York. While there he had accepted the position of Professor of Microbiology in the newly founded Australian National University. Since this university, located in Canberra, had no laboratories, Burnet had offered him the use of two laboratory rooms in the Walter and Eliza Hall Institute. In New York he had worked on the bacteriology of tuberculosis and had carried on with this work when he returned to Australia early in 1950. However, his heart was in virology, and in January 1951 he saw the immense potential interest presented by myxomatosis: what happens when a very virulent virus spreads through a very susceptible host species, on a continental scale?

The next entry into Burnet's diary was for 1 February 1951, and begins: 'Fenner anxious to take on myxoma as a major job, stimulated somewhat by Jean Macnamara I suspect'[56].

The encephalitis scare

In 1917–18 an outbreak of human encephalitis, called 'X-disease', occurred widely over south-eastern Australia. At first regarded as an unusual form of poliomyelitis, Cleland *et al.* (1918) produced encephalitis by intracerebral inoculation of monkeys, sheep, a foal and a calf, the latter animals being regarded as resistant to poliovirus. They maintained the virus for five serial passages in monkeys, but at that time no other studies were possible. The unusual meteorological conditions that allowed myxomatosis to spread over south-eastern Australia as it did in 1951 were also propitious for the spread of X-disease. Early in February 1951 outbreaks of encephalitis occurred in Mildura, a fruit-growing town on the lower River Murray, and in the Shepparton district of Victoria. Since it was a matter of considerable public health importance, Burnet asked one of his virologists, E.L. French, to investigate the outbreak. French soon isolated a virus from the brain of a fatal case (Anon., 1951a, b; French, 1952), and showed that it was serologically related to the well-known Japanese encephalitis virus. As was then usual with arbovirus infections, the new virus was given the name of the locality from which it was isolated: Murray Valley encephalitis virus.

Since both myxomatosis and encephalitis were both novel diseases and were occurring at the same time and in the same part of the world, the public was apprehensive[57], and numerous newspaper commentators linked the human and the rabbit diseases. The causative virus having been recovered in the Walter and Eliza Hall Institute, Burnet issued reassuring statements, but as he relates in his autobiography (Burnet, 1968), public disquiet persisted and the Minister for CSIRO (R.G. Casey) was concerned that the rumours be put to rest. Then early in March Burnet was challenged by the Chairman of the Mildura Hospital Board to test the harmlessness of myxoma virus on himself and the Minister, Mr Casey.

Burnet consulted with Fenner (Fig. 6.11), as the virologist working on myxomatosis, and the two agreed that such a measure was without risk. Both knew that many people in the Americas must have been bitten by mosquitoes carrying myxoma virus and that Ratcliffe and his team must also have been bitten, and that the human encephalitis in Mildura was caused by a very different virus. Further, two officers of the Wildlife Section had accidentally inoculated themselves with the virus, without ill-effects[58]. So Burnet and Fenner inoculated each other intradermally (to mimic a mosquito bite) with what might be reasonable mosquito-sized doses: 1, 10, and 100 rabbit-infectious doses. When Clunies Ross heard about this he insisted that he should also be included in the experiment, since myxomatosis was a CSIRO project. In the event, none of them got more than a slight reddening at the inoculation sites, with no antibody response. On 8 March 1951 the Minister (Casey) made a general statement about myxomatosis to the House of Representatives[59], including the comment that two research workers had been inoculated with the virus with no ill-effects. Since this announcement did not completely allay public anxiety, he subsequently announced that the persons who had been inoculated, without ill effect, were Burnet, Clunies Ross and Fenner[60]. Subsequent history has confirmed the very high host species specificity of myxoma virus, and a serum survey of many children in California who were exposed to infected mosquitoes failed to find any serological evidence of infection with myxoma virus, although many rabbits in an adjacent rabbitry were dying (Jackson *et al.*, 1966).

Fig. 6.11. Frank John Fenner (1914–). After graduating in medicine at the University of Adelaide in 1938 and working as an intern for a year, Fenner joined the Royal Australian Army Medical Corps and served in Palestine, Africa and the South-West Pacific Theatre, for the last two and a half years as a malariologist. After the war he worked with F.M. Burnet at the Walter and Eliza Hall Institute until mid-1948, when he went to the Rockefeller Institute in New York for a year. In 1949 he was appointed foundation Professor of Microbiology in the John Curtin School of Medical Research in the newly established Australian National University. From 1951 to 1965 he worked on various aspects of the virology of myxomatosis. In collaboration with F.N. Ratcliffe he wrote a book *Myxomatosis,* which was published in 1965. In 1967 he became Director of the John Curtin School of Medical Research and in 1973 Director of the Centre for Resource and Environmental Studies. He was involved in the successful campaign by the World Health Organization to eradicate smallpox, work for which he was awarded the Japan Prize. He was elected a Fellow of the Royal Society in 1958 and received its Copley Medal in 1995.

The recurrence of myxomatosis in spring 1951

In June 1951, when the epizootic appeared to have come to an end, CSIRO Head Office issued a press release[61], which contained the caution:

> While ... there is room for optimism, one sobering fact must be recognized and emphasized. There is an enormous amount of rabbit infested country in Australia regarding which it is impossible to hold out any hope that myxomatosis may supplant other methods of destruction, or indeed be of any real value in control.

Many areas that had reported what was called a Grade I kill (a mortality of over 90%) had been largely repopulated during the winter, mostly by immigration of rabbits from nearby unaffected areas. Then in July and August, before inoculation campaigns had been commenced, reports of sick rabbits came from scattered areas throughout south-eastern Australia. Practically all of these local outbreaks occurred outside the areas in which the summer epizootics had occurred, often well away from them. There were very few winter sightings of sick rabbits in southern Queensland, where the summer epizootics had been severe. Some were localized around swamps or hillside seepages, suggesting that they might be associated with local concentrations of mosquitoes. Others often showed a marked association with patches of thistles[m]. Surprisingly, some outbreaks occurred in distant places like Claraville, north-west of Alice Springs, in Central Australia, some thousand kilometres from the nearest known outbreak in New South Wales[62]. In an appreciation of the situation in October 1951, Ratcliffe noted that:

> the winter persistence and 'working' of the disease has been of negligible proportions from the point of view of rabbit control. It should, however, be of great importance epidemiologically. It is now apparent that the virus is widely 'seeded' over southeastern Australia, and it should be able to assume epidemic form as soon as conditions favouring rapid transmission develop in any region.

[m]Mostly saffron thistles (*Carthamus lanatus*). Later Dyce (1961) demonstrated experimentally that myxomatosis could be transmitted on the spines of thistles.

Although by August–September there were in many areas large populations of various species of *Aedes* mosquitoes (for humans, the most annoying species), these did not initiate outbreaks of myxomatosis[n]. In October–November, however, with the appearance of the mosquito *Anopheles annulipes*, intense outbreaks occurred throughout the Murray–Darling basin, accentuated in the summer by the presence of *Culex annulirostris*. Aided by inoculation campaigns, by the next winter the disease had spread throughout most of south-eastern Australia. Thus 33 years after Aragão had first made the suggestion that myxomatosis should be used to control rabbits in Australia, myxomatosis had been demonstrated to be a highly effective means of biological control of the rabbit. How effective and for how long will be discussed in Chapters 7 and 8.

Reasons for the Failure to Use Myxoma Virus Earlier

With hindsight, it seems that myxomatosis could have been introduced into the Australian rabbit population much earlier than it was, perhaps in 1943, just after the last Point Pearce study. Why was it delayed for so long? Four factors appear to have played a part: concern about specificity, economic factors, political factors and psychological factors.

Concern about specificity

With all agents proposed for the biological control of a pest, one of the most important questions to be answered is whether it will be specific, i.e. for agents designed to control insect pests or plants, will the proposed biological agent have an adverse effect on insects, or plants, of commercial or ecological value? Of course, the same question is asked about other methods of pest control such as the use of chemical insecticides or herbicides. One of the great advantages of biological pest control is that

[n]Subsequent investigations showed that *Aedes* spp. were not attracted to rabbits (see p. 167).

it is likely to be more specific than chemical methods, but against this must be considered the fact that the biological agent is likely to be self-perpetuating. This is usually a great advantage if the agent is highly specific for the target host, but is an added cause for concern if it is not.

Before quarantine authorities give authority for the release of agents proposed for the biological control of a vertebrate pest, especially a virus, evidence is required that the agent is highly host-specific, and will not adversely affect (usually, will not infect) any animal other than the target species, whether it be a domestic or companion animal, a native species, or human beings.

In general, viruses fall into one of two groups as regards host range. If testing is carried out in mature animals (because very immature animals may be much more susceptible than adults), a few viruses are found to have a very wide host range, some affect several species, often in the same genus, and some are highly host-species-specific. After fairly extensive testing in Brazil (Aragão, 1943), England (Martin, 1936) and Australia (Bull and Dickinson, 1937), and many years of field experience in California and South America, myxoma virus appeared to be highly host-specific, affecting only a very few species within the family Leporidae. There appeared to be no need to delay its introduction into Australia on the grounds of safety. However, partly because the Director-General of Health was sceptical about its possible value, he insisted on stringent conditions for the field experiments, including the requirement that they should be carried out in areas remote from human habitation. Such areas were usually arid, and hence lacked mosquitoes. If a good case had been made in 1943, permission to use it in a situation that would have permitted mosquito transmission may have been granted, although the response of the quarantine authorities to Bull's suggestion that investigations should be carried out by the chief veterinary officer of New South Wales (see below) makes this by no means certain. In the event, the virus was held in quarantine until it escaped in 1951.

Economic factors

In the early days of the European exploration and subsequent colonization of large parts of the world, in the 16th to 19th centuries, the Europeans had no inhibitions in moving the animals that they regarded as being useful in their home surroundings to any part of the world. Indeed, intending colonists always brought domestic and companion animals with them. As soon as Europeans became reasonably well established in 'new' countries like the Americas, Australia or New Zealand, they set up 'acclimatization societies' designed to encourage the importation and establishment of most of the animals and plants with which they were familiar at home. Such animals and plants were placed in a situation where they were freed of the predators and agents of disease that controlled their population growth in their original homes. It is for this reason that the most extreme examples of plant and animal pests occur in countries opened up to European exploitation in the past few centuries (the Americas, Australia, and oceanic islands), although there are also some examples of pest movements from the New World to Old World.

However, it took only a few years before the European rabbit was recognized as a major agricultural pest, and from about 1880 onwards farmers and their organizations exerted great pressure on politicians to control them. Although there was a substantial trade in rabbit skins and carcasses, on the whole the pressure from the agricultural lobby to destroy the pest was much stronger than that from the skin and carcass trade. After the Second World War, with the explosion in rabbit numbers and the knowledge that myxomatosis might be a 'magic bullet', this pressure was accentuated. Thus there were no substantial economic factors inhibiting the earlier introduction of myxomatosis.

Political factors

Quite apart from the political will of different governments to control a perceived pest, sometimes factors that may be called 'political' can lead to delays in action even when all parties concerned recognize the seriousness of a particular pest problem. This was the case when investigations on the control of rabbits by myxomatosis ceased in 1943. Bull explained the delay in the following terms[22]:

> CSIR was a research organization. We sought to gain knowledge by scientific research. From our Division [Animal Health] new knowledge was extended and applied by the several [State] Departments of Agriculture. As a research organization we had no statutory obligation in the control of rabbits. This was carried by vermin authorities in the several States. We took the view that myxomatosis might prove successful, even if its range were restricted, if used in regions where mosquitoes and possibly other winged insects would probably act as vectors. We thought that this might be determined by the State vermin authorities as it was impracticable for us to test a sufficient number of areas. To this end, I wrote to Hindmarsh who was chief veterinary officer in New South Wales. He agreed to carry out the follow up work and applied to the Director-General of Health for permission to use the virus. This was refused. Thus a dead end was reached.

At that time Australia was engaged, for the first time in its history, in a major war in which enemy planes were bombing Australian cities and there was a real threat of invasion. Most young men were in the armed services; likewise scientists and technical officers who until then had been working on 'civilian' problems were recruited for war-time tasks. One consequence of this was that during the closing years of the war the rabbit problem became even more severe, for there were no young men left in country areas to carry out whatever rabbit control was possible. Combined with favourable seasonal conditions for the multiplication of rabbits, the consequence of this neglect was that immediately after the war the rabbit problem was more severe than it had ever been. There was now a political imperative to do something about it, leading to the establishment of the Wildlife Survey Section of CSIRO, with a specific remit to study ways of controlling rabbits.

Psychological factors among scientists

In the earliest days of the proposal to use myxoma virus for rabbit control there was an understandable reluctance among quarantine officers to authorize the importation of an agent which they regarded as a potential threat to the health and possibly the lives of domestic and native animals, and perhaps humans. Nowadays, as well as quarantine considerations (which relate to danger to animals other than the target species), the public reaction concerning cruelty to animals must be taken into consideration before approval is given to import an agent to be used for the control of a mammal or bird (see Chapter 11). In the period immediately after the Second World War, when the use of myxomatosis in the field came up for serious consideration, the rabbit problem was recognized as so crucial to Australia's primary industries that the matter of the cruelty of any method of control was not seriously raised. However, when the outbreaks occurred in early 1951, many people not closely connected with primary production saw rabbits infected by myxomatosis, and it is an ugly disease (see illustrations in Chapter 5). Reports appeared in the press suggesting that the Royal Society for the Prevention of Cruelty to Animals would take legal action against CSIRO[63], but no prosecution occurred. The economic and ecological cost of rabbit infestations in Australia outweighed ethical sensitivities.

What might be called 'psychological' factors were also of some importance in the early attitude of the leading scientists. Bull, who had demonstrated that both fleas and mosquitoes could transmit myxomatosis, recognized the possible importance of flea transmission but was curiously reluctant to believe that mobile vectors such as mosquitoes might spread the disease between warrens[64]. Ratcliffe was reluctant to put the major energies of his small, inexperienced and newly established Section into what his senior colleague Bull regarded as a project of doubtful value, and he dismissed mosquitoes because it was not possible to alter their distribution or their numbers (see p. 134). The attitude of Bull and Ratcliffe was strengthened by their irritation with the insistent and somewhat strident pressure by Dame Jean Macnamara, both through the media and politically, to make myxomatosis the highest priority of the Wildlife Survey Section[65]. On the other hand, Macnamara, besides being deeply concerned by the land degradation caused by rabbits, was undoubtedly influenced by her perception that the use of myxomatosis was 'her' proposal. She also made the point that, as a worker in preventive medicine, she had learned 'that enthusiasm is not evoked by an occasional press statement, but by bringing active guidance to the people by a person aided by modern equipment such as visual aids, wireless, talks, the film, ...' which helps to explain her persistence. It is pleasing to note that the report in the journal *Nature* on the spread of myxomatosis in south-eastern Australia (Ratcliffe *et al.*, 1952) concludes with the sentence: 'It remains only to record that the enthusiastic and persistent advocacy by a Melbourne specialist in virus diseases, Dame Jean Macnamara, was a material factor in the decision to initiate experiments in 1933, and to resume field studies in 1950.'

Endnotes

[1]Basser Library Archives MS143/5/Myx 2. Letter from H.B. Aragão to A. Breinl, 15 January 1919. It is curious that he mentions Argentina, since at that time rabbits were not a pest there, although hares were.

[2]CSIRO Archives, Series 3, WA6/3/3/1. Letter from H.B. Allen to G. Lightfoot, Secretary of the Commonwealth Institute of Science and Industry, 27 September 1919.

[3]Basser Library Archives MS143/5/Myx 2. Letter from E.N. Robinson, Secretary of the Commonwealth Institute of Science and Industry, to H.B. Aragão, 19 December 1919.

[4]Basser Library Archives MS143/5/Myx 2. Letter from H.R. Seddon, Director of Veterinary Research, Glenfield, to Chief Veterinary Surgeon, New South Wales, 6 September 1929.

[5]However, '... surprisingly, since late 1966 the disease has been enzootic on the island, although the proportion of animals affected is low ...' From Infestation Control 1965–67, p. 55. Published 1969, Her Majesty's Stationery Office, London.

[6]This was the strain of virus provided by Dr R.E. Shope and subsequently used for the introduction of myxomatosis in Australia, where it was designated as the Standard Laboratory Strain.

[7]National Library Archives, MS 2399/1/133. Letter from Macnamara in New York City to her family, 6 January 1933.

[8]National Library Archives, MS 2399/3/4. 'Infectious myxomatosis of rabbits: Preliminary memorandum', for the Australian Minister in London, Mr Stanley Bruce. Sent from New York by Dr Jean Macnamara, April 1933. There was no final memorandum. It was subsequently sent by Bruce to A.C.D. Rivett, with a copy to H.J.L. Cumpston.

[9]CSIRO Archives, Series 3, WA6/3/3/2 Part I. Letter from Rivett to F.L. McDougall, 9 October 1933.

[10]CSIRO Archives, Series 3, WA6/3/3/3 Part I. Letter from Martin to Rivett, 28 September 1933.

[11]CSIRO Archives, Series 3, WA6/3/3/3 Parts I and II. Correspondence, mainly between Martin and Rivett, from 28 September 1933 and 13 April 1939.

[12]CSIRO Archives, Series 3, WA6/3/3/3 Part I. Letter from Martin to Rivett, 28 September 1933.

[13]CSIRO Archives, Series 3, WA6/3/3/10 Part V (I). Memorandum from L.B. Bull to Rivett, 6 November 1935. 'Rabbit myxomatosis: Observations by Sir Charles Martin and Recommendations for (1) Specificity Trials, and (2) Field Trials in Australia.'

[14]CSIRO Archives, Series 3, WA6/3/3/10 Part V (I). Memorandum from Bull to Rivett, 25 February 1936. 'Observations on the Projected Study of *Myxomatosis cuniculi*.'

[15]CSIRO Archives, Series 3, WA6/3/3/10 Part V (I). Letter from Cumpston to Rivett, 10 February 1936.

[16]CSIRO Archives, Series 3, WA6/3/3 Part I. Letter from Director of Agriculture, Tasmania, to Rivett, 27 August 1934.

[17]CSIRO Archives, Series 3, WA6/3/3/5 Part II. Letter from Acting Premier of Tasmania to Acting Prime Minister of Australia.

[18]Australian Archives Series A1928/1, file 225/9, Section 1. Memorandum from Cumpston to Rivett; letters from Rivett to Vermin Boards in Western Australia; and letters to Vermin Boards in Western Australia and from Prime Minister to H.K. Nock, MHR.
[19]CSIRO Archives, Series 3, WA6/3/3/10 Part IX (IV). A note on the possible use of virus myxomatosum in the control of rabbit populations, by L.B. Bull, 22 August 1938.
[20]CSIRO Archives, Series 3, WA6/3/3/10 Part V (II). Letter from Cumpston to Rivett, 26 March 1940.
[21]Australian Archives Series A1928/1, file 225/9, Section 2. Interim report on myxomatosis investigations, by L.B. Bull, February 1941.
[22]CSIRO Archives, Series 2, file 31/11. Letter from Bull to Rivett, 10 March 1942.
[23]CSIRO Archives, Series 3, WA6/3/3/10 Part IX. Letter from Rivett to Bull, 6 March 1942.
[24]CSIRO Archives, Series 379, file ICR 23/1. Several letters between Ratcliffe and Clunies Ross, written between 29 March 1947 and 15 July 1947.
[25]CSIRO Archives, Series 3, file WA6/3/2/7 Parts IV and V. Numerous letters written between January 1945 and December 1948.
[26]CSIRO Archives, Series 3, file WA6/3/2/7 Part V. Letter from the Secretary, The Federal Graziers' Council of Australia to CSIR, 19 March 1948.
[27]CSIRO Archives, Series 379, file ICR 23/1. Letter from Ratcliffe to Clunies Ross, 6 June 1948.
[28]CSIRO Archives, Series 379, file 18/8. Letter from Ratcliffe to CSIR Executive, 13 September 1946.
[29]CSIRO Archives, Series 379, file 18/8. Memorandum from G.A. Cook (Secretary, CSIR) to J.J. Dedman, Minister-in-Charge of the Council for Industrial and Scientific Research.
[30]CSIRO Archives, Series 379, file 18/8. Letter from Ratcliffe to CSIR, 17 September 1948.
[31]CSIRO Archives, Series 3, file WA6/3/3/10 Part II (I). Myxomatosis – Summary of information, by F.N. Ratcliffe, 5 August 1949. Part IV. Comments by Bull on Ratcliffe's proposals for field trials, 7 November 1949.
[32]CSIRO Archives, Series 3, file WA6/3/3/10 Part IX. Letter from Ratcliffe to Bull, 10 February 1949; Ratcliffe decides to '... take [myxomatosis] off the ice for one more trial.'
[33]CSIRO Archives, Series 3, file WA6/3/3/7 Part XVII. From an unpublished letter by Ratcliffe written in response to Macnamara's letter in *Stock and Land*, 8 June 1949.
[34]CSIRO Archives, Series 3, file WA6/3/3/2 Part II. Letter from R.E. Shope to K.B.L. Kesteven, February 1949.
[35]CSIRO Archives, Series 3, file WA6/3/3/2 Part II. Letter from Clunies Ross to Kesteven, 30 March 1949.
[36]CSIRO Archives, Series 3, file WA6/3/3/10 Part IX (VIII). Letter from Clunies Ross to Bull, 2 May 1949.
[37]Basser Library Archives, MS143/1/D5. Letter from Bull to Fenner, 13 December 1961.
[38]CSIRO Archives, Series 3, file WA6/3/3/6 Part I. Letter from Ratcliffe to Clunies Ross, 13 July 1949.
[39]CSIRO Archives, Series 643, ACAP, Session 2, Item 9(A). The rabbit problem. A survey of research needs and possibilities with some suggestions regarding practical policy, by F.N. Ratcliffe, Melbourne, 1951.
[40]CSIRO Archives, Series 3, file WA6/3/3/10 Part II(I). *Myxomatosis cuniculi* – summary of information, by F.N. Ratcliffe, 5 August 1949.
[41]CSIRO Archives, Series 3, file WA6/3/3/10 Part II. Letter from Ratcliffe to Bull, 31 October 1949.
[42]Basser Library Archives, MS143/1/D6. In a note to Fenner dated 14 November 1996, B.V. Fennessy noted that in conditions like those at Gunbower, Ratcliffe saw myxomatosis as a possible labour-saving alternative to fumigation.
[43]Basser Library Archives, MS143/1/D6. Letter from Fennessy to Fenner, 1996.
[44]CSIRO Archives, Series 3, file WA6/3/3/7 Part I. Press release: 'New Experiments on Rabbit Control', 1 March 1950.
[45]CSIRO Archives, Series 3, file WA6/3/3/7 Part I. Press release: 'Myxomatosis Fails Against Rabbits', 31 July 1950.
[46]CSIRO Archives, Series 3, file WA6/3/3/10 Part II. Letter from Ratcliffe to Bull, 12 September 1950.
[47]Letter from Ratcliffe to Clunies Ross, 5 October 1950. Quoted by O'Dea (1997), p. 292.
[48]Basser Library Archives, MS 143/1/D2. Letter from K. Myers to Fenner, 10 September 1996: 'Looking back on it, I am of the opinion that the *Anopheles annulipes* which supported the infection at Rutherglen (the only site with *Anopheles*) ... were flying out from the Murray River flats 10 miles to the north, and were equally likely to be carrying the virus back to the river in a spotty fashion. Since the inoculations were made on 7 September [1950] [and again on 27 October], individual mos-

quitoes could have been seeding down the river [for at least seven weeks and possibly] for ... three and a half months before we became aware of it.'
[49]CSIRO Archives, Series 379, file ICR 18/8. Myxomatosis of the rabbit: Some methods that can be used for the spread of the disease in rabbit populations. Information Circular No. 20. CSIRO Division of Animal Health and Production.
[50]CSIRO Archives, Series 3, file WA6/3/3/10. Letter from Macnamara to Clunies Ross, 2 February 1951. Macnamara's views were supported in a letter to the editor of the Adelaide *Stock and Station Journal* by R.R. Pennefather on 14 October 1952.
[51]CSIRO Archives, Series 3, file WA6/3/3/10 Part IX (X). Letter from Euston Young to Clunies Ross, 7 March 1951.
[52]CSIRO Archives, Series 379, file 18/8. Minutes of Conference on Myxomatosis in Rabbits, 15 February 1951.
[53]'£80,000 For State's War On Rabbits', *Sydney Morning Herald,* 1 May 1951.
[54]CSIRO Archives, Series 3, file WA6/3/3/6 Part VI. Wildlife Survey Section Circular Memorandum No. 1, 24 October 1951.
[55]University of Melbourne Archives. Australian Science Archives Project. A Guide to the Records of Frank Macfarlane Burnet, Series 2, 2/15. Diary entries, 31 January and 1 February 1951 (copies in Basser Library Archives, MS143/25/5A).
[56]As far as I (Fenner) can recall, I had had no contact with Macnamara at that time. I was working on *Mycobacterium ulcerans*, but I was anxious to move back into virology. The lack of contact with Macnamara is confirmed by a letter from her to Pemberton dated 8 February 1951, in which she presses him to appoint 'someone well-grounded in virology' to a team to be formed in Victoria to study the disease. CSIRO Archives, Series 379, file 18/8. Myxomatosis remained my principal research interest for the next 15 years.
[57]CSIRO Archives, Series 3, file WA6/3/3/10 Part IX (X). Letter from Dr E.S. Morris (Director-General of Public Health, New South Wales) to Clunies Ross, 27 February 1951, and reply, Clunies Ross to Morris, 5 March 1951.
[58]Basser Library Archives, MS143/1/D6. Letter from Fennessy to Fenner, 26 July 1991.
[59]Australian Parliamentary Debates, House of Representatives, 8 March 1951, 212, 171–173.
[60]Australian Parliamentary Debates, House of Representatives, 11 October 1951, 214, 523.
[61]CSIRO Archives, Series 379, file ICR 18/8. CSIRO Press Release, 7 June 1951.
[62]CSIRO Archives, Series 3, file WA6/3/3/10 Part II (II). Letter from Administrator of the Northern Territory to Secretary, Department of Territories, 13 September 1951.
[63]CSIRO Archives, Series 3, file WA6/3/3/13. Memorandum from Ratcliffe to all staff of the Wild Life Survey Section and cooperating officers, 16 October 1951: 'Myxomatosis: Possible Legal Action by RSPCA.'
[64]Basser Library Archives 143/25/5A. Letter dated 2 May 1996, from W.I.B. Beveridge to R. Humphreys, then working on a biography of Ian Clunies Ross. At the time of which he writes, Beveridge was Professor of Animal Pathology in the University of Cambridge. 'When he [Clunies Ross] visited me in Cambridge [in 1950] he said that Dame Macnamara had been rousing public opinion about CSIRO doing more to get myxomatosis going in the rabbits. Ian [Clunies Ross] said that he felt sure that it was a useless exercise because Lionel Bull was quite definite about it. But I pointed out to Ian that Bull was wrong. When I was in America some years previously there was an outbreak of myxo in farmed rabbits in California [in 1938] which was spread by mosquitoes. I had written to Bull about this and suggested myxo should be released in places and times where there were mosquitoes. But Bull was stubborn and ignored this. So I told Ian about this and said I believed myxo would spread if released under the right conditions. And, as you know, that is what happened. Bull was not receptive to other people's suggestions and was stuck with his own pet idea about rabbit fleas.' See also Beveridge (1997).
[65]Basser Library Archives 143/1/D1. Letter from Ratcliffe to D. Stewart (publisher of book by E. Rolls), dated 28 February 1968, giving Ratcliffe's considered opinion on the relations between Ratcliffe, Bull and Macnamara.

References

AAAS (1921) Report of the Fifteenth Meeting of the Australasian Association for the Advancement of Science. Australasian Association for the Advancement of Science, Sydney, pp. xxxi–xxxii.

Andrewes, C. (1981) Richard Edwin Shope. *Biographical Memoirs, National Academy of Sciences* 50, 353–375.

Anon. (1919) Rabbit myxoma. *Science and Industry* 1(3), 187; 1(4), 205.

Anon. (1942) A mechanical device for the spread of disease agents among rabbits. *Journal of the Council for Scientific and Industrial Research* 15, 82–83.

Anon. (1951a) Use of myxomatosis cuniculi against rabbits. *Medical Journal of Australia* 2, 652–653.

Anon. (1951b) Encephalitis and rabbit myxomatosis in Australia. *Lancet* 2, 794.

Aragão, H.B. (1943) O virus do mixoma no coelho do mato (*Sylvilagus minenses*), sua transmissão pelos *Aedes scapularis* e *aegypti*. *Memorias do Instituto Oswaldo Cruz* 38, 93–99.

Beveridge, I. (1997) *Fighting Diseases: My Varied Scientific Career.* W.I.B. Beveridge, Wentworth Falls, NSW, 105 pp.

Borg, K. (1962) Om myxomatos. *Särtryck ur Medlemsblat för Sveriges Veterinärförbund* 4, 1–10.

Bull, L.B. and Dickinson, C.G. (1937) Specificity of the virus of rabbit myxomatosis. *Journal of the Council for Scientific and Industrial Research* 10, 291–294.

Bull, L.B. and Mules, M.W. (1944) An investigation of *Myxomatosis cuniculi*, with special reference to the possible use of the disease to control rabbit populations in Australia. *Journal of the Council for Scientific and Industrial Research* 17, 79–93.

Burnet, M. (1968) *Changing Patterns. An Atypical Autobiography.* Heinemann, Melbourne, pp. 106–112.

Burnet, M. (1971) *Walter and Eliza Hall Institute 1916–1965.* Melbourne University Press, Melbourne, pp. 103, 133.

Chick, H. (1956) Charles James Martin 1866–1955. *Biographical Memoirs of Fellows of the Royal Society* 2, 173–208.

Cleland, J.B., Campbell, A.W. and Bradley, B. (1918) The Australian epidemics of an acute polio-encephalomyelitis (X disease). Report of the Director-General of Public Health, 1917, Sydney, pp. 150–280.

Clunies Ross, A.I. (1977) Ian Clunies Ross. *Records of the Australian Academy of Science* 3(3/4), 86–108.

Coman, B. (1998) Francis Ratcliffe, pioneer conservationist. *Quadrant* 42, 20–26.

Currie, G. and Graham, J. (1966) *The Origins of CSIRO: Science and the Commonwealth Government, 1901–1926.* Commonwealth Scientific and Industrial Research Organization, Melbourne, 203 pp.

Dyce, A.L. (1961) Transmission of myxomatosis on the spines of thistles, *Cirsium vulgare* (savi) Ten. *CSIRO Wildlife Research* 6, 88–90.

Fenner, F. (ed.) (1995) *The Australian Academy of Science: The First Forty Years.* Australian Academy of Science, Canberra, pp. 132–133.

French, E.L. (1952) Murray Valley encephalitis: isolation and characterization of the causative agent. *Medical Journal of Australia* 1, 100–105.

French, E.L. (1993) Bull, Lionel Bately (1889–1978). *Australian Dictionary of Biography* 13, 293–294.

French, E.L. and Stewart, D.F. (1983) Lionel Bately Bull 1889–1978. *Historical Records of Australian Science* 5(4), 90–110.

Hvass, J. and Schmit-Jensen, H.O. (1939) *Report of a Visit to the Dufeke Estate in Skaane (Sweden) on 27 April 1939, for the Purpose of Studying the Effects of Infection by Myxomatosis among Wild Rabbits.* Report to the Danish Ministry for Agriculture and Fisheries, Copenhagen.

Jackson, E.W., Dorn, C.R., Saito, J.K. and McKercher, D.G. (1966) Absence of serological evidence of myxoma virus infection in humans exposed during an outbreak of myxomatosis. *Nature* 211, 313–314.

Lockley. R.M. (1940) Some experiments in rabbit control. *Nature* 145, 767–769.

Lockley, R.M. (1955) Failure of myxomatosis on Skokholm Island. *Nature* 145, 906–907.

Mackerras, I.M. (1971) Francis Ratcliffe (1904–1970). *Search* 2(3), 74–75.

Martin, C. J. (1936) Observations on *myxomatosis cuniculi* (Sanarelli) made with a view to the use of the virus in the control of rabbit plagues. *Council for Scientific and Industrial Research Bulletin* No. 96, 28 pp.

Morison, P. (1986) Martin, Sir Charles James (1866–1955). *Australian Dictionary of Biography* 10, 423–425.

Mules, M.W. (1940) Notes on the life history and artificial breeding of the Australian 'stickfast' flea, *Echidnophaga myrmecobii* Rothschild. *Australian Journal of Experimental Biology and Medical Science* 12, 385–390.

Myers, K. (1954) Studies in the epidemiology of infectious myxomatosis of rabbits. II. Field experiment, August–November 1950, and the first epizootic of myxomatosis in the riverine plain of south-eastern Australia. *Journal of Hygiene* 52, 47–59.

Notini, G. (1952) The wild rabbit problem. In: Notini, G., Forselius, S., Bramford, S. and Mellström, B. (eds) Vildkaninen på Gotland (The rabbit problem in the Gotland Island). *Bulletin of the Royal School of Forestry, Stockholm* Nr 9, pp. 103–109.

O'Dea, M.C. (1997) *Ian Clunies Ross. A Biography.* Hyland House, Melbourne, 371 pp.

Ratcliffe, F.N. (1931) The Flying Fox (*Pteropus*) in Australia. *CSIR Bulletin* No. 53, Melbourne, 81 pp.

Ratcliffe, F.N. (1936) Soil Drift in the Arid Pastoral Areas of South Australia. *CSIR Pamphlet* No. 64, Melbourne, 84 pp.

Ratcliffe, F.N. (1938) *Flying Fox and Drifting Sand.* Angus & Robertson, Sydney, 332 pp.

Ratcliffe, F.N., Myers, K., Fennessy, B.V. and Calaby, J.H. (1952). Myxomatosis in Australia. A step towards the biological control of the rabbit. *Nature* 170, 7–11.

Roe, M. (1981) Cumpston, John Howard Lidgett (1880–1954). *Australian Dictionary of Biography* 8, 174–176.

Schedvin, C.B. (1987) *Shaping Science and Industry. A History of Australia's Council for Scientific and Industrial Research, 1926–49.* Allen and Unwin, Sydney, 374 pp.

Schedvin, C.B. (1993) Clunies Ross, Sir William Ian (1899–1959). *Australian Dictionary of Biography* 13, 449–451.

Schmit-Jensen, H.O. (1939) Summary of the experiments carried out in Vejro by the State Veterinary Serum Laboratory on the extermination of rabbits by myxomatosis virus. *Report to the Danish Ministry for Agriculture and Fisheries,* Copenhagen.

Shope, R.E. (1932) A filterable virus causing a tumor-like condition in rabbits and its relationship to virus myxomatosum. *Journal of Experimental Medicine* 56, 803–822.

Smith, A.G. (1968) Macnamara, Dame Annie Jean (1899–1968). *Australian Dictionary of Biography* 10, 346–347.

Tyrrell, M., Fenner, F. and Piesse, R.D. (1971). Some tributes to Francis Ratcliffe, O.B.E. A pioneer of conservation in Australia. *ACF Newsletter*, February 1971.

White, H.C. (1929) Observations on rabbit myxoma. *NSW Department of Agriculture Veterinary Research Report* No. 5, 1927–28, 46–47.

Zwar, D. (1984) *The Dame. The Life and Times of Dame Jean Macnamara; Medical Pioneer.* Macmillan Australia, Melbourne, 168 pp.

7

Myxomatosis in Australia: 1952–1966

Overview

Since myxomatosis had survived the winter of 1951 and become active again the following spring in many parts of the Murray–Darling Basin, it was clear that the disease was now enzootic in the rabbit population of south-eastern Australia. Its activity in the summer of 1952 exceeded expectations. Instead of being confined to the vicinity of the rivers and creeks, it spread along and between them. In many areas carriage of infected rabbits set up new centres of natural spread, and in all States natural spread was supplemented by inoculation campaigns, which were carried out extensively over subsequent years, especially in Victoria and New South Wales.

Initially, the major activities of the scientists of the Wildlife Survey Section and field workers of the State departments responsible for rabbit control were to elucidate the relative importance of various vectors (mainly mosquitoes) in different parts of Australia, and to make detailed studies of the performance of the disease in selected study sites, notably Lake Urana. Laboratory studies of the virulence of specimens of myxoma virus recovered by field workers and of changes in the genetic resistance of wild rabbits were carried out in Fenner's laboratory in the Australian National University. However, in 1954 laboratory studies of genetic resistance were begun in the CSIRO Division of Genetics, and in 1958 field work by the Victorian Department of Crown Lands and Survey was supplemented by laboratory studies under the direction of Douglas. The Victorian work became of major importance after 1965, when Fenner's laboratory had ceased work on myxomatosis and the Keith Turnbull Research Institute at Frankston had taken responsibility for continuing studies of virus virulence and rabbit resistance.

By the end of the 1954–55 season, myxomatosis, by natural spread supplemented by inoculation campaigns, had reached practically every part of Australia in which rabbits occurred. The year 1955 marked the beginning of a relatively stable situation in which overt myxomatosis activity would wax and wane in response to changes in vector activity and the fluctuating numbers of susceptible rabbits. By then, however, it was clear that most enzootic strains of virus were somewhat attenuated, and the possibility that the increased survival rates might lead to the development of genetic resistance had been raised. Because so many trained observers were involved in vector studies and extensive surveys, meetings of Federal and State authorities were convened annually between 1951 and 1956, at which reports of the season's myxomatosis activity in each State were presented. After 1956 the collection and collation of field data on an Australia-wide basis almost ceased, so that the more recent history of myxomatosis can be presented only in general terms or for particular study sites, except in Victoria.

Myxomatosis had brought incalculable benefit to Australia. It had rendered a previously impossible situation manageable, and in areas of high economic value follow-up measures like poisoning, ripping burrows and providing properties with rabbit-proof fencing had ensured that in some places the rabbit problem was under control. As time went on, however, it became clear that the increasing genetic resistance of rabbits would reduce the efficacy of myxomatosis as a method of rabbit control.

Spread of Myxomatosis: Spring 1951 to Winter 1955

Between 1951 and 1956 the staff of the Wildlife Survey Section was augmented by scientists transferred temporarily from other CSIRO divisions and by officers of State authorities engaged in rabbit control. Their prime concern was to observe the occurrence of myxomatosis and determine its epidemiology in different parts of Australia. It was clear that the initial explosive spread was due to flying vectors, and attention was concentrated on mosquitoes. *Anopheles annulipes,* a wide-ranging, adaptable mosquito which had been evident in only one of the trial sites, Rutherglen, during the spring of 1950 (Myers, 1954), became abundant during October–November 1951, and intense outbreaks of myxomatosis occurred along the Murray River, between Victoria and New South Wales. Then during the summer months, January to March 1952, *Culex annulirostris*, which had been the principal vector in the summer of 1951, took over from *An. annulipes* in southern Australia as well as in the basin of the River Darling in southern Queensland, where it had been responsible for the extensive spread in 1951. In contrast to the 1950–51 season[a], the disease occurred many kilometres back from the rivers. As soon as farmers realized the efficacy of myxomatosis, they carried infected rabbits to their farms, and the natural spread of myxomatosis was further aided by inoculation campaigns carried out by State authorities, which started in 1951 and were considerably expanded in subsequent years. By March 1952 myxomatosis was occurring over a huge area of undulating and hilly country, extending across the centre of Victoria and northward, in New South Wales, over the foothills and western slopes of the Great Dividing Range, with many localized outbreaks well beyond the margin of the epizootic zone.

The situation in the winter of 1952 was similar to that of 1951, but the localized winter outbreaks were more numerous and extensive, paving the way for intense epizootic activity in the spring and the 1952–53 summer. In the spring of 1952 the highly mobile simulid vector *Austrosimulium furiosum* was common and assisted in the widespread dispersal of the disease, which by the end of the 1952–53 season had become virtually coextensive with the distribution of rabbits in Australia (Fig. 7.1).

Already, however, observations had been made that heralded problems for the continuing efficacy of myxomatosis. In February 1952 Mykytowycz (1953; Fig. 7.2), working in the Australian Capital Territory, captured a wild rabbit which was recovering from myxomatosis. The causative virus was passaged serially in wild rabbits, by contact infection; none of eight rabbits died naturally, although most rabbits infected by mosquito probing died from a rather atypical disease. The observation that some field strains had diminished virulence was confirmed by Fenner (1953a), who recovered several attenuated strains of myxoma virus from pools of mosquitoes captured by Myers in November–December 1952, during the second epizootic at Lake Urana, New South Wales, one year after virulent virus was released in the area (see p. 156).

The pattern in 1953–54 was reviewed at the Myxomatosis Conference in July 1954[1].

[a]The period from spring to autumn, when most cases of myxomatosis occurred in Australia, was termed the 'myxomatosis season'.

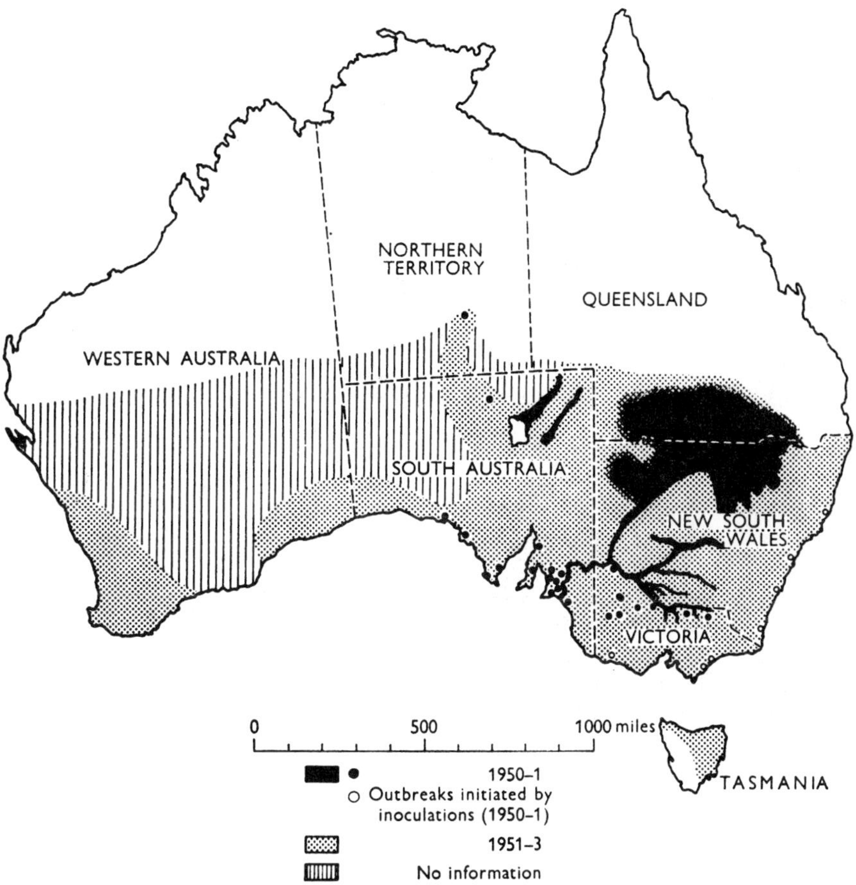

Fig. 7.1. Map of Australia showing the extent of myxomatosis at the end of 1953. From Fenner and Ratcliffe (1965), with permission.

It was more difficult to assess the impact of myxomatosis than in previous years because the rabbit population was then so much smaller and its distribution was patchy, there were few invasions of new territory, and because the novelty of myxomatosis had now worn off there were fewer reports from landholders. The rabbit infestations in many parts of New South Wales were now at their lowest in living memory. In contrast to previous years, there were few signs of persistence through the winter and there was a very late start to summer outbreaks. There were good kills along the Murrumbidgee and adjacent rivers, closely confined to the river frontages and extending along the River Murray from Albury to the Murray mouth in South Australia. Heavy summer rains at the end of January in northern Victoria and southern and central New South Wales were followed by outbreaks, sometimes locally very effective, in country well away from the rivers, where a remarkable and widespread development of myxomatosis in autumn and early winter provided the main kill of the season.

In August 1955 a series of papers on scientific aspects of myxomatosis in Australia were read at the meeting of the Australian and New Zealand Association for the Advancement of Science[2]. These papers contained, ahead of formal publication, information on the position at the time,

Fig. 7.2. Roman Mykytowycz (1921–1996). Born in the Ukraine and educated in Veterinary Science in Leipzig and Munich, Mykytowycz migrated to Australia in 1949 and joined the Wildlife Survey Section as a technical assistant in mid-1950, later becoming a Research Officer and rising to be Senior Principal Research Officer in the CSIRO Division of Wildlife Research before his retirement in 1986. As well as early work on myxomatosis, he made outstanding studies on rabbit behaviour, especially on the importance of the sense of smell.

including data on vectors, the appearance of attenuated strains and the mechanism by which they were selected, and the appearance in the field of rabbits that were more resistant than in earlier years.

The situation in different states

Since vermin control in Australia is a State responsibility, it is useful to look at the situation in the winters of 1953 and of 1955, at the end of the third and fifth myxomatosis seasons, on a State by State basis[3,4].

New South Wales

By winter 1953 myxomatosis had reached every part of the State, although its occurrence had been patchy on the northern tablelands. The extensive rangelands of the Western Division had been virtually cleared of rabbits.

The overall results in the 1954–55 season were even better than during the previous two seasons, the disease being active throughout the State, in some areas spreading in a spectacular fashion reminiscent of 1951–52. It was effective in several areas where it had performed poorly in previous years and was often active in low-density rabbit populations. Most outbreaks appeared to arise from enzootic infections, but a number of landholders believed that the disease did not spread until after inoculations. A total of 254,200 doses of myxoma virus were distributed, 212,150 to Pasture Protection Boards and 42,050 direct to landholders; the number of rabbits inoculated was probably about 10% of these figures.

Victoria

By mid-1953 myxomatosis had been recorded over most of the State, and the first extensive outbreaks had been recorded in the hilly country in Gippsland, in the east, in the Mallee, the low-rainfall area in the north-west, and in the Western District, where most outbreaks had occurred in late autumn and early winter.

Myxomatosis performed very satisfactorily during the 1954–55 season, and by the end of June rabbit numbers were at their lowest for many years. Much of the disease activity appeared to arise from the flaring up of persisting infection. Although interest in inoculation field days was less than in previous years, due to the scarcity of rabbits or the recognition that the disease was already active locally, there were many areas in which inoculation seemed to initiate outbreaks. In the northern part of the State myxomatosis occurred in the summer, due to the activity of *An. annulipes* and *Cu. annulirostris,* whereas in the south, where *Cu. annulirostris* did not occur, the peak activity occurred in late autumn and winter, and a variety of other mosquitoes acted as vectors.

South Australia

In the 1952–53 season myxomatosis was seen throughout most of the areas where observations were possible; the mortalities tending to be high in the good agricultural areas, but low to moderate in the low rainfall country of the northern part of the State. Myxomatosis epizootics occurred in the areas where the trials of Bull and Mules in 1942–43 had shown no tendency to spread (see p. 127), probably because climatic conditions were then unfavourable.

By 1954 the situation in South Australia had become more complex than in the eastern States, due in part to the migration of rabbits from the drier hinterland to the coastal belt or onto river frontages in the summer. There appeared to be more rapid build-up of rabbits in areas that had experienced moderate to good myxomatosis kills in previous seasons, but the general picture at the end of the 1954–55 season was of a very substantial overall reduction in rabbit numbers.

Queensland

Because much of Queensland lies within the tropics, rabbits thrived only in the southern part of the State. Myxomatosis had occurred throughout this area in the initial outbreaks of the summer of 1951, and in the following spring there was an extension into sparsely settled desert country around Lake Eyre, in South Australia.

In winter 1955, in the far south-western areas, rabbits remained at the low levels to which they had been reduced in 1951. Elsewhere there had been a slight build-up in numbers during 1954, but the 1954–55 season was one of the best on record, with intermittent flooding in rabbit-infected areas, and a resulting abundance of mosquitoes and sandflies. More than in other States, large areas of previously devastated pasture and farm land were coming back into production, and myxomatosis appeared to have 'all but wrecked' the rabbit industry (skins and carcasses) in Queensland. Although the authorities encouraged inoculation, particularly in areas away from the watercourses, by 1954 graziers in all areas had become apathetic, relying on good seasons to effect a widespread dissemination of myxomatosis.

Western Australia

Because of the absence of large, permanently flowing streams, combined with the normal dry summer season, the most effective mosquito vectors, *An. annulipes* and *Cu. annulirostris*, rarely occurred in large numbers in this State. Only two extensive epizootics occurred in the first decade after inoculation campaigns were begun in 1951, in the south-western corner of the State extending as far north as Geraldton, and for some distance into the country adjacent to the south coast as far as the South Australian border (Calaby *et al.*, 1960). The first of these was in the 1954–55 season, when heavy rain fell in mid-February over most of the rabbit-infested areas of the State. Disease activity started along the watercourses, later extending into the hills.

Tasmania

Initially the State Government legislated against the introduction of myxomatosis, but after an illegal importation in 1951 some outbreaks occurred. An inoculation campaign organized during the summer of 1952–53 resulted in establishment of the disease throughout the settled parts of the island, which were roughly coextensive with habitats suitable for rabbits. However, the authorities favoured poisoning with sodium fluoroacetate ('1080') rather than myxomatosis as their main weapon for rabbit control, and by 1954 myxomatosis had been completely overshadowed by poisoning. Unspectacular outbreaks occurred in a number of areas, in late summer or autumn.

Detailed studies at Lake Urana

Lake Urana (see Fig. 8.1, p. 183, for its location) was selected by the Wildlife Survey Section as a major and continuing study site for investigations of the epidemiology of myxomatosis. The first of these studies, by a combined CSIRO–Australian National

University team, was a detailed analysis of two successive epidemics over the period September 1951 to January 1953 (Myers *et al.*, 1954). This provided field evidence of two of the most important features of myxomatosis in Australian wild rabbits; its extreme lethality in genetically unselected animals, and the emergence, during its first winter in the area, of less virulent strains of the virus. Although there was a large rabbit population there, myxomatosis had not occurred at Lake Urana in 1951, probably because it was some distance from the river systems near which disease activity was concentrated. The lake (hardly recognized as such by persons who are accustomed to lakes always containing water) is a flat depression of just over 8000 hectares, an area of internal drainage which is flooded in wet years and at other times provides good grazing (Fig. 7.3). The lake bed has a rim of sandhills which provide an excellent habitat for rabbits.

Observations were made in two locations; Fig. 7.4 illustrates the results at one of these. The investigation involved initial population counts of rabbits[b], then some hundreds of rabbits were captured, inoculated with the Standard Laboratory Strain of myxoma virus, and released in the same site. Population counts of healthy and sick rabbits and counts of mosquitoes[c] were carried out at weekly and then monthly intervals for the next 12 months. Larval and adult mosquitoes were collected and identified, and when epizootics of

[b]Population counts of rabbits were made by sight counts on standard transects under uniform weather conditions, usually towards sunset, when rabbits were feeding and still visible (see Myers, 1954).

[c]Sixteen species of mosquitoes occurred in the region; the counts shown in Fig. 7.4 illustrate only the most numerous mosquitoes seen at different periods.

Fig. 7.3. Lake Urana, New South Wales, one of the major CSIRO study areas for investigations of myxomatosis and rabbit biology. In summer the lake bed was usually dry; rabbits lived in warrens in the sandhills surrounding the lake (foreground).

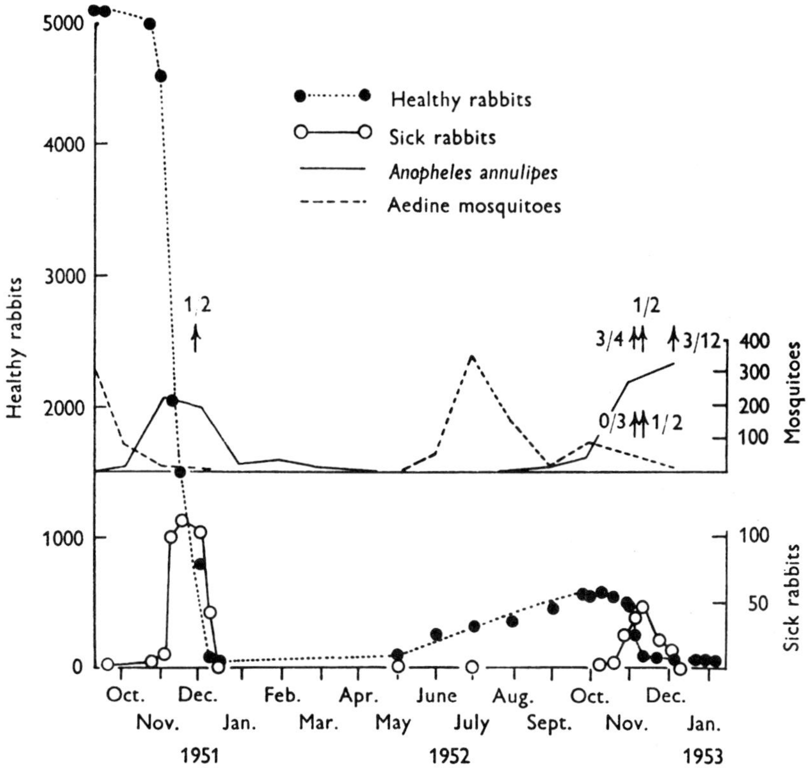

Fig. 7.4. Diagram illustrating the results at the study area at Lake Urana of two successive outbreaks of myxomatosis over the period September 1951 to January 1953, showing population counts of normal and sick rabbits and of adult *Aedes* spp. and *Anopheles annulipes* mosquitoes, and the results of assays of batches of mosquitoes for the presence of myxoma virus. From Myers *et al.* (1954), with permission.

myxomatosis occurred batches of about 50 mosquitoes of identified species were tested for the presence of virus. Viruses recovered from batches of adult mosquitoes were tested for their virulence by standard procedures[d].

Standard population counts during September, before the inoculations, averaged 5000. The first cases of myxomatosis in uninoculated rabbits were seen late in October, and within four weeks the population count had fallen from 5000 to 50. Serological tests showed that three-quarters of the survivors had escaped infection, so that the case-fatality rate was an extraordinary 99.8%. Very similar figures were observed at the second site, but the landholder killed off all survivors and it was not possible to proceed with the experiment there. Myxomatosis persisted at a very low level of transmission throughout the winter of 1952, at least one diseased rabbit being seen each month from October 1951 to December 1952, except for June, August and September.

The greatly reduced population was augmented in June 1952 by an influx of adult rabbits which immigrated to the area because of flooding of the lake floor,

[d]Groups of five laboratory rabbits were inoculated intradermally with small doses of virus that had been passed once in laboratory rabbits. Since at this time most strains killed all inoculated rabbits, mean survival times were used as a measure of virulence (see p. 94).

increasing the population to about 150. Young rabbits started to emerge from the warrens in July and rabbit numbers rose steeply to reach 550 in October, when 50 rabbits were captured, inoculated with the Standard Laboratory Strain of myxoma virus, and released. However, an epizootic commenced before cases could have occurred due to transmission from the inoculated rabbits, reaching a peak in November and reducing the population to about 60 by December, a level maintained for the rest of the summer. In contrast to the situation a year earlier, some two-thirds of the rabbits counted in January showed obvious signs of recent recovery from myxomatosis. Sera collected from the survivors showed that all were immune, and by excluding the few survivors from the first epizootic, the case-fatality rate among susceptible rabbits for the second epizootic was calculated to be 90%. This was still a very high death rate, but left 10% rather than less than 1% of survivors. Clearly, it was possible for breeding from such a population base to repopulate an area rather quickly, and since the vast majority of breeding rabbits had survived myxomatosis, the progeny might well be genetically more resistant than the original population. Over the next several years weanling rabbits were collected each spring from Lake Urana and tested for their genetic resistance to myxomatosis, and the site was used for many other field investigations.

Providing Information to the Public

It was clear that as soon as the Australian farming community heard about myxomatosis they were very interested in it as a means of controlling their worst pest animal (see pp. 125 and 140). This interest extended to the general public when the suspicion arose that infection with myxoma virus might cause encephalitis in humans. When myxomatosis escaped from the trial sites and spread over the rabbit-infested parts of Australia in the summer of 1950–51, it was essential that up-to-date scientific information should be provided. This was needed by State Departments of Agriculture, Crown Lands and Survey, and the like that had responsibilities for rabbit control and by the landholders, so that they could be advised how to carry out inoculations and how to capitalize on the reductions in rabbit numbers. Interested Federal and State government officials were put in the picture by holding conferences on myxomatosis that included the scientists and representatives of all interested government departments. Landholders were kept informed by frequent press releases aimed especially at the rural press.

Conferences on rabbit control

As soon as it became clear, early in 1951, that myxomatosis was killing large numbers of rabbits, a conference was held in Melbourne, attended by scientists working on myxomatosis (members of the Wildlife Survey Section and virologists from the Australian National University) and the relevant representatives of Commonwealth and New South Wales and Victorian State government departments[5]. This was followed by a meeting of a technical committee, to work out details concerning the preparation and distribution of myxoma virus for inoculation campaigns[6]. At the same time the Wildlife Survey Section organized a procedure for distributing information on the rapidly changing situation by means of Progress Reports[7], five of which were issued between 26 February and 25 May 1951, by which time the initial epizootic had come to an end. These reports collated all information available to the Section and distributed it to all scientists involved and to officials in interested State departments. With the expansion of the Wildlife Survey Section by the establishment of a unit in Perth and the co-option of officers from other branches of CSIRO to assist in work on myxomatosis, the Progress Reports were replaced by Circular Memoranda, designed to keep all officers informed on staff matters and scientific progress, and for distribution to interested outside groups[8].

Subsequently, at the end of each myxomatosis season, Conferences on Myxomatosis were held which were attended by scientists and representatives of Federal government departments and of relevant departments from all States[9]. Myxomatosis long remained a major focus of work of the Wildlife Survey Section, but with the developing assistance of relevant State departments, the frenetic activity of 1951–52 was replaced by a more orderly arrangement of larger and less frequent conferences, which were ultimately broadened into Vermin Control Conferences[10] and Vertebrate Pest Control Conferences[11]. Myxomatosis figured prominently in all the earlier conferences, but was later reduced to a minor component of the agendas, as ripping and poisoning with sodium fluoroacetate became more important topics for discussion. Over the years the proceedings of these conferences provide an interesting picture of changing attitudes to the role of myxomatosis in rabbit control, and of growing understanding of the changes that had occurred in both the virus and its host.

Press releases

Each Myxomatosis Conference was followed by a long press release distributed by CSIRO Headquarters[12], and State departments with responsibilities for rabbit control also issued press releases for the rural press in their respective States. Selected items from several of the Myxomatosis and Pest Control Conferences were reported in the CSIRO non-technical periodical publications *Rural Research*[13] and *Ecos*[14].

Inoculation Campaigns

Responsibility for inoculations

From the time of the first investigations of myxomatosis in Australia by Bull, farmer's and grazier's organizations had clamoured for supplies of myxoma virus so that landholders could introduce it on their properties. At the governmental level, State governments were responsible for pest control. After the dramatic spread of the disease in the summer of 1951, some officially sponsored inoculations were carried out before the end of that summer, in regions to which the disease had shown no signs of spreading. It became clear that inoculation would have to be undertaken on a State-wide basis, and at a meeting between Commonwealth and State representatives in June 1951 it was decided that the distribution of the virus should be the responsibility of the State authorities[15].

In a memorandum distributed by CSIRO[19], Ratcliffe made it clear that the field officers of the Wildlife Survey Section would be unable to assist in inoculation campaigns except by supplying advice to State and Pastures Protection Board officers, if this could be done without interfering with their investigational work.

Supply of myxoma virus for inoculation campaigns

Except for a few batches prepared at research institutions in Adelaide and Brisbane for use in 1951 in South Australia and Queensland respectively, all material for inoculation in the field was prepared either by the Veterinary Research Station at Glenfield (for use in New South Wales) or the Commonwealth Serum Laboratories[e] in Melbourne (for all other States). The virus used at Glenfield was an early field isolate from Dubbo, which became known as the Glenfield strain (see below). At the Commonwealth Serum Laboratories the Standard Laboratory Strain was used until 1961, when it was replaced by the Glenfield strain, which was in turn replaced by the Lausanne strain in 1974.

[e]The Commonwealth Serum Laboratories, located in Melbourne, were established by the Commonwealth Government during the First World War to ensure that needed vaccines, for both human and veterinary use, were available to the Australian public. Although various State and private institutions also produced veterinary vaccines, it was the institution to which governments turned for needed biological products. However, since pest control was a State responsibility, myxoma virus for use in inoculation campaigns could also be produced in laboratories under State control.

Initially the Walter and Eliza Hall Institute for Medical Research in Melbourne, where Fenner had worked on myxomatosis from early 1951 until late 1952, supplied myxoma lesions to the Commonwealth Serum Laboratories, from which they prepared the freeze-dried and ampouled product. Between 1957 and 1961 G.W. Douglas, of the Victorian Vermin and Noxious Weeds Destruction Board, provided this material, and after 1961 the Commonwealth Serum Laboratories produced the virus in microbiologically secure animal facilities on their own premises. Although demand decreased greatly during the 1970s, CSL Ltd. (as it became after the Commonwealth Serum Laboratories were privatized) continues to supply freeze-dried Lausanne strain myxoma virus, on request, to State pest control authorities and to individual farmers.

The Glenfield strain

Investigators at the Veterinary Research Station at Glenfield, in New South Wales, concluded in 1951 that the strain Aust/Dubbo/2–51/1 of Fenner and Marshall (1957) was more virulent than the Standard Laboratory Strain, and decided to use it for inoculation campaigns in that State. It became known as the Glenfield strain. In 1961 tests in Victoria in genetically resistant wild rabbits confirmed its greater lethality (97% case-fatality rate compared with 71% and 75% case-fatality rates in two trials; Douglas (1962); see also Table 8.3, p. 197). It continued to be used by the Glenfield Veterinary Research Station until 1980, when it was replaced by the Lausanne strain until production there ceased in 1982. In 1962 the Commonwealth Serum Laboratories switched from the Standard Laboratory Strain to the Glenfield strain; in 1969 this was replaced by a cloned derivative, GV/194.

The Lausanne strain

This strain, which had been used to introduce myxomatosis into France in 1952 (see p. 213), was found by tests in genetically resistant rabbits to be substantially more virulent than either the Glenfield or the Standard Laboratory strain (see Table 8.3, p. 197). In 1972 the Commonwealth Serum Laboratories undertook production with the Lausanne strain as well as the cloned Glenfield virus (GV/194). Initially supplies of Lausanne strain in Victoria were restricted to use by the Department of Crown Lands and Survey because they were using it for experiments on transmission by the European flea, but from 1975 only Lausanne strain virus was produced by the Commonwealth Serum Laboratories and it was supplied to State pest control authorities and to individual farmers as requested.

Methods of inoculation

Two methods of inoculation of rabbits were used (Fig. 7.5). Landholders were advised to infect rabbits by rubbing the reconstituted virus suspension vigorously on the rabbit's conjunctiva or on a scarified area of skin. This method could also be used to transfer pus from the eye of a diseased rabbit to that of a healthy rabbit, and especially in the early days landholders often carried out such transfers, often with rabbits brought from distant sites. Officers of rabbit control agencies usually infected captured rabbits by subcutaneous inoculation (Douglas, 1962). After infection, rabbits were either released, or held in wire mesh cages where they would be readily accessible to mosquitoes. In all States, but especially in Victoria, efforts were made to time inoculation campaigns to coincide with the expected appearances in different parts of the State of mosquitoes known to be good vectors.

To simplify inoculations by landholders, Sobey *et al.* (1967) developed a method of preparing dried virus mixed with an abrasive powder and packaging this in a folded polythene bag within an Alfoil sachet. This material was infective when placed under the eyelid, and was relatively heat stable. It was prepared and made available by the Glenfield Veterinary Research Station to landholders in New South Wales and Queensland (via the Queensland Lands Department). In a letter to the Commonwealth Serum Laboratories,

COMMONWEALTH OF AUSTRALIA

Department of Health

COMMONWEALTH SERUM LABORATORIES

LIVING MYXOMA VIRUS (DRIED)

Prepared for the initiation of myxomatosis in rabbits.

The Living Myxoma Virus (Dried) contained in this package has been prepared by the Department of Bacteriology of the Australian National University as a 10% suspension of the tissues of rabbits suffering from Myxomatosis. The suspension has been dried and sealed in vacuum by the Commonwealth Serum Laboratories and these instructions for use have been drawn up in conjunction with the Commonwealth Scientific and Industrial Research Organization.

STORAGE.

In the dried form the virus is reasonably resistant and provided it is kept away from the light in a cool place should retain its virulence long enough for practical purposes. For the best results obtain the virus fresh and use soon after receipt. If temporary storage is necessary keep it in a refrigerator or ice chest if possible.

DIRECTIONS FOR USE.

Opening the Ampoule: To do this make a deep scratch with the small file provided on the narrow neck of the ampoule. Make a clean break by bending away from the file mark.

Reconstitution: Add 10 c.cs. of cold water to the ampoule. This will fill the ampoule to about ¾ inch below the shoulder. The dried material should resuspend in the water immediately but slight shaking may facilitate this. 10 ccs. is sufficient to inoculate 100 rabbits. **Reconstituted Virus** should not be kept for use next day. It must be used as soon as possible and during use should be kept away from heat and direct sunlight.

Inoculation of Rabbits: After reconstitution of the Virus material dip into it a match stick around the end of which has been wrapped a small piece of steel wool or cotton wool to form a swab. Then rub the wet swab on the inside edge of the lower eyelid of the rabbit. September, 1952

SUBSEQUENT CARE OF RABBITS.

(1) The inoculated rabbits may be returned to the warrens from which they were taken: or

(2) They may be kept in netting cages or enclosures in the vicinity of well populated warrens. This is a practicable means of maintaining an infected area when insect carriers (e.g., mosquitoes) are prevalent. Freshly infected rabbits should be introduced into the cages or enclosures every seventh day in order to replace rabbits which die and to maintain a continuous supply of infected animals.

(3) Lastly, rabbits may be inoculated by transferring pus from the eyes of infected rabbits by means of match swabs. It takes about seven days from the time of the original infection until the development of pus in the eyes.

Renown Press, Carnegie, UM 2264, Vict.

Fig. 7.5. Information and instructions for use on ampoules of freeze-dried myxoma virus produced for the initiation of myxomatosis in wild rabbits.

the Queensland Department of Lands noted that the sachet virus was much preferred to the ampouled virus ‘owing to the simplicity of infecting the rabbit with eye powder and the better keeping capability of the powder’[16]. However, the Commonwealth Serum Laboratories decided not to produce sachet virus. In the early 1970s producers at the Glenfield Station replaced sachets by freeze-drying the diatomaceous earth/myxoma virus mixture in ampoules containing a glass bead[17].

Inoculation campaigns, State by State

The State body responsible for inoculations and the intensity of inoculation campaigns varied in different States, so a short account will be given of each.

New South Wales

Rabbit control was the responsibility of local bodies known as the Pastures Protection Boards, to which Glenfield strain (from 1980 Lausanne strain) virus was distributed as requested. Some Boards acted merely as retailers of the virus, passing vials to landholders, some inoculated rabbits brought in by landholders on advertised dates, and a few sent a rabbit inspector around the district in a specially equipped vehicle, not only inoculating rabbits for landholders but also catching, infecting and releasing rabbits at strategic points.

Virus distribution to Pastures Protection Boards continued until 1982, when production in the Glenfield Veterinary Research Station ceased. In the early and mid-1950s, some tens of thousands of rabbits were inoculated and released annually; subsequently the numbers fell to some thousands each year, fairly well distributed around the State. Organized distribution of virus ceased in 1982, although landholders are still able to purchase freeze-dried material from CSL Ltd.

Victoria

Rabbit control was the responsibility of the Vermin and Noxious Weeds Destruction Board, Department of Crown Lands and Survey. As soon as myxomatosis escaped in the summer of 1951, Dame Jean Macnamara ensured that the Board followed up the epizootic by widespread inoculation, and the Victorian government set up and supported the most enthusiastic and prolonged inoculation campaigns of any State (Table 7.1). Special credit for this should be given to G.W. Douglas (Fig. 7.6). Besides profiting from his earlier work with the Wildlife Survey Section, Douglas maintained close liaison with the Commonwealth Serum Laboratories, and for some years provided them with myxoma lesion material for the preparation of freeze-dried virus.

Demonstration experiments were carried out regularly in the field to assess the mortality in samples of regional rabbit populations, and popular interest in inoculations was stimulated by the issue of pamphlets and other forms of publicity (Douglas, 1958a,b, 1962; Fig. 7.7) Most inoculations were performed by the Lands Department's 140 inspectors, each of whom had several centres in his district at which he carried out inoculations on advertised dates, and in addition several hundred ampoules of virus were purchased each year by Victorian landholders directly from the Commonwealth Serum Laboratories. Each annual campaign was planned to fit in with the seasonal pattern of epizootics in different parts of the State and with the indications of probable vector activity. In general, the pattern was as follows. In the north, along the River Murray and its tributaries, inoculations

Table 7.1. Details of the use of myxoma virus for inoculations in Victoria, 1951–1966[a].

Year	Total	Rabbits inoculated		Inoculation centres	Farmers using centres
		For landholders	On Crown Land		
1950–51	5,000[b]	—	—	—	—
1951–52	40,000	—	—	—	—
1952–53	12,000	—	—	—	—
1953–54	30,000	—	—	—	—
1954–55	25,000	—	—	—	—
1955–56	15,000	—	—	—	—
1956–57	26,000	—	—	—	—
1957–58	30,000	—	—	—	—
1958–59	12,761	12,761	1346	416	1901
1959–60	11,739	10,366	1373	354	1816
1960–61	6,959	5,457	1502	217	1055
1961–62	24,817[c]	19,605	5212	520	3040
1962–63	24,645	19,097	5548	672	3165
1963–64	18,432	14,356	4076	594	2696
1964–65	23,003	17,026	5977	606	3095
1965–66	25,413	19,494	5919	736	3545

[a]From Douglas (1965, 1968).
[b]Approximate, no other details available.
[c]Well-publicized introduction of Glenfield strain for inoculations.

Fig. 7.6. Geoffrey William Douglas (1925–1985). After graduating in Agricultural Science in the University of Melbourne in 1951, Douglas joined the Department of Crown Lands and Survey and worked on myxomatosis in Victoria for 27 years, including a secondment to the Wildlife Survey Section from 1954 to 1957. He was appointed Deputy Chairman of the Vermin and Noxious Weeds Destruction Board in 1959 and Chairman in 1972. He played a major role in the establishment in 1967 of the Keith Turnbull Research Institute in Frankston, which undertook continuing research on various aspects of myxomatosis, especially changes in virus virulence and in the genetic resistance of rabbits, between then and 1982. He was also responsible for planning the intensive and continuing programme of introductions of myxoma virus into the field.

were arranged in early spring, whereas in the central and southern districts they were usually carried out in November or December and again in January or February. However, this timetable was modified if seasonal conditions varied. In a dry year, when the activity of *An. annulipes* would be suppressed, inoculations in the north were postponed until November or December to take advantage of the expected presence of *Cu. annulirostris*. On the other hand, in a wet year inoculations might be advanced to September. In addition, to exploit conditions that promised to be exceptionally favourable for transmission, a mobile unit worked through various districts, inoculating and releasing rabbits that were captured at night by spotlighting. Field staff also carried out inoculation campaigns in what they regarded as the most difficult rabbit areas, such as north-western Victoria, where there were few landholders, or where it appeared that rabbits were increasing rapidly. Sometimes they used their own initiative, sometimes they consulted research staff.

WHAT CAN WE EXPECT FROM MYXOMATOSIS?

• *"MYXO" HAS DONE A GOOD JOB. **Every effort is needed to strike a knock out blow** while the rabbit is down.*

When myxomatosis was first released, warnings were given that the disease would only be one of the methods used to control rabbits.

This same warning is given today. In spite of the continual war against this persistent enemy, the rabbit remains one of our greatest pests.

In 1950 the introduction of the myxomatosis virus provided welcome reinforcement to conventional methods of control. Over the seven years that followed, rabbits were reduced from plague proportions to light infestations with comparatively little effort. This eased the problem for the time being.

But with stocking rates now at record levels, many farmers are asking "How much can we expect from myxomatosis in the future?" Before answering this, let us look at the changes that have occurred since myxomatosis was released.

by G. W. DOUGLAS, B.Agr.Sc., Co-ordinator of Rabbit Control in Victoria

ACQUIRED IMMUNITY

In the early outbreaks from 1951 to 1953, myxomatosis usually swept through the dense rabbit populations in a few weeks, leaving only a small fraction of the original population untouched. This reduction in rabbit numbers was dramatic and unfortunately, the disease was hailed as the complete answer to the problem.

Later on it was quite evident that some of the rabbits which survived the initial outbreaks had recovered from the disease.

So this was the first change noticed — that rabbits recovered from myxomatosis and acquired an immunity.

WEAKER VIRUS

These rabbits survived because they had been infected with the field strain of virus which was weaker than the standard laboratory strain released in 1950-51.

Collecting blood from a rabbit to test for immunity.

Recently the rabbit flea has been imported from England to assist in spreading myxomatosis. If tests are satisfactory, the flea may be released in the field.

THE FUTURE

We can expect the virus to be a valuable weapon in the fight against the rabbit for some time yet. However we do not know to what extent the virus in the field will drop in killing power or how the change in genetic resistance in the rabbit will affect myxomatosis in the future.

But one thing is certain – Myxomatosis outbreaks should be followed up with an intensive poisoning campaign, particularly with 1080 (the newest rabbit poison), warren ripping and fumigation.

From past experience we know what the rabbit can do, and we realize that our control measures did not cope with the pest before 1950. Surely the stakes are so high that we must make every effort to hit the rabbits hard while they are down with every means at our disposal.

Although myxomatosis is the spearhead of the attack, conventional methods must still be used to support the campaign.

Fig. 7.7. First and last pages of a four-page publicity pamphlet issued to landholders in Victoria by the Vermin and Noxious Weeds Destruction Board in 1958. Notice the emphasis on the need for follow-up poisoning, warren ripping and fumigation.

Queensland

Inoculations were organized by the Coordinating Board, a body responsible to the Minister for Public Lands for vermin and noxious weeds control. For the first season (1951–52) virus was prepared in Brisbane; subsequently it was usually purchased from the Commonwealth Serum Laboratories; sometimes, as sachet virus, from the Glenfield Veterinary Research Station. The Department of Agriculture and Stock made its field staff available to inoculate rabbits brought in to advertised points in rabbit-infested areas, but after a few years the landholder response was so poor that inoculations were carried out only after requests from landholders.

South Australia

In the early 1950s there was no effective centralized rabbit control authority in South Australia, and only one organized inoculation campaign was conducted (Lines, 1952). In the spring of 1951 the local Department of Agriculture advertised dates at which landholders could bring in rabbits for inoculation to some 50 centres in the agricultural belt, from Eyre Peninsula in the west to Mount Gambier in the south-east. Virus was prepared in Adelaide, and because there was some trouble with the early batches, the western inoculation centres were reopened in December and January. Some 20,000 rabbits were inoculated in 1951, but organized inoculation campaigns were not conducted in subsequent years, although some landholders carried out their own inoculations.

Western Australia

Inoculations were organized by the Chief Vermin Control Officer in the Department of Agriculture (Calaby *et al.*, 1960). In the early spring of 1951 and during the next two years, a total of 21 inoculation centres were set up as a network covering that part of the State in which the rabbit was an economic problem. The sites were selected with regard to rabbit abundance and potential mosquito breeding places. In dry areas, sites were selected where rabbits were infested with stickfast fleas (*Echidnophaga* spp.). Each centre was in the charge of an attendant, who initially maintained infection in caged rabbits from which captured rabbits were infected, and subsequently carried out inoculations of ampouled virus. The inoculated rabbits were either kept in cages near streams or released in warrens. The centres were closed down for the season when an outbreak occurred or if it became apparent that there was little chance of a local outbreak. All the centres were closed down in 1954, after a total of some 27,000 rabbits had been infected and released. After 1952 ampoules of the Commonwealth Serum Laboratories virus were sold to individual landholders, but few were purchased after 1955.

After the centres had been closed, many thousands of rabbits were inoculated by mobile units and in drives conducted by the more active local Vermin Boards; these operated throughout the year. Although no officers were employed full time on rabbit inoculations after 1958, inoculations continued at a reduced level until the early 1980s.

Knowing his State and its climate, the Chief Vermin Control Officer, A.R. Tomlinson, did not expect dramatic results, but hoped to get the virus seeded throughout the rabbit-infested south-western part of the State, so that it would be present in most regions if conditions were favourable, a strategy that was successful, although unfortunately epizootics occurred only if there was exceptional and unseasonable rain. Because of the sporadic and unpredictable effects of myxomatosis, poisoning by 1080 was introduced in 1954–55 as the principal official method of rabbit control (Calaby *et al.*, 1960).

The value of inoculation campaigns

The natural spread of myxomatosis during the period 1951–1953 demonstrated its potential for dispersal if there were adequate numbers of susceptible rabbits and conditions favoured the activity of mobile vectors. It was difficult to evaluate the influence of inoculations in the areas covered by the initial spread, but clearly some of the early localized outbreaks in the coastal belt of New South Wales and Victoria were initiated in this way. By the third season, 1952–53, three-quarters of strains isolated from field cases were of Grade III virulence (Marshall and Fenner, 1960). The only 'marker' then available for inoculated virus, as distinct from enzootic virus, was that it should be of Grade I virulence. Such strains were occasionally recovered during the next six years (16% of strains tested in 1953–54 and 1954–55) and they probably represent virus derived from cases arising from inoculation campaigns.

In September 1954 experiments were set up at Lake Urana (see Fig. 8.1, p. 183) to test the efficacy of the strain of myxoma virus that had been used to introduce myxomatosis into Europe in 1952, because at the time this highly virulent strain (Lausanne) appeared to be maintaining its high virulence over several seasons (Fenner *et al.*, 1957). Providentially, they also provided an opportunity to compare the kill and persistence of viruses of virulence Grades I and III. The results at one site are illustrated in Fig. 7.8. Rabbits were captured, inoculated with the Lausanne strain, marked, and released over a period of six weeks in November–December 1954. Early in November, one of the rabbits captured for inoculation proved to be suffering from myxomatosis caused by the enzootic strain, of Grade III virulence. Thus the study was transformed into a comparison of the epizootic pattern and long-term persistence of two strains of differing virulence.

A major epizootic began in late November, and as many unmarked sick rabbits as possible were captured for testing whether they had been infected with the Lausanne or the local enzootic strain of myxoma virus. Analysis of the results was

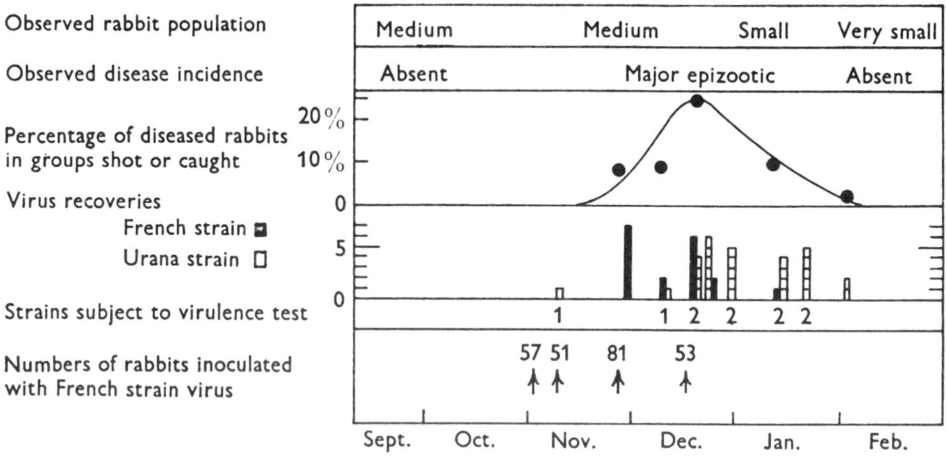

Fig. 7.8. The epizootic of myxomatosis that followed the introduction of the Lausanne strain of myxoma virus (French strain) into the wild rabbit population at Lake Urana in October–November 1954, showing the schedule of inoculation with the Lausanne strain, the appearance of a local strain early in November 1954, the incidence of myxomatosis, the type of virus as determined by the intradermal screening test and the size of the rabbit population. From Fenner *et al.* (1957), with permission.

simplified by the fact that intradermal inoculation of material from the captured rabbits on the shaved backs of laboratory rabbits produced quite different lesions – a large protuberant lesion in the case of the Lausanne strain and a flat lesion by the enzootic strain, so that inoculation of a single rabbit could provide data on several isolates. The sharp outbreak which began late in November was over by early February. The majority of sick rabbits captured during the first three weeks proved to be infected with the Lausanne strain; during the next two weeks strains of both kinds were found, and as the epizootic faded, only Australian enzootic strains were found. Significantly, only the latter survived through the winter and caused an intense epizootic in the following spring. A similar picture was observed at the other two sites where similar experiments were carried out.

The conclusion from these experiments was that in the presence of enzootic myxomatosis caused by slightly attenuated strains of myxoma virus, inoculation of virulent virus was probably not worth the organizational effort. However, after consideration of the whole scene, Ratcliffe concluded that even though the effort might be wasted in a season when vector activity was not at a high level, it was worth doing. Experience over the first five years had shown that a substantial part of the total benefit was due to the summation of many small, localized epizootics; furthermore, introductions of virulent virus provided the only strategy available for slowing down the development of genetic resistance in rabbits (Sobey, 1960).

Field Studies of Vectors[f]

Immediately after the dramatic spread of myxomatosis along the streams throughout the Murray–Darling Basin in early 1951, Ratcliffe arranged for an intensive

[f]Much of the information in this section is derived from Fenner and Ratcliffe (1965). The data derive from observations by CSIRO field team (from the Wildlife Survey Section and the Division of Entomology) comprising J.H. Calaby, A.L. Dyce, B.V. Fennessy, K. Myers, R. Mykytowycz and E.J. Waterhouse; G.W. Douglas of the Victorian Vermin and Noxious Weeds Destruction Board and D.J. Lee, of the Sydney School of Public Health and Tropical Medicine.

investigation of insect vectors[18], and in parallel Fenner and M.F. Day of the CSIRO Division of Entomology undertook detailed experimental investigations of the mechanism of mosquito transmission of myxomatosis (see p. 80). As Aragão (1943) and Bull and Mules (1944) had shown, transmission was mechanical, and tests with a wide range of mosquitoes, fleas, biting flies, ticks, lice and mites showed that any insect that bit two rabbits in succession was able to transmit the disease. Following his experimental studies, Fenner (1953b) suggested that mosquitoes could be regarded as 'flying pins', to draw attention to three important features of transmission: (i) in contrast to classical arbovirus infections[g], there was no vector specificity; (ii) there was no extrinsic incubation period (the period during which arboviruses multiplied in the vector); and (iii) the importance of the concentration of virus in the superficial cells in the skin lesions of infected rabbits and the irrelevance of viraemia in contaminating the mouthparts of vector insects.

The geographic–ecological background

In 1896 Spencer proposed a zoogeographic subdivision of Australia into three broad regions: Torresian, Bassian and Eyrean Provinces (Fig. 7.9). Much more precise subdivisions have been developed since then (see Thackway and Cresswell, 1995), but Spencer's subdivision is drawn at an ecological scale useful for considering the distribution of mosquito vectors of myxomatosis. Rabbits are a pest in Australia primarily in the temperate southern half of the continent, in the Bassian and Eyrian provinces. Most investigations on vectors of myxomatosis were carried out in Victoria and New South Wales, and to a lesser extent in south-western Western Australia. In the eastern States, the Great Dividing Range delimits the watersheds of streams flowing eastwards into the Pacific and into the Murray–Darling Basin to the west or north. As the Great Dividing Range turns westwards in Victoria it separates the northward-flowing tributaries of the River Murray from the shorter southward-flowing rivers. From the point of view of mosquito breeding, we can differentiate between the coastal region, on the seaward side of the Great Dividing Range, the Bassian Province, from the mountainous/hilly area of the Great Dividing Range itself and the relatively flat inland Eyrian Province. The moist temperate areas of South Australia and Western Australia are ecologically comparable to the Bassian Province.

The Eyrian Province

In the eastern States the Eyrian Province largely coincides with the Murray–Darling basin. The southern part of it, called the Riverina, is a relatively well-watered level or gently undulating plain. To the east the Riverina is bounded by the western slopes and foothills of the Great Dividing Range. Further inland it merges into the arid zone, where few entomological investigations were made. Except in more hilly areas of the Riverina, even moderately heavy falls of rain leave local accumulations of water that may persist for weeks. Temporary streams empty into swamps which last for months, and permanent streams flowing across this flat land have developed on their flood plains a system of lagoons (locally called billabongs) in many of which the water persists for very long periods. Waters in the Riverina are exploited by two dominant groups of mosquitoes with completely different ecological strategies. The main breeders in fluctuating waters (casual water and the edges of large swamps and billabongs), are *Ae. sagax* and *Ae. theobaldi,* which lay resistant eggs in the mud at the water's edge and develop in any water that persists for 10 days. Further east they are accompanied by *Ae. alboannulatus,* with similar

[g]Arbovirus infections are those in which the virus multiplies in the insect vector, the period between acquisition of virus by a blood meal and its presence in the salivary fluid in quantities adequate to infect a vertebrate host being called the 'extrinsic incubation period'. The 'intrinsic incubation period' of such infections is the interval between the infective bite and the first occurrence of symptoms.

Fig. 7.9. Scheme of zoogeographic provinces of Australia originally proposed by Spencer (1896).

patterns of behaviour. When conditions are favourable (when water rises after rains or the rivers flood) mass hatches of the eggs of these species make them the most important day-biting mosquitoes of the inland and the most obtrusive pests for man and stock. However, for several reasons, they transmit myxomatosis poorly; they usually prefer larger hosts than the rabbit, they are very persistent in obtaining a blood meal (and are therefore less likely to bite two rabbits in quick succession), and the life spans of individual Aedine mosquitoes are relatively short. The most abundant mosquitoes breeding in the deeper, more stable water, usually with emergent vegetation, are *An. annulipes, Cu. annulirostris* and *Cu. pipiens australicus.* All three species lay their eggs on the surface of suitable water, bite from dusk throughout the night, and are important vectors of myxomatosis.

The Bassian Province

On the coastal side of the Great Dividing Range in the eastern States, and in ecologically comparable areas in South Australia and Western Australia, the country is generally more hilly and the tree cover often much denser than in the Eyrean Province. The annual rainfall is higher but breeding waters are less extensive and the general level of mosquito abundance is much lower. Many species of mosquito may be found in any particular area, but except near coastal lagoons it is rare for any one species to reach overwhelming abundance. One or sometimes both of the main vectors of inland Australia (*An. annulipes* and *Cu. annulirostris*) occur on the seaward side of the Divide. In these areas it was often difficult to identify the main vector; several species of mosquito could be found feeding on rabbits but the onset of outbreaks of myxomatosis rarely coincided with peak activity of any one of them. When present, *An. annulipes* and *Cu. annulirostris* undoubtedly played a role, but other species were also probably involved in most outbreaks.

Culex annulirostris. This mosquito is a warm-country species and is common throughout most of the Eyrian Province, but absent from Tasmania, in Victoria it is rare south of the Divide, and it is virtually absent from south-eastern South Australia, although common in south-western Western Australia. It is a dusk and night biter that readily attacks humans, and precipitin tests of engorged adults showed that cattle and rabbits were the preferred hosts (Lee *et al.*, 1954).

Cu. annulirostris breeds in permanent or persistent water, especially in the river-frontages, swamps and lagoons of the Murray–Darling river system, from the plains of southern Queensland to the Murray mouth in South Australia. In Victoria it also breeds in stockwater dams and irrigation channels. In the north it may be active at any time of the year if conditions are favourable. Elsewhere it is usually active in mid- to late summer, which in southern Australia is the dry season. Under these conditions it is usually found in the vegetation fringing its breeding grounds, since it is sensitive to low atmospheric humidity. However, in exceptional seasons, when there is water lying in many areas away from the rivers in summer, it breeds in any persistent groundwater and becomes wide-ranging in its flight behaviour. Thus in the initial spread of myxomatosis in the summer of 1951, disease activity was confined to the vicinity of streams in the southern States, but in Queensland, where there had been exceptionally heavy summer rains, myxomatosis occurred throughout the area inhabited by rabbits.

In the vicinity of its permanent breeding grounds *Cu. annulirostris* is the most reliable and probably the most important vector. Even in its most favoured habitats, it is almost invariably associated with *An. annulipes* and *Cu. pipiens australicus*. However, since these three species attain peak abundance at different times, it will usually be obvious which is the main vector in any particular outbreak.

ANOPHELES ANNULIPES. This mosquito occurs Australia-wide and is probably the single most important vector of myxomatosis. Its vector capacity was demonstrated by the isolation of myxoma virus from numerous batches of mosquitoes, in December 1951 and November–December 1952 (Myers *et al.*, 1954). In a general survey of the first two years of myxomatosis, observations in the Eyrian region by the Wildlife Survey Section team (Ratcliffe *et al.*, 1952), noted that the presence of *An. annulipes:*

> as revealed by the discovery of adults in warrens or local breeding, [was] the one factor common to all those parts of the affected area in which we [were] able to make adequate and timely observations.

This observation was repeatedly confirmed in many parts of Victoria and New South Wales and in Western Australia.

An. annulipes is a dusk and night biter and tests on blood meals showed that rabbits, cattle and sheep were its preferred hosts. It breeds in swamps and billabongs, in the extensive surface waters present in the Riverina after heavy rain, and even in the slowly moving vegetated water on the margins of rivers and creeks. Adults are attracted to areas carrying dense rabbit populations and it habitually exploits rabbit burrows for daytime harbour. Using an ingenious cone-trap, Myers (1956) showed that it often attacked rabbits while underground, especially under arid conditions. Additional studies in the Riverina and in Victoria showed that it was possible to correlate the onset of epizootics not merely with the abundance of this vector but with the commencement of its large-scale attacks on rabbits.

It is a strong flier, and has been found up to 20 km from the nearest breeding site, and it is long-lived, adults sometimes being found in warrens two months after local breeding waters had dried up. Adult *An. annulipes* are present at all times of the year, with numbers peaking in spring and autumn. Along the Murray and its tributaries, where both *An. annulipes* and *Cu. annulirostris* are abundant in season, if conditions favour a massive spring emergence of *An. annulipes* an intense epizootic was likely to develop in October–November, leaving few susceptible rabbits for *Cu. annulirostris* to infect when it appeared in large numbers in December. In some southern parts of Victoria, beyond the southern limit of *Cu. annulirostris,* the autumn wave of *An. annulipes* was generally responsible for the extensive outbreaks, and disease activity often continued well into the winter, presumably because of continuing, low-intensity mosquito transmission.

Culex pipiens australicus. This species is co-extensive with the rabbit, on the mainland and in Tasmania, but is virtually absent from the more arid areas. Precipitin tests on blood from engorged females reveal that rabbits and poultry are preferred hosts. Evidence of its vector activity is less precise than with the mosquitoes just described, and derives from the information on its preferred hosts and from the correlation of its local abundance with disease outbreaks, especially in western and north-western Victoria. It is active only at and after dusk.

Cu. australicus breeds in many of the habitats favoured by *Cu. annulirostris* and *An. annulipes.* Where lagoons went through a regular seasonal cycle, becoming stabilized after winter floods, it tended to start breeding some weeks before *Cu. annulirostris.* Like *An. annulipes*, it showed spring and autumn peaks of abundance, and in all but the most humid parts of its range it sheltered in rabbit burrows during the day and occasionally fed on rabbits underground.

Other mosquitoes. Several other species of mosquitoes are probably vectors in different parts of Australia and in particular seasons; a detailed account of these is provided by Fenner and Ratcliffe (1965).

Other blood-sucking diptera

Two species of Simuliidae, *Simulium melatum* and *Austrosimulium furiosum,* are proven vectors, and although of minor importance compared with mosquitoes, they are capable of initiating and maintaining quite intense localized epizootics. Early in 1951, in a locality on the River Murray, Mykytowycz (1957) noticed that rabbits which exhibited very large swellings on one or both ears were being bitten on the ears by *Simulium melatum.* Experiments with caged rabbits confirmed that this insect bit them on the ears, producing similar signs. *Austrosimulium furiosum* was also demonstrated to be a vector, and it appeared to be the principal vector in a number of localized outbreaks in stands of timber away from rivers in the spring of 1952. Further, it has an exceptional flight range, and was probably responsible for widespread dispersal of myxoma virus, notably in the 1952–53 summer.

A few other small diptera, notably the ceratopogonids and species of *Phlebotomus,* are confirmed rabbit feeders and are probably occasional vectors.

Fleas and other ectoparasites

Bull and Mules (1944) had demonstrated experimentally that the stickfast flea (*Echidnophaga myrmecobii*) and several species of mosquitoes could transmit myxomatosis mechanically, and in some of their early field experiments they obtained evidence of the probable importance of stickfast fleas as vectors. In subsequent correspondence with Ratcliffe, Bull repeatedly mentioned the possibility that the stickfast flea might be a vector, but downplayed the vector potential of mosquitoes.

When myxomatosis escaped in 1950–51 it was obvious that a highly mobile insect vector was involved in its transmission, and *Cu. annulirostris* and *An. annulipes* were soon demonstrated to be efficient vectors. The timing of this epizootic points to *Cu. annulirostris* as the main vector. Subsequently localized outbreaks occurred in many parts of Australia, but in none of them could stickfast fleas be incriminated as vectors, even when the rabbits were heavily infested (Calaby *et al.*, 1960).

Three other ectoparasites commonly infest rabbits in Australia, the louse *Haemodipsus ventricosus* and the mites *Cheyletiella parasitovorax* and *Listrophorus gibbus.* The first two ectoparasites have been shown to be capable of transmitting infection, but it is unlikely that either could originate or maintain a large outbreak (Mykytowicz, 1958). Occasionally, in Tasmania and in the Australian Capital Territory, outbreaks appear to be limited or delayed by netting fences, suggesting that transmission must be predominantly by exchange of ectoparasites. Winter outbreaks occur when rabbits are breeding (fighting, mating, nuzzling), a time when exchange of ectoparasites might be expected to increase.

Proposal to Introduce the European Rabbit Flea

Because of Bull's repeated comments on the vector capability of the stickfast flea, Ratcliffe had early discussed with his colleagues the possibility of introducing the European rabbit flea, *Spilopsyllus cuniculi,* and distributing it amongst Australian wild rabbits[19]. Reports of the effectiveness of *S. cuniculi* as a vector in England (see p. 228) revived the idea of introducing fleas, in the hope that they might spread myxomatosis in some places where mosquitoes had not been effective.

However, in response to a request from Ratcliffe in May 1955, the Director-General of Health refused permission for their importation from England[20]. After intensive lobbying and with the unanimous support of the Australian Agricultural Council, CSIRO succeeded, in February 1957, in reversing this decision, but the Director-General of Health insisted that the fleas should be kept in a laboratory in Canberra, under strict quarantine. Under these conditions A.L. Dyce (Fig. 7.10) was unable to get the fleas to breed. Subsequently Mead-Briggs and Rothschild in England showed that breeding of *S. cuniculus* depended on gravid female fleas obtaining blood meals from a pregnant rabbit (see p. 85). European rabbit fleas were again imported into Australia in 1966; work on their breeding and release is summarized in Chapter 8 (see p. 182).

Fig. 7.10. Alan Lindsay Dyce (1923–). After serving in the Royal Australian Navy for 4 years during the Second World War, Dyce graduated with a BScAgric (Hons I) from the University of Sydney in 1952, majoring in entomology. In 1952 he joined the CSIRO Division of Entomology, and was immediately transferred (initially on loan, later confirmed) to the Wildlife Survey Section to work on the insect vectors of myxomatosis. He carried out important work on the biology and vector potential for myxomatosis of mosquitoes, simulids, phlebotomine sandflies and ceratopogonids in New South Wales and Queensland. In 1961 he transferred to the Division of Tropical Animal Production, where he worked on *Culicoides*, the vectors of bluetongue and ephemeral fever viruses. Since retirement in 1987 he has worked as a CSIRO Honorary Fellow on the taxonomy and systematics of *Culicoides*.

Myxomatosis in Victoria: 1957–1966

After 1956 the collection and collation of field data on an Australia-wide basis almost ceased, so that the history of myxomatosis between 1956 and 1966 can be presented in some detail only for the

State of Victoria[h]. The picture there (Douglas, 1965) provides a reasonable overview of the Australia-wide scene in the better agricultural land, but does not portray the situation in the more arid rangelands of the interior. Douglas estimated that by the mid-1960s the rabbit population over Victoria as a whole fluctuated around 20% of the pre-myxomatosis level, but varied greatly from region to region. In most areas outbreaks of low intensity occurred every year, with widespread epizootics every four years or so. In most parts of the State organized poisoning on a substantial scale was carried out to deal with infestations that built up over years of low myxomatosis activity.

Predation, mainly by foxes and eagles, became much more important after the emergence of attenuated strains of myxoma virus. Although such strains often did not kill rabbits, they made the animals so sick and helpless that they fell easy victims to the predators which congregated in areas of myxomatosis activity[21].

Tests on the Virulence of Field Isolates, 1951–1967

The discovery of attenuated strains of myxoma virus among rabbits naturally infected in 1952 and early 1953 (Fenner, 1953a; Mykytowicz, 1953) clearly posed problems as to whether myxomatosis would continue to be an effective method of rabbit control. It was essential that regular assessment of changes in the virulence of field strains should be instituted, and this work was undertaken initially by the team at the Australian National University[i] and later, for Victoria only, by scientists at the Keith Turnbull Research Institute[j].

For a virus being used for the biological control of a vertebrate pest, 'virulence' is best equated with capacity to kill (lethality). Testing viruses of rabbits for this attribute thus posed a problem, for it was impossible to inoculate and hold large enough groups of rabbits to provide statistically reliable estimates of virulence based on case-mortality rates (such as might be done with mice). A search was made for properties that might be correlated with lethality as determined by animal inoculation, but none was found. It was therefore decided to use survival time, as measured on groups of five or six rabbits, with statistical adjustments if there were one or two survivors, as a surrogate for the case-fatality rate (Fenner and Marshall, 1957). As described in Chapter 5, five (later six) 'virulence grades' were established, which have been used ever since.

It took a few years to organize systematic collection of large numbers of specimens of lesion material from infected rabbits, but between 1955 and 1959 over 100 strains of myxoma virus were tested by the ANU team each year (Marshall and Fenner, 1960). Lesion material from sick or dead rabbits was collected by officers of the Wildlife Survey Section of CSIRO and the Pastures Protection Boards of New South Wales, and inspectors of the Department of Crown Lands and Survey of Victoria. For most seasons between 1959–60 and 1966–67 testing was limited to strains collected in Victoria and carried out under

[h]Under pressure from Dame Jean Macnamara, Victoria had developed a very strong State rabbit control organization. The work there was directed by G.W. Douglas. Special laboratories for the experimental study of myxomatosis were set up (the Keith Turnbull Research Institute in Frankston), and valuable data continued to be reported at the Vermin Control and Pest Conferences that replaced the earlier conferences on myxomatosis, and in the scientific literature (see Chapter 8).

[i]From 1951 to 1965 the team working on myxomatosis at the Department of Microbiology at the Australian National University comprised F. Fenner, I.D. Marshall, G.M. Woodroofe and two or three laboratory assistants.

[j]Laboratory work on myxomatosis in Victoria was begun by G.W. Douglas in 1958, and was undertaken at the Keith Turnbull Research Institute from 1967 to 1982. The principal scientists at the Institute were R.C.H. Shepherd and J.W. Edmonds, with support from I.F. Nolan and A. Gocs.

the direction of G.W. Douglas. The results of tests from 1951 to 1966–67 (Marshall and Fenner, 1960; Fenner and Woodroofe, 1965; Douglas, 1968) are shown in Table 7.2.

Although myxoma virus may be carried for long distances by infected mosquitoes, there is no doubt that strains of reduced virulence soon appeared independently in many widely separated areas, all over Australia. Experience in Tasmania and Western Australia has been strikingly similar to that in south-eastern Australia, with strains of Grade III virulence becoming established as the dominant group by about 1955 and strains of very high virulence becoming uncommon after that date. The mechanism by which this change occurred is explored in Chapter 14.

Changes in the Genetic Resistance of Rabbits, 1953–1966

Apart from the presence of less lethal viruses, recovery from myxomatosis could be attributed to increased resistance, due to physiological or environmental factors or an increase in genetic resistance. An obvious physiological reason for such increased resistance could be the presence of maternal antibodies at the time of infection. Experiments in the laboratory indicated that passive immunity was effective in mitigating the severity of the disease in rabbits that it did not completely protect, but only in kittens less than two months old (Fenner and Marshall, 1954); this would account for only a small proportion of milder cases. The principal environmental factor to affect the fate of myxomatous rabbits was high ambient temperatures (Marshall, 1959), but rabbits were recovering in places and at times when this factor was not operating. The emergence and within a few years the dominance of field strains of myxoma virus that allowed some 10% of rabbits to survive would increase the number of survivors compared with what had been observed during the early outbreaks, but

Table 7.2. The virulence of strains of myxoma virus isolated from Australian wild rabbits between 1951 and 1966–67 (expressed as percentage of isolates tested).

Virulence Grade	I	II	III	IV	V	
Degree of virulence	Extreme	Very high	Moderate	Low	Very low	
Mean survival time (days)	<13	13–16	17–28	29–50	—	
Case-fatality rate (%)	99.5	95–99	70–95	50–70	<50	Number of strains tested
1950–51[a]	>99					
1951–52[a]	33	50	17	0	0	6
1952–53[a]	4	13	74	9	0	23
1953–54[a]	16	25	50	9	0	12
1954–55[a]	16	16	42	26	0	19
1955–56[a]	0	3	55	25	17	155
1956–57[a]	0	6	55	24	15	165
1957–58[a]	3	7	54	22	14	112
1958–59[a]	0.5	20	57	14	8	179
1959–60[b]	0	22	45	33	0	18
1960–61[b]	2	10	60	19	9	126
1961–62[b]	2	8	69	18	3	111
1962–63[b]	3	19	53	22	3	140
1963–64[b]	0	1	57	35	7	142
1964–65[b]	0	1	51	45	3	135
1965–66[b]	1	0	77	21	1	125
1966–67[b]	0	0	62	37	1	191

[a]Data from Victoria and New South Wales (Marshall and Fenner, 1960).
[b]Data from Victoria, provided by R.C.H. Shepherd.

this did not account for the increasing numbers of survivors that were being seen. However, the survival to breed of some 10%, rather than less than 1%, of infected rabbits foreshadowed the selection of rabbits with a greater genetic resistance to the disease. Experiments were therefore undertaken to test wild rabbits for their genetic resistance.

Tests for genetic resistance in wild rabbits

The plan was to catch a large number of young rabbits just after they emerged from their burrows in early spring, before myxomatosis had occurred, in a number of different areas (Marshall and Fenner, 1958). They were raised for about two months in mosquito-proof accommodation before being challenged with myxoma virus. This was to be repeated for several successive years. When these investigations were initiated it was expected that the acquisition of a high level of genetic resistance might take several decades. In order to detect lower levels of resistance, it was decided to inoculate them with a small dose (about five rabbit-infectious doses) of a strain of Grade III virulence. A large batch of ampoules of such a suspension was prepared, titrated and stored in dry ice. To ensure that the captured rabbits were free of maternal antibody, they were kept until they were about four months old and weighed at least 1.2 kg before being challenged. Groups of rabbits whose parents had never been exposed to myxomatosis were used as controls, and gave uniform case-fatality rates and survival times each year, throughout the experiment. The results, summarized in Table 7.3, were an unpleasant surprise[k] as far as continued reliance on myxomatosis as the principal method of rabbit control was concerned. It was thought that the unexpectedly skewed results in the rabbits tested in 1955 might have been due, in part, to the fact that the highly virulent Lausanne strain had been released in the area the previous season and was credited with killing some 70% of the population (Fenner *et al.*, 1957), possibly imposing an even more stringent selection for resistant rabbits as the progenitors of the generation that was tested in 1955.

Marshall and Douglas (1961) extended these experiments by testing rabbits collected in 1957 from Maryvale and Ouyen in the State of Victoria (both sites that had experienced annual epidemics since 1951–52), and Lake Urana in 1958 (two years after the last collection listed in Table 7.3). Subsamples of the Victorian rabbits were tested in the Canberra animal house (temperature 20–23°C) and in field enclosures at Maryvale and Ouyen

[k]From the time of its escape in 1950, Burnet (1952) and Fenner (1953b) had realized that the very high case-fatality rate was unlikely to be sustained, because of attenuating mutations in the virus and/or selection for genetic resistance in the rabbits. However, both scientists thought that both of these processes would take decades to have a noticeable effect, hence the surprise.

Table 7.3. The severity of myxomatosis in groups of Australian wild rabbits inoculated intradermally with a small dose of a moderately attenuated strain of myxoma virus (Grade III virulence)[a].

Group	Number of epizootics to which population had been exposed	Symptomatology		
		Severe (including fatal)	Moderate	Mild
Wild rabbits before myxomatosis	0	93%	5%	2%
Lake Urana, 1953	2	95	5	0
Lake Urana, 1954	3	93	5	2
Lake Urana, 1955	4	61	26	13
Lake Urana, 1956	5	75	14	11

[a]Data from Marshall and Fenner (1958).

Table 7.4. The mortality rates from myxomatosis due to highly and moderately virulent strains of myxoma virus in non-immune rabbits from areas where there had been six successive annual epidemics of myxomatosis[a].

Group	Conditions	Virus	Number of rabbits	Mortality rate (%)
Controls[b]	Animal house	Grade III[d]	70	67
Ouyen	Animal house	Grade III	84	21
Ouyen	Field enclosure	Grade III	106	6
Maryvale	Animal house	Grade III	52	38
Maryvale	Field enclosure	Grade III	103	7
Controls[c]	Field epidemics	Grade I[e]	many	>99
Ouyen	Field enclosure	Grade I	43	86
Maryvale	Field enclosure	Grade I	20	90

[a]Based on Marshall and Douglas (1961).
[b]Domestic rabbits.
[c]From Myers *et al.* (1954).
[d]Slightly less virulent than strain used for Table 7.3.
[e]Standard Laboratory Strain.

(maximum afternoon shade temperatures between a low of 16°C and a high of 38°C). The results (Table 7.4) show that the resistance of the rabbits had increased still further, and that the mortality rate was even lower among the rabbits in the field enclosure than in those in the animal house. When rabbits of a third subsample of each of the Victorian groups were challenged with the highly virulent Standard Laboratory Strain (which had a mortality of almost 100% in both laboratory and genetically unselected wild rabbits), their mortality rates were 86% and 90%. Clearly, annual exposure of groups of non-immune rabbits from areas that had experienced six successive annual epidemics of myxomatosis had a greatly increased resistance to the disease.

Breeding for resistance

The early recognition that wild rabbits were becoming resistant to myxomatosis led CSIRO to recruit W.R. Sobey (see Fig. 8.8, p. 196) to their Animal Genetics Section to study this phenomenon, and in May 1953 J.M. Rendel (Chief of the Section) and Sobey met at CSIRO Headquarters to discuss this research, which comprised a breeding programme with laboratory rabbits that had recovered from myxomatosis[22]. Initially attenuated strains were used for challenge, but as resistance increased virulent strains were used. Over the years Sobey reported progress findings to the annual conferences on rabbit control; his definitive results, published in 1969, are discussed in the next chapter.

The Proposal to Vaccinate Rabbits in Commercial Rabbitries with Fibroma Virus

Prior to the spread of myxomatosis, the only use of domestic rabbits in Australia was for research and diagnosis in medical laboratories and as pets. In 1951–52 several cases of myxomatosis occurred in laboratory rabbits in hospital laboratories in Melbourne and Sydney. Investigations concerning preparation of a vaccine were therefore made, using strains of rabbit fibroma virus provided by Dr R.E. Shope (Fenner and Woodroofe, 1954). An effective vaccine was produced using the Boerlage strain of fibroma virus (see p. 109). Between 1955 and 1961 79 ampoules of freeze-dried vaccine, each containing 100 doses, were issued by the Commonwealth Serum Laboratories to scientific organizations wishing to protect their laboratory rabbits[23].

Since it was essential to ensure that fibroma virus was not transmitted to wild rabbits, experiments were conducted concurrently on the efficiency of mosquito transmission (Day *et al.*, 1956). Although the titre of fibroma virus in skin lesions was high, it did not appear to be accessible to probing mosquitoes; only five positives were recorded from 1627 bites by 619 mosquitoes. Subsequently Dalmat and Stanton (1959) showed that fibromas in *Sylvilagus floridanus* (the natural host of fibroma virus) were infective because the tumours normally progressed to an advanced histological stage in which there were mature fibroma cells and large cytoplasmic inclusions in the epithelium overlying the tumour. Fibromas in *Oryctolagus cuniculus* never attain this stage, and become infectious for probing mosquitoes, except in infant rabbits or in adult rabbits treated with whole body X-irradiation or inoculated with 1,2,5,6-dibenzanthracene prior to inoculation with fibroma virus.

With the collapse of the wild rabbit carcass and fur industry due to myxomatosis, a few domestic rabbit farms were established in New South Wales, but not in other States. In New South Wales the issue of permits to keep domesticated rabbits was restricted to the Moss Vale Pastures Protection District until 1960, but was then relaxed somewhat. However, very few farms were established: 34 in 1959, 81 in 1960 and 352 in 1961[24]. Requests for fibroma virus to vaccinate these rabbits in commercial rabbitries were first made in 1960, and these were granted under stringent controls[25].

Early in 1961 an application from a commercial rabbit farmer to have fibroma virus registered for sale in South Australia aroused great concern in Victoria[26], leading to extensive negotiations beween CSIRO and the governments of Victoria and New South Wales. As a result, Dr R.E. Shope, the American virologist who had discovered the virus in 1932, was invited to Australia to report on the use of fibroma vaccine in commercial rabbitries. After visiting the eastern States for four weeks in March–April 1962, Shope issued a report[27] in which he concluded that:

> The indiscriminate and inadequately controlled use of fibroma virus in protecting domestic rabbits in commercial rabbitries against myxomatosis is considered to be a hazardous procedure. The possibility that fibroma virus used in this way might escape into the wild rabbit population and thus interfere with the effectiveness of rabbit control by myxomatosis is of definite but unknown magnitude.

Other aspects of Shope's report were critical of the changing emphasis of the Wildlife Survey Section's work from myxomatosis to other methods of rabbit control.

In July 1962 the Commonwealth Minister for Health issued a press statement in which he stated that he had instructed the Commonwealth Serum Laboratories to restrict its sale of fibroma virus to scientific laboratories[28]. On 22 October 1962 a conference was held at CSIRO Headquarters to discuss Shope's report, amongst other matters[29]. There was considerable criticism of the report on scientific grounds, but it was agreed that no further assistance should be provided by CSIRO to the commercial rabbit industry unless it was requested by the State government concerned. Shope's report was bitterly resented by Ratcliffe[30], who considered that it was scientifically inaccurate in relation to the potential danger of fibroma virus, had belittled the work of Australian scientists and especially of the Wildlife Survey Section, and saw in the whole exercise the hands of his old opponent, Dame Jean Macnamara. However, the legal position in the various States after this meeting was that sale of fibroma for use as vaccine was restricted to laboratory rabbits except in New South Wales, where its use in commercial rabbitries was permitted but strictly controlled. Nevertheless, the domestic rabbit industry collapsed, the number of permits falling from a maximum of 352 in 1961 to 47 in 1966 and 12 in 1977.

Effects of Myxomatosis on Agricultural Production

Rabbit control was introduced in Australia in order to mitigate what was regarded as Australia's major agricultural pest, the European rabbit. During the period covered by this chapter observations of the effects of myxomatosis were restricted to its influence on agricultural productivity, which were very considerable. Reid (1953) estimated that the value of increased wool and meat production due to myxomatosis in 1952–53 was £34 million ($A590 million in 1990 dollars), most of the increased production coming from New South Wales, South Australia and Victoria. A comparison of five-year averages before and after myxomatosis showed that in New South Wales wool production and the numbers of sheep and cattle slaughtered increased by 26%, 25% and 26% respectively (Waithman, 1979). Figures collected by Fennessy (1966) on a property basis in New South Wales and South Australia demonstrated increases of 26–50% in stock-carrying capacity following the removal of rabbits. Myers noted that landholders who used to spend some $2000 over a period of 4–6 weeks each year on rabbit control prior to 1951 were by 1960 spending some $100–$200 for 2–10 days work annually. Williams *et al.* (1995) provide a more detailed analysis of the economic impact of rabbits on pastoral and agricultural production.

Endnotes

[1]CSIRO Archives, Series 3, file WA6/3/3/9 Part VII. Conference on Myxomatosis, 28 July 1954.
[2]Proceedings of Symposium on 'Myxomatosis'. Journal of the Australian Institute of Agricultural Science (1955), 21(3), 130–151; 21(4), 250–253. Contains reviews by M.F. Day, F. Fenner, K. Myers, F.N. Ratcliffe and W.R. Sobey.
[3]CSIRO Archives, Series 3, WA6/3/3/9 Part VI. Conference on Myxomatosis, 8 July 1953.
[4]Basser Library Archives 143/5/T22. Review of Myxomatosis – 1954/55, by F.N. Ratcliffe.
[5]CSIRO Archives, WA6/3/3/9 Part I. Conference on Myxomatosis of Rabbits, 15 February 1951.
[6]CSIRO Archives, WA6/3/3/9 Part II. Technical Committee on Myxomatosis, 12 June 1951.
[7]CSIRO Archives, WA6/3/3/6 Part IV. Wildlife Survey Section Progress Reports No. 1–No. 5; 26 February 1951 to 25 May 1951.
[8]Wildlife Survey Section Circular Memoranda, 24 October 1951 and 6 December 1851.
[9]CSIRO Archives, WA6/3/3/9 Parts III, IV, VI, VII, VIII. Conferences on Myxomatosis, 31 July 1951, 17 June 1952, 8 July 1953, 28 July 1954, 18 September 1956.
[10]Australian Vermin Control Conferences: Perth, 1957; Hobart, 1960; Canberra, 1964; Melbourne, 1968; Canberra, 1973.
[11]Australian Vertebrate Pest Control Conferences: Canberra, 1978; Dubbo, 1983; Coolangatta, 1986.
[12]CSIRO Archives, WA6/3/3/7 Part I. Press releases on myxomatosis, 16 February 1951 to 12 December 1952.
[13]*Rural Research in C.S.I.R.O.* has been published quarterly by CSIRO since 1952, to provide a summary of recent research of practical significance, designed especially for the use of agricultural extension officers. It has included the following articles on myxomatosis:
March 1957: 'Myxomatosis Campaign Intensified'.
June 1959 'Under Review: Rabbit Control'.
June 1962: 'Rising Rabbit Numbers in the Riverina'.
September 1966: 'New Look at Myxoma Virus'.
[14]*Ecos* is a popular magazine produced by CSIRO since 1974, to cover its work on matters of environmental interest. It has included the following articles on myxomatosis:
Ecos 18, November 1978 'On the Trail of the Rabbit'.
Ecos 53, Spring 1987 'New Strategies in the Rabbit War'.
[15]CSIRO Archives, Series 3, file WA6/3/3/10 Part II (II). Memorandum: 'The Supply and Distribution of Myxomatosis Virus', 21 September 1951.
[16]CSL Registry, file 80/100. Letter from Acting Secretary of the Queensland Land Administration Commission to the Director, Commonwealth Serum Laboratories, 16 January 1970.
[17]Basser Library Archives 143/25/5A. Letter from Ted Batty, Technical Manager of the Virology Section of the Elizabeth Macarthur Agricultural Institute, to Fenner, 10 October 1997.

[18]CSIRO Archives, WA6/3/3/6 Part IV. Wildlife Survey Section Progress Report No. 3, 21 March 1951.
[19]CSIRO Archives, Series 3, file WA6/3/3/6 Part I. Letter from Ratcliffe to Clunies Ross, 13 July 1949.
[20]CSIRO Archives, Series 3, file WA6/3/3/4 Part VI. Letter from A.J. Metcalfe (Director-General of Health) to Ratcliffe, 22 November 1955.
[21]See Fenner and Ratcliffe, 1965, pp. 300, 303; observations by A.L. Dyce, G.W. Douglas and K. Myers.
[22]CSIRO Archives, file WA6/3/3/4 Part V. Letter from J.M. Rendel to CSIRO Headquarters outlining plans for Sobey's work, 9 April 1953.
[23]CSL Archives series 047. File 58/2419. Letter from P.L. Bazeley, Director of Commonwealth Serum Laboratories to F.W.G. White, Chairman of CSIRO, 28 February 1961.
[24]Basser Library Archives 143/5/Myx 3A. Letter from R.M. Watts, Director-General of Agriculture, New South Wales, to Fenner, 7 February 1978.
[25]CSIRO Archives, Series 3, file WA6/3/3/9 Part IX. History of fibroma (extracted from CSIRO files). (50 pages)
[26]Basser Library Archives 143/5/Myx 3A. Letter from G.W. Douglas to Fenner, 26 July 1961.
[27]Basser Library Archives 143/5/Myx 3A. The Shope Report. *Liaison Notes* No. 19, October 1962.
[28]Press release by Senator H.W. Wade, Commonwealth Minister for Health, 15 July 1962.
[29]CSIRO Archives, Series 3, file WA6/3/3/9 Part IX. Conference at CSIRO Headquarters on 22 October 1962, to discuss CSIRO rabbit research programmes.
[30]Basser Library Archives 143/5/Myx 3A. Letters from Ratcliffe to Shope, 5 June 1962. See also letter from Ratcliffe to Douglas Stewart, 28 February 1968 (Basser Library Archives 143/1/D1).

References

Aragão, H.B. (1943) O virus do mixoma no coelho do mato (*Sylvilagus minensis*), sua transmissão pelos *Aedes scapularis* e *aegypti*. *Memorias do Instituto Oswaldo Cruz* 38, 93–99.

Bull, L.B. and Mules, M.W. (1944) An investigation of *Myxomatosis cuniculi*, with special reference to the possible use of the disease to control rabbit populations in Australia. *Journal of the Council of Scientific and Industrial Research* 17, 79–93.

Burnet, F.M. (1952) Myxomatosis as a method of biological control against the Australian rabbit. *American Journal of Public Health* 42, 1522–1526.

Calaby, J.H., Gooding, C.G. and Tomlinson, A.R. (1960) Myxomatosis in Western Australia. *C.S.I.R.O. Wildlife Research* 5, 89–101.

Dalmat, H.T. and Stanton, M.F. (1959) A comparative study of the Shope fibroma in rabbits in relation to transmissibility by mosquitoes. *Journal of the National Cancer Institute* 22, 593–615.

Day, M.F., Fenner, F., Woodroofe, G.M. and McIntyre, G.A. (1956) Further studies on the mechanism of mosquito transmission of myxomatosis in the European rabbit. *Journal of Hygiene* 54, 258–283.

Douglas, G.W. (1958a) Some recent trends in myxomatosis. *Livestock Digest,* Spring 1958.

Douglas, G.W. (1958b) What can we expect from myxomatosis? *Dairyfarming Digest* 5, 23–26.

Douglas, G.W. (1962) The Glenfield strain of myxoma virus. Its use in Victoria. *Journal of Agriculture Victoria* 60, 511–516.

Douglas, G.W. (1965) A review of myxomatosis in Victoria, 1950–1965. *Journal of Agriculture Victoria* 63, 557–562.

Douglas, G.W. (1968) Observations on the virulence of field strains and on genetic resistance in wild rabbits in Victoria. In: *Australian Vermin Control Conference 1968.* Standing Committee on Agriculture of the Australian Agricultural Council, Melbourne, pp. 86–97.

Fenner, F. (1953a) Changes in the mortality rate due to myxomatosis in the Australian wild rabbit. *Nature* 172, 228.

Fenner, F. (1953b) Host parasite relationships in myxomatosis of the Australian wild rabbit. *Cold Spring Harbor Symposium on Quantitative Biology* 18, 291–294.

Fenner, F. and Marshall, I.D. (1954) Passive immunity in myxomatosis of the European rabbit (*Oryctolagus cuniculus*): the protection conferred on kittens born by immune does. *Journal of Hygiene* 52, 321–336.

Fenner, F. and Marshall, I.D. (1957) A comparison of the virulence for European rabbits (*Oryctolagus cuniculus*) of strains of myxoma virus isolated from the field in Australia, Europe and America. *Journal of Hygiene* 55, 149–191.

Fenner, F. and Ratcliffe, F.N. (1965). *Myxomatosis.* Cambridge University Press, Cambridge, 379 pp.
Fenner, F. and Woodroofe, G.M. (1954) Protection of laboratory rabbits against myxomatosis by vaccination with fibroma virus. *Australian Journal of Experimental Biology and Medical Science* 32, 653–668.
Fenner, F. and Woodroofe, G.M. (1965) Changes in the virulence and antigenic structure of myxoma virus recovered from Australian wild rabbits between 1950 and 1964. *Australian Journal of Experimental Biology and Medical Science* 43, 359–370.
Fenner, F., Poole, W.E., Marshall, I.D. and Dyce, A.L. (1957) Studies in the epidemiology of infectious myxomatosis of rabbits. VI. The experimental introduction of the European strain of myxoma virus into Australian wild rabbit populations. *Journal of Hygiene* 55, 192–206.
Fennessy, B.V. (1966) The impact of wildlife species on sheep production in Australia. *Proceedings of the Australian Society of Animal Production* 6, 148–156.
Lee, D.J., Clinton, K.J. and O'Gower, A.K. (1954) The blood sources of some Australian mosquitoes. *Australian Journal of Biological Science* 7, 282–301.
Lines, E.W.L. (1952) Some natural factors affecting rabbit populations in Australia. *South Australian Naturalist* 27, 28–30.
Marshall, I.D. (1959) The influence of ambient temperature on the course of myxomatosis in rabbits. *Journal of Hygiene* 57, 484–497.
Marshall, I.D. and Douglas, G.W. (1961) Studies in the epidemiology of infectious myxomatosis of rabbits. VIII. Further studies on the innate resistance of Australian wild rabbits exposed to myxomatosis. *Journal of Hygiene* 59, 117–122.
Marshall, I.D. and Fenner, F. (1958) Studies in the epidemiology of infectious myxomatosis of rabbits. V. Changes in the innate resistance of Australian wild rabbits exposed to myxomatosis. *Journal of Hygiene* 56, 288–302.
Marshall, I.D. and Fenner, F. (1960) Studies in the epidemiology of infectious myxomatosis of rabbits. VII. The virulence of strains of myxoma virus recovered from Australian wild rabbits between 1951 and 1959. *Journal of Hygiene* 58, 485–488.
Myers, K. (1954) Studies in the epidemiology of myxomatosis of rabbits. II. Field experiments, August–November 1950, and the first epizootic of myxomatosis in the riverine plain of south-eastern Australia. *Journal of Hygiene* 52, 47–59.
Myers, K. (1956) Methods of sampling winged insects feeding on the rabbit *Oryctolagus cuniculus* (L.). *C.S.I.R.O. Wildlife Research* 1, 45–58.
Myers, K., Marshall, I.D. and Fenner, F. (1954) Studies in the epidemiology of myxomatosis of rabbits. III. Observations on two successive epizootics in Australian wild rabbits on the riverine plain of south-eastern Australia 1951–1953. *Journal of Hygiene* 52, 337–360.
Mykytowycz, R. (1953) An attenuated strain of myxomatosis virus recovered from the field. *Nature* 172, 447–448.
Mykytowycz, R. (1957) Transmission of myxomatosis by *Simulium melatum* Wharton (Diptera: Simuliidae). *C.S.I.R.O. Wildlife Research* 2, 1–4.
Mykytowycz, R. (1958) Contact transmission of infectious myxomatosis of the rabbit *Oryctolagus cuniculus* (L). *C.S.I.R.O. Wildlife Research* 3, 1–6.
Ratcliffe, F.N., Myers, K., Fennessy, B.V. and Calaby, J.H. (1952) Myxomatosis in Australia. A step towards the biological control of the rabbit. *Nature* 170, 7–19.
Reid, P.A. (1953) Some economic results of myxomatosis. *Quarterly Review of Agricultural Economics* 6, 93–94.
Sobey, W.R. (1960) Myxomatosis: the virulence of the virus and its relation to genetic resistance in the rabbit. *Australian Journal of Science* 23, 53–55.
Sobey, W.R., Conolly, D. and Adams, K.M. (1967) Myxomatosis: the preparation of myxoma virus for inoculation via the eye. *Australian Journal of Science* 30, 233–234.
Spencer, W.B. (1896) *Summary of the Zoological, Botanical and Geological Results of the Expedition. Report of the Horn Expedition to Central Australia.* Aulau, London.
Thackway, R. and Cresswell I.D. (eds) (1995) *An Interim Biogeographic Regionalisation for Australia: a Framework for Establishing the National System of Reserves, Version 4.0.* Australian Nature Conservation Agency, Canberra.
Waithman, J. (1979) Rabbit control in New South Wales – past, present and future. *Wool Technology and Sheep Breeding* 27, 25–30.
Williams, K., Parer, I., Coman, B., Burley, J. and Braysher, M. (1995) *Managing Vertebrate Pests: Rabbits.* Australian Government Publishing Service, Canberra, pp. 57–65.

8

Myxomatosis in Australia: 1967–1997

Overview

By the mid-1960s myxomatosis was endemic throughout rabbit-infested parts of Australia and had become part of the rabbit scene. When inoculated in unselected laboratory rabbits, most strains of virus recovered from infected wild rabbits produced infections with mean survival times that equated with case-fatality rates of 70–90%. However, in adult wild rabbits in the field the disease was milder and the mortality rate much reduced because of their increased genetic resistance. Inoculation campaigns ceased in most States in the late 1960s, but continued in Victoria until the mid-1970s and after dwindling in Queensland until the mid-1980s, were then revived, only to be discontinued after the escape of the rabbit haemorrhagic disease virus in 1995.

By the late 1960s rabbit populations were beginning to increase substantially, and several of the State organizations responsible for rabbit control recruited new staff to deal with the re-emerging rabbit problem. After 1966 the major changes in the epidemiology and impact of myxomatosis resulted from the introduction of the European rabbit flea, *Spilopsyllus cuniculi*. Especially where mosquitoes had not been very effective vectors, the introduction of *S. cuniculi* changed the seasonal incidence of myxomatosis, producing protracted winter and spring outbreaks rather than the sharp summer outbreaks produced by mosquito transmission.

S. cuniculi did not persist in rabbit populations in the hot, dry rangelands, where the rabbit remained a serious environmental threat. After investigations lasting about ten years, in 1993 the Spanish rabbit flea, *Xenopsylla cunicularis,* was introduced into some of these areas. It has survived well there and is spreading slowly through the rabbit populations; it is still too soon to assess its effects.

After testing the virulence of field strains on groups of five or six laboratory rabbits for many years, several problems were identified that sometimes compromised their reliability. In tests on laboratory rabbits, but not in wild rabbits, infections with less virulent strains were commonly associated with secondary bacterial infections; their incidence was conditioned by the pre-existing bacterial flora of the rabbit's respiratory tract. Since many such infections were fatal, it became harder to ascribe deaths to viral virulence. Other factors that were more readily controlled were the temperature of the animal house and the use of stored lesion material for inoculations. A more difficult problem that was first clearly identified in the 1990s was that the mean survival time of groups of five or six rabbits sometimes appeared to be an unreliable predictor of what the case-fatality rate would have been in larger groups of rabbits.

Examination of field strains from Victoria suggested that most strains collected between 1960 and 1980 were of Grade III virulence, as judged by the inoculation of laboratory rabbits. In areas where the genetic resistance of the rabbits was higher a few strains of higher virulence (Grades I and II) were found. In the 1990s, most of the few strains that were collected in New South Wales and Victoria were Grade I strains, resembling the Standard Laboratory Strain in symptomatology and mean survival times.

Observations on the genetic resistance of wild rabbits showed that rabbits in the Victorian Mallee had become so resistant that their mortality from challenge with the virulent Standard Laboratory Strain had fallen to about 70% by 1961–66 and 60% by 1976–81; corresponding figures from cooler parts of Victoria were 94% and 79%. Sobey obtained a similar rapid increased resistance in laboratory rabbits bred from resistant parents, the heritability of resistance being 35–40%. Subsequent analysis of Sobey's data suggested that the sires of resistant rabbits contributed more than the does to resistance, and observations in field enclosures also suggested that a 'sire factor' might contribute to resistance.

Comparisons of myxomatosis in rabbits in cold areas and during hot summers confirmed the laboratory observations that high ambient temperatures diminished and low temperatures exacerbated the severity and lethality of myxomatosis.

Field observations in New South Wales suggested that some outbreaks might originate from the reactivation of latent infections, with the production of lesions from which myxomatosis was transmissible by insect vectors.

Although there was a general impression that myxomatosis had ceased to have much effect on rabbit numbers by the mid-1970s, careful investigations demonstrated that in fact myxomatosis continued to exert substantial control. Immunization experiments showed that myxomatosis reduced populations by a factor of about 10. Nevertheless, growing populations of rabbits highly resistant to myxomatosis stimulated new approaches to biological control. Initiatives were launched involving the use of myxoma virus as a vector for immunocontraception, and another highly host-specific virus that had recently devastated rabbits in Europe, the calicivirus that causes rabbit haemorrhagic disease, was introduced. In 1992 a Cooperative Research Centre for the Biological Control of Vertebrate Pests was established, with its headquarters in Canberra.

Inoculation Campaigns

Although organized inoculation campaigns had been a major activity of most State rabbit control authorities in the 1960s, enthusiasm for them fell off during subsequent decades. However, releases of Lausanne virus with fleas were undertaken for some years in Victoria and New South Wales, and some landholders have continued to purchase myxoma virus from CSL Ltd. In Western Australia inoculations by local Vermin Boards continued at a reduced level until the early 1980s. In Queensland, after almost ceasing to be used during the 1970s, inoculation campaigns were recommenced in 1983, using Lausanne virus in early spring instead of late spring/summer[1]. Inoculations were carried out in all rabbit problem areas, especially in southern Queensland from the far south-west to the Darling Downs-Moreton Bay Rabbit Board. Between 1983 and 1987 some 2400 rabbits were inoculated annually. The success of the early spring campaigns was indicated by the fall in the numbers captured for inoculation between 1988 and 1994 to about 1400 annually. Organized inoculations ceased in 1995, after the escape of rabbit haemorrhagic disease virus. Currently a few vials are purchased each year while the best management strategy for using inoculations of rabbit haemorrhagic disease virus and myxoma virus to complement each other is being developed.

Introduction of the European Rabbit Flea (*Spilopsyllus cuniculi*)

Myxomatosis was introduced into France in June 1952 and was first recognized in

England in October 1953 (see pp. 212 and 225). It was soon apparent that the principal vector in Britain was the rabbit flea (*Spilopsyllus cuniculi*), an ectoparasite that had not been introduced (with the rabbit) into Australia or New Zealand. From the early days of work on myxoma virus, Bull and later Ratcliffe had been impressed with the possibility of using rabbit fleas as vectors. They thought that it should be possible to breed them in the laboratory and then make experimental introductions under controlled conditions (see p. 134). To explore this possibility, European rabbit fleas had been imported under quarantine in 1957, but attempts to breed them were unsuccessful[2]. In the early 1960s the breeding cycle of *S. cuniculi* was worked out by Mead-Briggs and Rothschild in Britain (see p. 85), and another effort was made to use the flea in Australia, since it seemed probable that it could spread myxomatosis in situations where there were few mosquitoes.

Breeding fleas on laboratory rabbits

With this new information, European rabbit fleas were again imported into Australia in 1966 and breeding was successfully established in an insecticide-free quarantine site near Sydney (Sobey and Menzies, 1969). As a precondition for release, the Commonwealth Director of Animal Health and State representatives, although aware of the specialized adaptation of *S. cuniculi* to rabbits, required that the susceptibility of certain native fauna to infestation should be tested. Five species of marsupials were used, all females. It was found that within 15 minutes all fleas (50–100 per animal), except for an odd straggler, had left each female, whereas none had left either a female rabbit or a female hare (*Lepus europaeus*). European rabbit fleas can only breed after feeding on pregnant does and then on her newborn kittens. After quarantine restrictions had been lifted in 1968 it was found that they will feed, but not breed, on cats (Sobey, 1977).

Methods were developed for large-scale breeding of fleas in an ordinary animal house (Sobey *et al.*, 1974, 1977). A pregnant doe was housed in a nest box with a removable bottom over a tray and fleas were seeded onto her a week before she was due to litter. When the kittens were 12 days old the nest box was replaced by a new one and the bottom of the nest box was placed on a rack above a 'collector' rabbit. The first emergence fleas, the majority of which were females (Shepherd and Edmonds, 1980), were collected on this rabbit, from which they could be combed off. Second emergence fleas remained as pupae until a disturbance broke their quiescent state; they could be activated when required and recovered in the same way. Average yields of 4000 fleas per litter were obtained and they could be stored in plastic bags in the refrigerator until required. Initially fleas were induced to bite through virus-rich rabbit lesions to make them infective (Sobey and Conolly, 1975); later, batches of fleas were shaken up in a suspension of myxoma virus and dried (Sobey, 1977).

Initial release in Australia

In 1968 fleas were released on three properties on the central and northern tablelands in New South Wales. Within two rabbit breeding seasons almost every rabbit shot within half a kilometre of each release site was infested. If rabbits were then infected with 'marked' strains of myxoma virus, or if fleas that had probed through myxomatous lesions were released, the 'marked' strain was recovered from several infected rabbits under conditions suggesting that the virus had been transmitted by fleas (Sobey and Conolly, 1971). Later studies elaborated on the dispersal and fluctuations in numbers of fleas under natural conditions in Australia (Williams and Parer, 1971; Williams, 1973).

Further releases of European rabbit fleas

Following the success of these preliminary studies, fleas were bred and released on a large-scale in several States, notably in Victoria (Shepherd and Edmonds, 1976, 1978a), New South Wales (Sobey and Conolly, 1971; Williams and Parer, 1971), South Australia (Cooke, 1983) and Western

Australia (King *et al.*, 1985; Wheeler *et al.*, 1985). In each State sites for intensive study of fleas and rabbits were established (Figs 8.1 and 8.4). In all areas the fleas spread slowly but effectively, especially if the fleas were released early in the rabbit breeding season. In one locality 95% of rabbits captured four years after the releases carried the flea, often in numbers of up to 500 per rabbit (Shepherd and Edmonds, 1976). The results of trials in various States were presented at the triennial Australian Vertebrate Pest Control Conferences.

Effects on the epidemiology of myxomatosis

CSIRO scientists introduced fleas in New South Wales and the Australian Capital Territory, both to determine the feasibility of such introductions and to determine their effect on rabbit mortality due to myxomatosis. Fullagar (1977) found that fleas had little effect on the spread of myxomatosis among wild rabbits in a 36 hectare enclosure near Canberra; Parer *et al.* (1981) reported similar results at Lake Urana. Parer and Korn (1989) compared the monthly reporting of myxomatosis in Pasture Protection Districts throughout New South Wales over the periods 1959–64 and 1980–85, i.e. before and after the introduction of the rabbit flea. Throughout the State and during both periods myxomatosis was more frequent in summer than in winter, especially in the

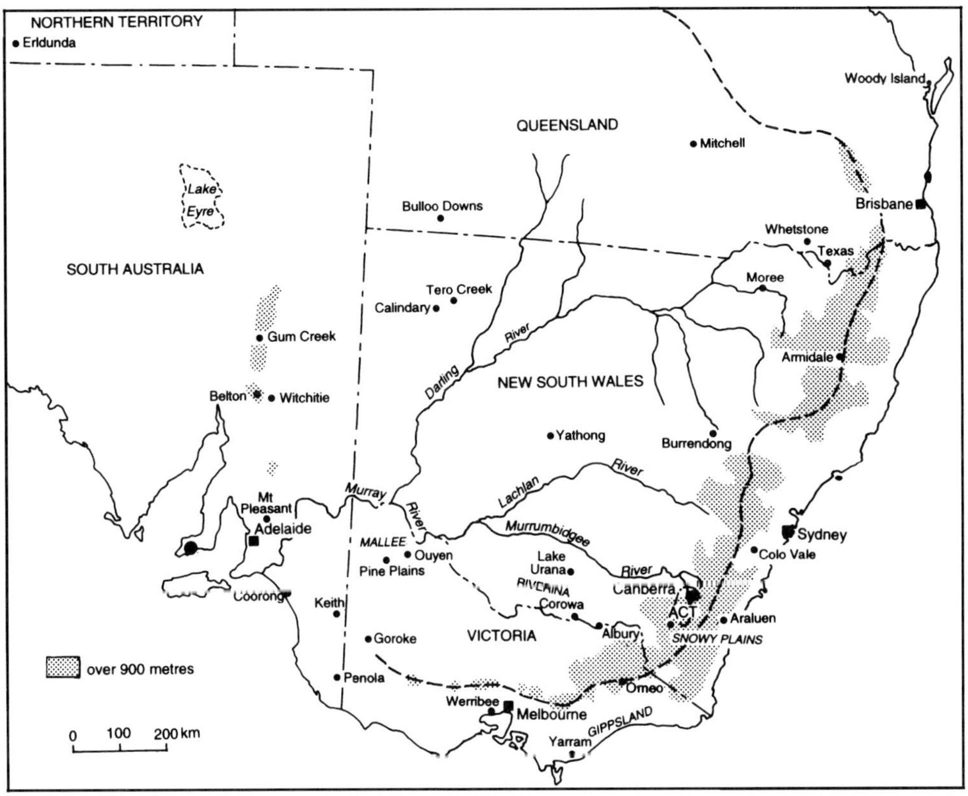

Fig. 8.1. Map of south-eastern Australia showing study sites for work on myxomatosis after its initial spread in 1951. Some of these sites were first used during the period covered in Chapter 7 (1952–1966); several more were developed to follow up the introduction of the European rabbit flea in 1967. Different sites began and ceased operation at different times; some of them are now being used as intensive study sites for rabbit haemorrhagic disease (see Fig. 11.10, p. 261).

drier inland areas. It was only in the tablelands, relatively near the coast, that myxomatosis was more frequent after introduction of the rabbit flea – 81% positive monthly reports for 1980–85 compared with 60% for 1959–64.

In most years since 1951, rabbits in the Mallee region of Victoria had been subjected to summer mosquito-transmitted epidemics of myxomatosis. In combination with late summer nutritional stress, these generally caused a reduction in population of about 80% among rabbits surviving from the previous breeding season, even though the enzootic strains of myxoma virus were of substantially reduced virulence and this population was the most resistant to myxomatosis in Victoria (Shepherd and Edmonds, 1978b; Shepherd *et al.*, 1978; Fig. 8.2). Three years after introduction of

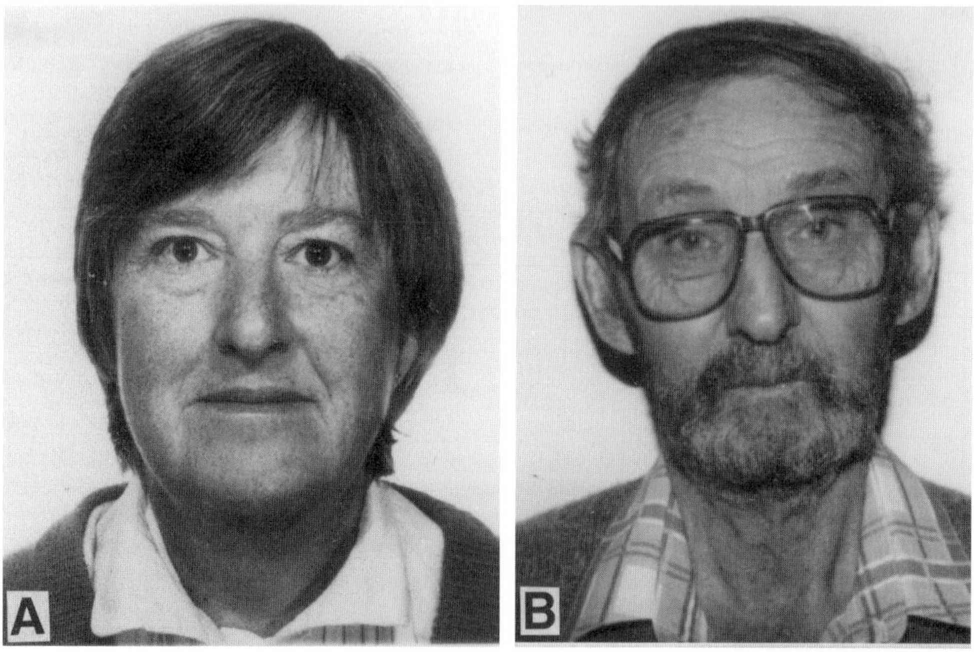

Fig. 8.2. (A) Rosamond Charmian Hollis Shepherd (1940–). Born and educated in Brisbane, Shepherd graduated with a BSc from the University of Queensland in 1962, majoring in entomology, and proceeded to a MSc degree in 1966. After working for several years in Brisbane and in England, she joined the Victorian Department of Crown Lands and Survey in 1969 and worked at the Keith Turnbull Research Station in Frankston on the European rabbit flea as a vector of myxomatosis. Over the next eleven years she studied the interactions of the ectoparasites of the rabbit, myxoma virus and rabbit ecology, focusing on rabbits in the Mallee region of Victoria. In 1985 she was awarded the degree of MAgrSc from the University of Melbourne for these studies. Her later years at the Keith Turnbull Research Institute were spent on the biological control of weeds. She retired in 1995 and works as a weed science consultant.
(B) John Wilson Eadie Edmonds (1924–). Born in Natimuk, Victoria, and educated in Clear Lake and Horsham, Edmonds served as a navigator and operations officer in the Royal Australian Air Force from 1943 to 1946. He graduated with a BAgrSc from the University of Melbourne in 1951 and worked on the family farm in the Wimmera until 1966, when he joined the Victorian Department of Crown Lands. He worked at the Keith Turnbull Research Station on the virulence of field strains of myxoma virus and the resistance of wild rabbits to myxomatosis. In 1974 he was awarded the degree of MAgrSc from the University of Melbourne for a thesis 'The Social Development of a Small Rural Community' and in 1979 the degree of MSc in Monash University for a study of immunoglobulin allotypes in wild rabbits of south-eastern Australia. He retired to a hobby farm in 1985 and now writes poetry, being awarded first prize in the C.J. Dennis Literary Awards in 1997.

rabbit fleas into the region in 1970, the fleas had spread sufficiently widely and bred up to adequate numbers to change the seasonality of myxomatosis, with spring rather than summer outbreaks each year from 1973 to 1975. These caused heavy mortality in kittens before they entered the observable population at about three months of age, so that in early summer few subadult rabbits were seen. Combined with lessened late summer nutritional stress, this led to an increase in the average age of the population. This effect was reversed in 1976, when dry conditions delayed rabbit breeding until late spring. Fleas bred poorly, there were relatively few cases of myxomatosis, and rabbit numbers built up rapidly. In the same region, it was found that the highly virulent Lausanne strain of myxoma virus introduced with fleas persisted and continued to kill rabbits for up to 16 weeks, despite competition with field strains (Shepherd and Edmonds, 1977a).

In the rich agricultural Western Plains region of Victoria, where climatic and topographic conditions were unsuited to the production of high densities of mosquitoes, myxomatosis occurred as an enzootic disease after the introduction of fleas, with diseased rabbits present throughout the winter (Tighe *et al.*, 1977).

The most systematic studies of the effects of fleas on the epidemiology of myxomatosis were carried out in South Australia, where Cooke (1983) used the ratio of young of the year to older rabbits, measured in summer and autumn, after the breeding season, as a measure of the mortality due to myxomatosis. The results were consistent in all study sites (see Fig. 8.1), which were located in areas with a Mediterranean-like climate, the average annual rainfall for different sites ranging from 240 mm to 655mm. The overall average of the ratio of the younger to older rabbits for the four years prior to the introduction of fleas was 9.68. This fell to 1.93 for the four years after fleas were introduced, implying that a high proportion of young rabbits were being removed from the population. At all sites, after the introduction of rabbit fleas the numbers of rabbits decreased and failed to increase in favourable seasons.

The spread of myxomatosis in Western Australia after its introduction in 1951 had been much less successful than in eastern Australia, probably because of the small numbers of vector mosquitoes, and in 1955 poisoning with 1080 was introduced as the principal official method of rabbit control. European rabbit fleas were introduced in various places in the south-west of the State in 1969 and periodically over the following decade (King *et al.*, 1985; Fig. 8.3). Detailed observations on rabbit populations were carried out at Chidlow and Cape Naturaliste (Fig. 8.4), at sites in farming land surrounded by native vegetation. Both places have a Mediterranean climate; rabbit breeding began in April–May (autumn) and ceased when pastures dried out in early summer. The fleas became established and dispersed widely and, as elsewhere, their numbers fluctuated seasonally and were highest on female rabbits during the breeding season. It was observed that although epidemics of myxomatosis remained irregular, after the introduction of fleas they occurred in winter and were more extensive than the summer epidemics which had occurred prior to the introduction of the fleas (King and Wheeler, 1985). Observations at the Cape Naturaliste study site showed that the mortality rate in two winter epidemics that occurred one and five years after the introduction of the fleas was substantially higher than in three summer epidemics that occurred during the five years preceding their introduction. Moreover, because the winter epidemics impacted heavily on young rabbits, they reduced the size of the population of breeding rabbits the next year more than did the occasional summer epidemic that had occurred before the introduction of fleas. Myxomatosis was so effective that the use of 1080 for rabbit control was discontinued.

In 1992–96, field experiments were conducted in Wellstead, in south-western Western Australia (Fig. 8.4) to study the effects of sterilization on the social

Fig. 8.3. (A) Stuart Hugh Wheeler (1944–). Born in the United kingdom, Wheeler graduated with a BSc in Zoology from the University of Durham in 1965. He came to the University of Adelaide, where he obtained a PhD degree in 1970 and then joined the Agriculture Protection Board of Western Australia in Perth. For the next 12 years he carried out long-term studies of the ecology and control of rabbits in study sites at Chidlow and Cape Naturaliste (see Fig. 8.4). Later he was involved in the management of donkeys and rodents in northern Western Australia. In 1988 he became Senior Research Officer in charge of the Vertebrate Pest Control Section. He has been involved in the Cooperative Centre for Vertebrate Biocontrol since its inception in 1992, more recently as ecology leader. With changed administrative arrangements he moved to Albany, and now works mainly in the Sustainable Rural Development Section as Pest and Weed Control Liaison Officer.
(B) Dennis Richard King (1942–). Born in Canada, King was educated in Vancouver and graduated with a BSc in zoology from the University of British Columbia in 1964 and an MSc in 1968. He then went to the University of Adelaide and worked on temperature regulation in the lizard *Varanus gouldi*, work for which he received a PhD in 1978. In 1973 he joined the Agriculture Protection Board of Western Australia and worked initially on the population dynamics and control of rabbits, including the effects of myxomatosis on rabbits in the south-western corner of the State. Later he undertook investigations of the tolerance of native fauna and introduced vertebrate pests to the poison 1080. He retired in 1996, and now works on the biology of lizards at the Western Australian Museum.

behaviour of rabbits (see p. 204). All rabbits on the trial sites were regularly bled and sera assayed by ELISA (Kerr *et al.*, 1998). Myxomatosis occurred as an annual spring–summer outbreak except in 1994, when there was no epidemic, and 1995–96. In 1995–96 cases of mild myxomatosis were seen from March until November 1995, when there was a large-scale epidemic which peaked in January 1996. The spring–summer epidemics, primarily due to mosquitoes, left more than 80% of susceptible rabbits immune; when combined with survivors from previous epidemics over 90% of rabbits on the sites were immune. This indicated that the spring–summer epidemics depended on the susceptible kittens entering the population each year.

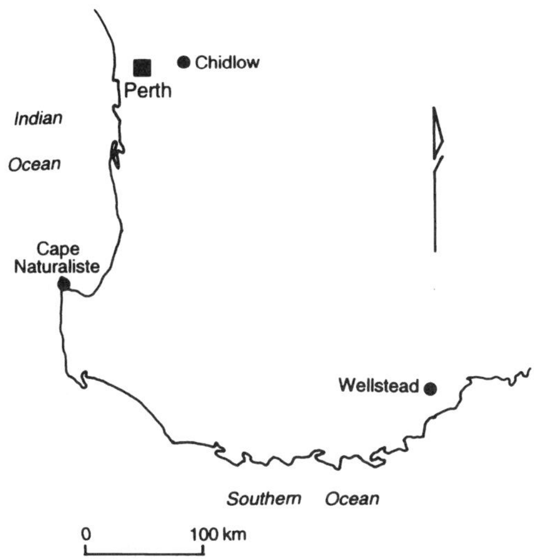

Fig. 8.4. Map of the south-western corner of Western Australia, showing the study sites used for investigation of the ecology of rabbits and the effects of rabbit fleas on the epidemiology of myxomatosis (Cape Naturaliste and Chidlow), and for the effects of sterilization of various proportions of the female rabbit population on rabbit numbers (Wellstead).

There were many susceptible rabbits present in 1994, but it was a very dry year and there were very few mosquitoes, although rabbit fleas appeared to be present in usual numbers. The absence of myxomatosis that year was probably due to the failure of the virus to be introduced into the trial sites by mosquitoes. The slowly progressive outbreak in 1995 was probably due to flea transmission, with a sharp epidemic due to mosquitoes in the summer.

At the 8th Australian Vertebrate Pest Control Conference, Cooke (1987; Fig. 8.5A) reviewed the overall effects on myxomatosis of the introduction of rabbit fleas 18 years earlier. In general, there was a clear inverse relationship between the apparent improvement in the effectiveness of myxomatosis following the release of the rabbit fleas and the general abundance of mosquito vectors. Thus in both Western Australia and South Australia, where mosquitoes had not been particularly effective, rabbit populations were often reduced in size and fewer young rabbits survived. There were also significant changes in the Victorian Mallee and the tableland areas of New South Wales, although mosquitoes were important vectors in these regions. In some areas where mosquitoes were effective every year, as at Lake Urana, the introduction of fleas did not change the pattern of the disease (Williams and Parer, 1972).

On the other hand, there are large areas of inland Australia where the climatic conditions are not suitable for the survival of European rabbit fleas because they die out during prolonged droughts, when rabbits do not breed (Cooke, 1984). These field observations were confirmed by laboratory studies which showed that flea larvae do not survive at relative humidities below 70%, whereas burrow humidities in dry years can fall well below this level (Cooke and Skewes, 1988), and in most of these arid areas mosquitoes were usually uncommon. There had also been a heightened appreciation of the great environmental damage caused by the largely uncontrolled rabbit populations in the arid rangelands. Summing up the position at a national workshop on Rabbits and their Control in September 1983, the chairman (Dr M.A.S. Jones), commented[3]:

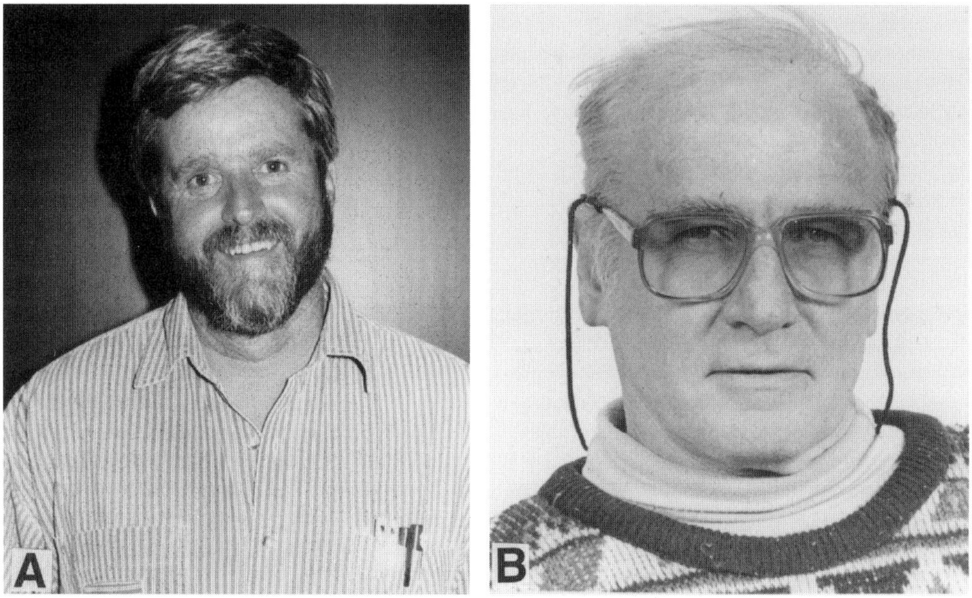

Fig. 8.5. (A) Brian Douglas Cooke (1944–). After graduating in Zoology in the University of Melbourne in 1965, Cooke joined the Vermin Control Branch of the South Australian Department of Lands and has continued to work with this group through changes in its name to Vertebrate Pests Control Authority and then Animal and Plant Control Commission. In 1970 he gained an MSc degree and in 1974 a PhD degree from the University of Adelaide, both for work on wild rabbits. He has been concerned with many aspects of rabbit biology, especially in arid inland Australia. After working on wild rabbits in Spain in the 1980s he became involved in the introduction of the Spanish rabbit flea as a potential vector for myxomatosis in arid parts of Australia, and subsequently with the introduction of rabbit haemorrhagic disease virus for rabbit control. Since September 1995 he has been seconded to the CSIRO Division of Wildlife and Ecology in Canberra to work on rabbit haemorrhagic disease virus as a method of rabbit control.
(B) Ian Patrick Parer (1940–). Born in Wewak, Papua New Guinea, Parer graduated in Agricultural Science with Honours from the University of Queensland in 1963, majoring in animal behaviour. He studied the ecology and behaviour of feral domestic fowls on a Barrier Reef island in 1964–65 before joining the CSIRO Division of Wildlife Research in 1966. At CSIRO he worked on the population dynamics of wild rabbits, the distribution of rabbits in relation to climate and soils and predator–prey interactions. He had a long-term interest in the epidemiology of myxomatosis and in the evolutionary changes that occurred in the rabbits and myxoma virus after the introduction of the virus into Australian wild rabbits in 1950. He retired in 1995.

> The present myxomatosis vectors, mosquitoes and rabbit fleas, have limited distribution in time and geophysical [space] respectively. There is an urgent need to examine whether a suitable flea biotype for the arid zone exists on the Iberian Peninsula.

As a result, funds were provided by the Meat Research Corporation to allow Cooke to go to Spain to obtain a strain of *S. cuniculi* better adapted to arid conditions. Cooke and Dr R. Soriguer, of the Consejo Superior de Investigaciones Cientificas, collected *S. cuniculi* from some of the driest areas in Spain, bred them in Seville, and sent pupae to Adelaide. After fulfilling quarantine protocols some were bred up, with difficulty, but physiological tests showed that they were no better adapted to aridity than the fleas already in use[4]. In 1988 Cooke went back to Europe, and after investigating the systematics of fleas that were specific to *Oryctolagus cuniculus*, he concentrated on two species, *Caenopsylla laptevi ibera* and *Xenopsylla cunicularis*. Both occurred commonly in the driest parts of Spain and tests showed that both were good vectors of myxomatosis[4]. Tests in Spain showed that although neither species required access to a pregnant rabbit for breeding, *C. laptevi ibera* was a seasonal breeder, with a prolonged diapause, whereas *X. cunicularis* bred year-round (Cooke, 1990).

The Introduction of *Xenopsylla cunicularis* from Spain

As a result of these investigations, *X. cunicularis* was imported from Spain and established in quarantine in Adelaide in August 1990. Then followed tests of host specificity, optimum temperatures and humidity for reproduction, capacity to transmit myxomatosis and interaction with other vectors of myxomatosis (Bartholomaeus, 1991). Approval was given for its release from quarantine in April 1993, and batches of pupae or newly emerged adults were released at over 500 sites across much of inland South Australia, New South Wales, Queensland and Northern Territory (Cooke, 1995). Despite low rainfall over much of inland Australia following the release, fleas were recovered from most of the release sites a year later. They appeared to be most successful in sandy and loamy soils, and spread at rates of up to 2 km per year. Following the introductions, unseasonal outbreaks of myxomatosis were reported from western Queensland which were confined to areas in which the fleas had been released (Robertshaw and Gould, 1995), and similar unseasonal outbreaks were observed in South Australia (Mutze, 1996). However, at their present rate of spread, which has been slowed by drought and outbreaks of rabbit haemorrhagic disease, it will take several years before *X. cunicularis* becomes widespread and numerous enough to modify the epidemiology of myxomatosis over large tracts of inland Australia. In the meantime, the rabbit haemorrhagic disease virus has spread over many parts of inland Australia (see Chapter 11), and it will be difficult to unravel the effects of these two methods of rabbit control.

Changes in Administrative Arrangements and Research Scientists

In 1966 the virology team from the Australian National University ceased working on myxomatosis. Investigations on viral virulence and host resistance were continued, for Victorian specimens, by G.W. Douglas, J.W. Edmonds, R.C.H. Shepherd and I.F. Nolan, who from 1967 worked in the newly established Keith Turnbull Research Institute near Frankston, where excellent facilities for laboratory work on rabbits were available. The gradual increase in rabbit numbers that became apparent in the late 1960s led other State governments to strengthen their rabbit control personnel, with the appointment of B.D. Cooke in South Australia and D.R. King and S.H. Wheeler in Western Australia. At about this time several of the CSIRO scientists who had been active in early studies of myxomatosis, including F.N. Ratcliffe, J.H. Calaby, K. Myers, R. Mykytowycz, A.L. Dyce and B.V. Fennessy, retired or moved to other activities. J.D. Dunsmore, P.J. Fullagar and R.T. Williams (Fig. 8.6B) worked on myxomatosis during the 1970s, and in 1976 W.R. Sobey and D. Conolly moved from the Division of Animal Genetics to the Division of Wildlife Research and were active in research on myxomatosis until their retirement in 1983. I. Parer (appointed in 1966), C.K. Williams (transferred from the Northern Territory in 1980) (Fig. 8.6A), and P.J. Kerr (Fig. 5.7, p. 104) continued working on myxomatosis through the 1990s.

By 1978 organized inoculation campaigns had ceased in all States, including Victoria, and the Victorian rabbit control authorities took over responsibility for stocks of freeze-dried myxoma virus from the Commonwealth Serum Laboratories. However, myxoma virus (Lausanne strain) continued to be made available by CSL Ltd to landholders who requested it. At a research level, instead of inoculating and releasing rabbits as a method of introducing myxoma virus, CSIRO teams in New South Wales and the Australian Capital

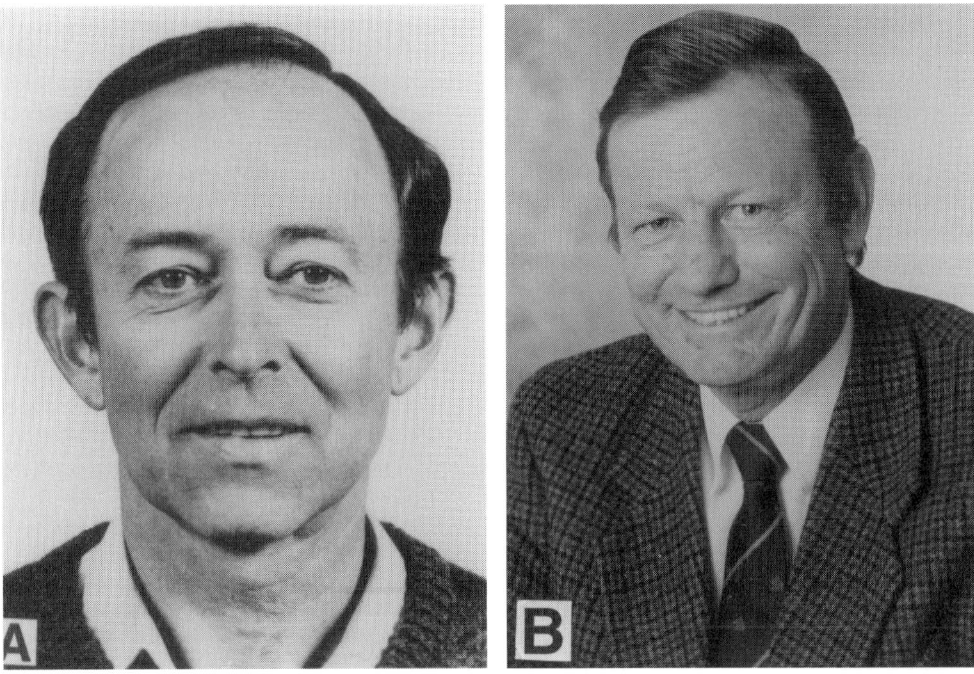

Fig. 8.6. (A) Christopher Kent Williams (1944–). Born in Fremantle, Western Australia, Williams graduated with a BSc (Hons) from the University of Western Australia in 1965, majoring in Zoology. He obtained the degree of PhD at the same university in 1971, and the same year Williams joined the CSIRO Division of Wildlife Research and worked in Darwin, Northern Territory, on the ecology of small native mammals and feral buffalo in the monsoonal lowlands. In 1980 he transferred to Canberra and began research on rabbits in 1983. He worked on processes of adaptation of the rabbit to Australian environments and regional variation in genetic resistance to myxomatosis before moving to studies of integrated conventional control. He was involved in planning the programme for assessing rabbit haemorrhagic disease virus as a method of biological control of the rabbit in Australia, and after its escape from Wardang Island he participated in the management of the introductions and follow-up research. Williams' current research is with the CSIRO Division of Wildlife and Ecology and the Vertebrate Biocontrol Cooperative Research Centre studying population responses of rabbits to the imposition of reproductive sterility.
(B) Richard Trevor Williams (1942–). Born in London, England, Williams graduated in Zoology with Honours from Durham University in 1964. In 1967 he obtained a PhD degree from the University of Nottingham for research on the water physiology of insect ectoparasites. In 1968 he came to Australia to work at the CSIRO Division of Wildlife Research on a five-year contract to study the impact of the introduction of the European rabbit flea on the epidemiology of myxomatosis. In 1974 he moved to Northern Territory to become the inaugural head of the Department of Biology at the Darwin Community College. In 1980 he returned to Canberra as a science policy analyst with the Commonwealth Government, most recently with the Bureau of Resource Sciences. He briefly re-entered the biological control debate in 1993–94 with a substantial contribution to discussions on the use of rabbit haemorrhagic disease virus.

Territory, and Edmonds and Shepherd in Victoria, introduced contaminated fleas into rabbit burrows. In Queensland, inoculations were banned in 1973–74, when the European flea was released, so that the effect of this vector on rabbit numbers could be evaluated. However, because rabbit numbers were increasing greatly, inoculations were recommenced in the summer of 1981–82. At this time field strains were already present, so from 1984 to 1985 releases were made in August–September, just prior to the build-up of vectors and before the arrival of field strains. This has continued on an annual basis and is considered an effective form of rabbit control[5].

Establishment of the Cooperative Research Centre for Biological Control of Vertebrate Pest Populations

Animal welfare groups, which have become more influential since the 1970s, have always pressed for more humane methods of pest control (Fisher and Marsh, 1996), and in 1990 Australian and New Zealand Federation of Animal Societies organized a Conference on Fertility Control in Wildlife[6]. Studies of the physiology of animal reproduction had raised the prospect of fertility control as a humane method of pest control (Tyndale-Biscoe, 1991). In response to the decrease in the effectiveness of myxomatosis as a lethal disease of rabbits, Tyndale-Biscoe suggested that myxoma virus might be used as a vector for a gene(s) which could effectively sterilize rabbits[7]. Potentially a similar method might be used for other pest species, such as foxes and mice in Australia and possums in New Zealand, and also for the control of populations of animals of high conservation value, if suitable viral vectors were discovered.

In 1990 the Australian Government introduced a new initiative to promote research and development by involving universities, CSIRO, State governments and where appropriate private industry in cooperative research and development projects, by establishing Cooperative Research Centres with guaranteed funding for five years, with the possibility of extensions. Seed money provided in 1989–90 had enabled Tyndale-Biscoe to develop the concept of immunocontraception to the stage at which it could be proposed for funding as the major programme of a Cooperative Research Centre. In 1992 the Division of Wildlife and Ecology of CSIRO, the Australian National University, the Agriculture Protection Board of Western Australia and the Department of Conservation and Land Management of Western Australia joined forces to form a Cooperative Research Centre for Biological Control of Vertebrate Pest Populations (Vertebrate Biocontrol Centre or VBC), based at the Division of Wildlife and Ecology in Canberra (Annual Reports, 1993–1994, 1994–1995, 1995–1996). The first Director was C.H. Tyndale-Biscoe, a distinguished reproductive physiologist (Fig. 8.7). The work of the Centre on rabbit control is discussed at the end of this chapter.

Changes in the Virulence of Myxoma Virus

Problems with virulence tests

Experience since the 1960s revealed several problems in obtaining results in laboratory tests that truly reflected the likely lethality of the virus in the field. Two of these were technical problems for which solutions were developed; the temperature of the animal house and the use of inocula that contained a large amount of inactive virus. Two other problems were more difficult to solve. One was an inescapable problem, the activation of inapparent bacterial infections in the rabbits used for the testing, and the other was the emergence in the field of strains of virus for which mean survival times in genetically unselected rabbits did not seem to be a satisfactory surrogate for lethality.

Temperature of the animal house

A potential complicating factor, as demonstrated experimentally by Marshall (1959) (see p. 107), was the temperature of the animal house. High temperatures reduced the severity of myxomatosis, whatever the virulence of the virus; low temperatures increased the severity. This complication was avoided in some but not all laboratories by suitable air-conditioning. However, environmental temperature did play a part in determining the kill in several field situations (see p. 199).

Use of concentrated suspensions of lesion material for inoculations

Sometimes lesion material took several weeks to be transferred from the field to the laboratory, or had been stored in glycerol in the laboratory (usually in a refrigerator) for some months before being tested. Suspensions of such material usually contained low concentrations of viable virus

Fig. 8.7. Cecil Hugh Tyndale-Biscoe (1929–). Born in Kashmir, India, Tyndale-Biscoe graduated in Botany and Zoology at Canterbury University College, Christchurch, New Zealand in 1950. He then joined the Department of Scientific and Industrial Research in New Zealand, working on the ecology of wild rabbits. He completed an MSc degree on the ecology of the brushtail possum in New Zealand in 1953, taught for three years in Peshawar, Pakistan, and then took a PhD in the University of Western Australia, on the reproductive physiology of the quokka. He joined the Department of Zoology at the Australian National University in 1962 and worked extensively on marsupial reproduction. In 1976 he was appointed leader of a newly-formed Marsupial Biology Group in the CSIRO Division of Wildlife Research and for the next 14 years led a programme on the endocrinology and biochemistry of marsupial reproduction. In 1989 the resources of the Group were redirected to investigate the possibility of controlling mammal pest species by immunocontraception, focusing initially on the rabbit and then extending to the fox and the mouse. Since 1992 this work has been carried out within the Cooperative Research Centre for the Biological Control of Vertebrate Pest Populations, of which Tyndale-Biscoe was the Director until his retirement in 1995. He was elected a Fellow of the Australian Academy of Sciences in 1986 and has received a number of prestigious scientific awards, and now works in the Research School of Biological Sciences of the Australian National University.

and presumably much inactivated virus. The survival times of rabbits inoculated with such suspensions was usually much longer than if the same low dose of viable virus was derived from fresh lesion material (Fenner and Woodroofe, 1965). This problem was overcome with Australian samples tested after 1959 (Marshall and Fenner, 1960; Edmonds *et al.*, 1975) and in British samples tested in Australia (Fenner and Chapple, 1965) by inoculating the lesion material in a rabbit and making the test inoculations either by needle prick of the lesion or by intradermal inoculation of a small dose of the passaged virus.

Secondary bacterial infections

The classical strains of Grade I virulence (Standard Laboratory Strain and strains derived directly from *Sylvilagus* spp.) killed rabbits quickly, with only slight signs of secondary bacterial infection of the respiratory tract in the few rabbits that survived longer than usual. However, less virulent strains caused severe immunosuppression, evidenced in infected rabbits by 'snuffles', with a profuse purulent nasal discharge. Snuffles was commonly seen in laboratory rabbits infected with moderately virulent (Grade III or IV) strains and it appeared to be the cause of death in some rabbits that had survived the acute disease. Experience in different laboratories showed that the incidence and severity of this complication varied from place to place and time to time, probably depending on the extent of inapparent infection of the test rabbits with *Pasteurella multocida*.

Australian wild rabbits also carry some potentially pathogenic bacteria in their respiratory tracts. After examining the upper and lower respiratory tracts of 181 wild rabbits, Cumming *et al.*[8], found *Bordetella bronchiseptica* in 15% of rabbit nares and 1% of rabbit lungs, compared with 41% for both sites in laboratory rabbits. Comparable figures for *P. multocida* were 2% of nares and 1% of lungs in wild rabbits, and 20% for both sites in laboratory rabbits. However, observers with considerable experience of seeing myxomatous rabbits in the field, in Victoria (Edmonds and Shepherd) and South Australia (Cooke) have said that they did not see snuffles in such rabbits[9].

Failure of survival times to measure lethality
The reliability of using mean survival times as a surrogate for case-fatality rates was questioned by Parer *et al.* (1994), who studied the resistance to several strains of myxoma virus of the offspring of rabbits collected between 1976 and 1980 in several widely separated parts of Australia. Instead of using laboratory rabbits as controls, they were able to use the offspring of wild rabbits imported from New Zealand, where myxomatosis had never become established. Using large numbers of New Zealand wild rabbits (a total of 240 rabbits for five strains of virus) they found that three strains that would have been graded I, II and IV on mean survival times actually killed all the inoculated rabbits, hence would have been classified as Grade I in terms of case-fatality rates. Two strains were classified as Grades II and IV respectively by both mean survival times and case-fatality rates. They pointed out that the original Grade I strains (Standard Laboratory Strain and Lausanne strain) killed rabbits rapidly by an uncharacterized effect, whereas infections with most field strains collected at that time caused immunosuppression when tested in laboratory rabbits. This phenomenon has been investigated in depth by Strayer *et al.* (1983); the consequent secondary infections often killed all the laboratory rabbits inoculated with these field strains, with much longer survival times than found with the original Grade I strains.

Parer *et al.* concluded that by the 1980s most Australian field strains were probably Grade I or II in virulence, as judged by case-fatality rates in unselected rabbits, but might be classified as Grade III or possibly Grade IV in terms of mean survival times in groups of five or six rabbits. However, in tests with offspring of genetically resistant Australian wild rabbits, these strains clearly differed in virulence from classical Grade I strains, as judged by case-fatality rates. To differentiate between strains of myxoma virus then circulating in Australia, Parer *et al.* suggested that field strains should be tested on progeny from a closed colony derived from genetically resistant wild rabbits, or as an alternative, the offspring of laboratory females impregnated with stored semen from genetically resistant wild rabbits. Such colonies of rabbits have not yet been established.

Changes in the virulence of field strains

From 1959 onwards Victorian investigators used the protocols of Fenner and Marshall (see p. 94) to test the virulence of field strains of myxoma virus. Ignoring the doubts about the validity of using survival rates as a surrogate for lethality, the Victorian findings, as reported (Edmonds *et al.*, 1975) and as communicated to the authors (as unpublished results), are summarized in Table 8.1. Because of the warmer, drier climate in the Mallee region and the higher resistance of rabbits in this region, the results from the Mallee and elsewhere in Victoria are tabulated separately. The trend observed during the period 1951 to 1966 (see p. 172) was maintained into the early 1980s. Strains of Grade III virulence continued to predominate, and in the Mallee region increasing numbers of strains of increased virulence (Grades I and II) were observed after 1970. In other parts of Victoria more Grade II strains and fewer Grade IV strains were observed among the few samples tested in 1984–85. However, if Parer *et al.* (1994) are correct, it is likely that many of the strains isolated after the mid-1970s were more lethal than estimated from the mean survival times of the inoculated rabbits.

The few assays of the virulence of field strains that have been made in the 1990s (P.J. Kerr, 1998, personal communication) indicate that most strains were of high virulence, as judged by both lethality and survival times when small doses were inoculated intradermally in laboratory rabbits. Clinically, and in their short survival times, most looked very like classical Grade I strains such as the Standard Laboratory Strain (Table 8.2). A few attenuated (Grade IV) strains were also recovered.

Table 8.1. The virulence of strains of myxoma virus recovered from the field in the Mallee region and elsewhere in Victoria between 1951 and 1985, calculated on the basis of mean survival times (expressed as percentages).

Virulence grade	I	II	III	IV	V	
Mean survival time (days)	9–13	14–16	17–28	29–50	—	Number
Presumed case-fatality rate (%)	>99	95–99	70–95	50–70	<50	of samples
Collection area: south-eastern Australia						
1950–51	>99					1
1952–55	13.3	20.0	53.3	13.3	0	60
1956–58	0.7	5.3	54.6	24.1	15.5	432
Collection area: Mallee region of Victoria[a]						
1959–63	0	4.3	57.1	34.3	4.3	70
1964–66	2.0	0	64.7	31.3	2.0	51
1967–69	0	0	68.1	31.9	0	51
1970–74	1.0	6.9	77.5	14.7	0	102
1975–81	3.0	5.8	67.8	23.4	0	121
1984–85	0	23.0	69.2	7.6	0	13
Collection area: Victoria except for Mallee region[a]						
1959–63	2.1	12.4	61.2	19.5	4.7	379
1964–66	0.4	0.4	63.5	34.5	1.2	255
1967–69	0	0	61.6	36.4	2.0	198
1970–74	0	1.4	69.4	29.2	0	72
1975–81	0	0	65.8	34.2	0	91
1984–85	0	14.2	80.0	4.7	0	21

[a]Data kindly provided by R.C.H. Shepherd and J.W. Edmonds.

Table 8.2. The virulence of field isolates of myxoma virus recovered from diseased rabbits in south-eastern Australia between 1991 and 1994, tested in genetically unselected laboratory rabbits[a].

	Rabbits inoculated		Survival times	
Virus isolate	Number	Survivors	of fatal cases (days)	Virulence grade
Cooma/2–94 (NSW)	6	0	9, 9, 10, 11, 11, 12	I
Bendigo/7–92 (Vic)	6	0	9, 10, 11, 11, 13, 13	I
Brooklands/4–93 (NSW)	6	0	10, 11, 12, 13, 13	I
Gungahlin/1–91 (ACT)	6	3	23, 24, 27	IV

[a]P.J. Kerr (personal communication, 1998).

Changes in the Resistance of the Rabbit

Apart from environmental factors such as ambient temperature, the risk of concurrent infection, starvation, or predation, three physiological factors can influence the fate of a rabbit infected with a standard small dose of a known strain of myxoma virus, namely active immunity, passive immunity, and genetic resistance.

Active immunity

Ordinarily, rabbits that had recovered from myxomatosis completely resisted reinfection, but occasionally, months after the initial infection, developed localized lesions at the site of challenge inoculation (see p. 105). To avoid the effect of active immunity in tests of wild rabbits for genetic resistance, only antibody-negative rabbits were used.

Passive immunity

Maternal antibodies may afford some protection to very young progeny. Experimentally, the very young offspring of does with maternal antibodies had a higher percentage recovery after challenge infection, extended survival times and a reduced chance of infection after probing by infective mosquitoes than the very young progeny of animals without maternal antibodies (Fenner and Marshall, 1954). In testing rabbits for genetic resistance, animals were held until they were about 16 weeks old, in order to avoid possible effects of passive immunity or age susceptibility (Marshall and Fenner, 1958; Sobey, 1969).

It was considered that maternal immunity was unlikely to be important in mosquito-transmitted epidemics, since these usually occurred in summer, when most kittens would be at least eight weeks old. However, after the introduction of fleas, infection with myxomatosis could occur at any time of the year. Investigations were therefore carried out to determine whether maternal immunity might affect mortality due to flea transmission. Sobey and Conolly (1975) found that kittens born to four different immune does in pens heavily infested with fleas that had been contaminated with the virulent Lausanne strain of myxoma virus contracted fatal myxomatosis when they were 2–3, 4–6, 5–6 and 6–8 weeks old. Rabbit kittens attract fleas from the first week after birth (Mead-Briggs, 1964), and myxomatosis rapidly kills very young rabbits without maternal antibody (Fenner and Marshall, 1954). This experiment suggested that kittens were protected from infection for such a short time that maternal antibody was unlikely to interfere with the efficacy of myxomatosis produced by the introduction of contaminated fleas into rabbit burrows.

Genetic resistance

The possibility that the very stringent selection imposed by such a lethal virus would lead to greater genetic resistance in the progeny of the rare survivors had been foreshadowed as early as 1953 (Fenner, 1953). It was assumed that such genetic resistance, evident after all maternal antibodies had been lost, would be due to natural selection for genes that enhanced the resistance of rabbits to myxomatosis. In the late 1980s there was evidence from analysis of earlier data that there might be an additional 'sire effect' (Sobey and Conolly, 1986; see below). Experiments on genetic resistance were commenced in wild rabbits in 1953 (see p. 173) and in laboratory rabbits in 1954.

Breeding laboratory rabbits for resistance

Sobey (Fig. 8.8) began experiments on breeding for resistance in laboratory rabbits in 1954 and continued them until 1974 (Sobey, 1969; Sobey and Conolly, 1986). Initially, four strains of myxoma virus were used, but the percentage of recoveries increased so rapidly with two attenuated strains that attention was concentrated on tests with two strains, the prototype strain of Grade III virulence (KM13), and as the resistance increased, the Standard Laboratory Strain (Grade I virulence).

The idea was to breed only from rabbits that had recovered from infection, but problems arose because surviving bucks were often sterile, or became fertile only after extended periods (Sobey and Turnbull, 1956). Since one fertile buck could service many does, a more important concern in the early years was that very few does survived. Two stratagems were used to overcome this problem: the use of an elevated environmental temperature for 24 hours three or four days after infection with the highly virulent strain, which increased the survival rate slightly (Sobey *et al.*, 1968), and breeding from unselected does. Since unselected as well as selected does were used, the results (death or survival) were recorded by 'grade' rather than generation. Unselected rabbits were graded 0; if a rabbit of grade 0 recovered it was graded 1 (or its previous grade plus 1), offspring of a 0 × 1 mating were graded 0.5, and an 0.5 grade rabbit which recovered became grade 1.5, and so on. Continued selection, from 1968 to 1974,

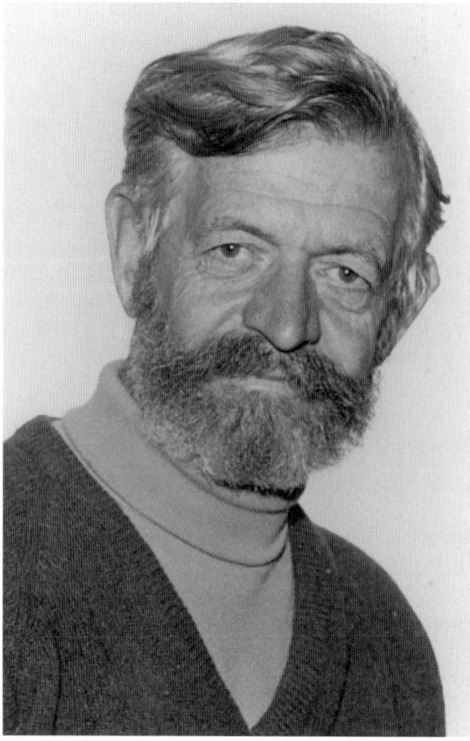

Fig. 8.8. William Rennie Sobey (1923–). Born in South Africa, Sobey served in the South African Air Force from 1941 until 1946, and then studied soil conservation at the University of Witwatersrand, graduating with a BSc in 1948. He then proceeded to the Department of Genetics of the University of Edinburgh, where he graduated with a PhD in animal genetics in December 1952. He was immediately recruited to the CSIRO Division of Animal Genetics, which was located in Sydney, and commenced studies on the genetic resistance of rabbits to myxomatosis. Later he was involved in the introduction of the European rabbit flea into Australia and its distribution among wild rabbits. In 1976 he transferred to the CSIRO Division of Wildlife Research in Canberra, where he worked on various aspects of rabbit control until his retirement in 1983.

was limited to three generations (Sobey and Conolly, 1986), and did not result in any increase in resistance. The results are summarized in Fig. 8.9. With both strains of virus the survival times of fatal cases were greatly extended as the selection process proceeded, and the percentage of recoveries rose steadily. The case-fatality rate in rabbits aged 17–53 weeks had fallen to 80% by selection grade 4, for KM13, over a period of nine years. The same drop in case-fatality rate was achieved by selection grade 6.5 for the Standard Laboratory Strain, after selection with that strain for a further five years. Intra-sire correlation, using survival time as an index or all-or-none probit analysis, gave an estimate of the heritability of resistance of about 35–40%.

Field observations of resistance

Studies on the increase of resistance in wild rabbits collected between 1953 and 1958 were described in Chapter 7. Subsequent studies were carried out by Victorian workers on rabbits collected from two parts of Victoria, Gippsland, a cool temperate area, and the Mallee, a hotter, drier area. Since by 1960 all Australian wild rabbits had become relatively highly resistant to the strain of moderate virulence which had been used in the earlier tests, strains of Grade I virulence were used for challenge. The results (Table 8.3) show that all animals became steadily more resistant and that the resistance of the Mallee rabbits was substantially greater than that of the Gippsland rabbits. The tests also demonstrated that there were substantial differences in the virulence of three strains that in unselected rabbits were all classified as of Grade I virulence, namely the Standard Laboratory, the Lausanne and the Glenfield strains.

The CSIRO Division of Wildlife and Ecology maintains a colony of wild rabbits at its headquarters in Canberra. Outbreaks of myxomatosis have regularly occurred in the ancestors of these rabbits over the past 40 years. In 1996 Kerr (personal communication, 1998) tested their genetic resistance using recently isolated strains, most of which had been shown to be highly virulent when tested in laboratory rabbits (see Table 8.2). All rabbits were serologically negative (by the sensitive ELISA) before being inoculated. They were inoculated with doses of 10^3 plaque-forming-units, i.e. ten times more than had

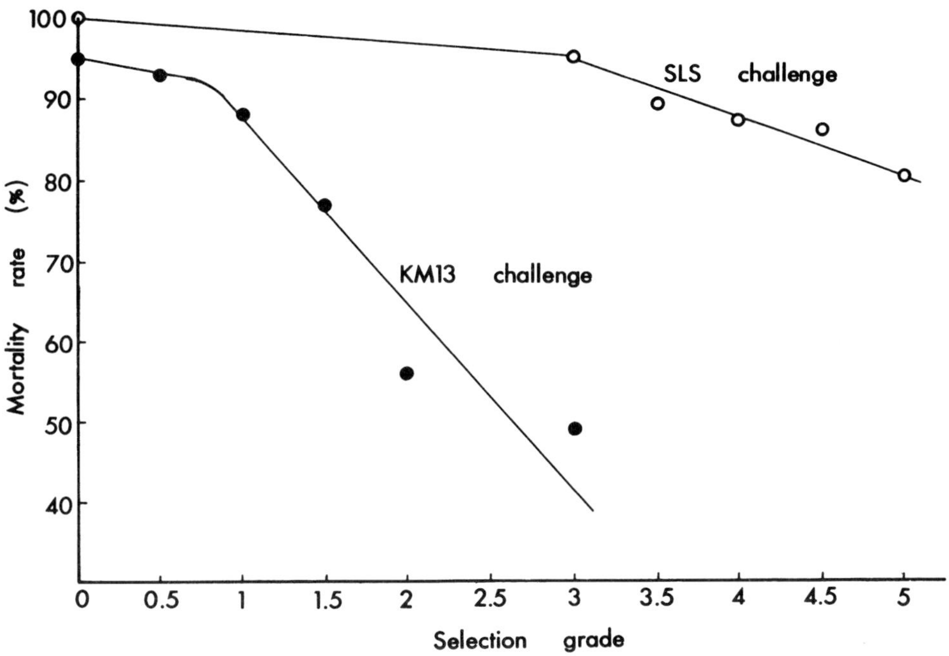

Fig. 8.9. Increases in the resistance of laboratory rabbits bred by mating males that had recovered from infection with two strains of myxoma virus with does that had recovered or were unselected. Selection Grade indicates the parentage of the rabbits under test: 0.5, one parent had recovered from myxomatosis; 1, both parents had recovered or male parent and grandparent had recovered, doe was unselected; 2, both parents and both grandparents had recovered, etc. Initial selection by inoculation with KM13 (Grade IIIA strain), from Selection Grade 3 by inoculation with Standard Laboratory Strain, which killed all unselected rabbits. Data from Sobey (1969).

Table 8.3. Case-fatality rates of non-immune rabbits challenged with strains of myxoma virus of Grade I virulence.

	Case-fatality rate (%)		
	Standard laboratory strain	Glenfield strain	Lausanne strain
Selective breeding experiments (laboratory rabbits)[a]			
grade 0 (no selection)	99		
grade 4	88		
grade 6.5	79		
Gippsland wild rabbits[b]			
1961–66	94	98	
1967–71	90	99	
1972–75	85	98	*c.* 100
1976–81	79	95	*c.* 100
Mallee wild rabbits[b]			
1961–66	68	98	
1967–71	66	94	
1972–75	67	96	*c.* 100
1976–81	60	91	98

[a]Data from Sobey (1969).
[b]Data kindly provided by R.C.H. Shepherd and J.W. Edmonds.

been used in the virulence tests in laboratory rabbits shown in Table 8.2. Using nine or ten wild rabbits in each test, there were some striking differences from those obtained in genetically unselected laboratory rabbits, and between strains of virus that gave very similar results in laboratory rabbits (Table 8.4). With each virus strain, some wild rabbits showed no clinical signs of myxomatosis except the local lesion at the site of inoculation. In no case did all rabbits show signs of severe disease, and case-fatality rates after inoculation with strains of Grade I virulence varied from zero to 50%. Severe cases that ultimately recovered, or survived for over three weeks before dying, would provide opportunities for mosquito transmisson; these were much more common in wild rabbits challenged with Grade I strains than with those challenged with the Grade IV strain (Gungahlin).

The 'sire effect' in resistance

It had been assumed that all the increase in resistance in laboratory and wild rabbits just described had a genetic basis, i.e. natural selection for progeny which had inherited genes favouring their resistance to myxomatosis. However, in re-analysis of their laboratory data, Sobey and Conolly (1986) observed that following inoculation with myxoma virus the survival times of offspring of genetically unselected does mated to bucks that had recovered from myxomatosis were extended beyond their expected range. The effect seemed to be greatest when the same strain of virus was used to challenge the offspring as had been used in the buck's challenge inoculation. Further, unselected does which had had a previous mating with a recovered buck appeared to confer a non-genetic survival advantage on their subsequent offspring.

Subsequently a similar effect was observed in studies of wild rabbit populations from arid, semi-arid, and subalpine climatic regions (Williams *et al.*, 1990; Williams and Moore, 1991). First generation progeny of rabbits from each climatic region were bred and reared in the laboratory, immunized with a highly attenuated strain of myxoma virus and mated randomly within regional groups. Random samples of about 50 members of second generation progeny of each regional group (approximately equal numbers of each sex) were inoculated at about one year of age with either sterile medium, a highly attenuated strain or the Lausanne strain of myxoma virus. Only Lausanne virus produced deaths from acute myxomatosis, but 12 of the 49 rabbits survived. Analysis of the results according to sire and dam and the length of time that had elapsed between prior immunization and birth showed that acquired 'paternal immunity' enhanced the resistance of progeny born within seven months of paternal infection with myxoma virus, reducing the risk of death by about 25%.

Table 8.4. Responses of wild rabbits to strains of myxoma virus isolated in the field in the 1990s[a].

Virus isolate	Virulence grade (laboratory rabbits)[b]	Clinical signs[c]	Severe disease	Lethality (deaths)[d]	Survival times of fatal cases (days)
Cooma/2–94 (NSW)	I	7/10	6/10	4/10	17, 24, 33, 37
Bendigo/7–92 (Vic)	I	5/10	4/10	0/10	—
Brooklands/4–93 (NSW)	I	8/9	6/9	5/9	25, 25, 25, 26, 35
Gungahlin (ACT)	IV	8/10	2/10	1/10	18
Lausanne	I	7/9	7/9	4/9	14, 21, 25, 25, 34

[a]P.J. Kerr (personal communication, 1998).
[b]See Table 8.2.
[c]Other than lesion at site of inoculation.
[d]Before 50 days.

Later Parer *et al.* (1995) tested the resistance of progeny of rabbits from five locations in Australia and from New Zealand, which differed in their resistance to myxomatosis (Parer *et al.*, 1994), by challenge with Lausanne virus and six Australian field strains of myxoma virus of differing virulence. The results were subjected to analysis of variance, taking account of all factors which could have influenced their survival. Male rabbits transmitted to their offspring some factor that increased by 22% the offspring's probability of surviving challenge infection with myxoma virus, if they had been born within 9 months of the sire challenge. The possible mechanism of this effect is discussed in Chapter 13.

Significance of the sire effect in the field

Parer *et al.* (1995) point out that the sire effect may have contributed to the observed initial rapid increase in the resistance of rabbits observed in the field (Marshall and Fenner, 1958; Marshall and Douglas, 1961) and in the animal house (Sobey, 1969). Its absence may also provide a partial explanation of epizootics that have resulted in unexpectedly high mortalities. In all outbreaks studied since the mid-1960s in which survival rates have been less than 15% (Dunsmore *et al.*, 1971; Williams *et al.*, 1972a, 1973b; Wheeler and King, 1985), the populations had not been exposed to myxoma virus for two or more years. In these circumstances there would be very few immune rabbits surviving, but in addition none of the susceptible rabbits would have had the added survival advantage produced by the sire effect. Edmonds and Shepherd have reported that non-immune rabbits from the Mallee are consistently more resistant to challenge infection with myxoma virus than those from the Gippsland district of Victoria (see Table 8.3). Although the rabbits from the two regions may have been genetically different and the Mallee rabbits have been subjected to more severe selection for genetic resistance because of the more frequent epizootics there, the sire effect may also have played a part in producing this difference.

Environmental Factors Affecting the Severity of Myxomatosis

It has long been known that in the laboratory the severity of myxomatosis is greatly ameliorated by holding infected rabbits at temperatures of 38–40°C (Parker and Thompson, 1942). This observation was confirmed by Marshall (1959), who also found that very low temperatures exacerbated the severity of the disease (see p. 108). Under field conditions rabbits are never exposed to temperatures as high as 40°C, because they shelter in burrows during the day. However, Mykytowycz (1956) found that wild rabbits infected with an attenuated strain of myxoma virus and maintained in unheated sheds were more likely to die in winter (monthly mean minimum temperatures between 0° and 5°C) than in summer. Observations in different parts of Australia showed that environmental temperature exerts an influence on mortality rates from myxomatosis in the field, especially in cold localities. In the course of long-term studies of rabbits at Snowy Plains, a subalpine region of south-eastern Australia, Dunsmore *et al.* (1971) observed an epidemic that infected all the rabbits in the area and killed some 80% of the population, spreading through the area in 3–4 weeks in February 1967. This speed of spread suggested transmission by a flying vector. No further cases were seen until May 1969, when a slowly progressive epidemic began in late May and ended in August 1969, the temperature being below freezing for most days from June to August. There were no mosquitoes or fleas in the area. Although the mite, *Listophorus gibbus*, was suspected as a vector, it does not probe through rabbit skin (Shepherd and Edmonds, 1977b). The case-fatality rate was 86%, and laboratory tests of the causative virus showed that it was of Grade III virulence (70–90% case-fatality rate in laboratory rabbits).

European rabbit fleas were introduced in November 1970 and myxomatosis recurred in the summer months, February to mid-April 1972. This outbreak appeared to be flea-transmitted and the case-mortality rate was only 33% (Dunsmore and Price, 1972). The high case-fatality rate of 86% during the winter epidemic was thought to be due to the extreme cold to which the rabbits were exposed.

The Source of Myxoma Virus in the Field, and the Question of Latency and Reactivation

Reviewing the results of tests for anti-myxoma antibodies in various groups of rabbits over a period of five years, Williams *et al.* (1973a) noticed that titres fluctuated over time, falling below measurable levels and later rising again, in the absence of an observable outbreak of myxomatosis. They speculated that the rising titres might be due to the reactivation of latent virus. In both of the epidemics at Snowy Plains, described above, and in winter epidemics in Canberra associated with concurrent presence of the Lausanne strain and field strains of myxoma virus (Williams *et al.*, 1972b; Fullagar, 1977), the authors were puzzled about the source of the field strains. No myxomatosis had been seen in the closely studied Snowy Plains region between February 1967 and May 1971, nor between August 1971 and February 1972. The region was surrounded by a belt of rugged timber country in which moderate numbers of rabbits found refuge but there was little breeding. Any flying vector would have had to cross for several kilometres over rugged high country (>2000 metres above sea level), at a time of year unfavourable for flying vectors. From observations on the scattered warrens, Dunsmore *et al.* (1971) considered that several different rabbits initiated sub-centres of infection and that these rabbits had survived a previous epidemic. One tagged rabbit had been trapped in September 1968 and gave a positive serological test for myxomatosis. It was a socially dominant male and was found at the same warren in mid-April 1969 suffering from myxomatosis. It was seen several times during the next few weeks with lesions, but finally recovered.

Williams *et al.* (1972a) then investigated the possibility that overwintering of myxomatosis was due to the presence of 'carrier' animals, in which virus persisted in a latent state, but could be reactivated with the production of infectious lesions, possibly only after severe stress. Their suspicions were heightened by observations of an epidemic in an enclosed rabbit population in Canberra in December 1969. The population comprised 42 adult rabbits with antibodies, 19 with no antibodies and the 120 offspring of these rabbits. Eleven cases of myxomatosis were noted in the enclosure on December 17, of which seven were adults that had previously shown circulating antibodies, two were susceptible adults and two were juveniles. If the virus had been introduced by a flying vector it must have selectively transmitted the virus to rabbits in which antibodies had previously been detected, since these were outnumbered by susceptibles 42:139 at the time. In an attempt to produce reactivation experimentally, they then placed 20 antibody-positive rabbits in individual cages in a temperature-controlled room. Over the next seven months these rabbits were subjected to two periods of heat stress and one of cold stress, and during one period of heat stress they were hormonally stressed (ACTH). One rabbit showed generalized signs of myxomatosis during a period of heat stress, producing characteristic lesions from which myxoma virus was isolated.

These observations and experiments have profound implications for the epidemiology of myxomatosis. They do not accord with what we know about poxvirus infections in general. Although long-lasting infectious lesions may occur in *Sylvilagus* rabbits infected with fibroma virus (Kilham and Dalmat, 1955; Dalmat and Stanton, 1959), latency and reactivation has not been described for poxvirus infections. However, the evidence assembled by Williams *et al.* is persuasive.

Overall Effectiveness of Myxomatosis

With the inevitable decrease in interest that accompanied the establishment of myxomatosis as part of the ecological scene, it became more and more difficult to determine the extent to which it was still exerting control over rabbit numbers. Casual observations attest to the much more frequent sightings of dead kangaroos than dead rabbits on country roads in agricultural areas of southern Australia, a dramatic reversal of the situation in the late 1940s, when a driver would rarely travel a kilometre without seeing several dead rabbits.

Myers (1970) made an interesting assessment of the continuing effects of myxomatosis over the first 20 years of its presence. He made accurate counts of rabbit numbers, warrens and active burrows in four areas where trial releases of myxoma virus had been made in 1950 (see Chapter 6). In all four areas, the decrease in all indicators of rabbit infestation was dramatic (Table 8.5).

Several investigations at the major CSIRO study site at Lake Urana have provided evidence of the position up to about 1980, including the effects of introductions of virulent virus. Williams and Parer (1972) systematically observed a 280 hectare rabbit-proof paddock each year from 1968 to 1971. European rabbit fleas were released there in June 1968 and were in all warren areas by November 1969. An epidemic occurred each year, but factors other than myxomatosis caused most of the unusual high mortality that occurred each year, especially in the younger rabbits. Transmission was rapid in 1968 and 1969, with *Anopheles annulipes* as the vector, and slower in 1970, when *Spilopsyllus cuniculi* was the vector, but all three epidemics peaked early in November. The 1971 epidemic peaked in December and transmission was rapid, but the vector was not determined. Overall, some 75% of the rabbits older than 3 months and between 16% and 55% of younger rabbits contracted myxomatosis. In contrast to the dramatic mortalities of over 99% and 90% seen in 1952 and 1953, only 40% of the infected older rabbits and 50% of the infected younger rabbits died of myxomatosis. Circumstantial evidence suggested the first three epidemics may have been initiated by the reactivation of latent virus in rabbits that were more than twelve months old at the start of the epidemic.

Many factors contribute to rabbit numbers. In an effort to focus on the influence of myxomatosis, Parer *et al.* (1985) examined the effect of introducing

Table 8.5. Long-term effects of myxomatosis on rabbit numbers at four CSIRO study sites in south-eastern Australia (Myers, 1970).

	Before myxomatosis (1950)	After myxomatosis (1970)
Rutherglen, Vic, 5.9 hectares		
Number of warrens	15	0
Number of rabbits	1000	3
Active burrows	850	4
Coreen, NSW, 7.7 hectares		
Number of warrens	26	0
Number of rabbits	400	2
Active burrows	600	2
Balldale, NSW, 7.7 hectares		
Number of warrens	150	5
Number of rabbits	300	20
Active burrows	1250	17
Urana, NSW, 154 hectares		
Number of rabbits	5000	11

an immunizing strain of myxoma virus into populations of flea-infested rabbits in one of the CSIRO study sites, Lake Urana, each year from 1978 to 1980. Mosquito-transmitted myxomatosis occurred annually at this site. The immunizing strain was a field strain (FS98) that had earlier been shown to kill some 30% of genetically unselected adult rabbits (Sobey *et al.*, 1983). However, all Lake Urana rabbits, both adults and kittens, survived inoculation with FS98 because they were genetically more resistant as a result of exposure to repeated outbreaks of myxomatosis. The 280 hectare site was divided by rabbit-proof fences into four pens with similar large populations of rabbits. From 1978 to 1980 the immunizing strain was introduced by placing fleas contaminated with that virus into the burrows on two pens and fleas contaminated with the virulent Lausanne strain into burrows on the other two pens. Each year cases of myxomatosis due to local field strains affected populations in all four pens. Within two years the numbers of adult rabbits in one of the immunized populations had increased by a factor of 8 and the other by a factor of 12, suggesting that even in animals with a high level of genetic resistance myxomatosis was still an important factor in suppressing rabbit populations.

By the mid-1980s, rabbits were under reasonable control in much of the good agricultural land of Australia, and myxomatosis was playing a part in maintaining this level of control. However, in the arid rangelands, which constitute such a large part of inland Australia, rabbits remained a major pest (see p. 260). Much of this land was unoccupied Crown land and pastoral holdings were very extensive, so that it was not economic to carry out rabbit control by poisoning or ripping. The occurrence of myxomatosis was erratic, because mosquitoes were rarely present in large numbers and the European rabbit flea was poorly adapted to the hot conditions. The main reason for introducing the Spanish rabbit flea was to increase the transmission of myxomatosis amongst these populations, and rangelands rabbits are the principal target of rabbit haemorrhagic disease virus (see p. 265).

New Initiatives: Immunocontraception for Rabbit Control

As described earlier, the Vertebrate Biocontrol Centre was established in 1992 primarily to investigate the possibility of controlling rabbits and foxes by immunocontraception; later mice were added. The concept of immunocontraception is illustrated in Fig. 8.10. Four research programmes relating directly to immunocontraception were established: ecology, virology, reproduction and immunobiology; their results to date are summarized in the Annual Reports of the Vertebrate Biocontrol Centre. All aspects of the problem were discussed at the Fourth International Conference on Fertility Control for Wildlife Management in Queensland on 8–11 July 1996 (Proceedings, 1997).

Feasibility of immunocontraception with a genetically-engineered virus

It was clearly desirable to demonstrate that the concept of immunocontraception by infection with a genetically-engineered virus was feasible. Since mice are much more convenient laboratory animals than rabbits, experiments were carried out by the Immunobiology Program with a natural pathogen of mice, ectromelia virus, which belongs to the same family (*Poxviridae*) as myxoma virus. Mice were infected with a recombinant ectromelia virus expressing a mouse zona pellucida glycoprotein (ZP3), a protein found on the outside of ova (Jackson *et al.*, 1997). Female mice infected with the recombinant virus were infertile for five to nine months following infection, but became fertile as the level of anti-ZP3 antibody fell. Reinfection with the recombinant virus boosted the anti-ZP3 response and restored infertility.

Research on rabbit reproduction

Although not discussed earlier in this book, research on the reproduction of the

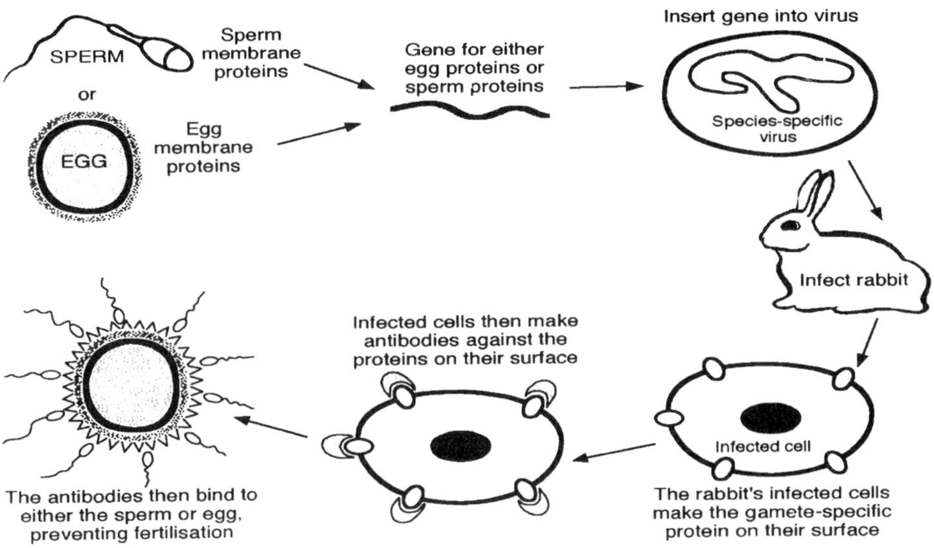

Fig. 8.10. The concept of virus-vectored contraception for rabbit control. Immunocontraceptive components of the sperm or egg are identified, the encoding gene is cloned and inserted into the myxoma virus genome. When the virus replicates in the rabbit the immunocontraceptive protein is produced and induces an immune response, which neutralizes the antigen on the gametes, thus destroying fertility. From Holland and Jackson (1994), with permission.

rabbit has been an important component of CSIRO research on rabbits since the 1960s (Myers *et al.*, 1994). By 1987 sufficient information was available to suggest that immunosterilization might be a more effective method of rabbit control than conventional mortality agents (Tyndale-Biscoe, 1997). In many wild animals, reproductive success is closely linked to high rank in the social hierarchy of the population. There is no evidence that dominant females suppress breeding by subordinates among wild rabbits, but the survival of the kittens of dominant females is significantly greater than those of subordinate females (Mykytowycz and Fullagar, 1973; Cowan, 1987). Thus sterilization of dominant animals could affect fecundity of the population disproportionately provided that the sterilized rabbits maintained or improved their status in the social hierarchy, and that a sufficiently high proportion of the population was sterilized. This ruled out the use of sex hormones for sterilization, but agents that affected gametes, fertilization or implantation could prove useful if they could be presented to the target animals in such a way as to induce a strong and persistent immune response. Use of recombinant myxoma virus encoding genes for proteins involved in fertilization or implantation presented a possible method for sterilizing rabbits without affecting their sexual activity or social status. Scientists of the Vertebrate Biocontrol Centre therefore faced five key questions (Tyndale-Biscoe, 1997):

1. What proportion of females in a wild population must be sterile in order to reduce to a useful extent the rate of growth of the population?
2. Can components of rabbit gametes or reproductive tract secretions provoke a long-lasting immunocontraceptive response in rabbits?
3. Can the encoding genes be inserted into the genome of myxoma virus in such a way that they effectively immunosterilize rabbits infected with the recombinant virus?

4. Can such a recombinant virus spread effectively in populations in which myxomatosis is enzootic?
5. Can this be achieved in such a way that it does not endanger other species of animals in Australia or rabbits in countries where they might be valued animals?

Research on population ecology and social behaviour

In 1992 two large field experiments were set up to answer the first question, what proportion of the females of that very fecund pest animal, the rabbit, needed to be sterilized to obtain a useful reduction in abundance. These trials ran for four years, one in New South Wales, near Canberra, and one in south-western Western Australia (Williams and Twigg, 1996). The results of the New South Wales experiments have not yet been fully analysed, but agree well with those in Western Australia, which were described by Twigg *et al.* (1999). In both experiments 12 sites with free-ranging rabbit populations of up to 100–200 rabbits were selected. Sites were assigned a level of female sterility of 0%, 40%, 60% or 80% of the adult female population of 25–100 rabbits (three replicates of each), by tubal ligation, which did not interfere with hormones and reproductive behaviour and, as controls, sham operation. In succeeding years a similar proportion of female recruits were sterilized on each site. There was a significant decrease in rabbit productivity with increasing sterility level (Fig. 8.11), although there was a compensatory increase in adult and kitten survival on the high sterility sites. In all sterility cohorts, sterile females lived longer than other rabbits. Overall, there was a marked decrease in the seasonal peaks of rabbit abundance at the high sterility sites. The experiment demonstrated that a small sustained, long-term reduction in rabbit abundance could be achieved if 60–80% of the female rabbits could be prevented from breeding. It was considered that this could be achieved by immunocontraception if the myxoma vector retained good transmissibility and all infected rabbits that survived remained sterile for life. These are difficult provisions to fulfil, and provide the principal targets of the virology and reproduction programmes.

It had been thought that the imposition of female sterility might alter the epidemiology of myxomatosis either by reducing the number of susceptible kittens or reducing the number of fleas because of the reduced numbers of pregnant rabbits. Although it was not deliberately introduced, myxomatosis occurred as annual spring–summer epizootic in 1991–92, 1992–93 and late 1993, with a delayed outbreak, probably flea-borne, occurring between March and June 1995 (see p. 186). Imposed levels of female sterility as high as 80% for three years had no significant effect on the proportion of seropositive rabbits (Kerr *et al.*, 1998).

Research on virology

To determine whether myxoma virus could be used as a vector for an immunocontraceptive gene, Kerr and Jackson (1995) inserted the influenza virus haemagglutinin gene into the myxoma virus genome and demonstrated that rabbits infected with this recombinant developed high plasma antibodies to the haemagglutinin. Plasmid transfer vectors were constructed which could be inserted in an intergenic region of myxoma virus. Recombinant myxoma viruses express the product in cell culture (Jackson and Bults, 1992), and in infected rabbits, without associated attenuation of viral virulence (Jackson *et al.*, 1996). Investigations have also begun on the spread among wild rabbits of a strain of myxoma virus which can be recognized by polymerase chain reaction (PCR) tests, in competition with enzootic strains (Robinson *et al.*, 1997), as a preliminary to studying the spread of a recombinant myxoma virus in the field.

The research of the Immunobiology Group complements that of the Virology Group by designing methods to enhance the immune response to the expressed protein, so that immunosterilization will be effective and prolonged. Since this is more likely to be achieved by high levels of antibodies in the reproductive tract,

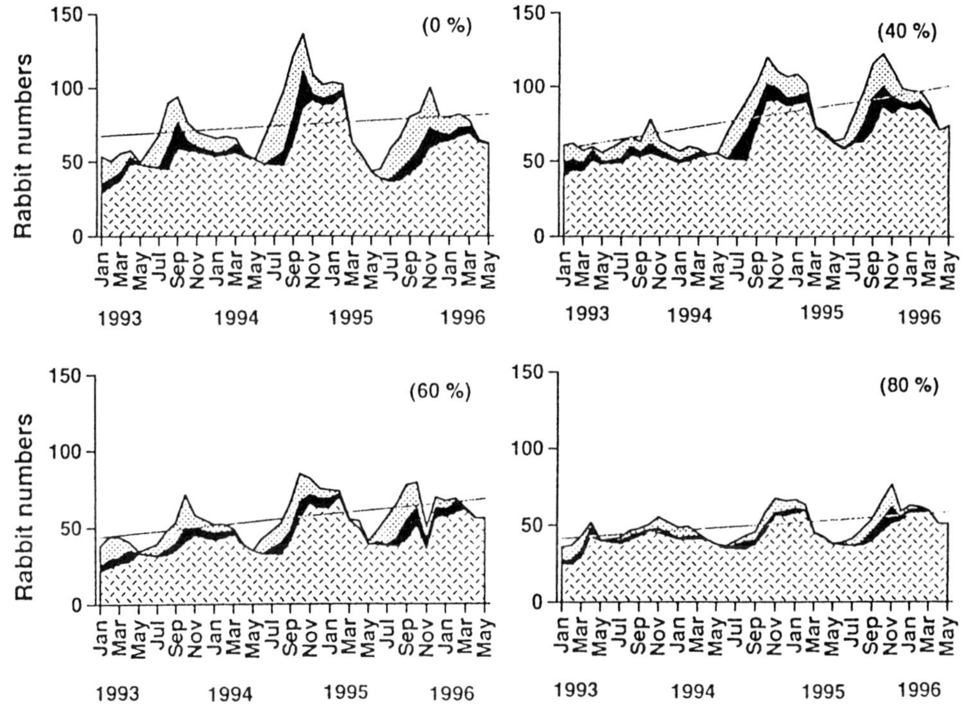

Fig. 8.11. Mean changes in the proportion of adult (hatching), subadult (solid) and kitten (stippled) rabbits in three sites in which various proportions of the female rabbits were surgically sterilized. From L.E.Twigg *et al.* (1999, unpublished results).

attention is focused on enhancement of mucosal immunity, both by the method of antigen delivery and the enhancement of the activity of the cytokines which act as B cell (antibody) adjuvants (Ramsay and Ramshaw, 1997).

Research on reproduction

The prime target of the Reproduction Programme was to determine which of several potential target antigens, reproductive hormones, membrane proteins of spermatozoa or proteins of the zona pellucida of the fertilized ovum, would be the most suitable. Reproductive hormones were excluded because of their lack of specificity and possible effects on reproductive behaviour. Attempts were therefore made to identify target molecules on the membrane of both spermatozoa and ova (Holland and Jackson, 1994).

Immunization of female rabbits with the sperm protein PH-20, inoculation of which had been demonstrated to render both female and male guinea-pigs infertile (Primakoff *et al.*, 1988), produced a strong systemic immune response but no detectable mucosal response in the vagina, even after inoculation via Peyer's patches (Holland *et al.*, 1997). Further studies showed that serum antibody did not enter the oviduct and uterus to a significant extent, and induction of a strong mucosal immune response within the female genital tract was difficult. Attention was therefore turned to ZPB protein, a component of the rabbit zona pellucida, since it had already been shown that serum antibody could enter ovarian follicles and access the developing oocyte. Approximately 70% of female rabbits immunized with ZPB produced as a recombinant protein in a vaccinia virus T7 system, and boosted twice with the protein in Freund's incomplete adjuvant, were rendered infertile and had high serum antibody titres. Similar results were obtained in rabbits infected with a recombinant myxoma virus expressing ZBP and boosted twice with protein, but only 25% of females infected with the recombinant virus alone were sterilized (Kerr *et al.*, 1998). Much more work is needed before this finding can be turned into a useful method for rabbit control.

Legal and ethical issues

Animal welfare groups strongly support immunosterilization as a humane method of controlling wild animal populations, whether they are feral animals that may be a pest or protected animals such as elephants or koalas whose numbers may become too large locally. However, there are other aspects to be considered if an infectious agent is used to produce the sterilization, which Tyndale-Biscoe (1997) discusses at some length. He points out that the use of infectious agents for immunosterilization has both national and international implications. Like the release of virulent viruses for pest control (see p. 250), the release of genetically-engineered viruses into the environment has potential consequences that need to be discussed publicly, so that informed decisions can be made about their future use. Fortunately, there is already something of a precedent, since baits containing recombinant vaccinia virus–rabies virus glycoprotein have been widely and successfully used since 1989 in Belgium and France, for the control and local eradication of fox rabies, and in the United States, for the control of raccoon rabies, with no adverse effects (Pastoret *et al.*, 1997). It was found that although vaccinia virus has a wide host range and the baits could potentially infect other animals, the infections that it caused (in foxes and raccoons) were not contagious.

In Australia, the main risk in the present proposals is that the recombinant virus may infect animals other than those of the target population. Once released, such a virus cannot be recalled. Myxoma virus has been spreading among rabbits in Australia and Europe for almost 50 years without any evidence of the infection of any other animals, except, very rarely, the European hare. However, rabbits, both wild and domesticated, are valued animals in some European countries, and myxoma virus can infect several species of *Sylvilagus* rabbits endemic in the Americas. For these reasons, the International Union for Conservation of Nature/Status Survey and Conservation Lagomorph Specialist Group has expressed strong reservations about the use of genetically engineered myxoma virus. Tyndale-Biscoe (1997) outlines the sort of risk assessment that would be required to address these concerns.

Endnotes

[1]Basser Library Archives MS143/25/5A. Letter from K.H. Strong to Fenner, 26 November 1997.

[2]Basser Library Archives MS143/25/5A. Letter from A.L. Dyce to Fenner, 13 January 1997.

[3]Report on conference 'Rabbits and their Control', organized by the Australian Meat Research Committee, September 1983.

[4]Basser Library Archives MS143/25/5A. Letter from B.D. Cooke to Fenner, 4 August 1997.

[5]Basser Library Archives MS143/25/5A. Letter from D. Berman to Fenner, 3 February 1998.

[6]Conference on Fertility Control in Wildlife, University of Melbourne, November, 1990. Program in Tyndale-Biscoe (1991); speaker's papers available from Australian and New Zealand Federation of Animal Societies, PO Box 1023, Collingwood, Victoria 3066, Australia.

[7]Tyndale-Biscoe (personal communication, 1997) says that the idea arose from what was almost a throw-away remark by Dr S.J. Robbins in an interview with journalists about early work on genetic engineering of myxoma virus. After discussing the possibility of introducing genes that might alter the virulence of myxoma virus, perhaps by introducing genes for bacterial toxins, he said (*The Canberra Times*, 17 February 1987): 'A more humane approach, which might accommodate the

objections of the animal-rights movement, might be to equip the virus with a mammalian gene that would synthesize some sort of natural contraceptive, so that rabbits would be unable to breed, and would simply die out naturally'.

[8]Basser Library Archives MS143/25/5A. Paper in preparation: Cumming, S.A., Graves, S., Adler, B. and Edmonds, J.W.E. Respiratory tract infections of wild rabbits (*Oryctolagus cuniculus*) in Victoria, Australia.

[9]Personal communications from J.W. Edmonds and B.D. Cooke to Fenner, June and August, 1997. See also Basser Library Archives MS143/25/5A. Letter from J.W. Edmonds to Fenner, 24 April 1998.

References

Annual Reports (1993–1994, 1994–1995, 1995–1996) *Annual Reports of the Cooperative Research Centre for the Biological Control of Vertebrate Pests.* CSIRO, Canberra.

Bartholomaeus, F.W. (1991) Rabbit Fleas and myxomatosis: Update on the Spanish connection. Unpublished data. In: *Working Papers; 9th Australian Vertebrate Pest Control Conference, Adelaide 1991*, pp. 101–105. Quoted with the author's permission.

Cooke, B.D. (1983) Changes in the age-structure and size of populations of wild rabbits in South Australia, following the introduction of European rabbit fleas, *Spilopsyllus cuniculi* (Dale), as vectors of myxomatosis. *Australian Wildlife Research* 10, 105–120.

Cooke, B.D. (1984) Factors limiting the distribution of the European rabbit fleas, *Spilopsyllus cuniculi* (Dale) (Siphonoptera) in inland South Australia. *Australian Journal of Zoology* 32, 493–506.

Cooke, B.D. (1987) The European rabbit flea in Australia. Unpublished data. In: *Working Papers; 8th Australian Vertebrate Pest Control Conference, Coolangatta, 1986,* pp. 81–83. Quoted with the author's permission.

Cooke, B.D. (1990) Notes on the comparative reproductive biology and laboratory breeding of the rabbit flea *Xenopsylla cunicularis* Smit (Siphonaptera:Pulicidae). *Australian Journal of Zoology* 38, 527–534.

Cooke, B.D. (1995) Spanish rabbit fleas, *Xenopsylla cunicularis* in arid Australia: a progress report. In: *Proceedings of the 10th Australian Vertebrate Pest Control Conference, Hobart, 1995*, pp. 399–401.

Cooke, B.D. and Skewes, M.A. (1988) The effects of temperature and humidity on the survival and development of the European rabbit flea *Spilopsyllus cuniculi* (Dale). *Australian Journal of Zoology* 38, 649–659.

Cowan, D.P. (1987) Group living in the European rabbit (*Oryctolagus cuniculus*) mutual benefit or resource localization? *Journal of Animal Ecology* 56, 779–795.

Dalmat, H.T. and Stanton, M. (1959) A comparative study of the Shope fibroma in rabbits in relation to transmissibility in mosquitoes. *Journal of the National Cancer Institute* 22, 593–615.

Dunsmore, J.D. and Price, W.J. (1972) A non-winter epizootic of myxomatosis in subalpine south-eastern Australia. *Australian Journal of Zoology* 20, 405–409.

Dunsmore, J.D., Williams, R.T. and Price, W.J. (1971) A winter epizootic of myxomatosis in subalpine south-eastern Australia. *Australian Journal of Zoology* 19, 275–286.

Edmonds, J.W., Nolan, I.F., Shepherd, R.C.H. and Gocs, A. (1975) Myxomatosis: the virulence of field strains of myxoma virus in a population of wild rabbits (*Oryctolagus cuniculus* L.) with high resistance to myxomatosis. *Journal of Hygiene* 74, 417–418.

Fenner, F. (1953) Host parasite relationships in myxomatosis of the Australian wild rabbit. *Cold Spring Harbor Symposia in Quantitative Biology* 18, 291–294.

Fenner, F. and Chapple, P.L. (1965) Evolutionary changes in myxoma virus in Britain. An examination of 222 naturally occurring strains obtained from 80 counties during the period October–November 1962. *Journal of Hygiene* 63, 175–185.

Fenner, F. and Marshall, I.D. (1954) Passive immunity in myxomatosis of the European rabbit (*Oryctolagus cuniculus*): the protection conferred on kittens born by immune does. *Journal of Hygiene* 52, 321–336.

Fenner, F. and Woodroofe, G.M. (1965) Changes in the virulence and antigenic structure of strains of myxoma virus recovered from Australian wild rabbits between 1950 and 1964. *Australian Journal of Experimental Biology and Medicine* 43, 359–370.

Fisher, P.M. and Marsh, C.A. (eds) (1996) *Humaneness and Vertebrate Pest Control.* Proceedings of the Seminar Held March 27th 1996. Agriculture Victoria, Department of Natural Resources and Environment, Melbourne, 65 pp.

Fullagar, P.J. (1977) Observations on myxomatosis in a rabbit population with immune adults. *Australian Wildlife Research* 4, 263–280.
Holland, M.K. and Jackson, R.J. (1994) Virus-vectored immunocontraception for control of wild rabbits: identification of target antigens and construction of recombinant viruses. *Reproduction, Fertility and Development* 6, 631–642.
Holland, M.K., Andrews, J., Clarke, H. and Hinds, L. (1997) Selection of antigens for use in a virus-vectored immunocontraceptive vaccine: PH-20 as a case study. *Reproduction, Fertility and Development* 9, 117–124.
Jackson, R.J. and Bults, H.G. (1992) A myxoma virus intergenic transient dominant selection vector. *Journal of General Virology* 73, 3241–3245.
Jackson, R.J., Hall, D.F. and Kerr, P.J. (1996) Construction of recombinant myxoma virus expressing foreign genes in different intergenic sites without associated attenuation. *Journal of General Virology* 77, 1569–1575.
Jackson, R.J., Maguire, D.J., Hinds, L.A. and Ramshaw, I.A. (1997) Infertility in mice induced by a recombinant ectromelia virus expressing mouse zona pellucida glycoprotein 3. *Biology of Reproduction* 58, 152–159.
Kerr, J.D. and Jackson, R.J. (1995) Myxoma virus as a vaccine vector for rabbits: antibody levels to influenza virus haemagglutinin presented by a recombinant myxoma virus. *Vaccine* 13, 1722–1726.
Kerr, J.D., Twigg, L.E., Silvers, L., Lowe, T.J. and Forrester, R.I. (1998) Serological monitoring of the epidemiology of myxoma virus to assess the effects of imposed fertility control of female rabbits on myxomatosis. *Wildlife Research* 25, 123–131.
Kerr, P.J., Jackson, R.J., Robinson, A.J., Swan, J., Silvers, L., French, N., Clarke, H., Hall, D.F. and Holland, M.K. (1999) Infertility in female rabbits (*Oryctolagus cuniculus*) alloimmunized with rabbit zona pellucida protein ZPS either as a purified recombinant protein or expressed by recombinant myxoma virus. *Biology of Reproduction* (in press).
Kilham, L. and Dalmat, H.T. (1955) Host-virus-mosquito relations of Shope fibromas in cottontail rabbits. *American Journal of Hygiene* 61, 45–54.
King, D.R. and Wheeler, S.H. (1985) The European rabbit in south-western Australia. I. Study sites and population dynamics. *Australian Wildlife Research* 12, 183–196.
King, D.R., Oliver, A.J. and Wheeler, S.H. (1985) The European rabbit flea *Spilopsyllus cuniculi,* in south-western Australia. *Australian Wildlife Research* 12, 227–236.
Marshall, I.D. (1959) The influence of ambient temperature on the course of myxomatosis in rabbits. *Journal of Hygiene* 57, 484–497.
Marshall, I.D. and Douglas, G.W. (1961) Studies in the epidemiology of infectious myxomatosis of rabbits. VIII. Further observations on changes in the innate resistance of Australian wild rabbits exposed to myxomatosis. *Journal of Hygiene* 59, 117–122.
Marshall, I.D. and Fenner, F. (1958) Studies in the epidemiology of infectious myxomatosis of rabbits. VI. Changes in the innate resistance of Australian wild rabbits exposed to myxomatosis. *Journal of Hygiene* 56, 288–302.
Marshall, I.D. and Fenner, F. (1960) Studies in the epidemiology of infectious myxomatosis of rabbits. VII. The virulence of strains of myxoma virus recovered from Australian wild rabbits between 1951 and 1959. *Journal of Hygiene* 58, 485–488.
Mead-Briggs, A.R. (1964) Some experiments concerning the interchange of rabbit fleas, *Spilopsyllus cuniculi* (Dale), between living rabbits. *Journal of Animal Ecology* 33, 13–26.
Mutze, G.J. (1996) *Release of Spanish Rabbit Fleas as Vectors of Myxomatosis in Inland Australia.* Final Report, Feral Pests Program: Pest Animal Control in Drought. Australian Nature Conservation Agency, 17 pp.
Myers, K. (1970) The rabbit in Australia. In: den Boer, P.J. and Gradwell, G.R. (eds) *Dynamics of Populations.* Centre for Agricultural Publishing and Documentation, Wageninen, pp. 478–506.
Myers, K., Parer, I., Wood, D. and Cooke, B.D. (1994) The rabbit in Australia. In: Thompson, H.V. and King, C.M. (eds) *The European Rabbit. The History and Biology of a Successful Colonizer.* Oxford University Press, Oxford, pp. 108–157.
Mykytowycz, R. (1956) The effect of season and mode of transmission on the severity of myxomatosis due to an attenuated strain of virus. *Australian Journal of Experimental Biology and Medicine* 34, 121–132.
Mykytowycz, R. and Fullagar, P.J. (1973) Effect of social environment on reproduction in the rabbit, *Oryctolagus cuniculus* (L.). *Journal of Reproduction and Fertility* Supplement 19, 503–522.

Parer, I. and Korn, T.J. (1989) Seasonal incidence of myxomatosis in New South Wales. *Australian Wildlife Research* 16, 563–568.

Parer, I., Conolly, D. and Sobey, W.R. (1981) Myxomatosis: the introduction of a highly virulent strain of myxomatosis into a wild rabbit population at Urana in New South Wales. *Australian Wildlife Research* 8, 613–626.

Parer, I., Conolly, D. and Sobey, W.R. (1985) Myxomatosis: the effects of annual introduction of an immunizing strain and a highly virulent strain of myxoma virus into rabbit populations at Urana, N.S.W. *Australian Wildlife Research* 12, 407–423.

Parer, I., Sobey, W.R., Conolly, D. and Morton, R. (1994) Virulence of strains of myxoma virus and resistance of wild rabbits, *Oryctolagus cuniculus* (L.), from different locations in Australasia. *Australian Journal of Zoology* 42, 347–362.

Parer, I., Sobey, W.R., Conolly, D. and Morton, R. (1995) Sire transmission of acquired resistance to myxomatosis. *Australian Journal of Zoology* 43, 459–465.

Parker, R.F. and Thompson, R.L. (1942) The effect of external temperature on the course of infectious myxomatosis of rabbits. *Journal of Experimental Medicine* 75, 567–573.

Pastoret, P.-P., Brochier, B., Aguilat-Setien, A. and Blancou, J. (1997) Vaccination against rabies. In: Pastoret, P.-P., Blancou, J., Vannier, P. and Verschueren, C. (eds) *Veterinary Vaccinology*. Elsevier Science, Amsterdam, pp. 616–628.

Primakoff, P., Lathrop, W., Woolman, L., Cowan, A. and Myles, D.G. (1988) Fully effective contraception in male and female guinea pigs immunised with the sperm protein PH-20. *Nature* 335, 543–546.

Proceedings (1997) Proceedings of the Fourth International Conference on Fertility Control for Wildlife Management. *Reproduction, Fertility and Development* 9, 1–186.

Ramsay, A.J. and Ramshaw, I.A. (1997) Cytokine enhancement of immune responses important for immunocontraception. *Reproduction, Fertility and Development* 9, 91–97.

Robertshaw, J.D. and Gould, W.J. (1995) Queensland Spanish rabbit flea – releases and results. In: *Proceedings of the 10th Australian Vertebrate Pest Control Conference, Hobart, 1995*, pp. 402–407.

Robinson, A.J., Jackson, R., Kerr, P., Merchant, J., Parer, I. and Pech, R. (1997) Progress towards using recombinant myxoma virus as a vector for fertility control in rabbits. *Reproduction, Fertility and Development* 9, 77–83.

Shepherd, R.C.H. and Edmonds, J.W. (1976) The establishment and spread of *Spilopsyllus cuniculi* (Dale) and its location on the host, *Oryctolagus cuniculus* (L.), in the Mallee Region of Victoria. *Australian Wildlife Research* 3, 29–44.

Shepherd, R.C.H. and Edmonds, J.W. (1977a) Myxomatosis: the transmission of a highly virulent strain of myxoma virus by the European rabbit flea *Spilopsyllus cuniculi* (Dale) in the Mallee Region of Victoria. *Journal of Hygiene* 79, 405–409.

Shepherd, R.C.H. and Edmonds, J.W. (1977b) Ectoparasites of the wild rabbit *Oryctolagus cuniculus* (L.) in Victoria: the occurrence of the mites *Leporacarus gibbus* (Pagenstecher) and *Cheyletiella parasitivorax* (Megnin) and the louse *Haemodipsus ventricosus* (Denny). *Journal of the Australian Entomological Society* 16, 237–244.

Shepherd, R.C.H. and Edmonds, J.W. (1978a) Myxomatosis: the release and spread of the European rabbit flea *Spilopsyllus cuniculi* (Dale) in the Central District of Victoria. *Journal of Hygiene* 83, 285–294.

Shepherd, R.C.H. and Edmonds, J.W. (1978b) Myxomatosis: changes in the epidemiology of myxomatosis coincident with the establishment of the European rabbit flea *Spilopsyllus cuniculi* (Dale) in the Mallee Region of Victoria. *Journal of Hygiene* 81, 399–403.

Shepherd, R.C.H. and Edmonds, J.W. (1980) Myxomatosis: the emergence of male and female European rabbit fleas *Spilopsyllus cuniculi* (Dale) from laboratory cultures. *Journal of Hygiene* 84, 109–113.

Shepherd, R.C.H., Edmonds, J.W., Nolan, I.F. and Gocs, A. (1978) Myxomatosis in the Mallee region of Victoria, Australia. *Journal of Hygiene* 81, 239–243.

Sobey, W.R. (1969) Selection for resistance to myxomatosis in domestic rabbits (*Oryctolagus cuniculus*). *Journal of Hygiene* 67, 743–754.

Sobey, W.R. (1977) Rabbit fleas. *Wool Technology and Sheep Breeding* 25, 37–40.

Sobey, W.R. and Conolly, D. (1971) Myxomatosis: the introduction of the European rabbit flea *Spilopsyllus cuniculi* (Dale) into wild rabbit populations in Australia. *Journal of Hygiene* 69, 331–346.

Sobey, W.R. and Conolly, D. (1975) Myxomatosis: passive immunity in the offspring of immune rabbits (*Oryctolagus cuniculus*) infested with fleas (*Spilopsyllus cuniculi* Dale) and exposed to myxoma virus. *Journal of Hygiene* 74, 43–55.

Sobey, W.R. and Conolly, D. (1986) Myxomatosis: non-genetic aspects of resistance to myxomatosis in the rabbit *Oryctolagus cuniculus. Australian Wildlife Research* 13, 177–187.

Sobey, W.R. and Menzies, W. (1969) Myxomatosis: the introduction of the European rabbit flea *Spilopsyllus cuniculi* (Dale) into Australia. *Australian Journal of Science* 31, 404–406.

Sobey, W.R. and Turnbull, K. (1956) Fertility in rabbits recovering from myxomatosis. *Australian Journal of Biological Sciences* 9, 455–461.

Sobey, W.R., Menzies, W., Conolly, D. and Adams, K.M. (1968) Myxomatosis: the effect of raised ambient temperature on survival time. *Australian Journal of Science* 30, 322.

Sobey, W.R., Menzies, W. and Conolly, D. (1974) Myxomatosis: some observations on breeding the European rabbit flea *Spilopsyllus cuniculi* (Dale) in an animal house. *Journal of Hygiene* 72, 453–465.

Sobey, W.R., Conolly, D. and Menzies, W. (1977) Myxomatosis: breeding large numbers of rabbit fleas (*Spilopsyllus cuniculi* Dale). *Journal of Hygiene* 78, 349–353.

Sobey, W.R., Conolly, D. and Westwood, N. (1983) Myxomatosis: a search for a strain of virus to immunize a wild population of rabbits, *Oryctolagus cuniculus. Australian Wildlife Research* 10, 287–295.

Strayer, D.S., Skaletsky, E., Cabirac, G.F., Sharp, P.A., Corbeil, L.B., Sell, S. and Leibowitz, J.L. (1983) Malignant rabbit fibroma virus causes secondary immunosuppression in rabbits. *Journal of Immunology* 130, 339–404.

Tighe, F.G., Edmonds, J.W., Nolan, I.F., Shepherd, R.C.H. and Gocs, A. (1977) Myxomatosis on the western plains of Victoria. *Journal of Hygiene* 79, 209–217.

Tyndale-Biscoe, C.H. (1991) Fertility control in wildlife. *Reproduction, Fertility and Development* 3, 339–343.

Tyndale-Biscoe, C.H. (1997) Immunosterilization for wild rabbits: the options. In: Kreeger, T.J. (ed.) *Contraception in Wildlife Management.* United States Department of Agriculture, APHIS Technical Bulletin 1853, U.S. Government Printer, Washington, pp. 223–234.

Wheeler, S.H. and King, D.R. (1985) The European rabbit in south-western Australia. III. Survival. *Australian Wildlife Research* 12, 213–225.

Wheeler, S.H., Oliver, A.J. and King, D.R. (1985) The European rabbit flea, *Spilopsyllus cuniculi,* in south-western Australia. III. Survival. *Australian Wildlife Research* 12, 227–236.

Williams, C.K. and Moore, R.J. (1991) Inheritance of acquired immunity to myxomatosis. *Australian Journal of Zoology* 39, 307–311.

Williams, C.K. and Twigg, L.E. (1996) Responses of wild rabbit populations to imposed sterility. In: Floyd, R.B., Sheppard, A.W. and De Barro, P. (eds) *Frontiers of Population Ecology.* CSIRO, Melbourne, pp. 547–560.

Williams, C.K., Moore, R.J. and Robbins, S.J. (1990) Genetic resistance to myxomatosis in Australian wild rabbits, *Oryctolagus cuniculus* (L.). *Australian Journal of Zoology* 37, 697–703.

Williams, R.T. (1973) Establishment and seasonal variation in abundance of the European rabbit flea *Spilopsyllus cuniculi* (Dale), on wild rabbits in Australia. *Journal of Entomology* (A)48, 117–127.

Williams, R.T. and Parer, I. (1971) Observations on the dispersal of the European rabbit flea *Spilopsyllus cuniculi* (Dale), through a population of wild rabbits, *Oryctolagus cuniculus* (L.). *Australian Journal of Zoology* 19, 129–140.

Williams, R.T. and Parer, I. (1972) The status of myxomatosis at Urana, New South Wales, from 1968 until 1971. *Australian Journal of Zoology* 20, 391–404.

Williams, R.T., Dunsmore, J.D. and Parer, I. (1972a) Evidence for the existence of latent myxoma virus in rabbits (*Oryctolagus cuniculus* (L.)). *Nature* 238, 99–101.

Williams, R.T., Fullagar, P.J., Davey, C.C. and Kogon, C. (1972b) Factors affecting the survival time of rabbits in a winter epizootic of myxomatosis at Canberra, Australia. *Journal of Applied Ecology* 9, 399–410.

Williams, R.T., Dunsmore, J.D. and Sobey, W.R. (1973a) Fluctuations in the titre of antibody to a soluble antigen of myxoma virus in field populations of rabbits, *Oryctolagus cuniculus* (L.) in Australia. *Journal of Hygiene* 71, 487–500.

Williams, R.T., Fullagar, P.J., Kogon, C. and Davey, C. (1973b) Observations on a naturally occurring winter epizootic of myxomatosis at Canberra, Australia, in the presence of rabbit fleas (*Spilopsyllus cuniculi*) and virulent myxoma virus. *Journal of Applied Ecology* 10, 407–427.

9
Myxomatosis in France

Overview

Following the dramatic spread of myxomatosis in Australia early in 1951, an illegal introduction for rabbit control was made on an estate near Paris in June 1952. The strain of virus used ('Lausanne') was recently derived from its *Sylvilagus* host in Brazil. It led to the establishment of the disease in France. Over the next decade myxomatosis spread to most of the countries of Europe in which rabbits are found, producing very high mortalities.

The public reaction to myxomatosis in France was dominated by deep concern by rabbit breeders for the safety of their domestic rabbits and by chasseurs for the destruction of an important hunting animal, the wild rabbit. On the other hand, foresters and most farmers welcomed the destruction of a major pest. Vaccination of domestic and wild rabbits was practised on a large scale.

The Lausanne strain was highly lethal and the skin lesions were much more protuberant than those produced by the virus used for the Australian introduction. Less virulent strains were recognized two years after its introduction, and tests nine and fifteen years later showed that a variety of strains of differing virulence were present, the percentage of highly virulent strains decreasing progessively to 2% by 1962. In 1980 a 'non-myxomatous' or 'respiratory' form of myxomatosis was observed in commercial rabbitries; it soon became the commonest form and was thought to be transmitted by close contact.

Initially the populations of wild rabbits were decimated, and even with the reduction in virulence of the virus and the increased resistance of the rabbits, by the 1990s myxomatosis appeared to have produced reductions in the wild rabbit populations to somewhat less than half those found before 1953.

Introduction into France

The dramatic outbreak of myxomatosis among rabbits in Australia in early 1951 received extensive media coverage in Europe. The first international responses to this demonstration of its effectiveness for rabbit control occurred in France.

Enquiries from the Institut Pasteur, Paris, January 1952

On 22 January 1952 Dr G. Remaudière, of the Service de Parasitologie vegetale of the Institut Pasteur, wrote to Francis Ratcliffe, the head of the Wildlife Survey Section of CSIRO, requesting reprints on myxomatosis, a specimen of the virus and details on its cultivation, and saying, amongst other things[1]:

> Les lapins constituant dans certaines parties de la France, un véritable fléau, l'Institut Pasteur désirerait entreprendre la lutte biologique contre ce Rongeur et essayer de développer une épidémie de myxomatose en

> Sologne. Cette région infestée de lapins semble, à priori, favorable à l'expansion de la maladie, car les Culicides et Simulides sont abondants.

Ratcliffe passed the letter to his colleague Fenner, who on 15 February 1952 sent Remaudière two vials of freeze-dried myxoma virus and full particulars of the method of preparation of large quantities for inoculation campaigns. Receipt of the letter and virus was acknowledged on 29 February, but no further information about the use of the virus was received in Australia or could be found in a recent search of archival sources in France.

Release of myxoma virus on Maillebois Estate, June 1952

Dr P.F. Armand Delille (Fig. 9.1) was a distinguished physician and bacteriologist who was 77 years old when he read in the Paris papers of the great epizootics of myxomatosis in Australia in 1950–51, and conceived the idea of eliminating wild rabbits from his estate at Chateau Maillebois (Eure-et-Loire) by the use of this virus. Lockley describes his visit to Maillebois in December 1953 in the following terms (Lockley, 1964):

> At that moment Dr Delille was a very worried man; proceedings against him had been begun by hunting and sporting interests in France, and he was engaged with his legal advisers in Paris preparing to defend his actions in the law courts. A test case was being brought against him, with the financial backing of the hunting clubs of France, by the local owner of a domestic rabbitry in which the rabbits had all died of myxomatosis. ...
>
> The Chateau Maillebois is a striking turreted medieval house lying in a beautiful wooded estate of 600 acres with a farm and small river, the whole enclosed with a high stone wall. This wall is broken only by certain entrance gates which had been rendered rabbit-proof before the introduction of the virus. Thousands of wild rabbits were devouring the farm crops and killing the tender forest trees, Dr Delille's son told us, when in June, 1952, two wild-caught rabbits were inoculated with myxoma virus. In six weeks about 98% of the wild rabbits were dead, but none of the domestic rabbits in the hutches on the estate was affected.

Fig. 9.1. Paul F. Armand Delille (1874–1963). Born in Fourchambault (Nièvre) in 1874, Delille graduated in medicine and became one of the leading paediatricians in Paris, carrying out distinguished research on tuberculosis and infectious diseases of children. He was appointed a professor in the Paris Medical School and served as vice-president of the Société de Biologie. During the First World War he was Chief of Bacteriology in l'Armée d'Orient and carried out valuable work on malaria, for which he was made a Commander of the Legion of Honour. He retired to a 300 hectare walled estate at Maillebois, outside Paris. His relevance to this book derives from the fact that in 1952, at the age of 78, he introduced myxomatosis to France and thence to Europe by inoculating two rabbits on his estate with virus derived from a type culture collection in Lausanne, an act for which he was reviled by rabbit breeders, furriers and hunters but praised by farmers and foresters.

> In October 1952, the disease was identified from a corpse picked up near Rambuoillet, the residence of the President of France, fifty kilometres from Maillebois.

The disease at Rambouillet was identified as myxomatosis by Jacotot and Vallée (1953), but at the time its source was unknown to the authorities or the public. Concerning the escape from Maillebois, Delille did not believe that the virus had been carried by mosquitoes, because his own domestic rabbits had not been affected. He thought that perhaps other landowners had broken in and captured diseased rabbits for distribution on their properties. About a year later, by which time some 35% of domestic rabbits and an estimated 45% of wild rabbits in France had died of myxomatosis, Delille made a public statement. On 14 June 1953 and again on 14 October 1953, he read papers to the Académie d'Agriculture, claiming that the method should be used systematically for rabbit control (Delille, 1953). At this time, one of us (F.F.) was in Paris, and could not but be amused by the controversy then raging in *Le Temps* about the height that a rabbit could jump, because this had been proposed as the way by which an infected rabbit had escaped from Maillebois.

While agriculturalists and especially foresters supported Delille, and were ultimately to award him a gold medal (Fig. 9.2), he was vigorously denounced by the powerful hunting organizations. The Ministry of Agriculture was in a difficult position, because in addition to its duties in relation to agriculture and forestry, it was responsible for the Conseil Supérieure de La Chasse, which derived its income from gun licence revenue paid by sportsmen who were primarily interested in hunting rabbits.

Bills were introduced into the French Assembly to make the introduction and use of myxomatosis illegal, but it was too late – the disease was established and spreading. With the financial support of the hunting clubs of France, a test case was brought against Delille by the owner of a local domestic rabbitry, all of whose rabbits had died of myxomatosis. The case failed on a technicality – it could not be proved that the virus had been introduced directly into the affected rabbitry by Delille (as indeed it had not).

Fig. 9.2. Medal presented to Dr P.F. Armand Delille by French agriculturists and foresters, at a ceremony organized by M. Chavet, Secrétaire général de la Fédération nationale des Producteurs de Bois et Reboiseurs francais.

The virus that Delille had used differed from those used previously for introductions into European rabbits, which were derived from the Moses strain and had been extensively passaged in laboratory rabbits. It was sent to Delille by his friend Professor Hauduroy, Director of the Centre de Collection de Types microbien in Lausanne, Switzerland[2]. Dr G. Bouvier, Director of Institut Galli-Valerio in Lausanne, has described the origin of this strain (Bouvier, 1954), which was derived from the *Sylvilagus* reservoir in Brazil with at most six passages in domestic rabbits before being sent to Delille. It has been called 'Lausanne' in scientific papers since a description of its properties by Fenner and Marshall (1957).

Attitude to Rabbits in France

Myxomatosis was a subject of great public interest in France, because there was a large commercial rabbit-breeding industry, many people had back-yard hutches ('clapiers') and there was a large and influential hunting fraternity for whom the wild rabbit was an important resource. On the other hand, foresters and farmers were very much aware of the importance of rabbits as an agricultural pest (Fig. 9.3). They shared with wild boars the distinction of being regarded as the major vertebrate pest, and in 1952, before the advent of myxomatosis, rabbit damage to agriculture and forestry had reached an estimated annual cost of 1000 million francs and 88 of the 90 départmentes in France had declared the rabbit a pest (Siriez, 1957).

The French Game Act of 1844, with later amendments, protects wild rabbits as game; predators which attack rabbits or

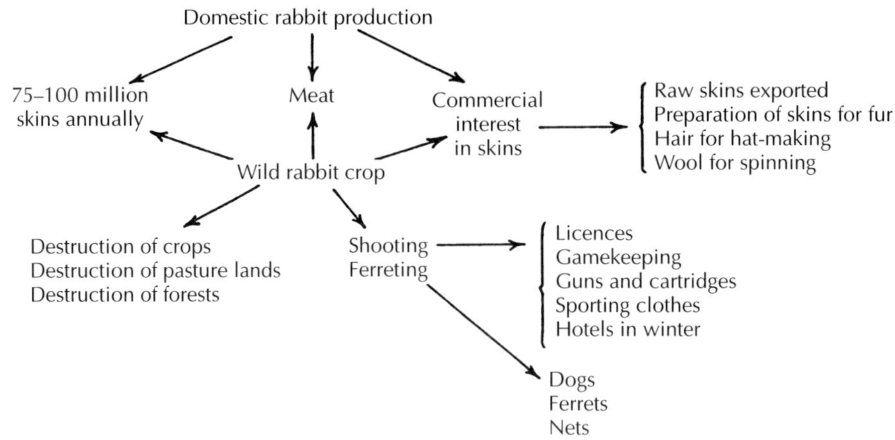

Fig. 9.3. The place of the rabbit (wild and domestic) in France as it was before myxomatosis. (From Fenner and Ratcliffe, 1965; modified from Barthélémy, 1953.)

game birds, such as foxes, stoats and weasels, are regarded as vermin. There is an open season for hunting rabbits between early September and early January, with possible extension to 31 March. In 1952, the year before the introduction of myxomatosis, 1,850,000 shooting permits were issued, of which some 80% were held by persons who were primarily rabbit hunters. The importance of adequate numbers of wild rabbits to these enthusiastic *chasseurs* is obvious (see Table 9.1, below), as well as their significance for the supporting industries indicated in Fig. 9.3. The major hunting organizations, the Conseil Supérieur de la Chasse and the St Hubert Club de France, made prolonged and vigorous protests about the destruction of rabbits by myxomatosis, and supported such counter-measures as the vaccination of wild rabbits with fibroma virus, the introduction of cottontails from the United States, and the introduction of resistant rabbits from Australia, all to no avail.

Besides the importance of wild rabbits for hunters, there was a large domestic rabbit industry. In 1950 some 140 million domestic rabbits were produced and consumed annually in France, and many retired workers were partly dependent upon rabbit raising for their livelihood. Because mosquitoes were major vectors and myxomatosis was sweeping through the wild rabbit population, many outbreaks of the disease occurred among domestic rabbits in the early 1950s.

Short books on myxomatosis soon appeared (Radot and Lépine, 1953; Virat-Pilet, 1954) and a comprehensive two-volume monograph was published in 1972–73 (Joubert *et al.*, 1972, 1973). The spread of myxomatosis in France prompted the organization of an International Symposium on Myxomatosis by l'Office Internationale des Epizooties in Paris in November 1953, and there were special sections on myxomatosis at the International Veterinary Congress in Stockholm in August 1953 and at the International Congress of Microbiology in Rome in September 1953.

Official Action on Myxomatosis

A Central Service for the Fight Against Myxomatosis was set up by the Conseil Superiéur de la Chasse on 16 July 1953, responsible to Dr F. Barthélémy, Engineer in Charge of Water and Forestry. It worked in close collaboration with the Veterinary Service, the Laboratory of Veterinary Research and the Institut Pasteur. Details of legislation providing for the control of myxomatosis are set out in the monograph by Joubert *et al.* (1973). The French Rural Code (RC) and Penal Code (PC) severely penalize deliberate importation and spread of infectious diseases of domestic and wild animals (PC articles 452, 454–1, 31 October 1955). Some articles of the Rural Code authorize the destruction of animal pests (RC articles 393–395) and specify the methods that can be used. Rabbits can be shot, trapped or poisoned. Sanitary regulations of rabbit farms are strictly policed, and farmers are required to maintain a register containing all information on diseases of their stock and vaccinations carried out.

No immediate official action was taken when myxomatosis was first recognized in October 1952. However, after the rapid spread of the disease in Spring 1953, decrees were published by the Ministry of Agriculture on 27 May 1953 and 27 June 1953. These made myxomatosis a notifiable disease and prohibited the movement of rabbits in an infected area. Appropriate measures were specified for the isolation and disinfection of domestic rabbitries, dead animals being incinerated and their remains buried in quicklime. It was further required that the sanitary status of rabbit farms had to be published fortnightly in the National Sanitary Bulletin (Ministerial decree of 9 July 1953). Persons guilty of 'the voluntary propagation of an epizootic' were liable to imprisonment for a period of one to five years. In the case of wild rabbits, areas at risk had to be indicated by notices bearing the words 'Myxomatose, maladie contagieuse du lapin', and all rabbits within the area were supposed to be destroyed.

However, strict as these regulations seem to be, it was impossible to eliminate the disease, and periodically outbreaks continue to occur among wild rabbits, and sometimes spread from them to domestic rabbits which are not protected by vaccination or insect-proof screening.

Clinical Features of Myxomatosis as Seen in France

Myxomatosis in France was initiated by the inoculation of two wild rabbits with the Lausanne strain of myxoma virus; no further inoculations of virulent virus were ever made in the field. Laboratory studies in Australia showed that the Lausanne strain produced a disease (Fig. 9.4A) with more protuberant lesions than those caused by the Standard Laboratory Strain, initially used in Australia for rabbit control. In experiments conducted in Australia, challenge infections of rabbits with increased innate resistance showed that the Lausanne strain was also

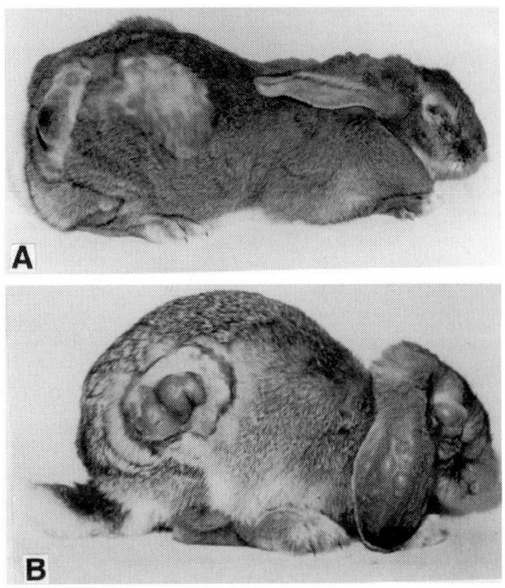

Fig. 9.4. Clinical signs of myxomatosis caused by strains of myxoma virus found in France during the early stages of the epizootic in France. (A) Lesions produced by the Lausanne strain. Rabbit photographed 10 days after inoculation; it died on the twelfth day. The primary lesion on the flank was very large, protuberant and deep purple in colour. The head was oedematous ('leonine facies') and the eyes completely closed with a profuse conjunctival discharge. The secondary lesions were also raised, purple in colour and not demarcated from the surrounding skin. (B) Lesions produced by the Loiret 55 strain. The photograph was taken on the 25th day and the rabbit died next day. About one-third of rabbits inoculated with this strain recovered. The primary lesion was very large and exuded serum. There were numerous secondary lesions all over the body, the eyes were completely closed and the ears hung down because of numerous lesions on the pinnae. From Fenner and Marshall (1957), with permission.

more virulent than the Standard Laboratory Strain, in that the case-fatality rate was about 99% when the Standard Laboratory Strain was associated with a case-fatality rate of 60–80%[3].

Attenuated strains of myxoma virus were first recognized in 1955, in the département of Loiret (Jacotot *et al.*, 1955; Fig. 9.5). The lesions produced by this strain ('Loiret 55') retained the protuberance characteristic of the Lausanne strain lesions (Fig. 9.4B) but the disease evolved more slowly. In more extensive tests the case-fatality rate was 65% and the mean survival time 33.1 days, with a range of 19 days to recovery. It was later designated as the prototype European strain of Grade IV virulence.

Fig. 9.5. Henri Jacotot (1896–1991). Born in Dijon in 1896, Jacotot served in France in the 1914–1918 war and was decorated with the Military Cross. Subsequently he graduated in veterinary science and was posted to the Pasteur Institute at Nhatrang, in Indo-China, in 1922, initially as chief assistant to Alexandre Yersin. In 1927 he was appointed Director of that institute. In 1948 he returned to the Pasteur Institute in Paris as head of the service of animal microbiology, a post which he occupied until he retired in 1965 as a Professor Emeritus. In Indo-China he studied the many veterinary and zoonotic diseases found in that country and was responsible for the construction of the building currently used by the Pasteur Institute at Nharang. After returning to Paris he took a leading role in the investigations of myxomatosis in France, describing the first recognized case in 1953, experimenting with insect transmission and studying the changes in virulence of the virus. A highly respected scientist and recipient of many prizes, he was a member of the Academies of Veterinary Science and of Medicine and served as President of the former.

Respiratory or non-myxomatous myxomatosis

In addition to the increase in the occurrence of attenuated strains characterized by the usual protuberant skin lesions but lower case-fatality rates (see below), in 1980 French scientists observed for the first time a 'respiratory or non-myxomatous' form of the disease (Brun *et al.*, 1981b; Joubert *et al.*, 1982). This syndrome was characterized by a longer incubation period (1–3 weeks), swelling of the eyelids accompanied by severe purulent conjunctivitis, genital lesions, and prominent nasal lesions accompanied by lacrimation and a mucopurulent nasal discharge (Fig. 9.6; see pp. 97 and 104). There were often pink or red spots on the ears, but no nodular skin lesions (Arthur and Louzis, 1988). Its epidemiology is described below (p. 220).

Although initially described in domestic rabbits raised by intensive husbandry and vaccinated with the SG33 attenuated myxoma virus vaccine (Chantal, 1981), this syndrome was also seen among rabbits kept under traditional husbandry and sometimes in wild rabbits. It was often accompanied by sterility and the abandonment of litters by farmed does.

The Spread of Myxomatosis in France

As commonly occurs when a dramatic 'emerging' disease breaks out, a number of fantastic stories about its mode of spread circulated during the early days of the French epizootic (Lockley, 1953). The consumption of grass that might have been contaminated by myxomatous wild rabbits

Fig. 9.6. Hutch rabbit showing signs of the 'respiratory' form of myxomatosis. Courtesy Dr A. Brun.

was thought to transfer myxomatosis to hutch rabbits, and motor cars which ran over infected rabbits were blamed for moving the disease to new districts. It was commonly reported that there was a black market for myxomatous rabbits, up to 5000 francs being paid by landholders anxious to get rid of a pest. However, it soon became apparent that mosquitoes were important vectors in France, and the presence of such a highly mobile and efficient vector provided an explanation for the movement of the disease to new areas (Fig. 9.7) and from wild to domestic rabbits. In the early days of the outbreak, there was often inadvertant transfer of disease between rabbit farms via infected rabbits still incubating the disease. Rabbit fleas are as common amongst wild rabbits in France as they are in Britain and undoubtedly contribute greatly to maintenance of enzootic myxomatosis throughout the year.

Epidemiology

With the knowledge that mosquitoes had been important vectors in Australia, it was

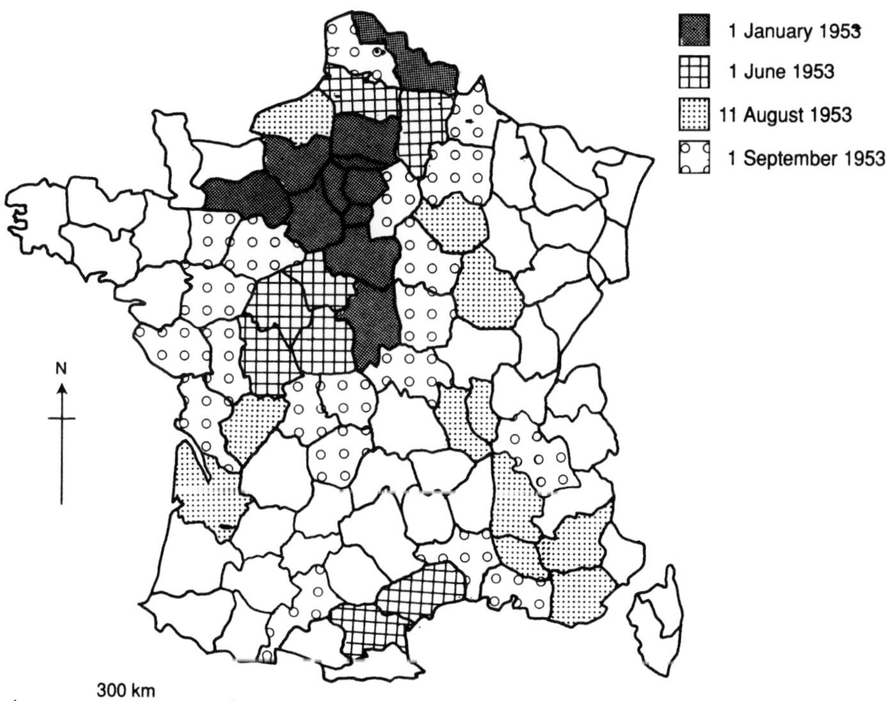

Fig. 9.7. The spread of myxomatosis in France between January and September, 1953. From Rogers *et al.* (1994), after Joubert *et al.* (1972), with permission.

soon shown that the prime suspect for transmission in the early epizootics in France[4], *Anopheles maculipennis atroparvus,* was an effective vector (Jacotot *et al.,* 1954). Subsequently, quiescent rabbit fleas were recovered from soil scrapings from deep burrows that had been abandoned by rabbits ten weeks earlier, following autumn epizootics of myxomatosis, and myxoma virus was recovered from these fleas (Joubert *et al.,* 1969). Arthur and Louzis (1988) postulated that eventually an enzootic–epizootic cycle involving reservoirs, vectors and susceptible rabbits developed. They suggested that the virus might overwinter in fleas in the soil of the burrow, in rabbits convalescing from infection, and in hibernating mosquitoes that might be contaminated with the virus. With the onset of warm weather, sharp mosquito-borne epizootics occurred among wild rabbits, rising to a peak in mid-summer and extending into autumn. These spread into domestic rabbits housed in non-mosquito-proof shelters.

Sightings of diseased rabbits fell when the wild rabbit population became greatly depleted in the mid-1950s, but after the attenuation of the virus and growing innate resistance in the rabbits in the late 1960s, the numbers of rabbits increased again. The seasonal and annual variations in myxomatosis then seen resulted from the interaction of the abundance of rabbits, their immune status (which was influenced by the age structure of the population) and the abundance of vectors. The occurrence of epizootics in southern France during summer and autumn was due to the concurrence of large numbers of aggressive mosquitoes and the presence of non-immune young rabbits. If there had been no myxomatosis the previous year, outbreaks often occurred in the spring and simulids and ceratopogonids were often involved (Joubert and Monnet, 1975). Spring outbreaks affected mainly young rabbits that had not been infected the previous summer, the resulting decrease in breeding females greatly depressing reproduction. Infection is maintained by rabbit fleas during autumn, winter and early spring, spread then being slow. Overwintering in the absence of obvious disease may be due to the persistence of virus on the mouth parts of fleas, which can remain in a quiescent state for some time in uninhabited burrows.

Impact on wild rabbits

Myxomatosis clearly had a major impact on the number of wild rabbits that were available for hunting, and the bag did not reach even 20% of its pre-myxomatosis level until the 1961–62 season, although the recorded figures may partly reflect the paucity of hunters as well as of rabbits (Table 9.1). In the Sologne, a favoured département for rabbit hunting, the mean numbers shot per season (compared with

Table 9.1. Numbers of rabbits killed each hunting season in 25 hunts in France, before and after the introduction of myxomatosis[a].

1950–51	64,439,000	
1951–52	37,723,000	
1952–53	55,133,000	
1953–54	6,721,000	of which 80% were from two unaffected hunts
1954–55	2,750,000	of which 87% were from one unaffected hunt
1955–56	2,404,000	of which 50% were from one hunt
1956–57	1,293,000	
1957–58	3,357,000	
1958–59	2,043,000	
1959–60	4,074,000	
1960–61	5,491,000	
1961–62	10,906,000	

[a]Data from Giban (1956) and personal communication to Fenner (1963).

pre-myxomatosis figures) increased from 1.5% in 1953–58 to 13.5% in 1967–72 and 17.6% in 1972–77 (Arthur *et al.*, 1980). However, after rabbits began to build up in numbers due to the presence of attenuated strains of virus and greater innate resistance, variations in the timing of major outbreaks of myxomatosis in relation to the hunting season caused large fluctuations in the numbers of rabbits shot. National surveys in France in 1977 and 1978 (Arthur *et al.*, 1980) showed that rabbit density varied from place to place depending on environmental factors, such as type of soil and agricultural practice, and was also affected by hunting pressure and predation. At that time rabbits were most abundant in north-western France and much less common in the south and west, where in contrast to pre-myxomatosis times they often lived only in isolated pockets. Twenty years later (1997) the picture has not changed.

After the first disastrous epidemics, a variety of measures were undertaken by hunting organizations to mitigate the impact of myxomatosis. Vaccination with fibroma virus was carried out in an attempt to preserve some rabbits for shooting and hopefully to build up 'barriers' to the movement of myxomatosis. Later it was suggested that, as with domestic rabbits, wild rabbits should first be injected with fibroma virus followed 6–8 weeks later by SG33, both being administered by jet injector (Joubert *et al.*, 1982; Fig. 9.8). Other early and unsuccessful methods for preserving rabbits as game animals included the introduction of cottontails (*Sylvilagus floridanus*), both as game animals and to hybridize with *Oryctolagus*, and a proposal to import rabbits from regions of Australia where innate resistance was high.

Fig. 9.8. Louis Joubert (1922–1989). Born in Grenoble in 1922, Joubert enrolled in the National Veterinary School at Toulouse in 1939, graduating with honours in 1945 and becoming assistant to the Professor of Infectious Diseases. He continued his studies in veterinary medicine and in pharmacy, and in 1948 was appointed project director at the Veterinary School in Lyon, where he advanced to become a professor in 1962. His early research was focused on the epidemiology of zoonotic diseases, and after Jacotot had retired he became the leading expert on myxomatosis in France, showing particular interest in unusual insect vectors and especially in the 'respiratory' or 'amyxomatous' form of the disease. He was the senior author of a major book on myxomatosis, published in Paris in 1972–1973.

Impact on domestic rabbits

Myxomatosis was spread by mosquitoes from wild to domestic rabbits and initially it devastated the rabbit-breeding industry; it was estimated in 1954 that 30–40% of the industry had been destroyed. Vaccination afforded some protection, and with the reduction of the size of the wild rabbit reservoir of myxomatosis in the late 1950s new outbreaks in rabbitries became less common. Initially fibroma vaccine was used on a very large scale, amounting to tens of millions of doses annually; later it was used on a smaller scale, primarily to protect breeding animals. It was only moderately effective, and after limited trials with an attenuated myxoma virus vaccine developed in California (Saito *et al.*, 1964), another attenuated vaccine (SG33) was developed in France (Saurat *et al.*, 1978). Given to rabbits at 2–3 months of age, it was effective for about a year after vaccination. However, the use of the SG33 vaccine was found to have an immunosuppressive effect, which among rabbits housed under conditions of poor hygiene led to a variety of secondary bacterial infections (Brun *et al.*, 1981; Godard, 1980). In such circumstances it was recommended that fibroma virus should be used for primary vaccination, followed by SG33 vaccine a month later.

The epidemiological cycle as seen in the 1980s differed according to the type of husbandry (Arthur and Louzis, 1988). Under conditions of traditional husbandry, in animal quarters with open contact with the outside world, the strains of virus were usually of the traditional nodular type, and were usually more lethal among the (genetically) unselected domestic rabbits than they were for wild rabbits. Outbreaks usually occurred in the autumn; there was a delay of 6–8 weeks between epizootic peaks in wild and domestic rabbits, the domestic rabbit peak coinciding with the arrival of mosquitoes within farm buildings (Puech, 1980). This form of myxomatosis could be prevented by efficient protection against the entry of mosquitoes by netting, unless animals harbouring the virus were brought in during the purchase of breeding animals.

Respiratory or non-myxomatous myxomatosis

Under intensive husbandry within closed mosquito-proof buildings, a non-myxomatous, pulmonary form of the disease, in which respiratory signs predominated, was most common (Brun *et al.*, 1981a; see p. 216). It appeared to be transmitted by the inhalation of infective particles, both experimentally (Chantal, 1981) and in the rabbit farms. The incubation period varied between one and three weeks, and infection was manifested by swelling of the eyelids, genital and nasal lesions, lacrimation and a mucopurulent nasal discharge. This occurred primarily in premises in which there was poor ventilation and a high frequency of respiratory disease, and episodes of this form of myxomatosis were likely to occur at any time of the year. Most outbreaks occurred after recent introductions of rabbits from farms which had experienced infection during the previous 2–3 months. Such animals may have been incubating the disease for a longer period than usual, or, it has been suggested, they were asymptomatic carriers, in which reactivation occurred after the stress of transfer to new premises.

Changes in the Virulence of the Virus

Attenuated strains of myxoma virus were first recognized in 1955, in the département of Loiret (see Fig. 9.4B). Other strains with a similar degree of attenuation were soon found elsewhere in France (Fenner and Marshall, 1955; Siriez, 1960); the

Table 9.2. The virulence of field strains of myxoma virus in France in 1953, 1962, and 1968, calculated on the basis of mean survival times and expressed as percentages[a].

Virulence grade	I	II	IIIA	IIIB	IV	V
Mean survival time (days)	9–13	14–16	17–22	23–28	29–50	—
Presumed case-fatality rate (%)	>99	95–99	90–95	70–90	50–70	<50
Year						
1953[b]	>99					
1962	11	19.3	34.6	20.8	13.5	0.8
1968	2.0	4.1	14.4	20.7	58.8	4.3

[a]From Joubert *et al.* (1972), p. 132.
[b]From field observations.

virulence of these strains was not altered by several serial passages in the rabbit testis (Jacotot *et al.*, 1956). Of ten French strains recovered in 1960–63, three were classified as of Grade IIIA virulence, three of Grade IIIB, three of Grade IV and one of grade V (Fenner and Ratcliffe, 1965). More recent data (Table 9.2) suggested that 15 years after the introduction of the virus a few highly virulent strains were still circulating, although the majority of strains were moderately or highly attenuated.

Endnotes

[1]Basser Library Archives, MS143/25/5A. Letters; G. Remaudière to F.N. Ratcliffe (22 January 1952), F. Fenner to Remaudière (15 February 1952) and reply (19 February 1952).
[2]Basser Library Archives, MS143/25/5A. Correspondence between G. Bouvier and Fenner about origins of 'Lausanne' strain, used to introduce myxomatosis into Europe.
[3]G.W. Douglas, personal communication to Fenner (1978), J.W. Edmonds and R.C.H. Shepherd, personal communication to Fenner (1982), published in Fenner, F. (1983). The Florey Lecture, 1983. *Proceedings of the Royal Society* B218, p. 268.
[4]In a communication to Jacotot *et al.* (1954), M.E. Roubaud, an entomologist with special knowledge of mosquitoes, commented: 'Au point de vue de la transmission de la myxomatose dans le milieu des clapiers, c'est cet anophèle *A. maculipennis s. lat.* qui apparait sans nul doute, en Europe, comme devant jouer le rôle principal. Dans les bois, dans les garennes, parmi les lapins sauvage, ce sont d'autres espèces culicidiennes qu'il convient d'incriminer, parmi lesquelles de nombreux *Aedes*, l'*Anopheles claviger* (*bifurcatus*) et surtout l'*A. plumbeus*'.

References

Arthur, C.P. and Louzis, C. (1988) A review of myxomatosis among rabbits in France. *Revue scientifique et technique de l'Office International des Epizooties* 7, 937–957, 959–976.

Arthur, C.P., Chapuis, J.L., Pages, M.V. and Spitz, F. (1980) Enquètes sur la situation et la répartition écologique du lapin de garenne en France. *Bulletin spécial scientifique and technique de l'Office National de la Chasse* December 1980, 37–86.

Barthélémy, F. (1953). Cahiers francais d'Information. 237, 4. Paris.

Bouvier, G. (1954) Quelque remarques sur la myxomatose. *Bulletin de l'Office International des Epizooties* 46, 76–77.

Brun, A., Godard, A. and Moreau, Y. (1981a) La vaccination contre la myxomatose: vaccins heterologue et homologue. *Bulletin de la Société Science Véterinaire et Médecine comparée,* Lyon 83, 251–254.

Brun, A., Saurat, P., Gilbert, Y., Godard, A. and Bouquet, J.F. (1981b) Données actuelle sur l'épidémiologie, la pathogénie et la symptomatologie de la myxomatose. *Revue de Médecine Vétérinaire* 132, 585–590.

Chantal, J. (1981) Comptes rendus des travaux effectués sur les accidents de vaccination en élévage de lapins domestique. *Convention 80 12, Ecole Nationale Vétérinaire de Toulouse-Office National de la Chasse,* Fevrier 1981.

Delille, P.F.A. (1953) Une methode nouvelle permettant à l'agriculture de lutter efficacement contre la pullulation du lapin. *Compte rendu hebdomadaires des Séances de l'Académie d'agriculture de France* 39, 638–642.

Fenner, F. and Marshall, I.D. (1955) Occurrence of attenuated strains of myxoma virus in Europe. *Nature* 176, 782–783.

Fenner, F. and Marshall, I.D. (1957) A comparison of the virulence for European rabbits (*Oryctolagus cuniculus*) of strains of myxoma virus recovered in the field in Australia, Europe and America. *Journal of Hygiene* 55, 149–191.

Fenner, F. and Ratcliffe, F.N. (1965) *Myxomatosis.* Cambridge University Press, Cambridge, p. 230.

Giban, J. (1956) Répercussion de la myxomatose sur les populations de lapin de garenne en France. *Terre et la Vie* 103, 179–187.

Godard, A. (1980) Apres le retrait du vaccin SG 33 contre la myxomatose. Les explications du fabricant. *Eleveur Lapin* 11, 17–20.

Jacotot, H. and Vallée, A. (1953) Un foyer de myxome infectieux chez des lapins de garenne dans la region de Rambouillet. *Annales de l'Institut Pasteur* 85, 448–450.

Jacotot, H., Toumanoff, C., Vallée, A. and Virat, B. (1954) Transmission expérimentale de la myxomatose au lapin par *Anopheles maculipennis atroparvus* et *Anopheles stephensi. Annales de l'Institut Pasteur* 87, 477–485.

Jacotot, H., Vallée, A. and Virat, B. (1955) Apparition en France d'un mutant naturellement attenuée du virus de Sanarelli. *Annales de l'Institut Pasteur* 89, 361–364.

Jacotot, H., Vallée, A. and Virat, B. (1956) Étude de quelque souches francaises de virus attenuée du myxome infectieux. *Annales de l'Institut Pasteur* 90, 779–783.

Joubert, L. and Monnet, P. (1975) Verification expérimentale du role des Simulies dans la transmission du virus myxomateux en Haute-Provence. *Revue de Médecine Vétérinaire* 12, 617–634.

Joubert, L., Chippaux, A., Mouchet, J. and Oudar, J. (1969) Entretien hiverno-vernal du virus myxomateux dans les terriers. Myxomatose d'inoculation par la puce du lapin et myxomatose du fouissement. *Bulletin de l'Académie vétérinaire de France* 42, 93–101.

Joubert, L., Duclos, Ph. and Tuaillon, P. (1982) La myxomatose des garennes dans le Sud-Est. La myxomatose amyxomatose. *Revue de Médecine vétérinaire* 133, 739–753.

Joubert, L., Leftheriotis, L. and Mouchet, J. (1972, 1973) *La Myxomatose.* Vols I and II. L'Expansion, Paris, 588 pp.

Lockley, R.M. (1953) *Myxomatosis in France. A Survey for the Nature Conservancy. Report of a visit to France, 5–19th December 1953.* The Nature Conservancy, London, 22 pp.

Lockley, R.M. (1964) *The Private Life of the Rabbit. An Account of the Life History and Social Behaviour of the Wild Rabbit.* Andre Deutsch, London, pp. 116–117.

Puech, M. (1980) Contribution aux recherches de cartographie épidémiologique: étude écologique des populations de Culicidés adult en rapport avec la myxomatose dans la Region Rhône-Alpes. Thèse 3e cycle, Université de Grenoble, 210 pp. Quoted by Arthur and Louzis (1988).

Radot, C. and Lépine, P. (1953) *La Myxomatose. Nouvelle Maladie des Lapins.* Flammarion, Paris, 122 pp.

Rogers, P.M., Arthur, C.P. and Soriguer, R.C. (1994) The rabbit in continental Europe. In: Thompson, H.V. and King, C.M. (eds) *The European Rabbit.* Oxford University Press, Oxford, p. 30.

Saito, J.K., McKercher, D.G. and Castrucci, G. (1964) Attenuation of the myxoma virus and use of the living attenuated virus as an immunizing agent for myxomatosis. *Journal of Infectious Diseases* 114, 417–428.

Saurat, P., Gilbert, Y. and Ganière, J.-P. (1978) Étude d'un souche de virus myxomateux modifié. *Revue de Médecine Vétérinaire* 129, 415–451.

Siriez, H. (1957) *La Myxomatose, Moyen de Lutte Biologique contre le Lapin, Rongeur Nuisible.* Société des Editions Pharmaceutique, Paris, 88 pp.

Siriez, H. (1960) *Lapins et Myxomatose. L'Evolution de la Maladie de 1956 a 1960 et Quelques Compléments a une Précédente Etude.* Société des Editions Pharmaceutique, Paris, 24 pp.

Virat-Pilet, J. (1954) *La Myxomatose du Lapin et du Lièvre.* Vigot Frères, Paris, 85 pp.

10
Myxomatosis Elsewhere in Europe

Overview

After an unsuccessful attempt to introduce myxoma virus in Scotland in 1952, between 1953 and 1961 myxomatosis spread from France to most of the countries of Europe in which rabbits occur. It produced very high mortalities in both wild and domestic rabbits in continental countries, but in Britain there was a high mortality in wild rabbits and not much disease in domestic rabbits. This difference reflected the great importance of the European rabbit flea (*Spilopsyllus cuniculi*) and the relative unimportance of mosquitoes as vectors in Britain, whereas in many countries on the Continent mosquitoes were major vectors in extensive summer epizootics while the flea maintained the disease at a low level during the winter.

Although the rabbit was recognized as an agricultural pest in the UK, the response of the general public there was strongly influenced by concern about its 'cruelty'. Inoculation of wild rabbits with virulent myxoma virus was forbidden.

Detailed studies in Britain showed that by 1962 the majority of strains were of intermediate virulence, although over 20% were still classified as highly virulent. By 1981 strains resembling the original Lausanne strain had disappeared, but 35% of strains were classified as highly virulent (Grade II). Tests of the resistance of wild rabbits in Britain showed that there was a progressive decrease in the case-fatality rate after challenge with a highly virulent virus (lethal in 98% of unselected wild rabbits) from about 92% in 1966–67 to about 45% in 1978–79.

Initially the populations of wild rabbits were decimated in all countries where myxomatosis occurred, and even with the reduction in virulence of the virus and the increased resistance of the rabbits, by the 1990s myxomatosis appeared to have produced reductions in the wild rabbit populations in most countries of Europe to somewhat less than half of those found before 1953.

Introduction of Myxomatosis into the Heisker Islands, Scotland, July 1952

Immediately after the great epizootic in Australia and independently of Delille's initiative (see Chapter 9), sheep farmers in Scotland whose grazings were being destroyed by rabbits requested samples of the virus. Before granting the request, scientists from the North of Scotland College of Agriculture, using virus supplied by Dr E.W. Hurst of ICI, Pty Ltd, carried out introductions on the rabbit-infested Heisker Islands, in the Atlantic Ocean 12 km west of Benbecula, in the Outer Hebrides (Shanks *et al.*, 1955). Groups of infected laboratory rabbits were released on the islands on eight occasions between July 1952 and May 1953. Myxomatosis spread to the wild rabbits and caused considerable mortality in 1953, but it did not become enzootic and by September 1954 the rabbit population was as large as ever. By that time myxomatosis had become established in England, and presumably the Scottish farmers used other methods to obtain the virus.

Spread of Myxomatosis from France

During the decade after its introduction into France, myxomatosis spread to every country in Europe where rabbits occur (Table 10.1), the last country to be reached being Sweden in 1961.

Cases first occurred in Morocco and Algeria (then French colonies) in

Table 10.1. The spread of myxomatosis throughout countries of Europe in which wild rabbits occur[a].

Country	Myxomatosis first reported	Rate and extent of spread
France	Inoculations June 1952 First case reported October 1952	Spread to half the départements by September 1953 extended to all rabbit-infested areas by 1957
Germany	August 1953	Cases in Berlin in 1955, across the Elbe in 1958
Belgium	August 1953	Spread throughout country during 1953
Holland	September 1953	Spread to all except the three north-eastern provinces by October 1954
Spain	September 1953	Spread through most of country by 1955, only Cantabria being unaffected by 1959
Luxembourg	1953	
Britain	October 1953	Spread throughout country by 1955
Switzerland	August 1954	Wild rabbits rare; scattered foci each summer among domestic rabbits
Czechoslovakia	1954	
Italy	August 1955	Spread rapidly throughout country
Austria	August 1955	A few outbreaks among wild and domestic rabbits each summer
Poland	1955	
Portugal	May 1956	Spread rapidly throughout the country
Denmark	September 1960	
Sweden	September 1961	Limited spread among wild rabbits in southern Sweden
Yugoslavia	?	Few wild rabbits; many domestic rabbits
Romania	?	
Morocco	September 1957	
Algeria	November 1957	

[a]Based partly on correspondence by Fenner with authorities in various countries during 1962; references to first sightings in Fenner, F. and Ross, J. (1994), p. 208.

September and November 1957. Published reports in most countries record little more than the occurrence or prevalence of the disease and sometimes accounts of the vaccination of domestic rabbits, but there is an extensive and continuing literature on myxomatosis in France and Britain.

Myxomatosis in the UK

Establishment and spread in southern England in 1953

Because of the damage caused to pastures and agricultural crops, there had been suggestions ever since Martin's demonstration of the virulence and specificity of myxomatosis that the virus should be introduced in Britain, 'but this method did not appeal to our national temper and received no official encouragement' (Thompson and Worden, 1956). After myxomatosis had spread through France it was recognized that it would probably cross the Channel. Myxomatosis was first recognized in wild rabbits in Britain in October 1953. The mechanism by which it crossed the Channel has never been determined with certainty. Sellers (1987) suggested that *Anopheles atroparvus* contaminated with myxoma virus may have been carried from France across the English Channel on the night of 11–12 August 1953, when meteorological conditions were favourable, leading to recognized cases by mid-October. On the other hand, Thompson (1994; Fig. 10.1), who was closely involved as an officer of the Ministry of Agriculture, Fisheries and Food, wrote: 'A local man brought a myxomatous rabbit from France and the first outbreak was confirmed near Edenbridge, Kent, on 13 October 1953, and a second outbreak on 27 October, in East Sussex'.

Efforts were made to eradicate the disease by the erection of rabbit proof fencing around outbreak areas and destruction of the rabbits within the enclosures, but nine further small outbreaks were observed in Kent, Sussex, Essex and East Suffolk by February 1954 (Armour and Thompson, 1955). When it became clear that the disease had survived the winter, the Advisory Committee on Myxomatosis concluded that further efforts to contain the

Fig. 10.1. Harry V. Thompson (1918–). After graduating in Zoology at the University of London in 1940, Thompson worked at the Bureau of Animal Population at the University of Oxford, where he came under the influence of the famous British ecologist C.S. Elton, with whom F.N. Ratcliffe worked in 1948, before coming back to Australia to set up the Wildlife Survey Section of CSIRO. In 1946 Thompson joined the Ministry of Agriculture, Fisheries and Food, where he became head of the department dealing with research on wild mammals and birds affecting agriculture. Inevitably, he became interested in rabbits, and was at the forefront of work on myxomatosis in rabbits in Britain. In 1959 he set up the Ministry's Worplesdon Laboratory at Guildford, Surrey, and remained its Director until leaving the Ministry in 1982 to become a private consultant. Besides serving on most committees dealing with rabbits and myxomatosis, and numerous national and international bodies concerned with wildlife and conservation, Thompson published numerous scientific papers and two important books on the rabbit, in 1956 and 1994.

disease would be useless, but that no attempt should be made to assist its spread or to introduce it into unaffected areas (Advisory Committee, 1954). However, the spread accelerated during the months June through to October 1954, and by November 408 foci had been recognized, distributed over an area of 375 square miles. The rate of linear spread in the Edenbridge area averaged 5.6 km a month from February to November 1954, but by November 1954 the disease had spread over a large part of Britain and had reached, or was taken to, the Orkney and Shetland Islands (Ritchie *et al.,* 1954). Mosquitoes appear to have been of limited importance in the spread from Edenbridge – the rabbit flea was regarded as the principal vector. The more distant spread was due to transfer of infected rabbits from one place to another by farmers and others for whom the rabbit was a major pest (Lockley, 1964), with the result that by the end of 1955 there were few areas of Britain unaffected by the disease and at this time it was calculated that over 90% of the wild rabbits on the island had died.

Attitude to rabbits in Britain

The only country in Europe other than France in which there have been extensive investigations of myxomatosis is Great Britain. However, the breeding of domestic rabbits occurs there on a much smaller scale than in France, and the attitude to wild rabbits is very different from that in France. In Britain wild rabbits had been bred in warrens for several centuries after their introduction from Normandy in the 11th and 12th centuries. Escapes from warrens occurred, but rabbits did not increase in numbers and distribution in the wild until the 19th century (Lloyd, 1981). Rabbits remained important game animals on large estates, both the Game Conservancy and the British Association for Shooting and Conservation maintaining records of annual shooting bags. Rabbits on game estates were protected by the control of their predators, and they overran adjacent farmland and tree plantations.

However, by the 20th century, pressure from farmers and foresters led to the classification of rabbits as pests. With the need for maximum agricultural production during the the Second World War responsibility for rabbit control on agricultural land was assumed by the Ministry for Agriculture and Fisheries. Reduction of the rabbit population was successful only in intensively farmed areas and reafforestation was difficult unless rabbit-proof netting was used. Indicating the need for drastic action, in 1948 a senior civil servant argued a case for virtually exterminating rabbits. Parliament prescribed measures for their destruction in the Prevention of Damage by Rabbits Act, 1939, the Agriculture Act, 1947, and the Pests Act, 1954 (Worrall, 1956).

The widespread occurrence of myxomatosis in the summer of 1954 led to a heated debate on the ethics of using myxomatosis as a means of rabbit control (Sheail, 1991). Britain is a densely populated island, wild rabbits were common and tame rabbits were often kept as children's pets – quiet, shy and cuddlesome. Myxomatosis presented a horrible aspect, as blind, sick rabbits stumbled on the local commons and across roads. Out of respect for public sentiment, the Advisory Committee on Myxomatosis that had been set up in November 1953 recommended that myxomatosis should not be spread by inoculation (Advisory Committee, 1954), and the animal lovers of Britain obtained an insertion into the Pests Act, 1954, which provided that any person who knowingly used or permitted the use of a rabbit infected with myxomatosis to spread it among uninfected rabbits thereby committed an offence and was liable to a fine. This provision of the Act was almost impossible to police (only two prosecutions had been made up to December 1962), but it was reinforced by an advertisement placed in British newspapers (Fig. 10.2), which vividly portrayed one aspect of the method by which it was spread (the boot of the motor car was an important accessory to the crime). However, after the spread of myxomatosis landholders cooperated in setting up 'rabbit clearance areas' and formed Rabbit Clearance

Societies, of which there were 750 in 1964, covering 46% of the farmed land in Britain (Thompson, 1994).

Epidemiology

The seasonal incidence of myxomatosis in Britain was much less pronounced than in France. Although rabbits were more numerous and cases of myxomatosis were more often seen during the summer months, analysis of records for England and Wales in 1961 (Table 10.2), showed that the number of counties reporting the presence of myxomatosis was almost constant throughout the year, that new outbreaks were reported to Rabbit Clearance Societies each month, and that positive tests for antigen were recorded every month. In addition, outbreaks often appeared to be restricted by fences, pockets of uninfected rabbits were often found adjacent to areas that were severely affected, and compared with the situation in France, cases among domestic rabbits were very rare. These findings suggested that the epidemiology in Britain was different from that found in France, where mosquitoes cause summer epidemics and movement of the disease over long distances, and infections of hutch rabbits were common, while the rabbit flea (*Spilopsyllus cuniculi*) played an important role in maintaining infection among wild rabbits during the winter months. Fleas carried by gulls were thought to be responsible for introducing myxomatosis to several Scottish islands (Lockie, 1956).

Because of their important role as vectors in Australia and France, the initial investigations at Edenbridge in Kent in 1954 focused on the role of mosquitoes. The most abundant mosquitoes there, *Aedes cantans* and *Aedes annulipes*, although biting humans viciously, were not attracted to rabbits, and sentinel rabbits suspended in cages above ground level were not infected (Muirhead-Thompson, 1956a). Late in the summer of 1954 heavy losses were reported among domestic and pet rabbits in Newhaven, on the Sussex coast. Infected *Anopheles atroparvus* were recovered from the rabbitry and laboratory experiments showed that they were efficient vectors and could carry the infection

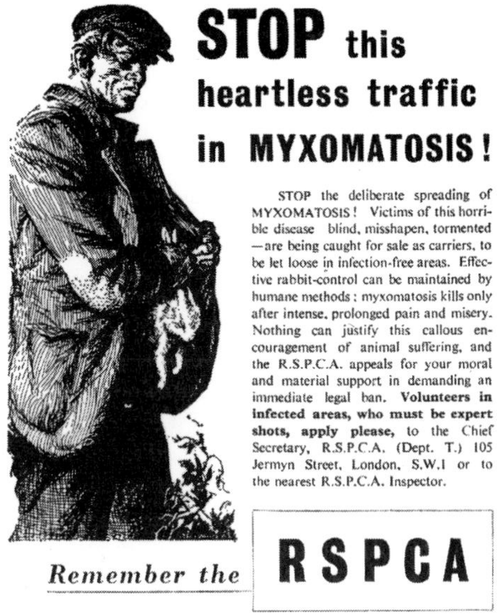

Fig. 10.2. Advertisement inserted in newspapers in Britain in 1954, by the Royal Society for the Protection of Cruelty to Animals. In spite of the feelings of the general public, during 1954 myxomatosis was distributed widely throughout Britain by carrying diseased rabbits from one place to another. From Fenner and Ratcliffe (1965), with permission.

Table 10.2. The seasonal incidence of myxomatosis in wild rabbits in England and Wales in 1961[a].

	Jan	Feb	Mar	Apr	May	Jun	Jul	Aug	Sep	Oct	Nov	Dec
Counties affected[b]	52	48	53	51	52	52	55	57	57	57	58	58
Antigen-positive samples[c]	12	12	7	19	9	29	57	23	26	36	38	11

[a]From Fenner and Ratcliffe (1965).
[b]Out of 61 counties. Data from monthly reports of Rabbit Clearance Societies.
[c]Gel-diffusion tests conducted at the Ministry of Agriculture, Fisheries and Food Laboratories.

over winter until the following spring (Muirhead-Thompson, 1956b; Andrewes *et al.*, 1956). Subsequent experiments were concentrated on the biology and vector behaviour of the rabbit flea, but in 1971, after interest in moving infected rabbits around as a means of spreading the disease had abated, Service (1971) found that some species of *Aedes* and *Anopheles* would feed on healthy rabbits and might be responsible for the sudden reappearance of myxomatosis in areas from which it has been absent for some time.

The importance of the rabbit flea as a vector
The year-round occurrence of cases of myxomatosis focused attention on the European rabbit flea (*Spilopsyllus cuniculi*) as the important vector in Britain. During their experiments on the introduction of myxomatosis into the Heisker Islands (Shanks *et al.*, 1955), Allan and Shanks had shown that the fleas would transmit the infection, an observation confirmed by Lockley (1954). Rothschild (1953), a world authority on fleas, emphasized the peculiar suitablity of the rabbit flea for maintaining infection through the winter. This was confirmed by demonstrations that rabbits released in deserted burrows 50 days after the inhabitants had died of myxomatosis became infested with fleas and died of myxomatosis (Brown *et al.*, 1956), suggesting that fleas could act as a reservoir of infection for several months after rabbits had deserted a burrow. The potential prolonged infectivity of fleas was confirmed by the observation that some fleas that had fed through lesions of a rabbit with myxomatosis and were then buried in the ground in glass tubes were infective for as long as 112 days (Chapple and Lewis, 1965). These observations led to intensive investigations into the biology of the rabbit flea, notably into its breeding cycle, which have been admirably reviewed by Mead-Briggs (1977) (see p. 83).

Impact on wild rabbits

The best summaries of the changes in the numbers of wild rabbits in Britain between early 1953 until the mid-1970s are provided by Lloyd (1970, 1981) and from then until 1985 by Trout *et al.* (1986). Before myxomatosis, rabbits were present in 94% of agricultural holdings over an acre in area in England and Wales, and 47% were heavily infested. Myxomatosis did not sweep over Britain as it had in 1953–54 in France, but was distributed gradually and in a patchy manner, by rabbit fleas, following human transfers of infected rabbits to new sites. However, by the end of 1955 myxomatosis had occurred throughout Britain, and it was estimated that in all areas through which myxomatosis had passed about 99% of the rabbits had died (Thompson and Worden, 1956). As late as 1957 rabbits were still few and widely separated, indeed rabbits had become so rare that 'in many parts of the country as few as 10 or so rabbits were regarded [by farmers] with considerable alarm and foreboding' (Lloyd, 1970). A survey of 5668 holdings in 1969 showed that there were rabbits on 60% of holdings and only 2.1% of farms were heavily infested. Outbreaks of myxomatosis were sporadic and generally occurred only when rabbits had increased in numbers locally – many such outbreaks probably resulted from deliberate introduction of the disease.

By 1962 moderately attenuated strains constituted some 64% of strains collected from sick rabbits in 80 counties of Great Britain (Fenner and Chapple, 1965). Seasonal and recurrent outbreaks occurred more commonly on estates in which rabbits were regarded as game than on farms, because the former supported larger populations containing a high proportion of adult immune rabbits. The outbreaks occurred in the summer, when the populations were reinforced by the emergence of young rabbits. The gradual increase in bags of rabbits on game-keepered estates, where predator control was persistent and vigorous, is illustrated in Fig. 10.3.

Despite the continued presence of myxomatosis, rabbit numbers slowly increased as moderately attenuated strains promoted the development of rabbits with greater innate resistance, so that by 1979 it was estimated that the over-wintering

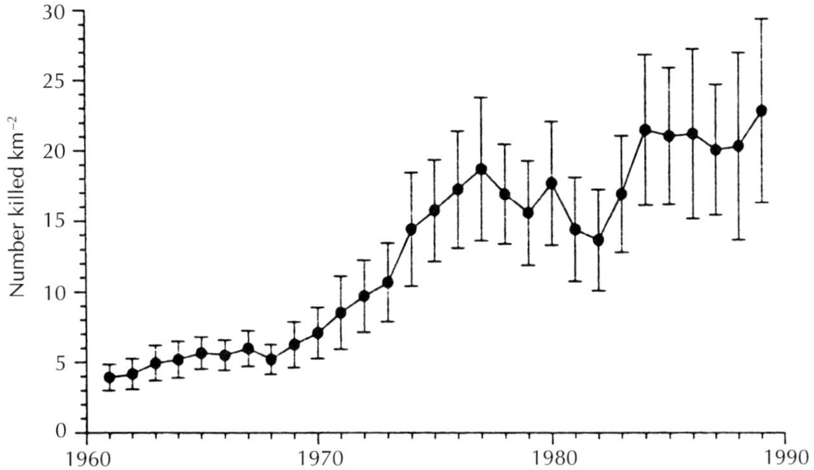

Fig. 10.3. Trends in the average bag of rabbits on game estates between 1961 and 1990. From Thompson (1994), with permission.

rabbit population had increased to about 20% of pre-myxomatosis numbers (Lloyd, 1981), although they were differently distributed, with few in Wales, which had formerly had large rabbit populations (Fig. 10.4).

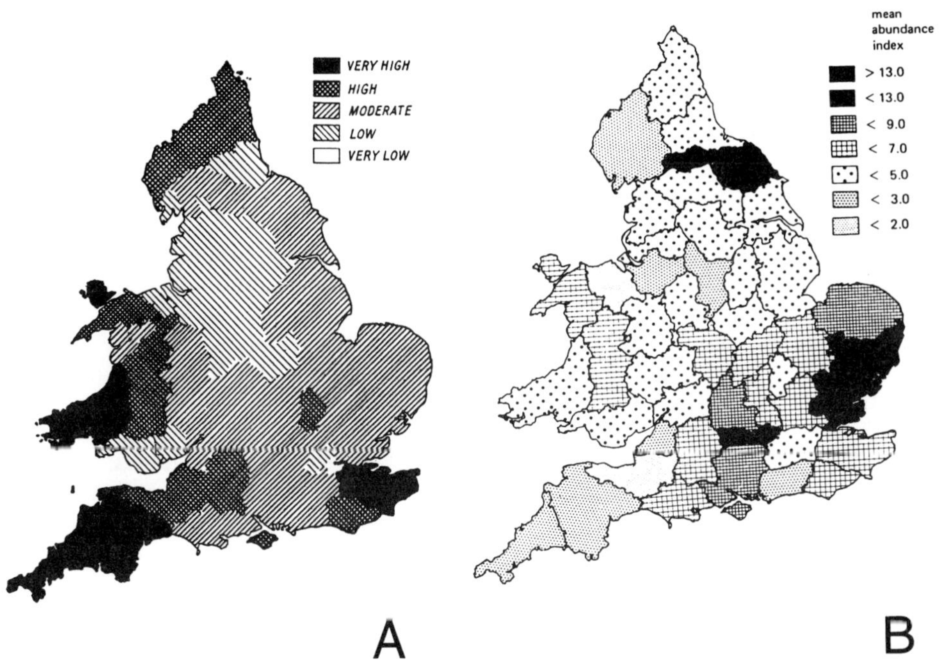

Fig. 10.4. Maps showing the distribution of rabbits in England and Wales, (A) before myxomatosis, and (B) 25 years after the introduction of myxomatosis. (A) From Thompson and Worden (1956), with permission. (B) From MAFF (1981), Crown copyright.

In Britain as in Australia, it was difficult to know whether myxomatosis was having an effect on rabbit numbers when rabbits had become more resistant and the very virulent strains of virus had disappeared. As Parer *et al.* (1985) observed in Australia, the only way to determine this was by manipulating the epidemiology of the disease, albeit in a different manner. Over a period of nine years from 1978 to 1986, flea numbers in study areas were periodically decreased by the application of insecticide and increased by the addition of fleas (Trout *et al.*, 1992). Changes in the prevalence of myxomatosis followed the changes in flea numbers, rabbit numbers increasing two- to threefold during the flea reduction periods, which eliminated myxomatosis, and falling when myxomatosis recurred in the reintroduction periods. Myxomatosis was clearly continuing to exert a controlling effect on rabbit numbers (Trout *et al.*, 1993); summarizing the situation in the early 1990s, Flowerdew *et al.* (1992) estimated that rabbit numbers had returned to about one-third of the pre-myxomatosis levels. In the mid-1980s rabbit damage was estimated to cost between £90 million and £120 million a year, and could rise.

More recently, Kolb (1994) assessed the impact of myxomatosis on rabbit infestation in Scotland, using the same methods as employed by the Department of Agriculture and Fisheries of Scotland in 1969, 1970, 1973 and 1974. In the East, North-east and Eastern Highlands 26.5% of farmers reported serious infestations in 1991, compared with 55.9% before myxomatosis reached Scotland (1954) and 1.5% in 1969–70. In other regions of Scotland serious rabbit problems disappeared after 1954 and are now at a level of only 5%. He calculated that the total annual loss attributable to rabbits was £11,790,000 in 1991, compared with a pre-myxomatosis level (at current values of the pound) of about £38,000,000.

Impact on domestic rabbits

Before myxomatosis, the breeding stock of domestic rabbits in Britain was about one million, and the annual turnover about 12 million (Thompson and Worden, 1956). They were not much affected by myxomatosis, because flying vectors were not important in Britain. However, in the initial outbreaks in 1953–55 an unknown number were killed, but after that many breeders followed advice to screen rabbit houses, and fibroma vaccine was made available in April 1954.

Changes in the virulence of the virus

Although not tested until much later (Fenner and Marshall, 1957), attenuated strains of myxoma virus were recovered from a rabbit in Sussex in September 1954 and from a pool of mosquitoes from Edenbridge in October 1954. The first reports of the occurrence of attenuated strains in Britain refer to material from Sherwood Forest, Nottinghamshire (Fenner and Marshall, 1955; Hudson *et al.*, 1955). Myxomatosis was introduced there in September 1954, and from April 1955 onwards an unusual number of rabbits was found to be surviving infection. Antibodies were recovered from 32 out of 64 sera and four attenuated strains of myxoma virus were recovered from surviving rabbits, although fully virulent virus had produced high mortalities a few miles to the north. Analysis of one of these strains by single-pock cloning revealed that it was a mixture of a highly virulent 'Lausanne-type' strain and a strain that produced protuberant lesions but killed few rabbits (Fenner and Marshall, 1957). The two substrains also differed in pock morphology.

Large-scale studies of the virulence of British strains of myxoma virus were initiated with a collection made from all over Great Britain in 1962 and analysed in Australia. By then moderately attenuated strains constituted some 64% of strains collected from sick rabbits in 80 counties of Great Britain (Fenner and Chapple, 1965; Table 10.3). In 1963 it had been decided to subdivide virulence Grade III (to which the majority of isolates tested in Australia between 1953 and 1959 had been allocated) into two subgroups, IIIA and IIIB, with mean survival times of 17–22 days and 23–28 days, and estimated case-

Table 10.3. The virulence of field strains of myxoma virus in Britain in 1962, 1975, and 1981, calculated on the basis of survival times (expressed as percentages).

Virulence grade	I	II	IIIA	IIIB	IV	V	
Mean survival time (days)	10–13	14–16	17–22	23–28	210–50	—	Number
Presumed case-fatality rate (%)	>99	95–99	90–95	70–90	50–70	<50	of samples
1953–55[a]	>99						
1962[b]	4.1	17.6	38.8	24.8	14.0	0.9	222
1975[c]	1.6	25.8	48.4	18.0	5.5	0.8	128
1981[c]	0	35.8	54.5	8.1	1.6	0	123

[a]From field observations.
[b]From Fenner and Chapple (1965).
[c]From Ross and Sanders (1987).

fatality rates of 90–95% and 70–90% respectively. In 1962 there were still a substantial number of highly virulent strains (22% Grades I and II) and the majority of the Grade III strains fell into the more virulent IIIA subgroup. Only 15% fell into the more attenuated Grades IV and V.

Follow-up studies in Britain, using similar protocols (Ross and Sanders, 1987), showed that Grade I strains of the Lausanne type had become rare by 1975 and had disappeared by 1981, but strains of Grade II were common (26% and 36% for 1975 and 1981 respectively), and the majority of strains were of Grade IIIA virulence (Table 10.3). Moderately and highly attenuated strains were rare or very rare. Apart from the absence of Grade I strains, virulence as judged by mean survival times in laboratory rabbits had moved steadily to the left of the spectrum over the period 1962 to 1981.

The strains used to initiate myxomatosis in Australia and Europe differed in their symptomatology, the strain used in Australia producing relatively flat skin lesions, whereas the Lausanne strain produced protuberant lesions (see Chapter 5). By 1962, 20% of the strains from Britain produced protuberant skin lesions, 45% produced flat lesions and the remainder were of intermediate morphology (Fenner and Chapple, 1965). There was little correlation between protuberance and lethality.

Changes in the innate resistance of rabbits

From analogy with what had occurred in Australia, authorities in Britain anticipated that as increasing numbers of rabbits survived because of the appearance of somewhat less virulent strains of virus, rabbits with greater innate resistance would be selected. Preliminary experiments had shown little difference between the case-fatality rates and mean survival times of non-immune rabbits collected between 1964 and 1967 from two localities (Kent and Norfolk) and wild rabbits from an island where there had been no myxomatosis (Vaughan and Vaughan, 1968). Extension of these studies for rabbits from Norfolk to 1976 (Ross and Sanders, 1977; Fig. 10.5) showed that there were slight increases in innate resistance between 1966 and 1969, then a substantial increase (tested on only 27 rabbits) in 1970 and a dramatic increase in the next large group of rabbits tested, in 1976. These tests were extended to rabbits from two other counties in Britain, Wiltshire and Angus (Ross and Sanders, 1987). The consolidated results (Table 10.4), demonstrate that by 1978 there was a considerable degree of increased resistance in each of the three widely separated rabbit populations. Comparison of the times of death of fatal cases and the case-fatality rates (Fig. 10.6) showed that while the ratio of deaths between 17 and 44 days and the case-fatality rates was linear, the percentage of rabbits dying early (17–28 days after inoculation) decreased more

Fig. 10.5. John Ross (1938–). After graduating in chemistry at the University of Aberdeen in 1961, Ross went on to obtain a PhD degree for studies on interferon in 1964. After postdoctoral experience at the University of Florida and the University of Birmingham, he was appointed in 1967 to the Ministry of Agriculture, Fisheries and Food in the United Kingdom to work on myxomatosis. From then until his retirement in 1996 he was the principal British scientist involved in studies of the epidemiology of myxomatosis in wild rabbit populations, the effects of the disease on rabbit numbers and the coevolution of virus and host.

quickly than the mortality rate, indicating that lengthening survival times were an early indication of increasing resistance.

Myxomatosis in Spain

In Spain, the ancestral home of the rabbit, it was never considered the pest that it was in France. Wild rabbits are the most important game animal in Spain, and before myxomatosis, about 10 million wild rabbits were shot annually by 1.5 million licensed hunters. Myxomatosis extended from France into the province of Gerona, in the north-east of Spain, near the border with France, in September–October 1953. From there it had spread to another 18 provinces by winter 1955 and in 1956 was reported from Portugal. Within five years of its introduction into north-east Spain it was observed that many wild rabbits captured there showed signs of the disease but were in comparative good health.

Spain had long had severe regulations prohibiting the deliberate spread of infectious diseases of animals. In December 1952, two months after the recognition of myxomatosis in France, a new law was introduced with four sections devoted to its control. With some difficulty, agreement was reached between the interests of foresters, to whom the rabbit was a pest, and those of hunters, commercial breeders and the leather and fur felt industries, the latter group objecting strongly to the damage to their industries caused by myxomatosis (Munos Goyanes, 1960).

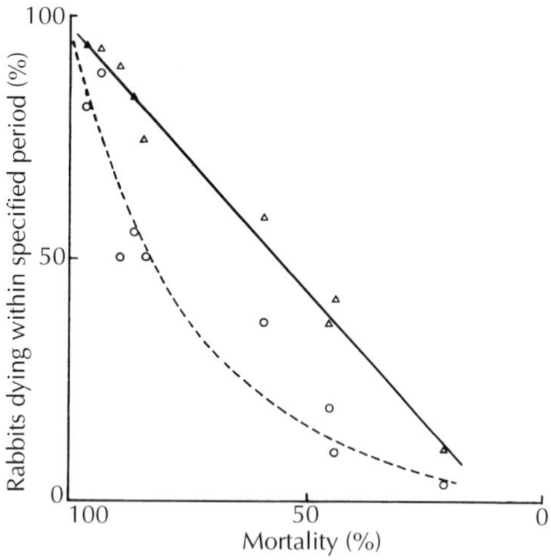

Fig. 10.6. Relationship between the mortality rate among rabbits infected with a strain of myxoma virus of intermediate (Grade IIIA) virulence and the percentages of rabbits dying between 17 and 44 days (solid line) and between 17 and 28 days (dashed line) after inoculation. From Ross and Sanders (1984), with permission.

Table 10.4. Changes in the innate resistance of wild rabbits from two locations in Britain, 1966–1978, after challenge with strains of different virulence[a].

Location	Year	Strain (virulence grade)	Number of rabbits tested	Case-fatality rate	Mean survival time of fatal cases (days)
Norfolk	1966	IIIA	41	90%	26.7
	1967		34	94%	23.2
	1968		71	86%	25.5
	1969		74	84%	25.9
	1970		27	59%	24.8
	1976		63	21%	35.9
Controls (each year)			9–18	100%	17.9–18.1
Wiltshire	1978	IIIA	71	45%	28.6
	Controls		18	100%	18
	1979	II	53	44%	29.4
	Controls		18	100%	16.3
	1980	I	44	52%	22.1
	Controls		9	100%	11.3

[a]Data from Ross and Sanders (1984); virulence checked on each occasion in laboratory rabbits.

Myxomatosis in Other Countries of Continental Europe

Italy

Italian rabbit farmers have long imported breeding stock from France, hence myxomatosis spread rapidly through rabbitries, especially in the north of the country. For a long time a locally produced attenuated vaccine (Brescia) was used, and may have interfered with vaccination with the SG33 vaccine, which was introduced in 1979. For whatever reason, use of the latter strain was attended by a high mortality, which led to some litigation.

The Netherlands

The first cases of myxomatosis were reported in autumn 1954, in the coastal dunes near the North Sea, where rabbits were abundant. In general, the public reaction was favourable to the disease, because of the harm rabbits caused to the dune vegetation.

References

Advisory Committee (1954) *Myxomatosis. Report of the Advisory Committee on Myxomatosis.* Her Majesty's Stationery Office, London, p. 7.

Andrewes, C.H., Muirhead-Thompson, R.C. and Stevenson, J.P. (1956) Laboratory studies of *Anopheles atroparvus* in relation to myxomatosis. *Journal of Hygiene* 54, 478–486.

Armour, C.J. and Thompson, H.V. (1955) Spread of myxomatosis in the first outbreak in Great Britain. *Annals of Applied Biology* 43, 511–518.

Brown, P.W., Allan, R.M. and Shanks, P.L. (1956) Rabbits and myxomatosis in the N.E. of Scotland. *Scottish Agriculture* 35, 204–207.

Chapple, P.L. and Lewis, N.D. (1965) Myxomatosis and the rabbit flea. *Nature* 207, 388–389.

Fenner, F. and Chapple, P.L. (1965) Evolutionary changes in myxoma virus in Britain. *Journal of Hygiene* 63, 175–185.

Fenner, F. and Marshall, I.D. (1955) Occurrence of attenuated strains of myxoma virus in Europe. *Nature* 176, 782–783.

Fenner, F. and Marshall, I.D. (1957) A comparison of the virulence for European rabbits (*Oryctolagus cuniculus*) of strains of myxoma virus recovered in the field in Australia, Europe and America. *Journal of Hygiene* 55, 149–191.

Fenner, F. and Ratcliffe, F.N. (1965) *Myxomatosis*. Cambridge University Press, Cambridge, p. 230.

Fenner, F. and Ross, J. (1994) Myxomatosis. In: Thompson, H.V. and King, C.M. (eds) *The European Rabbit*. Oxford University Press, Oxford, p. 208.

Flowerdew, J.R., Trout, R.C. and Ross, J. (1992) Myxomatosis: population dynamics of rabbits (*Oryctolagus cuniculus* Linnaeus, 1758) and ecological effects in the United Kingdom. *Revue Scientifique et Technique de l'Office Internationale des Epizooties* 11, 1110–1113.

Hudson, J.R., Thompson, H.V. and Mansi, W. (1955) Myxoma virus in Britain. *Nature* 176, 783.

Kolb, H.H. (1994) Rabbit *Oryctolagus cuniculus* populations in Scotland since the introduction of myxomatosis. *Mammal Review* 24, 41–48.

Lloyd, H.G. (1970) Post-myxomatosis rabbit populations in England and Wales. *European Plant Protection Organization, Publication Series A* 58, 197–215.

Lloyd, H.G. (1981) Biological observations on post-myxomatosis rabbit populations in Britain 1955–1979. In: Myers, K. and McInnes, C.D. (eds) *Proceedings of the World Lagomorph Conference – 1979*. University of Guelph, Ontario, pp. 623–628.

Lockie, J.D. (1956) More light on myxomatosis. *Scottish Agriculture* 36, 44–45.

Lockley, R.M. (1940) Some experiments in rabbit control. *Nature* 145, 767–769.

Lockley, R.M. (1954) The European rabbit flea, *Spilopsyllus cuniculi*, as a vector of myxomatosis in Britain. *Veterinary Record* 66, 434–435.

Lockley, R.M. (1964) *The Private Life of the Rabbit. An Account of the Life History and Social Behaviour of the Wild Rabbit*. Andre Deutsch, London, pp. 116–117.

MAFF (1981) Agricultural Science Service Research and Development Reports, *Mammal and Bird Pests*. HMSO, London.

Mead-Briggs, A.R. (1977) The European rabbit, the European rabbit flea and myxomatosis. *Applied Biology* 2, 183–261.

Muirhead-Thompson, R.C. (1956a) Field studies of the role of *Anopheles atroparvus* in the transmission of myxomatosis in England. *Journal of Hygiene* 54, 472–477.

Muirhead-Thompson, R.C. (1956b) The part played by woodland mosquitoes of the genus *Aedes* in the transmission of myxomatosis in England. *Journal of Hygiene* 54, 461–471.

Munos Goyanes, G. (1960) *Anverso y reverso de la mixomatosis*. Diression General de Montes, Casa y Pesca Fluvial, Madrid, 153 pp.

Parer, I., Conolly, D. and Sobey, W.R. (1985) Myxomatosis: the effects of annual introductions of an immunizing strain and a highly virulent strain of myxoma virus into rabbit populations at Urana, NSW. *Australian Wildlife Research* 12, 407–423.

Ritchie, J.N., Hudson, J.R. and Thompson, H.V. (1954) Myxomatosis. *Veterinary Record* 66, 796–804.

Ross, J. and Sanders, M.F. (1977) Innate resistance to myxomatosis in wild rabbits in England. *Journal of Hygiene* 79, 411–415.

Ross, J. and Sanders, M.F. (1984) The development of genetic resistance to myxomatosis in wild rabbits in Britain. *Journal of Hygiene* 92, 255–261.

Ross, J. and Sanders, M.F. (1987) Changes in the virulence of myxoma strains in Britain. *Epidemiology and Infection* 98, 113–117.

Rothschild, M. (1953) Notes on the European rabbit Flea. *Report to the Myxomatosis Advisory Committee*, 6 December 1953.

Sellers, R.F. (1987) Possible windborne spread of myxomatosis to England in 1953. *Journal of Hygiene* 98, 1110–1125.

Service, M.W. (1971) Reappraisal of the role of mosquitoes in the transmission of myxomatosis in Britain. *Journal of Hygiene* 69, 105–111.

Shanks, P.O., Sharman, G.A., Allan, R., Donald, L.G., Young, S. and Marr, T.G. (1955) Experiments with myxomatosis in the Hebrides. *British Veterinary Journal* 111, 25–30.

Sheail, J. (1991) The management of an animal population: changing attitudes to the wild rabbit in Britain. *Journal of Environmental Management* 33, 189–203.

Thompson, H.V. (1994) The rabbit in Britain. In: Thompson, H.V. and King, C.M. (eds) *The European Rabbit*. Oxford University Press, Oxford, pp. 64–107.

Thompson, H.V. and Worden, A.N. (1956) *The Rabbit*. Collins New Naturalist, London, 240 pp.

Trout, R.C., Tapper, S.C. and Harradine, J. (1986) Recent trends in the rabbit population in Britain. *Mammal Review* 16, 117–123.

Trout, R.C., Ross, J., Tittensor, A.M. and Fox, A.P. (1992) The effect on a British wild rabbit population (*Oryctolagus cuniculus*) of manipulating myxomatosis. *Journal of Applied Ecology* 29, 671–686.
Trout, R.C., Ross, J. and Fox, A.P. (1993) Does myxomatosis still regulate numbers of rabbits (*Oyryctolagus cuniculus* Linnaeus, 1758) in the United Kingdom? *Revue Scientifique et Technique de l'Office Internationale des Epizooties* 12, 35–38.
Vaughan, H.E.N. and Vaughan, J.A. (1968) Some aspects of the epizootiology of myxomatosis. *Symposium of the Zoological Society of London* 24, 281–309.
Worrall, V. (1956) Legal aspects. In: Thompson, H.V. and Worden, A.N. *The Rabbit.* Collins New Naturalist, London, pp. 191–210.

11

The Use of Rabbit Haemorrhagic Disease Virus for Rabbit Control

Overview

Rabbit haemorrhagic disease virus (RHDV) is a highly virulent calicivirus first observed in domestic rabbits in China in 1984. It was subsequently reported in many countries in Europe and other parts of the world. It caused havoc in commercial rabbitries until an inactivated vaccine prepared from infected rabbit tissues became available. RHDV appears to be a mutant form of a virus that has long been enzootic in commercial and wild rabbits in Europe, causing inapparent infections. Experience in Europe and China showed that it had a very restricted host range and was difficult if not impossible to grow in tissue culture. It is highly lethal for mature rabbits, killing them within one to three days, with signs of disseminated intravascular coagulation. Very young rabbits are much less susceptible to its lethal effects, and if infected they often recover and acquire lifelong immunity to reinfection. If a doe has high titre antibodies, maternal immunity may protect her progeny for a short time after the waning of this physiological resistance.

In commercial rabbitries it is spread by contact, and spread between rabbitries and between countries appears to be due to movement of infected rabbits, rabbit meat or fomites. In wild rabbit populations the virus appears to be spread by insects that become contaminated with the virus by feeding on the internal organs of dead rabbits, and also by contact via virus excreted in the urine or faeces. Contaminated insects transfer the virus mechanically, both locally, and under favourable meteorological conditions, over long distances.

After observations of outbreaks among wild rabbits in arid areas of Spain in 1988, it was decided to investigate the possibility that RHDV could be used for the biological control of rabbits in Australia. Authorities in New Zealand also expressed an interest and provided financial support for further investigations of the virus.

A strain of the virus was imported under quarantine into the Australian Animal Health Laboratory in 1991. Tests on a colony of wild rabbits confirmed its lethality and the apparently 'quiet' deaths of infected rabbits. Investigations on a range of domestic and native animals confirmed the high species specificity of the virus. After conferences involving animal welfare groups as well as scientists and rabbit control authorities, it was agreed by all governments in Australia and New Zealand to proceed to field trials on Wardang Island in Spencer Gulf, South Australia. Elaborate precautions were taken to maintain high quarantine security in the trial site. However, it was impractical to enclose the whole of the trial site within insect-proof netting, and in October 1995 the virus escaped from the trial site to the adjacent mainland and soon spread over distances of several hundred kilometres.

While this work was in progress, studies were carried out to provide the relevant Commonwealth government authorities with information to fulfil requirements for the release of the virus, as required by various Acts. In view of the escape of the virus, and having satisfied all legal requirements, in October 1996 releases were made at nominated sites throughout Australia, at each of which monitoring by State rabbit control authorities had been organized. The efficacy of the virus was high in the hotter arid regions, during the winter months. Spread in wetter, temperate regions was most effective in the warmer months.

In July 1997 the New Zealand government decided not to release the virus in that country, but in August rabbits that had died from RHDV were detected in rabbit-infested country in South Island and soon after that in other parts of New Zealand. Initially the Government attempted to limit the spread of the virus, but in September it agreed to allow farmers to spread the virus.

It is too early to determine how effective RHDV will be, and for how long, in either Australia or New Zealand. Attenuation of the virus and the consequent evolution of resistance in rabbits, as occurred in myxomatosis, depends primarily on whether transmissibility is enhanced by attenuation. If the principal source of virus available for transmission (by contaminated insects) is that in the internal organs of fatal cases, there may be selection for maintaining virulent strains of virus.

The Discovery and Spread of Rabbit Haemorrhagic Disease Virus

The People's Republic of China is the world's largest exporter of domestic rabbit meat, exports rising from 308 tonnes in 1975 to 53,200 tonnes in 1983 (Feng-Yi, 1990). If one adds to this the large local use of rabbit meat, from both large commercial rabbitries and small 'back-yard' hutches, it is clear that by the 1980s there was a very large population of domestic rabbits in China. In 1984 a hitherto unknown disease was seen in Angora rabbits that had been imported a few days earlier from the German Democratic Republic into China (Liu *et al.*, 1984; Xu, 1991). Except for suckling animals, almost all the rabbits in the affected rabbitry had died within a week, and in less than nine months the disease had spread through commercial rabbitries in an area of about 50,000 square kilometres (Xu, 1991), and soon spread to Korea (Park *et al.*, 1991). After failing to cultivate the causative virus in many primary rabbit cells and several cell lines, it was claimed that it could be grown, with difficulty, in a cell line derived from primary rabbit kidney cells (Ji *et al.*, 1991). An inactivated vaccine was developed, using livers of infected rabbits (Huang, 1991), and it was claimed that by the end of 1986 the disease had been brought under control in China (Xu, 1991).

The disease was recognized in Italy in 1986 (Cancellotti and Renzi, 1991) and soon spread to most countries in Europe (Morisse *et al.*, 1991). Chinese investigators realized that the disease was of viral origin, although they initially suspected the agent to be a parvovirus rather than a calicivirus. When the disease occurred in Europe in 1986 it was at first regarded as being due to a toxin, or to fallout from the Chernobyl disaster. In 1987–88 the link was established with the 'haemorrhagic pneumonia' of rabbits in China, and with knowledge of its viral origin vigorous research was initiated on this new and very important disease of an animal of commercial value for both meat production and hunting.

By what appears to be a remarkable coincidence, a very similar disease of hares, also caused by a calicivirus, had been causing significant losses in farmed and wild hares in northern Europe since 1980 (Gavier-Widén and Mörner, 1991). Retrospective testing has shown the existence of hares seropositive for the causative virus since 1971 (Moussa *et al.*, 1992). Although related (see below), the two viruses are distinct; they do not cross-protect (Chasey *et al.*, 1992; Lavazza *et al.*, 1996) and the hare disease occurred in countries such as Great Britain before rabbit haemorrhagic disease had occurred there (Chasey and Duff, 1990).

In May 1989 the Office International des Épizooties designated these diseases of lagomorphs 'viral haemorrhagic disease'. The disease of rabbits was called rabbit haemorrhagic disease (RHD) and the disease of hares the European brown hare syndrome (EBHS), and in 1991 the Office devoted an issue of its journal to these two diseases (OIE, 1991). Although initially the causative agent of RHD was suspected to be a parvovirus (Xu, 1991; Gregg *et al.*, 1991), virologists in Germany, Italy and Spain demonstrated that the causative viruses of both diseases were caliciviruses (Ohlinger and Thiel, 1991), and soon after this the genome of RHDV was completely sequenced (Meyers *et al.*, 1991).

The rabbit disease was first described outside China and Korea when it appeared among domestic rabbits in Italy in 1986, and by 1988 it had been reported among domestic rabbits in many countries in Europe and the Russian Federation, in the Middle East and parts of Africa, and in Cuba, Mexico, the United States of America, India and Reunion Island (Morisse *et al.*, 1991). It was probably spread by trade in rabbit meat, or by shipment of consignments of rabbits among which some were infected before despatch. Mortality rates among rabbits in many commercial rabbitries in Europe were exceedingly high until an inactivated vaccine was introduced. The virus was imported into Mexico with a shipment of 18 tonnes of rabbit meat exported from China to a supermarket outside of Mexico City in December 1988, but was successfully eradicated (Gregg *et al.*, 1991), an accomplishment made possible by the absence of wild European rabbits in Mexico. However, in Europe the virus spread quickly from domestic to wild rabbits and was soon established in wild rabbit populations of most countries in Europe.

Origin of the virus

Since viruses do not appear by spontaneous generation, three possible sources were considered: transfer of the virus of European brown hare syndrome to rabbits (which leaves unsolved the origin of that virus), change in the properties of a pre-existing non-pathogenic virus of rabbits, or transfer of a hitherto unknown virus from another animal species, such as was documented for myxomatosis. The first possibility was eliminated when it was shown that the viruses causing disease in rabbits and hares were related but distinctly different caliciviruses. The second possibility was raised because antibodies that cross-reacted with RHDV were detected in rabbit sera collected in the Czech Republic 12 years before the first outbreaks (Rodak *et al.*, 1990) and Capucci *et al.* (1994, 1997; Fig. 11.1) showed that seroconversion occurred in asymptomatic rabbits in some rabbitries where RHD had never been seen.

By ingenious manipulation of breeding rabbits and their progeny before and after being moved to different cages for fattening, Capucci *et al.* (1996) recovered a calicivirus that caused no symptoms in rabbits but produced seroconversion and protected the rabbits against infection with RHDV. A serological survey of rabbits in an industrial rabbitry showed that infection probably occurred immediately after weaning (Capucci *et al.*, 1997). The non-pathogenic virus, for which they proposed the name rabbit calicivirus (RCV), was more closely related to RHDV than to the European brown hare syndrome calicivirus, when tested by serology and sequence comparisons of the capsid proteins (Capucci *et al.*, 1996). The occurrence of the inapparent infection in some rabbitries but not in others explained the patchy nature of outbreaks of RHD in some commercial rabbitries in Europe. Its discovery dispenses with the need to consider the third alternative, namely that RHDV came from some other host than the European rabbit. Given that the European brown hare syndrome was also apparently a 'new' disease, three intriguing questions remain unanswered: how long ago did the separation of the rabbit and hare caliciviruses occur, and what were the

Fig. 11.1. (A) Lorenzo Capucci (1954–). Born in Cremona, Italy, Capucci took a degree in biology at the State University of Milan. In 1983 he joined the staff of the Istituto Zooprofilattico Sperimentale della Lombardia e dell'Emillia, where he still works. After working on foot-and-mouth disease, he commenced work on the calciviruses of lagomorphs in 1989. In 1991 his laboratory was designated by the Office Internationale des Epizooties (OIE) as the OIE Reference Laboratory for Rabbit Haemorrhagic Disease Virus. In 1996 he identified a non-pathogenic calicivirus in rabbits which is closely related to the rabbit haemorrhagic disease virus. Since 1997 he has been collaborating with Australian scientists on the laboratory diagnosis of rabbit haemorrhagic disease.
(B) Brian Coman (1944–). Born in Kyneton, Coman graduated with a BVSc from the University of Melbourne in 1967 and obtained a PhD degree in 1976, on parasitology. From 1968 to 1991 he worked at the Keith Turnbull Research Institute at Frankston, primarily on the ecology and economic significance of introduced predators. Since 1991 he has worked as a private consultant on projects associated with rabbit control in Australia and New Zealand, such as coauthorship of books on managing foxes and rabbits produced by the Bureau of Resource Sciences, the production of the draft environmental impact assessment of rabbit calicivirus and the development of a RHDV bait.

genomic changes responsible for the transition from RCV to RHDV, and did EBHSV also originate from a non-pathogenic hare calicivirus?

Classification and Properties of Caliciviruses

Viruses of the family *Caliciviridae* (Ohlinger and Thiel, 1991) are small round viruses with a characteristic appearance in electron-micrographs (Fig. 11.2) and a genome comprising a single positive sense RNA strand (Ohlinger *et al.*, 1990; Parra and Prieto, 1990).

Genome comparisons

The genome of rabbit haemorrhagic disease virus (RHDV) has been completely sequenced and consists of 7437 nucleotides, excluding the poly(A) tail (Meyers *et al.*, 1991; Rasschaert *et al.*, 1995; Gould *et al.*, 1997). The genome of the European brown hare syndrome virus (EBHSV) has also been sequenced. It is 7442 bases long; alignment of the sequences of RHDV and EBHSV shows 71% nucleotide identity (Gall *et al.*,

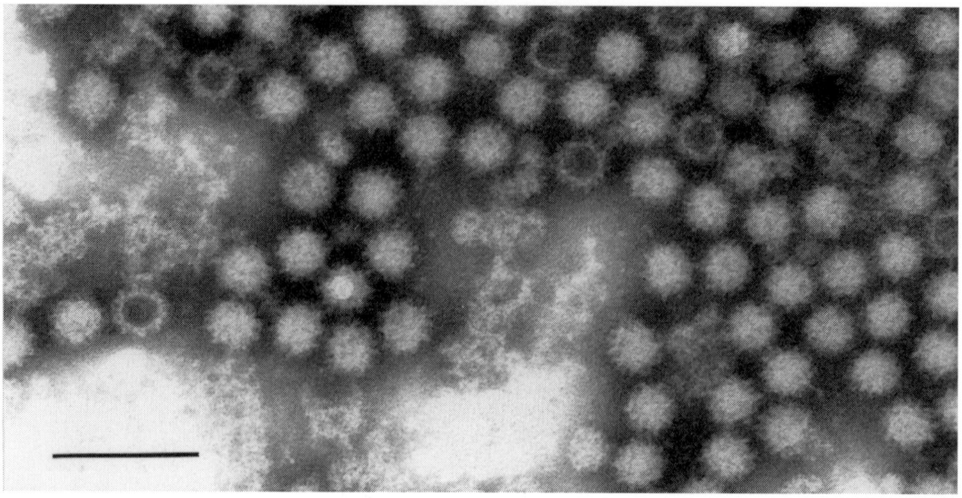

Fig. 11.2. Electron-micrograph of rabbit haemorrhagic disease virus, a calicivirus (Latin *calix*, cup or goblet). Bar = 100 nm. Caliciviruses are small round particles characterized by 32 cup-shaped surface depressions arranged in T = 3 icosahedral symmetry. Courtesy Dr H.A. Westbury, Australian Animal Health Laboratory.

1996). Comparisons of partial nucleotide sequences of the capsid protein gene of representatives of all known groups of caliciviruses show that those of lagomorphs, for which the generic name *Lagovirus* has been proposed, cluster together (Fig. 11.3). Comparisons have also been made of the relatedness of partial nucleotide sequences of the capsid protein gene of 44 strains of RHDV isolated from cases in several European countries, Mexico, China and Korea between 1989 and 1995 (Nowotny *et al.*, 1997; Fig. 11.4); they show between 89% and 100% identity. Comparisons between 19 strains of EBHSV showed similar homology within that species but there was only 53–60% homology between strains of RHDV and EBHSV.

Host range

Some caliciviruses, notably San Miguel sealion virus (which produced vesicular exanthema in swine), and the group of serologically related viruses that infect marine mammals, have a wide host range and are readily grown in cell culture. However, like the human caliciviruses, for which the only other susceptible host appears to be the chimpanzee (Wyatt *et al.*, 1978) and which cannot be grown in tissue culture (Kapikian *et al.*, 1995), both of the lagomorph viruses are highly species-specific and cannot be grown in tissue culture. There was a weak antibody response (but no disease and no protection), when hares were inoculated with RHDV or rabbits with EBHSV (Lavazza *et al.*, 1996), but no other animals among those that had been tested by a number of workers were susceptible. Further host range studies carried out to determine whether it was safe to release the virus among wild rabbits in Australia are described below (p. 249).

Although there have been two claims that RHDV could be cultivated, with difficulty, in cell lines derived from rabbit kidneys (Ji *et al.*, 1991; Nawwar *et al.*, 1996), other investigators have been unable to grow it in tissue culture. For example, Gregg *et al.* (1991) were unable to grow RHDV in any of 32 cell lines, in cultured cells prepared from many organs of a rabbit infected 24 hours earlier or from foetal rabbits, or by inoculation of embryonated chicken eggs. However, by carefully preparing suspensions of hepatocytes by

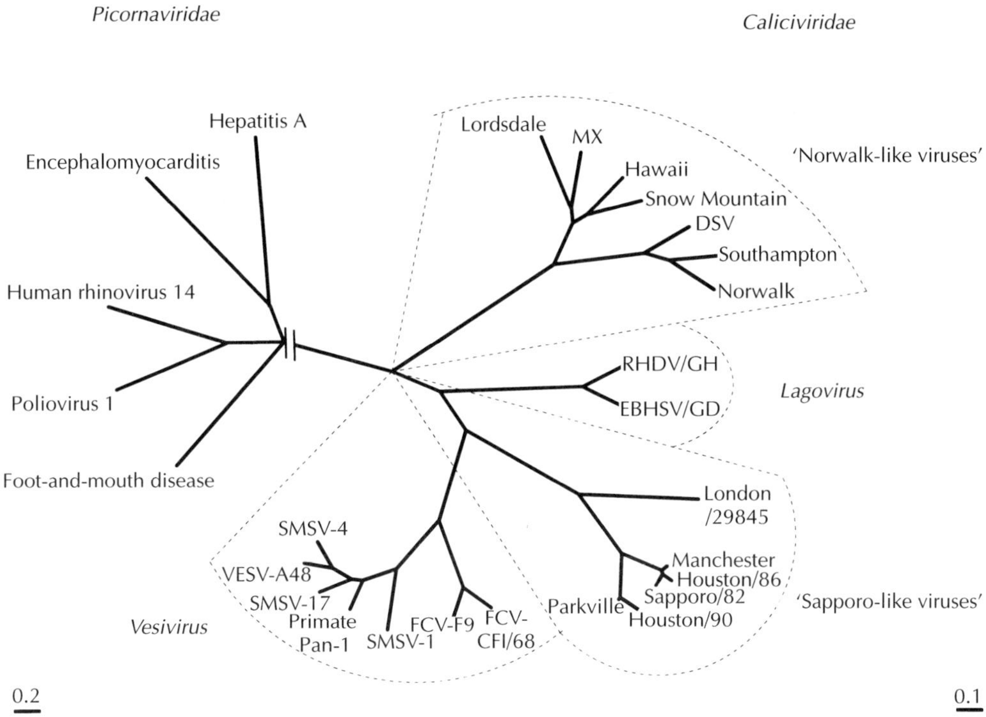

Fig. 11.3. Genetic relationships of caliciviruses recovered from a variety of animal hosts, based on comparisons of the hypervariable sequences of the capsid protein gene. There are four groups: two groups isolated from cases of human diarrhoea (Norwalk-like and Sapporo-like), the proposed genus *Vesivirus*, comprising the caliciviruses of pinnipeds, vesicular exanthema virus and feline calicivirus, and the proposed genus *Lagovirus*, containing rabbit haemorrhagic disease virus and European brown hare syndrome virus. The family *Picornaviridae* is distantly related to the *Caliciviridae*. Courtesy of Dr T. Berke.

collagenase treatment of rabbit liver, König *et al.* (1998) were able to infect these cells *in vitro*.

Clinical Features of Rabbit Haemorrhagic Disease

Infection of adult rabbits with RHDV leads to peracute or acute clinical disease in one to three days, rabbits infected by the oral route surviving on average for one day longer than animals infected by inoculation. No significant clinical signs are seen in peracute cases, but acutely affected cases appear quiet but have an increased body temperature and respiration rate and die within 12 hours (Marcato *et al.*, 1991). Haematuria and/or vaginal haemorrhage and foamy discharge from the nostrils are occasionally seen, and occasionally infected animals develop signs of central nervous system disease. No animals recover from the peracute disease and the few animals that recover from the acute disease may exhibit jaundice and die a few days later. Virus is found in all secretions and excretions of diseased rabbits.

Studies at the Australian Animal Health Laboratory (Lenghaus *et al.*, 1994) showed that after infection with a standard dose of RHDV, Australian wild rabbits died sooner than laboratory rabbits, within 20–24 hours rather than 30–36 hours, and apparently with minimal distress. They commented that:

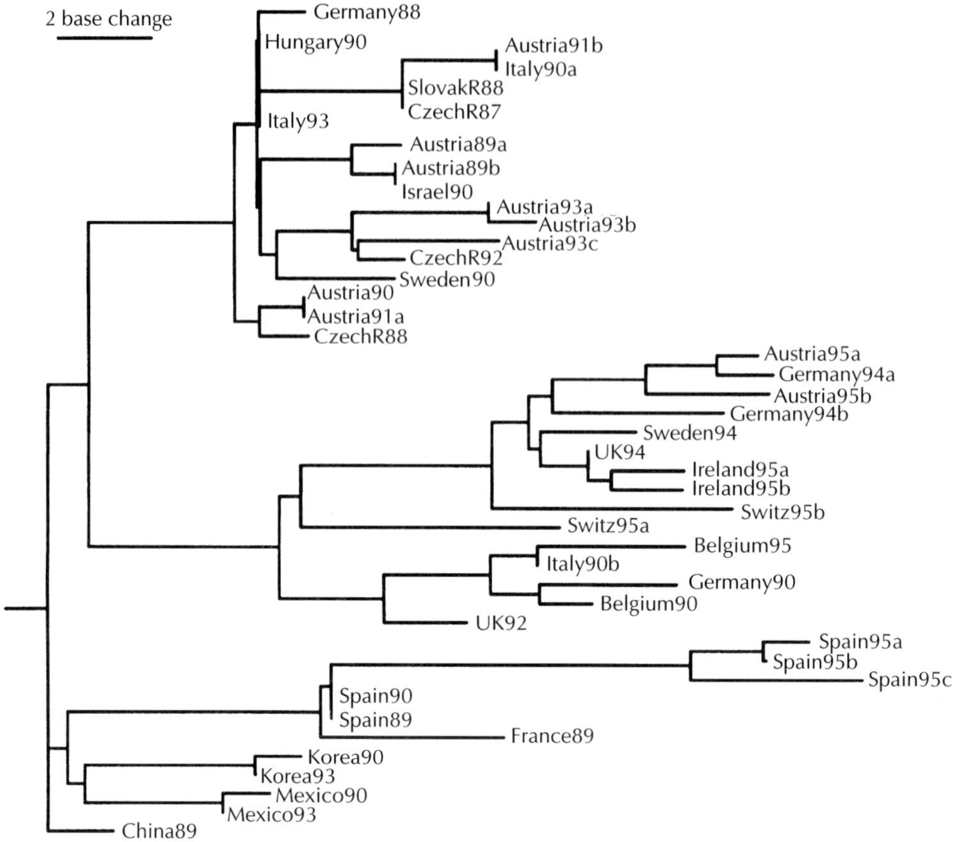

Fig. 11.4. Dendrogram showing the genetic relationships of 44 samples of RHDV from various parts of the world, based on a partial nucleotide sequence of the capsid protein gene. Although all samples are closely related, there are three main branches. From Nowotny *et al.* (1997), with permission.

> The haemorrhages that featured prominently in some overseas reports were of such a minor significance in our studies that the name 'Rabbit Haemorrhagic Disease' was seen to be seriously misleading – 'Rabbit Quiet Death Syndrome' or 'Big Spleen Disease' would be far less emotive and more accurate descriptors for the disease syndrome encountered here.

They noted that 'Rabbit Calicivirus Disease' (RCD) had also been suggested as an alternative name for RHD, and after the meeting in September 1993 at which the report by Lenghaus *et al.* was discussed, the Australian Government authorities adopted this name for the disease and 'rabbit calicivirus' as the name for the virus[a]. However, since Capucci *et al.* (1996) gave this name to the avirulent precursor virus, and since RHDV is still used internationally for the virulent virus, we have continued to use the latter name in this book.

In the Wardang Island trial (see p. 253) Cooke *et al.* (1997) found that inoculated wild rabbits died on average 42 hours later (range 21–48 hours), usually in their burrows. Behavioural changes were seen in

[a]This decision was influenced by the public disquiet about 'haemorrhagic' diseases, influenced no doubt by the publicity given to Ebola haemorrhagic fever and similar diseases.

some animals 12 hours before death; few infected rabbits came above the ground and those which emerged stayed unusually close to their warren. In later experiments Cooke found that female rabbits died about 12 hours earlier than males and that ambient temperature did not affect survival time.

Reactions of immature rabbits

European observers (Morisse *et al.*, 1991; Rodak *et al.*, 1991) reported that even in the absence of maternal antibodies, domestic rabbits less than four weeks of age did not develop clinical signs or pathological lesions, although they were infected and developed lifelong immunity. They became more susceptible as they got older. Lenghaus *et al.* (1994) confirmed this with laboratory-bred Australian wild rabbits, finding that although all baby rabbits (less than 10 days old) survived, only six of 13 five-week old and two of 18 seven- to nine-week old inoculated with RHDV survived. However, the baby rabbits were infected and they excreted enough virus to infect a sentinel adult rabbit housed in an adjacent cage. As adults they had high antibody levels, so that not only would they be immune if infected again, but maternal immunity from the female rabbits could protect their progeny for a few weeks after the waning of their physiological resistance. Experiments by Cooke at Gum Creek, in South Australia, showed that maternal antibodies persisted in progeny for between 5 and 11 weeks, depending on the maternal titre, and that even low levels of antibody could sometimes prevent death.

Pathology of Rabbit Haemorrhagic Disease

The most consistent pathological lesion in adult rabbits infected with RHDV is a necrotizing hepatitis which affects the hepatocytes in the peripheral areas of the liver lobules most severely (Marcato *et al.*, 1991; Fuchs and Weissenbock, 1992). Other characteristic lesions, not seen in all cases, are lymphocyte depletion and necrosis of the spleen. Disseminated intravascular coagulation produces fibrinous thrombi within small blood vessels in most organs, especially in the lungs, heart and kidneys, in which there are occasionally striking haemorrhages, from which the disease derived its name. Liver, spleen and blood contain high concentrations of virus (Xu, 1991). Lenghaus *et al.* (1994) noted that the damage to the liver and spleen in immature rabbits which survived infection did not progress beyond scattered small foci of lytic necrosis, and the extensive blood coagulation seen in adult animals was entirely absent. Titred by PCR (polymerase chain reaction) (see below), the concentration of virus in the livers of adult rabbits was a million times higher than in those of baby rabbits.

Clinical Diagnosis

Apart from epidemiological features (an acute disease with a high mortality in adult rabbits), the presence of gross lesions of acute hepatitis, a swollen spleen and congested and haemorrhagic lungs suggests rabbit haemorrhagic disease. The histopathological lesions in the liver, spleen, kidneys, heart and lungs are highly suggestive (Fuchs and Weissenbock, 1992).

Laboratory Diagnosis

Although it is difficult or impossible to grow the virus in tissue culture, electron microscopic examination of sections or smears of liver or spleen show very numerous small round virus particles, and immunofluorescence or enzyme-linked immunoassay (ELISA) with suitable polyclonal or monoclonal antibodies is diagnostic.

Advances in molecular biological technology have provided an exquisitely sensitive technique for detecting known nucleic acid sequences, the polymerase chain reaction (PCR). Guittré *et al.* (1995) developed a reverse transcriptase PCR for a conserved part of the capsid protein and showed that it was 10,000 times more sensitive than ELISA testing for the detection of RHDV. Using similar primers, Gould *et al.* (1997) were able to obtain positive results with material from infected liver at dilutions up to 10^{-10}. This test is now used routinely to detect whether various species of insect carry RHDV. Unfortunately, it does not distinguish between live and inactivated virus; in the absence of a method for culturing the virus this requires the inoculation of rabbits with suspensions of potentially contaminated insects.

Serological tests

Several serological tests can be used to demonstrate past infection in rabbits that survive infection, including haemagglutination-inhibition and several types of ELISA. These tests also reveal infection with the avirulent rabbit calicivirus. The most useful test for use in surveys of wild rabbits is the competition ELISA, using polyclonal or monoclonal antibodies (Capucci *et al.*, 1991, 1995). Using suitable anti-isotype antibodies, it is possible to distinguish between IgG, IgM and IgA (Capucci *et al.*, 1997). This is useful in epidemiological studies, since it can be used to distinguish between past infection with RHDV (IgG, some IgA and possibly IgM in adult rabbits), recent recovery from infection (high titre of IgM, with IgA and IgG at lower titres) and maternal antibody (IgG only, in rabbits less than 10 weeks old).

Development of Vaccines

Being such a devastating disease for the rabbit industry, Chinese scientists proceeded with the development of an inactivated tissue vaccine even before the causal agent had been characterized (Huang, 1991). They subsequently refined the procedure, using virus obtained from livers and spleens of infected rabbits inactivated with formalin. In Europe also, attempts to produce a vaccine were undertaken as soon as the disease was recognized; the large literature has been reviewed by Argüello Villares (1991). Formalin-inactivated vaccine produced from the liver and other organs of infected rabbits is usually administered to rabbits at about ten weeks of age. A single dose gives protection for the productive life of the rabbit, and immunity can be boosted by revaccination if necessary (as in pet or pedigree rabbits).

Since it has not been possible to grow the virus in cultured cells, other methods have also been used to produce vaccines. These include the production of virus-like particles from capsid protein obtained from insect cells infected with a genetically modified baculovirus (Plana-Duran *et al.*, 1996). Suspensions of these particles were immunogenic by the oral route as well as by injection, and could therefore be used in baits to immunize wild rabbits. Vaccines have also been produced from the vaccine strain of myxoma virus (SG 33; see p. 220) into which the gene for the coat protein of RHDV has been inserted, so that domestic rabbits can be simultaneously immunized against both myxomatosis and RHD (Bertagnoli *et al.*, 1996).

Epidemiology of Rabbit Haemorrhagic Disease

Resistance to environmental temperature is important in considering the epidemiology of RHD. Experiments at the Australian Animal Health Laboratory showed that purified virus survived between 28 and 35 days at 22°C but only 3–7 days at 37°C and about 15 minutes at 56°C (Smid *et al.*, 1991; Westbury, 1996). Viable virus was detected for as long as 105 days after RHDV has been dried on cloth and kept at about 20°C (Rodak *et al.*, 1991).

RHD may be spread by oral, nasal or parenteral transmission. Virus is excreted in the urine and faeces, which may be infectious for up to four weeks after infection if rabbits survive that long (Gregg *et al.*, 1991). Since the virus is moderately resistant to environmental temperature, transmission of infection in commercial rabbitries in temperate climates could readily occur via contaminated foodstuffs or other fomites, and the history of its spread from China to Mexico shows that meat from infected rabbits could serve as a means of long-distance movement of the virus. Insects of various kinds can also serve as vectors, transporting virus on their mouthparts or in their gut (see below).

In domestic rabbits

When first introduced into a susceptible population of domestic rabbits RHD spread rapidly, with mortality rates of up to 100% in rabbits older than two months (Ohlinger *et al.*, 1993). When the disease became endemic the morbidity and mortality rates fell, because of the presence of cohorts of animals that were infected previously as young rabbits and were therefore immune.

The initial outbreaks among domestic rabbits in various countries in Europe occurred in widely separated areas, and were probably initiated through the movement of rabbits in the domestic rabbit industry. The epidemiology in domestic rabbits was complicated initially by the presence in some rabbitries but not others of the avirulent rabbit calicivirus (see above), and later by that factor and the widespread vaccination that was carried out. Nevertheless, it was believed by people in the rabbit industry that 100–200 million domestic rabbits were killed in Italy in the big epidemic in 1988–90[1].

In wild rabbits

The disease was found in wild rabbits in Europe soon after it had been recognized in domestic rabbits, and an outbreak occurred in Spain in 1988. It has persisted in wild rabbit populations in Europe; in southern Spain the disease appears to break out every second year if sufficient susceptible rabbits are present. The first systematic study of the epidemiology of the disease among wild rabbits was carried out in an epizootic that occurred in the Donana National Park in Spain in April–May 1990 (Villafuerte *et al.*, 1994). The rabbit population in a section of the Park had been carefully monitored by capture–recapture, radiotracking and direct observation since October 1988. The disease was first recognized in the Park on 30 March 1990, and the epizootic spread through the population under observation between 21 April and 23 May. Only adult animals were found dead.

The mode of spread among wild rabbits living close to one another is almost certainly by the faecal–oral route, either directly or by contamination of the burrow or of food. The floor of burrows and adjacent soil contaminated with faeces, urine or infected organs can remain infective for several days, depending on the environmental temperature. No calicivirus of vertebrates is known to replicate in an insect vector, but with the experience of myxomatosis in mind, experiments were carried out in the Australian Animal Health Laboratory to see whether it might be mechanically transmitted by insects (Lenghaus *et al.*, 1994). Viraemia in rabbit haemorrhagic disease reaches a high titre, and as expected mechanical transmission could be demonstrated with fleas or mosquitoes which were transferred to naïve rabbits shortly after feeding on RHDV-infected rabbits.

The potential for mechanical transmission has also been demonstrated in the laboratory for the Australian bush fly (*Musca vetustissima*) and several other non-biting species of flies, which presumably acquire virus by feeding on blood or tissue exudates after the rabbit's death. Using PCR, Westbury (1996) demonstrated the presence of RHDV on a number of other species of flies and mosquitoes caught in traps, and in maggots obtained from rabbit carcasses. In field experiments, Cooke (1997) collected a variety of insects from study sites in an arid part of South Australia in which epidemics of RHD had occurred over the period November 1995 to February 1997. Several features of interest emerged. Virus in maggots did not persist through the larval and pupal stages. Adult blowflies and mosquitoes were readily collected near rabbit burrows, hence could well have fed on infected rabbits or their carcasses. PCR techniques showed that virus was retained in the bodies of bush flies for up to seven days and in blowflies (*Calliphora stygia*) for up to nine days. Fly spots, produced by regurgitation, are also infective, and flies may contaminate pasture plants and soil by both regurgitation and defecation. Cooke found that eight different species of flies occurring in arid parts of Australia were contaminated with RHDV. Flies may be important in spreading virus between warrens and also in long distance transfer of the virus, since bush flies and blowflies can fly rapidly and can be carried long distances by wind. Meteorological studies described below suggest that the initial spread from the Wardang Island trial site in Australia was due to carriage of contaminated flies by the wind.

Experience in both Spain and Australia shows that RHDV can persist locally for long periods, since almost immediate recurrence of outbreaks has been observed whenever conditions became favourable. The virus can survive for several weeks in carcasses and skin. Since flies carrying the virus have been detected in arid regions in Australia during mid-summer, it may persist by circulating at undetectable levels in local populations.

Control of RHD

Control in rabbitries is accomplished by attention to quarantine, hygiene and vaccination (Chasey, 1997), and measures designed to control the movement of potentially infected rabbits were instituted in Europe. Inactivated vaccines are highly effective for the productive life of commercial rabbits, and new methods of immunizing rabbits against both myxomatosis and RHDV are being developed.

Shortly after the virus was imported into Mexico it was decided to launch an eradication campaign there, based on quarantine of known and suspected infected premises, prohibition of the movement or sale of rabbits from infected zones, and slaughter of sick animals, with cleaning and disinfection of affected premises. This was feasible because there are no wild European rabbits in Mexico, and the effort was initiated soon after the disease was recognized there. Repopulation of the rabbitries was preceded by the use of sentinel animals to assess the effectiveness of these measures, and in 1993 Mexico was declared to be free of the disease.

Proposal to Use RHDV for Biological Control in Australia and New Zealand

Historically, there were some similarities, but enormous differences, between the procedures followed for the introduction and release of myxoma virus in Australia over the period 1936–1951 and the introduction and release of rabbit haemorrhagic disease virus during the 1990s. The similarities were that both viruses were found to be innocuous to all Australian native animals that were tested, that Wardang Island was used as the site for the first field trial, and that each virus ‘escaped’ from a trial site. The differences related to the regulatory requirements and the degree of consultation with the public. With myxomatosis, the only regulatory authority concerned was the Chief Quarantine Officer and the public in general was not consulted, although pastoralists, who were not ‘consulted’ but had heard of the virus, pushed strongly for its use. In the 40 years since the introduction of myxomatosis there has been a great increase in public interest in the environment, and the importance of regarding rabbits as a major issue for environmental conservation as well as an agricultural problem had emerged during the 1980s (see Chapter 12). Since the 1960s there had also been a substantial strengthening of groups concerned with animal welfare and cruelty to animals.

The Commonwealth Government responded to these public concerns by the establishment of new legislative and regulatory arrangements. Four different regulatory authorities were involved in the discussions that preceded the release of RHDV in Australia, and there was substantial consultation with the public and bodies concerned with animal welfare and cruelty to animals, as well as agriculture and conservation. New Zealand had already experienced the differences between the 1950s and the 1980s in its two attempts to introduce myxomatosis, in 1951 and 1987 (see p. 27); for Australian scientists and administrators RHDV was their first challenge since the change in public attitudes.

Cooperative arrangements between Australia and New Zealand

The Australian Quarantine Inspection Service and the New Zealand Ministry of Agriculture and Fisheries had established a mechanism for the harmonization of quarantine requirements in 1964. This was strengthened with the signing of the Australia/New Zealand Closer Economic Relations Trade Agreement in 1983, as a result of which in 1991 the Australian Agricultural Council (consisting of relevant Commonwealth and State Ministers) became part of the Agriculture and Resource Management Council of Australia and New Zealand, and the [Australian] Council of Nature Conservation Ministers became part of the Australian and New Zealand Environment and Conservation Council. The proposal to introduce RHDV for rabbit control, which was initiated through the Council of Nature Conservation Ministers, was the first proposal to introduce a biological control agent for a vertebrate pest since the introduction of myxoma virus. As it developed, it was the first proposal for any kind of biological control agent to be considered jointly by the Australian and New Zealand governments.

Investigations of RHD in Europe

Rabbit haemorrhagic disease had been causing havoc in commercial rabbitries in China since 1984 and Europe since 1986. Dr B.D. Cooke, of the South Australian Animal and Plant Control Commission, who was investigating the possibility of introducing the Spanish rabbit flea for use in arid parts of Australia (see p. 189), witnessed the first outbreak of a new and lethal disease in wild rabbits in Spain in July 1988. After his return to Australia in September, Cooke heard from his Spanish colleague, Dr R. Soriguer, that the cause of the epidemic had been confirmed as RHDV. He suggested to Dr B.H. Walker, Chief of the CSIRO Division of Wildlife and Ecology, that the virus might have potential for the biological control of the rabbit in arid areas in Australia[2]. At an *ad hoc* committee meeting at the CSIRO Division of Wildlife and Ecology in June 1989 it was suggested that Dr H. Westbury, of the Australian Animal Health Laboratory[b], should visit laboratories in Europe and the United States to obtain up-to-date information about the virus and the disease it caused. It was also suggested that Cooke, who planned to return to Spain to finalize arrangements for the importation of the Spanish rabbit flea, should be asked to make a preliminary study of the epidemiology of RHDV.

Presentations on RHDV and the need for rabbit control in arid regions were made by Cooke and Mr L. Best, of the South Australian National Parks and Wildlife Service, to a meeting of the Council of Nature Conservation Ministers in Nelson, New Zealand in July 1989. Having been briefed by Walker, Mr Barry Jones, the Commonwealth Minister for Science, recommended that funding should be provided for further investigations of RHDV. As a result, a Joint Working Party was established by the Australian Agricultural Council and the Council of Nature Conservation Ministers, with Dr C.H. Tyndale-Biscoe, of the CSIRO Division of Wildlife and Ecology, as chairman. In November 1989 the Joint Working Party agreed with the suggestion that Westbury and Cooke should make further studies of RHDV overseas. It was decided to seek the support of the Australian Agricultural Council for a combined approach of all Australian Governments (Commonwealth

[b]The Australian Animal Health Laboratory is a high security laboratory operated by CSIRO specifically to investigate exotic diseases of livestock. In addition to laboratories, it has facilities for handling large and small experimental animals behind its security barriers.

and State) and the New Zealand Government to investigate the use of RHDV for rabbit control.

The meeting of the newly established Australian and New Zealand Environment and Conservation Council in July 1990 considered the reports by Cooke[3] and Westbury[4] sufficiently promising to appoint a Working Party to investigate further the potential use of RHDV. A budget of $750,000, of which New Zealand provided $100,000 per annum, was allocated for a three-year laboratory investigation in the Australian Animal Health Laboratory. The Working Party consisted of representatives from State pest control agencies, the CSIRO Division of Wildlife and Ecology, the Australian Quarantine and Inspection Service, the Australian Animal Health Laboratory, the Bureau of Resource Sciences, and representatives from the New Zealand Ministry of Agriculture and Fisheries. Its terms of reference were:

1. to assess the potential of RHD for rabbit control;
2. to assess the possible adverse effects of RHD, with particular reference to non-target species, rabbit-based industries, animal welfare, and ecological impact;
3. to recommend whether RHD virus should be released for rabbit control in Australia and/or New Zealand;
4. to recommend the necessary conditions and timing for possible future release; and
5. to recommend steps for dealing with the accidental escape of RHD in Australia or New Zealand.

The first result, in August 1991, was the issue by the Australian Quarantine and Inspection Service of a permit to CSIRO to import RHDV into Australia, so that research investigations could be carried out under quarantine at the Australian Animal Health Laboratory, where a small breeding colony of wild rabbits had been established. In September 1991 RHDV strain CAPM was imported from the Veterinary Research Institute, Brno, Czech Republic, into the microbiologically secure facility at the Australian Animal Health Laboratory. The results of the work carried out over the next few months were so promising that in August 1992 the Working Party met again.

Workshop, Canberra, 25–26 August 1992

Meeting at the CSIRO Division of Wildlife and Ecology in Canberra, the Working Party of the Australian and New Zealand Environment and Conservation Council, augmented by representatives of animal welfare societies, considered the preliminary findings at the Australian Animal Health Laboratory to be sufficiently encouraging to plan the next steps. These were to develop a test protocol to determine the susceptibility of non-target species, to select which domestic, feral and native animals should be tested at the Australian Animal Health Laboratory and to begin consideration of possible ecological impacts of sustained reductions in rabbit numbers, especially in the semi-arid rangelands. There was also an obligation to increase public awareness of the proposal, and a number of national and State conservation and industry public interest groups that should be involved in the consultative process were identified.

Workshop, Geelong, 16–18 September 1993

A three-day workshop on RHD at the Australian Animal Health Laboratory, Geelong, involving a wide range of organizations, was organized by the Bureau of Resource Sciences and CSIRO, with the support of the New Zealand Government. Besides abstracts of the papers, a 127-page discussion paper that analysed the steps that would need to be taken prior to the granting of authority to release the virus (Bureau of Resource Sciences, 1993) was available to delegates. The results of the Workshop were summarized in a report (Munro and Williams, 1994) designed to disseminate the information presented and contribute to the ongoing assessment of RHDV as a biological control agent.

There was agreement at this time that the parts of Australia in which it was most urgent to improve rabbit control were the arid and semi-arid regions, which constituted some 70% of the mainland, and that rabbit haemorrhagic disease was potentially a powerful new agent for control in such areas. However, it was thought that in Australia steps would need to be taken to ensure that the essential requirements of the Quarantine Act (1908), the Wildlife Protection Act (1982), the Biological Control Act (1984), the Environment Protection (Impact of Proposals) Act (1974) and the Endangered Species Protection Act (1992) were fulfilled. Subsequently it was agreed that the Endangered Species Protection Act could be removed from the list. Following receipt of a report on the environmental impact of the proposed use of rabbit calicivirus (Coman, 1996), the Environment Protection Authority decided not to call for an Environmental Impact Statement, processing under the Biological Control Act being considered adequate for purposes of the Environment Protection (Impact of Proposals) Act.

In addition, the Workshop agreed that it would be necessary to continue to raise public awareness of the serious environmental impact of rabbits, to meet long-term animal welfare concerns, in terms of nett (minimal) suffering, to recognize the needs of Aboriginal Australians who might use rabbits as a food resource, and to convince land managers that if the virus was used, it was to be seen as a component of integrated pest and land management, not as a cure-all.

Recommendations were also made on further research at the Australian Animal Health Laboratory and on a trial release, under quarantine, on a safe island site, in order to obtain further information of behaviour of the virus under field conditions. As well as providing information on the epidemiology of the disease under Australian conditions, such a trial release would assist the relevant regulatory authorities and animal welfare interests in their assessments.

Laboratory Tests on Australian and New Zealand Native Fauna

Studies in Europe, China and the United States of America (summarized in Bureau of Resource Sciences, 1996) revealed that none of 26 species of animal (other than European rabbits) showed clinical signs after inoculation with RHDV in various doses and by various routes. Nor had there been any reports of illness or disease in humans or domestic animals that were in close contact with sick rabbits in commercial or backyard rabbitries, or RHDV-infected pet rabbits, in China or various countries in Europe during the past decade. Taken in conjunction with the great difficulty experienced by most laboratory workers in growing the virus in cultured cells, it appeared that RHDV was a highly host-specific virus.

Nevertheless, it was essential that its effects should be examined in a range of Australian native species. Tests were therefore undertaken in the Australian Animal Health Laboratory on four to six individuals of 13 introduced mammal species and two imported bird species, and three native rodents, six species of marsupial, two monotremes, three birds and one lizard. One species of bat and one species of bird (kiwi) from New Zealand were also tested. The species of Australian native animals were chosen from those found in association with rabbits, the list being approved by the Australia and New Zealand Environment and Conservation Council. The results are summarized in a report under the Biological Control Act (Bureau of Resource Sciences, 1996). Subsequently, in April 1996, the Minister for Primary Industries and Energy commissioned a study on the susceptibility of the echidna, koala and wombat. None of these animals showed signs of disease, nor could signs of infection be detected by a range of techniques, which included gross pathology, histopathology, and virus detection by the highly sensitive polymerase chain reaction. Serological tests were uniformly negative except for the kiwis, which exhibited a rising titre over the five-week test period, a result attributed to the very high dose (300,000 rabbit lethal doses) used (Buddle *et al.*, 1997).

Committees to Oversee Field Testing and Release

On 29 April 1994 the joint Ministerial Councils of the Australia and New Zealand Environment and Conservation Council and the Agriculture and Resource Management Council of Australia and New Zealand established a group to evaluate the virus and approved a $3 million budget. Later that year three committees were set up to oversee the field testing of RHDV (under quarantine) and subsequent release and monitoring: a Government Assessment Facilitation Committee, a Proponent Committee and a Research Management Committee. The proposed timetable is shown in Table 11.1.

The role of the Government Assessment Facilitation Committee was to develop and facilitate an efficient and effective process for the assessment of RHDV which met legislative requirements in Australia and New Zealand, so as to lead to a decision in the national interest. Its members were drawn from the Australian Quarantine and Inspection Service, the Environment Protection Agency and the Australian Nature Conservation Agency (now subsumed into Environment Australia), the Australian and New Zealand Environment and Conservation Council, the Agriculture and Resource Management Council of Australia and New Zealand, the New Zealand Ministry of Agriculture and Fisheries, with an observer from the Proponent Committee. It was initially chaired by Dr P. O'Brien (Bureau of Resource Sciences); the secretary was Ms S. Thomas.

The Proponent Committee was concerned with programme objectives, strategic planning, the preparation of documents for legislative approval, budget settings, communication with public groups and non-governmental organizations, intergovernmental arrangements and agency interactions. Its membership included representatives of the Australian Nature Conservation Agency, the Meat Research Corporation (which also represented the International Wool Secretariat), the Australian and New Zealand Environment and Conservation Council, the Agriculture and Resource Council of Australia and New Zealand, the New Zealand Ministry of Agriculture and Fisheries and CSIRO (represented by the Chiefs of Animal Health (Dr M. Rickard) and Wildlife and Ecology (Dr B.H. Walker) and the Director of the Vertebrate Biocontrol Centre (Dr C.H. Tyndale-Biscoe)). It was chaired by Dr Walker, and in August 1994 Mr N. Newland (Fig. 11.5A) was appointed as Program Coordinator, to bring together the scientific, planning, financial, legislative and communication aspects of the Program in liaison with the Australian and New Zealand governments. The Research Management Committee, drawn principally from CSIRO, was responsible for planning and monitoring research activities, and reported to the Proponent Committee. It was co-chaired by Dr K. Murray (Australian Animal Health Laboratory) and Dr C.H. Tyndale-Biscoe (CSIRO Wildlife and Ecology). With the escape of the virus from quarantine in October 1995, the programme outlined in Table 11.1 was

Table 11.1. Proposed timetable for release of rabbit haemorrhagic disease virus in Australia.

Stage	Activity	Period
1	Laboratory assessment	1992–1994
2	Limited pen trials on island	1994–1995
3	Whole-island trials[a]	1996–1997
4	Public consultation and approval	1994–1997
5	Release on mainland	1997–1998
6	Monitoring and assessment	from mid-1997; ongoing

[a]It was later recognized that whole-island trials were not practicable and Stage 3 was amended to comprise larger field trials within the quarantine site.

Fig. 11.5. (A) Nicholas Paul Newland (1948–). Born and educated in Adelaide, Newland graduated from Roseworthy Agricultural College in 1967. He initially worked for the CSIRO Division of Soils and then the Vermin Control Branch of the South Australian Department of Lands, where he gained an interest in rabbit control. After country postings with that department's Land Settlements Division, he undertook a post-graduate diploma in rural extension at Hawkesbury College and subsequently completed a MEnvStud degree at the University of Adelaide. He held several senior positions in the South Australian Environment Department involving nature and land conservation, rangeland and coastal management prior to becoming the Program Coordinator of the Australia New Zealand Rabbit Calicivirus Disease Program from 1994 until the end of 1996.
(B) Anthony John (Tony) Robinson (1943–). After graduating with a BVSc from the University of Queensland in 1966, Robinson went into veterinary practice in New Zealand for a short time, then worked with viruses at Wallaceville Animal Research Centre, and obtained a Dip. Microbiol. from the University of Otago in 1970. He graduated with a PhD in virology at the Australian National University in 1974, then worked in London, England, and Corvallis, Oregon, for three years before becoming Senior Lecturer in Virology at Massey University in New Zealand, in 1978. From 1983 until 1993 he was Director of the Medical Research Council Virus Research Unit in Dunedin. His principal research in New Zealand was on orf virus, a poxvirus on which he became the world authority. In 1993 he joined the CSIRO Division of Wildlife and Ecology as Virology Program Leader, and he has taken a major role in directing research on the use of myxoma virus as a vector for immunocontraception and on RHDV, as Chairman of the Australian RCD Program Science Subcommittee.

accelerated, and in July 1996 the Proponent Committee was replaced by a Management Group.

Shortly after its establishment, the Proponent Committee created the Australia and New Zealand Rabbit Calicivirus Disease Program, initially designed to operate over a four year period and, subject to the necessary approvals, to oversee a planned mainland release in 1998. It was a consortium of different organizations, reflected in the range of funding bodies: the Meat Research Corporation, CSIRO, the Vertebrate Biocontrol Cooperative Research Centre, New Zealand Ministry of Agriculture and Fisheries, Australian Nature Conservation Agency, Australian and New Zealand Environment and Conservation Council, the Agriculture and Resource Management Council of Australia and New Zealand and the International Wool Secretariat. The RCD Program established a

Scientific Subcommittee, of which Dr A.J. Robinson (Fig. 11.5.B) was appointed Chairman, and undertook extensive publicity operations to keep the public and land managers informed about various aspects of RHDV (Fig. 11.6).

Environmental impact assessment

In November 1995 the Australia and New Zealand Rabbit Calicivirus Disease Program commissioned Dr B.J. Coman to prepare a draft environmental impact assessment, for consideration by the Commonwealth Minister of the Environment as to whether an Environmental Impact Statement was required. A document of some 100 pages (Coman, 1996) thoroughly canvassed issues of real and possible adverse effects, and concluded that:

> the benefits accruing from the strategic release of RCD as part of a wider program of integrated control of rabbit damage throughout Australia far exceed the possible damage caused by the expected high mortality of rabbits. In commercial terms, the ratio of benefits to costs certainly exceeds 100:1 and is likely to be much higher. Potential environmental benefits are enormous and outweigh any possible deleterious effects. The disease is species-specific, highly effective as a control agent and likely to persist far into the future. It will greatly complement the existing rabbit control measures available. Perhaps most importantly, the strategic release of this virus will significantly enhance the prospects for long-term sustainablity of millions of hectares of rabbit-prone country in Australia.

Controlling wild rabbits:
need for integrated control strategy

Background on rabbit calicivirus:
history and basic facts about the virus

Strategic release of rabbit calicivirus:
What, when, how and why?

Animal welfare and rabbit calicivirus:
reducing suffering through effective rabbit control

Field effects of rabbit calicivirus:
What are you likely to see?

Getting the most out of rabbit calicivirus:
field monitoring & surveillance essential

Native fauna and rabbit calicivirus:
predicting and managing the impact of reduced rabbit numbers

Attention rabbit owners:
pet and farmed rabbit need protection from rabbit calicivirus

Fig. 11.6. Titles of leaflets on RHDV produced by the Australia and New Zealand Rabbit Calicivirus Disease Program and distributed to the general public, on request and through schools, Land-Care groups, at field days and on other similar occasions. Each leaflet comprised one to four pages outlining all available information about the relevant topic, expressed in non-technical language.

This draft environmental impact assessment was considered by the Environmental Protection Authority, which made recommendations to the Biological Control Authority regarding the release of RHDV.

Assessment by the Biological Control Authority

The process for release of biological control agents in Australia requires an assessment process under one of the Biological Control Acts existing in either the Commonwealth or States/Territories, following a unanimous request from the Agriculture and Resource Management Council of Australia and New Zealand. Although the Act was enacted in 1984, it had not been used before. The planned procedure was to carry out field testing of the virus on a remote island and, if this showed promise, to proceed with a series of steps (see Table 11.1), which were set out in detail in the draft discussion paper submitted to the 1993 Workshop and published in the report of its proceedings (Munro and Williams, 1994). Because of the escape of RHDV from Wardang Island to the mainland in October 1995, the planned programme had to be accelerated; the actual sequence of events is shown in Table 11.2.

In March 1996 the Agriculture and Resource Management Council of Australia and New Zealand nominated rabbits as target organisms and RHDV as an agent organism. An assessment was commenced by the Biological Control Authority, which sought submissions from the public in support of, or opposition to, the use of RHDV (Bureau of Resource Sciences, 1996). It received 472 submissions, of which 78% supported the proposal, based mainly on the benefits that would follow for agriculture and the environment if RHDV substantially increased the effectiveness of rabbit control. However, many submissions raised concerns about possible adverse effects on the environment or people from the release of the virus or the control of rabbits. The nature of these concerns are described later in this chapter.

Field Test on Wardang Island

The Australian Animal Health Laboratory was given responsibility for selecting a site for a field trial. During late 1993 and early 1994 a search was undertaken for a suitable island, critical criteria, among others, being size, the presence of reasonable numbers of rabbits, occurrence of myxomatosis, remoteness, accessibility, proximity to other islands, accommodation, ownership, presence of predators, other wildlife and livestock, and last but not least, cost (Robinson and Westbury, 1997). After investigations led by Dr A. Newsome, Wardang Island, an island in Spencer Gulf, South Australia (Fig. 11.7A), which had been used for the first myxomatosis field trial in 1937–39 (Bull and Mules, 1944), was chosen, and in December 1994 the Australian Quarantine and Inspection Service gave permission for a field trial to be conducted there. The objective (Cooke, 1996) was to find out whether the virus would spread among Australian wild rabbits living in natural warrens, and to evaluate (a) the immediate impact of the disease in terms of the humaneness of death and the rates of mortality, (b) rates of transmission, (c) effect of season on rates of transmission, and (d) persistence of the virus.

After further discussions between scientists of the Australian Animal Health Laboratory and the CSIRO Division of Wildlife and Ecology, a quarantine enclosure of some 50 hectares was built, surrounded by an electrified rabbit- and cat-proof fence and containing two pens, one containing six and the other four smaller sites, each one hectare in area and surrounded by two rabbit-proof fences (Fig. 11.7B). Elaborate quarantine protocols were set up by the Australian Animal Health Laboratory and approved by the Australian Quarantine and Inspection Service to minimize spread by fomites or insect vectors. Staff carrying out experiments washed their boots in footbaths containing glutaraldehyde and changed clothing and footwear four times on crossing barrier fences between the outer predator-proof perimeter and the inner sites where rabbits were kept.

Table 11.2. Actual sequence of events leading to the release of rabbit haemorrhagic disease virus in Australia and its occurrence in New Zealand.

Date	Action
July 1989	Council of Nature Conservation Ministers suggest use of RHDV for biological control of the rabbit, and commissions B.D. Cooke and H.A. Westbury to investigate the situation in Europe
July 1990	Australia and New Zealand Environment and Conservation Council appoints Working Party to investigate potential use of RHDV
August 1991	Australian Quarantine and Inspection Service issues permit for importation of RHDV by the Australian Animal Health Laboratory
August 1992	Workshop at CSIRO, Canberra develops protocol for laboratory tests on non-target species in the Australian Animal Health Laboratory
October 1992	Non-target species testing commenced
September 1993	Draft protocol for assessment produced by Bureau of Resource Sciences Workshop held at Australian Animal Health Laboratory
April 1994	Ministerial Councils establish group to evaluate RHDV and approve $3 million budget
August 1994	Proponent, Government Assessment and Facilitation and Research Management Committees appointed
September 1994	Proponent Committee creates Australia and New Zealand Rabbit Calivivirus Disease Program
December 1994	Australian Quarantine and Inspection Service gives permission for field trial (under quarantine) on Wardang Island
March 1995	Trials commence on Wardang Island
October 1995	RHDV escapes from Wardang Island to mainland and spreads widely
November 1995	Australia and New Zealand Rabbit Calivivirus Disease Program commissions an environmental impact assessment, which was published in February 1996 Advertisements in newspapers calling for public comment about use of RHDV Attempts at controlling RHD on mainland stopped
February 1996	National RCD Monitoring and Surveillance Program established. Started in July 1996, intended to operate until June 1998
March 1996	Agriculture and Resource Council of Australia and New Zealand nominates rabbits as target organisms and RHDV as an agent organism
April 1996	Minister for Primary Industries and Energy commissions a Human Health Study and orders tests on echidna, koala and wombat
July 1996	Proponent Committee replaced by a Management Group
August 1996	Publication of report on rabbit calicivirus disease, including analysis of public comments, produced under Biological Control Act Agriculture and Resource Management Council of Australia asked to approve release of RHDV
September 1996	RHDV registered as pest control agent by the National Registration Authority for Agricultural and Veterinary Chemicals Governor-General in Council lifts quarantine restrictions Declaration of rabbits as target organism and RHDV as agent organism under the Commonwealth Biological Control Act
October 1996	Rabbits inoculated with RHDV at selected sites in all mainland States and Territories
January 1997	Rabbits inoculated with RHDV at selected sites in Tasmania
August 1997	Wild rabbits infected with RHDV discovered in South Island, New Zealand
September 1997	Quarantine restrictions on RHDV lifted in New Zealand

Consideration was given to enclosing the 50 hectare quarantine enclosure under fly-proof netting, but this judged to be far too expensive. Following advice from the CSIRO Division of Entomology, all fleas on rabbits within the quarantine enclosure were killed with insecticide, the experimental burrows were sprayed with Deltamethrin (a residual insecticide), saline swamps were treated with *Bacillus thurin-*

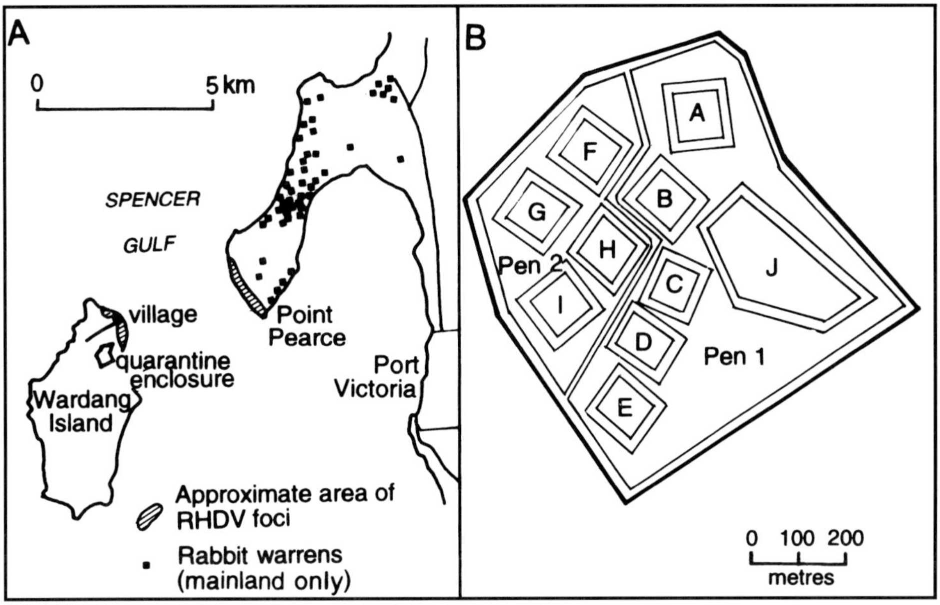

Fig. 11.7. (A) Map showing the location of Wardang Island (see also Fig. 6.4, p. 126) and the site of the experimental pens and Point Pearce, and of rabbit warrens on the island and adjacent parts of the mainland. (B) Arrangement of the quarantine enclosure on Wardang Island, showing locations of experimental sites. Each of the small enclosures contained a rabbit warren. Redrawn from Cooke (1996), with permission.

giensis israelensis to control mosquito breeding and fly traps were used to trap large numbers of flies.

A rabbit-free zone 300 metres wide was maintained around the perimeter of the quarantine area and rabbits elsewhere on the island (see Fig. 11.7A) were kept under surveillance. All experimental rabbits were fitted with radio collars, and were observed from hides at dawn and dusk. The risk of scavenging birds spreading the virus was reduced by infecting only a few rabbits at a time and removing any rabbits which died above ground as quickly as possible at dawn. Rabbits which died underground were precisely located by the radio collars and removed from the warren by digging vertically into the burrows. Carcasses never remained for more than one day.

Tests commenced in March 1995. Each of the small one hectare enclosures was stocked with four male and six female rabbits, and trials were usually initiated by inoculating two rabbits on a site and observing the animals for three weeks before killing all remaining rabbits. The spread of infection from inoculated to contact rabbits in individual sites varied from nil to rapid killing of seven of the eight contact rabbits, the interval between successive deaths varying from two to seven days, suggesting that virus could persist in a warren for at least five days. However, when a site was restocked seven weeks after an earlier experiment, none of the rabbits was infected. Sites being prepared for later experiments were regarded as 'sentinel' sites and monitored to detect any spread of the virus.

On 1 July virus was spreading in experimental sites D and E but suddenly appeared in 'sentinel' site H (Table 11.3), possibly associated with scavenging by a raven. Rabbits from sites D, E and H were trapped and killed and further experiments delayed until it had been confirmed that the virus had not spread further. Tighter precautions were taken to prevent scavenging and no spreading beyond the experimental site occurred in an experiment in mid-August.

Table 11.3. Chronology of events indicating escape of virus from experimental and quarantine sites between June and October 1995[a].

Date	Significant events
20 June	Rabbits inoculated on sites D and E
1 July	Rabbit infected in 'sentinel' site H
13 September	Rabbits inoculated on sites G and I
15 September	Inoculated rabbits on sites G and I die
18, 19 September	Contact rabbits on site I die
22, 23 September	Rabbits die of RHD on 'sentinel' site B
24 September	Virus present on blowflies collected outside quarantine area
25 September	Dead rabbits found on 'sentinel' sites F and J
29 September	Dead rabbits found on 'sentinel' site A; first infected rabbit found outside quarantine area
30 September	Rabbit control started on Wardang Island
1 October	Probable spread of virus to Point Pearce
12 October	First infected rabbit found at Point Pearce
12–14 October	Likely spread of virus to Yunta and Blinman
17 October	Rabbit control started at Point Pearce
19 October	Last known case of RHD on Wardang Island
25, 26 October	Dead rabbits found at Yunta and Blinman
28 October	RHDV identified from Yunta rabbit; further rabbit control at Point Pearce abandoned

[a]Based on Cooke (1996).

However, in an experiment beginning on 13 September virus again spread to two 'sentinel' sites, and on 29 September a rabbit that had died from RHD was found 900 metres north-east of the quarantine area and foci developed in the warrens in the area (see Fig. 11.7A). Accidental spread by staff was ruled out and spread by ravens or other scavengers was considered unlikely; spread by insects could not be excluded. As part of the contingency plans for an escape from quarantine, poisoning with 1080 and ripping of burrows in that area was carried out; nevertheless cases of RHD continued to appear on the island until 19 October.

Overall, in the trials on Wardang Island only 13 contact rabbits became infected despite 14 being inoculated, suggesting that RHD spread rather poorly among wild rabbits living in natural warrens when the mechanism of spread was by contact between infected rabbits or brief contact with cadavers.

Escape and spread from Wardang Island

Following the discovery of infected rabbits in 'sentinel' sites on 22 September and outside the quarantine area on 29 September, extensive searches were made on rabbit-infested areas on the adjacent mainland from 3 October onwards. On 12 October a dead rabbit subsequently shown to have been infected with RHDV was found at Point Pearce, on the mainland about 5 kilometres from the quarantine area, and cases were found over an area of some 20 hectares (see Fig. 11.7A), in a place that was 'buzzing with flies and other insects' even on cool, windy days. In accordance with a pre-planned programme agreed with the South Australian goverment ('Operation Garter'), rabbit control procedures were begun at Point Pearce on 17 October (Table 11.3). However, it was already too late; RHDV was identified from a dead rabbit collected at Yunta, some 360 kilometres away from Point Pearce, on 28 October. On 26 October dead rabbits, subsequently identified as being infected with RHDV, were found near Blinman, some 390 kilometres from both Point Pearce and Yunta. Subsequent observations[5] showed that the Yunta and Blinman outbreaks (see Fig. 11.8) occurred within 24 hours of each other, and were probably spatially distinct events, which occurred in

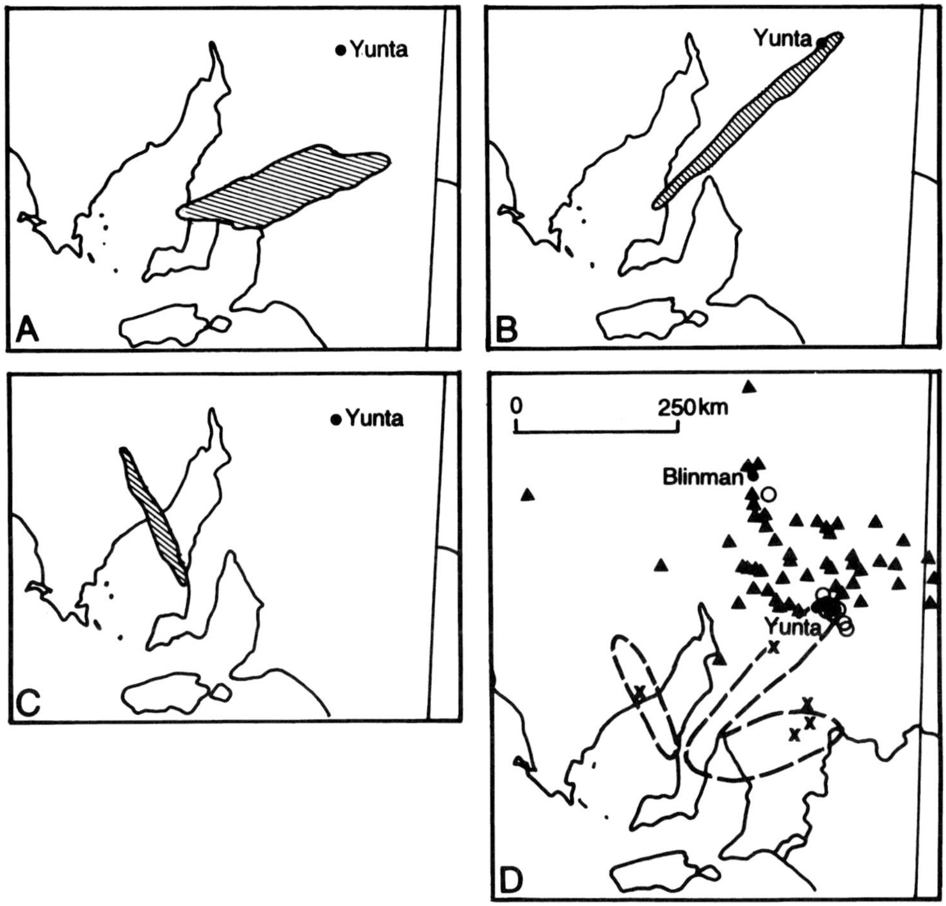

Fig. 11.8. Maps of part of South Australia showing the wind pattern from Wardang Island on (A) 12 October, 1995, between 0545 and 1815 hours, (B) 13 October 1995, between 0545 and 1830 hours, and (C) 14 October 1995, between 0900 and 1830 hours. Map D shows where RHDV-infected wild rabbits were found in October, November and December 1995. Redrawn from Cooke (1996), with permission, using meteorological details from Wardhaugh and Rochester (1996).

semi-arid areas of traditionally high rabbit density, probably beginning 'at about the same time as the Point Pearce outbreak'. Nucleotide sequencing of RNA from virus samples showed that viruses taken from site J, inside the quarantine area, from Wardang Island outside the quarantine area, and from Yunta were identical.

All the indications were that both short- and long-distance movement of the virus was effected by flying vectors. As with myxomatosis, transmission must have been mechanical, since caliciviruses of vertebrates do not replicate in arthropods. In contrast to myxomatosis, on this occasion at least, flies seem more likely than mosquitoes to have been the vectors. Both bushflies (*Musca vetustissima*) and blowflies (mainly *Calliphora augur* and *C. dubia*) were common on Wardang Island, especially from September onwards, whereas mosquitoes were rare (Cooke, 1997). In tests conducted at the Australian Animal Health Laboratory, no fewer that 13 of 16 pools of insects collected in areas where RHD was active proved positive for the virus by the polymerase chain reaction (Westbury, 1996). Virus has been demonstrated in eight species of flies, in two species of mosquitoes and in rabbit fleas. There is now ample evidence that under suitable meteorological conditions many kinds of flying insects can be carried for long distances. Under a contract between the Proponent Committee and the CSIRO Division of Entomology, simulations by Wardhaugh and Rochester (1996) of possible aerial movement of flying insects over the relevant period were carried out (Fig. 11.8). The likely trajectories of flies during the period 12–14 October accurately describe the areas in which new foci of disease were found over the succeeding two months. The areas where RHD was first noticed (Yunta and near Blinman) were in a region where rabbits were abundant and therefore sites at which infection was more likely to be established and seen than the farming areas north of Adelaide and in Eyre Peninsula (where there were relatively few rabbits), which would be in the trajectories for 12 October and 14 October.

Subsequent Spread of RHDV and Planned Releases

The spread of RHDV in Australia after its escape distorted the planned series of steps to be taken prior to release (see Table 11.1), and necessitated urgent action. Administratively, it was realized that it would be necessary to fulfil the provisions of the Agricultural and Veterinary Chemicals Code Act (1994) for registering pest control agents before deliberate release could be undertaken, and this was clearly necessary earlier than had been planned. The relevant Acts governing the release of the virus in New Zealand were the Animals Act (1967) and the Hazardous Substances and New Organisms Bill (at the time before Parliament).

Following a decision of the Consultative Committee on Exotic Animal Diseases that the virus could be regarded as endemic in Australia, attempts to control the spread of the virus at Point Pearce were stopped early in November 1995. Later that month notices were published in national newspapers and in the Commonwealth of Australia Gazette (Fig. 11.9) calling for public comment on the use of RHDV for rabbit control.

Newland, as RCD Program Coordinator, took a major role in reorganizing activities. State pest control authorities, especially in South Australia and New South Wales, intensified their monitoring of rabbits in rangelands locations. To ensure the provision of information to the public, from May 1995 a newsletter, *Rabbit Calicivirus Update,* was produced and widely distributed, and a 'Rabbit' site was set up on the Internet. Both contained general information about rabbit control and maps of Australia showing where confirmed cases of RHD had occurred, and in the numbers for November and December 1996, the release sites.

Australia-wide releases

Following the receipt of submissions from the public early in January 1996, an environmental impact document was produced, followed by a detailed report under the Biological Control Act, in which all submissions from the public were considered. Having also received a document advising that no adverse effects on human health were anticipated, in August the Agriculture and Resource Management Council of Australia and New Zealand agreed unanimously to approve the release of RHDV. The final steps were taken in September 1996, when the virus was registered by CSIRO as a pest control agent under the Agricultural and Veterinary Chemicals Code Act, the Minister for Primary Industries and Energy gave final approval, the Governor-General in Council lifted quarantine restrictions, and the declarations of rabbits as the 'target organisms' and RHDV as the 'agent organism' were published in the Commonwealth Gazette.

During 1996, the Management Group established an RCD Monitoring and Surveillance Program and an RCD

CONTROL OF RABBITS WITH RABBIT CALICIVIRUS DISEASE (RCD) AND THE ENVIRONMENTAL IMPACTS

CALL FOR PUBLIC COMMENT

Governments have been investigating the possible use of RCD as a biological control agent for European rabbits (*Oryctolagus cuniculus*). In October 1995, RCD accidentally escaped from experimental trials on Wardang Island and is now found in areas of South Australia and New South Wales. Governments now need to consider if there should be deliberate releases of RCD for the biological control of rabbits, particularly in areas not currently affected. RCD is intended to control rabbits by reducing their numbers.

The Commonwealth Ministers for the Environment, Sport and Territories and Primary Industries and Energy, and the State/Territory Ministers with agriculture and environment responsibilities have agreed to co-operate in their assessments. They are now seeking:-

- public comment into the development of guidelines for preparation of an Environmental Impact Statement; and
- submissions from any persons who object to, or support, rabbits being declared a target organism and/or RCD being declared an agent organism under the *Biological Control Act*.

Responses should be in writing and may include any issues which should be considered in an Environmental Impact Statement and/or particulars of the grounds for support or objection to declarations of rabbits and RCD. Responses should be forwarded by 10 January 1996 to the:-

- Biological Control Authority
  GPO Box 858
  Canberra ACT 2600.

The Environment Protection Agency will take the responses into account in determining the guidelines for the preparation of an Environmental Impact Statement.

Background papers on the requirements for preparation of Environmental Impact Statements, the problems caused by rabbits, and on RCD are available free of charge from the Biological Control Authority. Requests for information should be made in the first instance to Mr Michael Hillier (tel 06 2716330, fax 06 2723372).

Alan Smart
First Assistant Secretary
Livestock and Pastoral Division
Department of Primary Industries and Energy
Authorised Delegate of the Biological Control Authority.

Fig. 11.9. Abbreviated reproduction of a notice in the Commonwealth of Australia Gazette, No. GN47, 29 November 1995, pp. 4446–4447, which was also published in the national newspapers, calling for public comment on RHDV and its potential environmental impact. An assessment of the comments received was published in the report under the Biological Control Act 1984 (see p. 266).

Epidemiology Program. State government agricultural and conservation agencies and their respective Commonwealth government counterparts contributed to the two-year Monitoring and Surveillance through the Agriculture and Resource Management Council of Australia and New Zealand Program and the Australian and New Zealand Environment and Conservation Council, on a 50:50 State:Commonwealth basis. The two-year Epidemiology Program was funded by a consortium including State agriculture and conservation agencies, the Meat Research Corporation,

Environment Australia, the International Wool Secretariat, and the Anti-Rabbit Research Foundation of Australia.

Under the Monitoring and Surveillance Program, twelve localities representing different bioclimatic regions on the Australian mainland were designated for intensive study by State agriculture and conservation agencies for the effect of RHD on rabbit populations, fauna, flora and soil conservation (Fig. 11.10A). On all except one site there were sufficient rabbit population data to allow comparison with subsequent population measurements. In addition, each State has a number of broad-scale sites (a total of 54 observation sites in all) that were to be studied at varying levels of intensity. Deliberate releases have been made on many more sites throughout rabbit-infested regions of Australia (Fig. 11.10B), starting in Wagga Wagga in New South Wales on 9 October 1996. Tasmania was at first reluctant, but agreed to release the virus in January–February 1997 (Fig. 11.11).

Months before the official release, the disease had spread locally and over long distances and had been introduced unofficially to other places such as Victoria and Western Australia (Fig. 11.12A, B). Further spread occurred after the releases; the situation in April 1998 is shown in Figure 11.12C. So far only two reports have appeared in the scientific literature of the effectiveness of RHD (Mutze *et al.*, 1998a; Lugton, 1999), although detailed publicly-available reports have been published every four months since April 1997 by the RCD Monitoring and Surveillance Program. The two published reports refer to the situation in two very different areas. Mutze *et al.* described the findings at Gum Creek, in the Flinders Ranges in South Australia, one of the places to which the virus spread at the time of its escape from Wardang Island (see Fig. 11.8), whereas Lugton carried out a retrospective cross-sectional study following 245 releases made in New South Wales between October 1996 and December 1997.

The July 1998 Monitoring and Surveillance Program Report indicated that at that time the disease was still having a major impact on rabbit populations in the arid and semi-arid regions. In the Flinders Ranges locality, where five sub-sites were being monitored, including Gum Creek, declines in populations of over 90% have been recorded in all since the first epidemic in 1995. Three successive naturally occurring outbreaks of RHD have occurred and populations have remained at or below 10% of pre-RHD levels. A similar picture has been seen at the Erldunda locality in the Northern Territory, at Lake Burrendong in New South Wales, at Muncoonie Lakes in Queensland following the arrival of the virus in early 1996 (all arid/semi-arid land sites) and in a northern Tasmanian site after the virus arrived there in early 1998. At the Hattah-Kulkyne National Park in Victoria, the Coorong site in South Australia and the Euchareena site in New South Wales, populations have declined by about 65% and have not substantially recovered.

In contrast, variable results have been seen on other sites. In one site in the Stirling Ranges in southern Western Australia an epidemic occurred in May 1996 but since that time the disease has not reappeared and populations have recovered to 50% of their pre-RHD levels. However, at a nearby National Park site the population has remained low following the first appearance of the virus. Mixed results have also been seen at sites on the western slopes of New South Wales, where despite confirmation of the presence of naturally-infected rabbits in December 1996, numbers of rabbits have not significantly declined on one site while on others rabbit numbers have remained low. In an assessment of 245 sites in New South Wales, Lugton (1999) found that outbreaks of RHD were associated with relatively low ambient temperatures, high flea infestations and moderate rainfall in the month of release. He also noted that for each week following the cessation of rabbit breeding there was a 7% decline in the odds of an outbreak occurring. A patchy effect of the virus has also been reported in regions of higher rainfall in Victoria.

Based on the observed spread since the escape of the virus from Wardang Island in

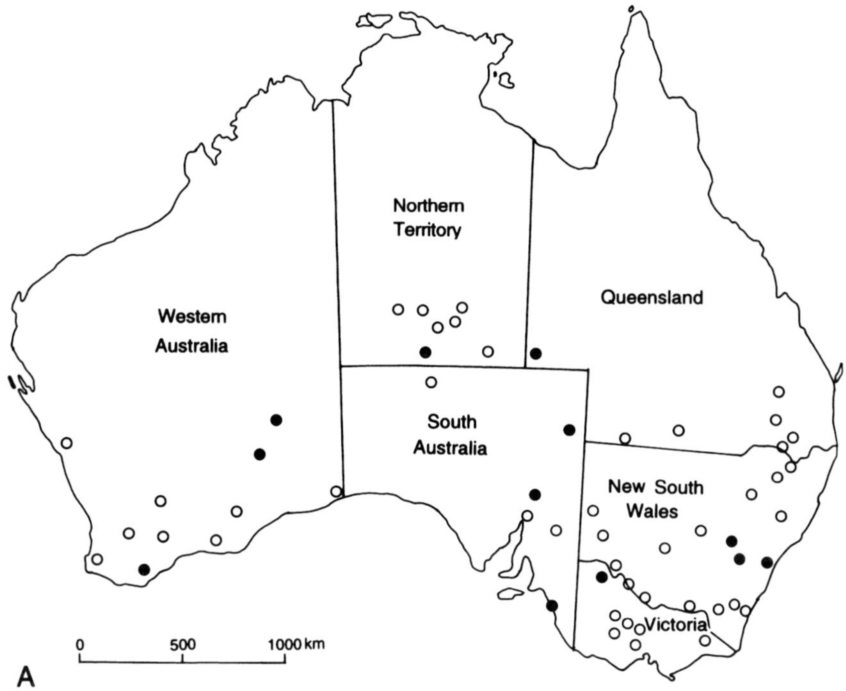

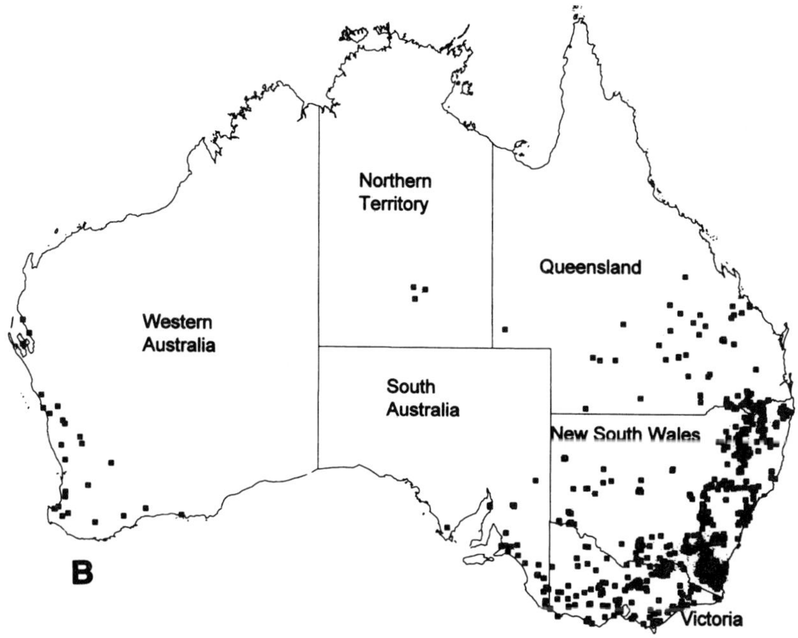

Fig. 11.10. Maps of mainland Australia showing (A) Monitoring and surveillance sites for RHD; dark circles, major monitoring sites; open circles, surveillance sites. (B) Release sites for RHDV. Data courtesy of Dr J. Kovalski and Mr M. Hillier.

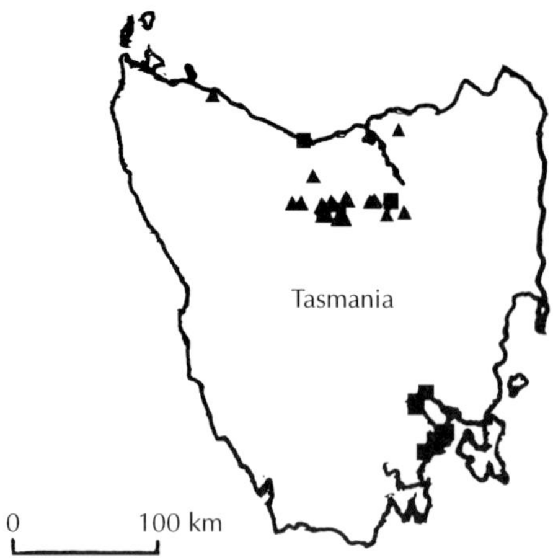

Fig. 11.11. Map of Tasmania showing release sites (square symbols) and sites where confirmed cases of RHD were found up to May 1998 (triangular symbols). Courtesy of Dr J. Kovalski.

October 1995 and the initiation of inoculations in October 1996, Smyth *et al.* (1997) produced a series of maps showing the likely spread of RHD in spring, summer, autumn and winter (Fig. 11.13). The maps were based on weather data from the Australian Meteorological Bureau, long-term weather predictions and the observed spread of the disease over the first 12 months since its escape. Summer air temperatures in the rangelands of northern South Australia commonly reach 45°C and soil surface temperatures can be 60°C or more for up to six hours of each day. Even in rabbit burrows, mid-summer ambient temperatures may exceed 32°C (Cooke, 1990). This explains the quiescence of RHD in the summer everywhere except in southern or highland areas (Fig. 11.13B), and the fact that in hot dry areas it spreads best in winter (Fig. 11.13C). Later observations may modify the picture, but on a seasonal basis the best times for further releases of the virus were as follows:

- Summer – Tasmania, southern Victoria, and eastern highlands and tablelands;
- Autumn – southern third of Australia except for the eastern highlands and tablelands;
- Winter – northern limits of the distribution of rabbits;
- Spring – southern third of Australia except for the eastern highlands and tablelands and the wetter regions.

Although the initial outbreak of RHD in Australia occurred in the spring, such sharp outbreaks can no longer be expected in the breeding season in areas where the disease has already occurred. In such circumstances not only are most breeding rabbits immune, but their young are protected against death by their physiological resistance (for about four weeks after birth) and by maternal antibody (for about eight weeks). Over summer, as the rate of spread of the virus slows due to the high environmental temperature, most young lose their maternal antibodies and may then die from RHDV as young adults when conditions for spread allow insect activity to resume in the autumn.

Effectiveness of RHD for rabbit control

In advising governments and landholders of the value of RHD for rabbit control, the scientists and administrators concerned were mindful of the history of myxomatosis (see Chapters 6–8). The initial impact of myxomatosis in Australia was so dramatic that it was difficult to persuade land managers that other methods of rabbit control were needed, in spite of repeated urging from experts like Francis Ratcliffe. On high value agricultural land, owners sometimes took advantage of the very small numbers of rabbits left after the first few outbreaks to eliminate these animals and build rabbit-proof fences around their property to prevent reinfestation. However, in the main other methods of rabbit control were neglected. In 1989, when it was first recommended that RHDV should be examined as a potential biological control agent for Australian rabbits, the relevant authorities were determined that full advantage should be taken of any resulting large-scale rabbit kills. As part of a comprehensive plan to develop up-to-date guidelines for the control of vertebrate pests, the Bureau of Resource Sciences of the

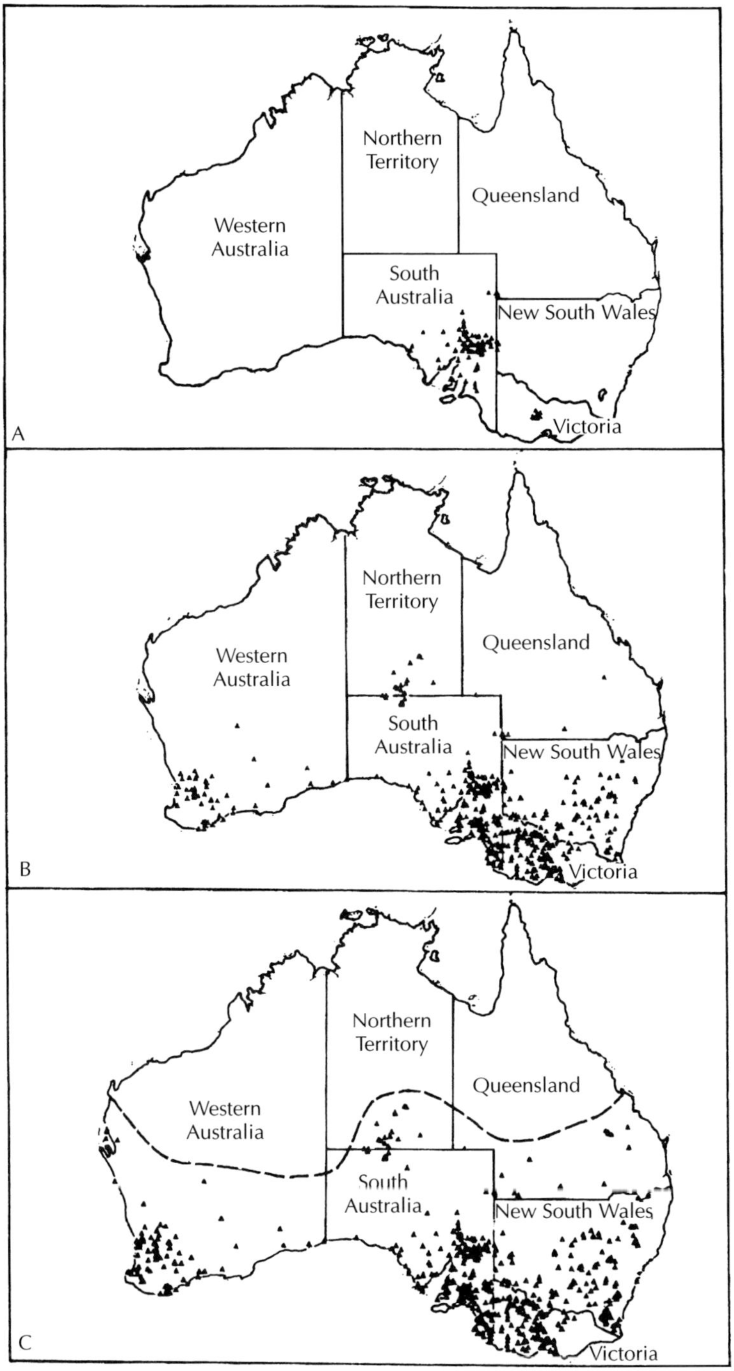

Fig. 11.12. Maps of mainland Australia showing (A) sites of confirmed cases in March 1996, five months after the escape from Wardang Island (the cases in Victoria occurred during February 1996); (B) confirmed cases in October 1996, just before official releases began; and (C) confirmed cases in April 1998. Courtesy Dr J. Kovalski.

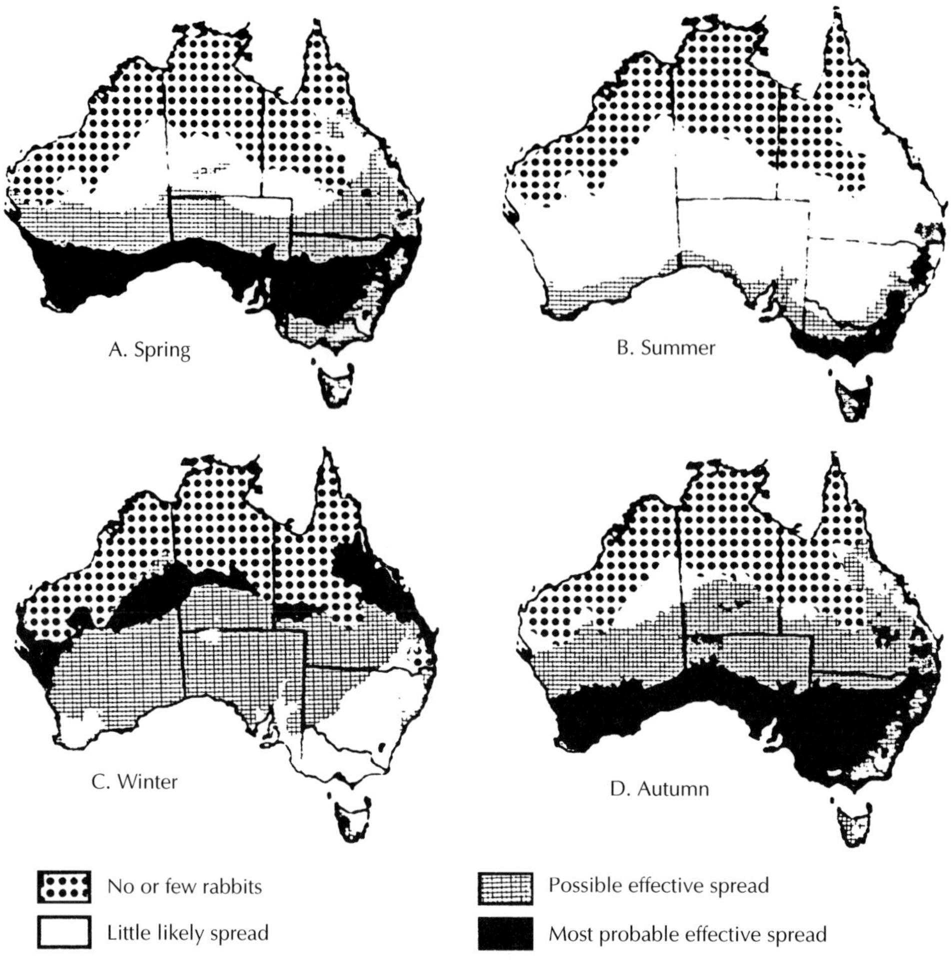

Fig. 11.13. Maps of Australia showing expected optimum seasons for spread of RHDV in different parts of Australia. Adapted from Smyth *et al.* (1997), with permission.

Commonwealth Government and the CSIRO Division of Wildlife and Ecology produced a detailed account of integrated rabbit control (Williams *et al.*, 1995). This was followed up in December 1997, when a booklet produced by the Anti-Rabbit Research Foundation of Australia (Coman, 1997) emphasizing the importance of follow-up was launched at the homestead at Barwon Park, the place where wild rabbits were introduced into Australia in 1859. The booklet is being distributed to landholders all over Australia.

The message from the scientists was that RHD, like myxomatosis, should be regarded as part, albeit a very important part, of a programme of integrated pest control. It is impossible to summarize all aspects of integrated rabbit control here; suffice it to say that the approach proposed was that any kill following the introduction of RHDV should be systematically followed up with measures of warren destruction, removal of surface cover in which rabbits might shelter, and destruction of residual rabbits by poisoning or fumigation. Although the Victorian government instituted a substantial programme to help farmers rip burrows after outbreaks of RHDV, most State and Territory governments have not provided funds to follow through the Australia and New Zealand Rabbit Calicivirus Program adequately, despite considerable pleading by the scientists and administrators involved.

There was some concern as to whether there would be any interference between myxomatosis and RHD. Observations at intensive monitoring sites in South Australia and New South Wales show that both diseases may be active at the same time, apparently without diminishing the lethality of either disease. It is still too early to draw general conclusions about the long-term effects of RHD on rabbit numbers and the consequent effects on native flora and fauna and agricultural production on a national scale. Bomford *et al.* (1998) reported on the overall picture in May 1998, about two and a half years after the escape of RHDV. At this time there had been significant regeneration of native shrubs in the Flinders Ranges in South Australia, where there had been outbreaks in October–November 1995, mid-August 1996 and May 1997, and at some monitoring sites in New South Wales. Earlier, there had been concern that with the decline in rabbits, cats and foxes might switch to killing more native animals, but there was as yet no evidence of a significant relationship between declines in rabbit numbers and changes in small mammal and reptile numbers at any of the monitoring sites. The numbers of raptors and other native predators had declined at some monitoring sites but not at others. At the Flinders Range site there was preliminary evidence that the endangered yellow-footed rock wallaby (*Petrogale xanthopus*) may have increased in numbers, and the sandy inland mouse (*Pseudomys hermannsbergensis*) was recorded there for the first time.

RHDV has been most effective in the semi-arid rangelands, where the rabbit problem was most acute. The changes have been most dramatic in Yunta–Blinman areas of the Flinders Ranges, where the first outbreaks occurred in October 1995, and where there had been cases in the spring of 1996 and 1997 (Mutze *et al.,* 1998a, b). In this area there has been a remarkable regrowth of native vegetation to an extent not seen within the memory of current landholders.

Introduction of RHDV into New Zealand

Although the New Zealand Government was a party to the investigations on RHDV from the outset, on 2 July 1997 it decided not to recommend its use. However, as was the case with myxomatosis in Britain (see p. 227), farmers in areas where rabbits were in plague numbers took the matter into their own hands. On 23 August 1997 the New Zealand Ministry of Agriculture confirmed that RHD had occurred on at least one farm in Central Otago, on South Island (Thompson and Clark, 1997). Tests with monoclonal antibodies confirmed the suspected identity of the virus with that found in Australia (O'Keefe *et al.,* 1998). Containment measures were started, but it soon became evident that the virus was also present in many other farms and that farmers were spreading it deliberately, usually by contaminating carrot bait with suspensions prepared from the entrails of rabbits that had died from the disease. In September 1997 the New Zealand Government bowed to the inevitable and made the possession and spreading of RHDV legal[6]. In mid-1998 a local biotechnology company was granted an experimental use permit to sell limited supplies of the virus on the open market, to be used on oat or carrot bait. The results were to be monitored in the hope of obtaining enough data to support full registration[7].

Public Concern about the Release of RHDV

From the earliest stages of the programme steps were taken to consult landholders, conservation interests, animal welfare groups and the general public and elicit their views on the use of RHDV. On his appointment as Coordinator of the Australian and New Zealand Rabbit Calicivirus Disease Program, Newland devoted a considerable effort to promoting public discussion of the use of RHDV. As well as appearing on television, he spoke to representatives of Australian Conservation Foundation, the South Australian Aboriginal Lands Trust and the Central Lands Council, the Royal Society for the Prevention of Cruelty to Animals, the Australia and New Zealand Federation of Animal Societies and the National Consultative Committee on Animal Welfare. A survey commissioned in 1994 showed that 23% of those interviewed had heard of the rabbit calicivirus and the majority of those interviewed believed that something should be done about the rabbit problem in Australia.

Representatives of the Royal Society for the Prevention of Cruelty to Animals and other 'special interest' groups were invited to all of the major workshops. In preparing the report for consideration under the Biological Control Act (Bureau of Resource Sciences, 1996) the views of government agencies, organizations with rural interests, animal welfare bodies and the general public were sought through a notice in the Commonwealth of Australia Gazette (see Fig. 11.9) and newspaper advertisements. These were designed primarily to elicit views about possible adverse effects of releasing RHDV or of controlling rabbits. In addition, some scientists in the United States of America had expressed views about the implications of RHDV for human health which obtained widespread publicity in Australia. A total of 472 submissions were received, 281 from individuals and the others from a variety of groups. An analysis of these views was included in the Bureau of Resource Sciences report, from which the following summary has been extracted. The numbers in brackets refer to submissions received in various categories:

government (2)
local government (8)
pest/rabbit control boards (6)
rural lands protection boards (17)
soil conservation boards (11)
pastoral companies (26)
community groups (31)
rabbit industries (2)
investment company, pet food supplier, native species recovery team, proponent group, pastoral board, each (1)
government agencies (20)
animal and plant control boards (16)
catchment and land protection boards (6)
farmers/graziers association (24)
landcare groups (23
mining companies (2)

Support for the use of RHDV was expressed in 78% of the submissions received, but many of these, as well as those which did not support use of the virus, expressed concern about possible adverse effects on the environment or people from release of the virus or the control of rabbits.

Possible Adverse Effects on People

Potentially, the release of RHDV could have direct adverse effects on human health or on the health of domestic animals, it could affect people involved in operating rabbit industries or hunting wild rabbits for sustenance, shooting rabbits as a sport, or keeping rabbits as pets. The presence of the disease in Australia might also have implications for international trade.

Direct effects on health of humans or domestic animals

Although the evidence from many countries in Europe over the last decade suggested that so far no species except the European rabbit had been affected, about

30% of the submissions to the Report expressed concern that RHDV might acquire a wider host range by mutation. This view was strongly influenced by the opinions of Dr D.O. Matson[8], an American expert on the *Caliciviridae*, and Dr Alvin Smith, of Oregon State University, which had received wide media publicity in Australia and overseas (Drollette, 1996). Smith had considerable experience with the pinniped caliciviruses (Smith and Boyt, 1990), which first attracted attention when one of them caused vesicular exanthema of swine in California in 1932. Over the next 20 years many outbreaks of the swine disease occurred, which were traced to garbage containing sealion or pork residues. Unlike the great majority of caliciviruses, this species has a wide host range and is readily grown in cell cultures. Smith believed that there was a substantial risk that RHDV would acquire the capacity to infect and cause disease in species other than the rabbit[9]. Most other virologists, on the basis of the high species-specificity of the human caliciviruses, some 14 years' experience of close contact between diseased rabbits and humans and their domestic animals in China and Europe, and the extensive testing of the susceptibility of other species to RHDV that had been carried out in Europe and Australia, regarded such a possibility as remote.

Rabbit calicivirus human health study group

This Study Group, consisting of representatives of the Commonwealth Department of Health and Family Services, the Bureau of Resource Sciences and independent infectious disease experts, was set up in April 1996 to examine the risk to human health. In addition to surveying the literature from Europe, 151 people who had been occupationally exposed to RHDV, 72% of whom had handled diseased rabbits with their naked hands, and 118 controls, were questioned concerning illnesses and tested by sensitive serological assays. None gave a history of illness that could be attributed to RHDV and none gave a positive serological response (Carman *et al.*, 1998). Responses to enquiries were received from some 50 overseas laboratories or groups working with RHDV, in 16 countries; these confirmed that human infection with the virus was unknown[10]. Based on this information, the Chief Medical Advisor to the Commonwealth Department of Health advised that he had 'no objections to the controlled release of rabbit calicivirus as a means of rabbit control'[11].

Effects on rabbit industries

At the time of the proposed release there were seven rabbit farms in Western Australia. Rabbit farming was not permitted in South Australia or Victoria but had again been made legal in New South Wales in 1995. It was regarded as only marginally profitable, and the cost of vaccination against RHDV would probably put some farms out of business. There was a small but more profitable wild rabbit trade, with an average annual value over the years 1988–89 to 1994–95 of about $3 million (Foster and Telford, 1996). Since this occurred mainly in the semi-arid rangelands where it was expected that RHDV would have its maximum impact, this industry would suffer. Aborigines living in Central Australia regarded rabbits as indigenous animals and used them as food, and also participated in the wild rabbit trade (Hetzel, 1978; Wilson *et al.*, 1992). The spread of RHD would affect them adversely, and its occurrence in Australia will effectively exclude Australia as a source of rabbit products for several overseas countries. Discussions with several Central Australian Aboriginal communities indicated that any substantial decline of rabbits in their country would substantially and irrevocably change their lives[12].

About 20,000 laboratory rabbits are used annually in Australia. Based on sales of rabbit food, there are thought to be about one million pet rabbits in New South Wales and somewhat fewer in the other States. Most laboratory and pet rabbits are located in cities where the natural spread of RHDV is less likely, but owners might nevertheless be put to the additional expense of vaccinating their rabbits.

Animal welfare issues

Wild rabbits in Australia die naturally by predation, disease (coccidiosis, myxomatosis) or starvation or thirst, or by such control techniques as fumigation, poisoning, ripping and shooting. In comparison, RHD is no less humane than other control methods and may be more humane.

Possible increased predation of livestock

Rabbits are a major source of food for both native and introduced predator species, especially foxes, and a sudden and substantial decrease in rabbit numbers due to RHD might lead to prey switching. There is some evidence that there might be increased predation of lambs and farmyard poultry, but any such increase would be more than offset by the benefits of more effective rabbit control.

Potential Adverse Effects on the Environment

The overall environmental effects of the occurrence of RHD among wild rabbits in the rangelands of Australia were expected to be favourable, being dominated by the increased growth of native plants and decreased erosion that would follow the destruction of rabbits. Possible adverse environmental effects of the use of RHDV are discussed in Chapter 12.

The Future of RHDV as a Biological Control Agent

It is too early to predict the potential of RHDV as an adjunct for rabbit control in Australia and New Zealand. So far, its performance has been patchy, with excellent results being obtained in some areas, especially in hotter, more arid parts of Australia, and poor spread in others. In contrast to myxomatosis, where until the European rabbit flea was imported effective spread could be predicted on the basis of the presence of large numbers of the more important mosquito vectors, the mechanical transmission of RHDV by flies of various species and their airborne movement over long distances is more difficult to monitor. As among wild rabbits in Spain, it appears to have become enzootic in some parts of Australia, recurring naturally in successive years in the Flinders Ranges, for example.

Over the longer term, its future depends on whether it retains its high virulence, and whether rabbits become genetically resistant. Although detailed investigations have not been conducted in Spain, eight years after its appearance there are no reports of either virus attenuation or increasing genetic resistance among wild rabbits. The physiological resistance of immature rabbits complicates forecasts of the future of RHD. If an epidemic occurs when there are many young kittens dependent on their mothers, this factor is of little importance, because unless they were survivors of an earlier epidemic, the mothers would die of RHD and the kittens would die from neglect. Physiological resistance would diminish the impact of RHD if there were many young rabbits which were independent of their mothers, since a proportion of such animals would survive and be resistant in future outbreaks. However, there would not be any elements of genetic resistance in such animals or their offspring; any resistance would be physiological rather than genetic, associated with their age, or due to maternal antibody.

As with myxoma virus, whether the virus retains its high virulence depends primarily on what governs its transmissibility. If transmission, and especially the spread of virus between warrens and between districts, depends primarily on the airborne movement of insect vectors, and such insects become contaminated by feeding on exposed internal organs of fatal cases, there could be a selective pressure for sustained virulence. However, if animals that recover due to physiological (age) resistance excrete virus for substantial periods, and if such virus is important for transmission, there would be no selection for high virulence, nor would there be any selection for viruses of reduced virulence.

Endnotes

[1]Basser Library Archives 143/25/5B Letter from L. Capucci to Fenner, 11 May 1998.
[2]Basser Library Archives 143/25/5B Letter from B.D. Cooke to Fenner, 4 August 1997.
[3]Basser Library Archives 143/25/5B Report by B.D. Cooke on viral haemorrhagic disease in wild rabbits in Spain, funded by the Committee of Nature Conservation Ministers of Australia, 1989.
[4]Basser Library Archives 143/25/5B Report by H.A. Westbury on viral haemorrhagic disease of rabbits, funded by the Committee of Nature Conservation Ministers of Australia, 1989.
[5]Basser Library Archives 143/25/5B Report on RCD outbreaks in the Yunta and Blinman regions of South Australia, by A. Newsome and G. Mutze, November 1995.
[6]Basser Library Archives 143/25/5B Rabbit calicivirus arrives in New Zealand. *Rabbit Calicivirus Update*, September 1997, p. 3.
[7]Basser Library Archives 143/25/5B Rabbit calicivirus disease update in New Zealand. In: *Australasian Wildlife Management Society Newsletter* (1998) 11(2), p. 12.
[8]Basser Library Archives 143/25/5B Addendum to Coman (1996); correspondence between Coman and Dr D.O. Matson.
[9]Basser Library Archives 143/25/5B Letter from Dr A.W. Smith to the Prime Minister of New Zealand, 16 December 1996, with comment by Professor M.J. Studdert, 16 December 1997. See also Smith *et al.* (1998) and correspondence in *Emerging Infectious Diseases*, April–June 1998.
[10]Basser Library Archives 143/25/5B *Rabbit Calicivirus and Human Health.* Report of the Rabbit Calicivirus Human Health Study Group, Canberra, 30 pp, with many attachments.
[11]Personal communication from Dr A.I. Adams to the Bureau of Resource Sciences, 1996. Quoted in Bureau of Resource Sciences (1996).
[12]Basser Library Archives 143/25/5B Report on the concerns and issues raised by Aboriginal communities in the southern Central Land Council region at meetings held in December 1995. Report to the Central Land Council, March 1996, by L. Baker, of Wallambia Consultants, Urunga, New South Wales.

References

Argüello Villares, J.L. (1991) Viral haemorrhagic disease of rabbits: vaccination and immune response. *Revue Scientifique et Technique de l'Office International des Épizooties* 10, 471–480.

Bertagnoli, S., Gelfi, J., Le Gall, G., Boilletot, E., Vautherot, J.-F., Rasschaert, D., Laurent, S., Petit, F., Boucrat-Baralon, C. and Milon, A. (1996) Protection against myxomatosis and rabbit viral haemorrhagic disease with recombinant myxoma viruses expressing rabbit haemorrhagic disease virus capsid protein. *Journal of Virology* 70, 5061–5066.

Bomford, M., Neave, H. and Conibear, M. (1998) Lessons from rabbit calicivirus disease. In: *Proceedings of the 11th Australian Vertebrate Pest Conference, Bunbury, 1998,* pp. 117–121.

Buddle, B.M., de Lisle, A.W., McColl, K., Collins, B.J., Morrissy, C. and Westbury, H.A. (1997) Response of the North Island brown kiwi, *Apteryx australis mantelli* and the lesser short-tailed bat, *Mystacina tuberculata* to a measured dose of rabbit haemorrhagic disease virus. *New Zealand Veterinary Journal* 45, 109–113.

Bull, L.B. and Mules, M.W. (1944) An investigation of *Myxomatosis cuniculi*, with special reference to the possible use of the disease to control rabbit populations in Australia. *Journal of the Council for Scientific and Industrial Research* 17, 79–93.

Bureau of Resource Sciences (1993) *Rabbit Haemorrhagic Disease. Draft Discussion Paper.* Bureau of Resource Sciences, Canberra, 127 pp.

Bureau of Resource Sciences (1996) *Rabbit Calicivirus Disease: a Report under the Biological Control Act 1984.* Bureau of Resource Sciences, Canberra, 127 pp.

Cancellotti, F.M. and Renzi, M. (1991) Epidemiology and the current situation of viral haemorrhagic disease of rabbits and the European brown hare syndrome in Italy. *Revue Scientifique et Technique de l'Office International des Épizooties* 10, 409–422.

Capucci, L., Scicluna, M.T. and Lavazza, A. (1991) Diagnosis of viral haemorrhagic disease of rabbits and the European brown hare syndrome. *Revue Scientifique et Technique de l'Office International des Épizooties* 10, 347–370.

Capucci, L., Lavazza, A., Civardi, A., Frigoli, G., Barbieri, B. and Rossi, C. (1994) Seroconversion in asymptomatic rabbits provides further evidence of putative RHDV-like virus. In: Schwyzer, M.

and Ackermann, M. (eds) *Immunobiology of Viral Infections*. Third Congress of European Society of Veterinary Virology, Interlaken, Switzerland. Fondation Marcel Merieux, Lyon, France, pp. 88–93.

Capucci, L., Frigoli, G., Ronshold, L., Lavazza, A., Brochi, E. and Rossi, C. (1995) Antigenicity of the rabbit hemorrhagic disease virus studied by its reactivity with monoclonal antibodies. *Virus Research* 37, 221–238.

Capucci, L., Fusi, P., Lavazza, A., Pacciarini, M.L. and Rossi, C. (1996) Detection and preliminary characterization of a new calicivirus related to rabbit hemorrhagic disease virus but nonpathogenic. *Journal of Virology* 70, 8614–8623.

Capucci, L., Nardin, A. and Lavazza, A. (1997) Seroconversion in an industrial unit of rabbits infected with a non-pathogenic rabbit haemorrhagic disease-like virus. *Veterinary Record* 140, 647–650.

Carman, J.A., Garner, M.G., Catton, M.G., Thomas, S., Westbury, H.A., Cannon, R.M., Collins, B.J. and Tribe, I.G. (1998) Viral haemorrhagic disease of rabbits and human health. *Epidemiology and Infection* 121, 409–418.

Chasey, D. (1997) Rabbit haemorrhagic disease – the new scourge of *Oryctolagus cuniculus*. *Laboratory Animals* 31, 33–44.

Chasey, D and Duff, P. (1990) European brown hare disease and associated virus particles in the UK. *Veterinary Record* 126, 623–624.

Chasey, D., Lucas, M., Westcott, D. and Williams, M. (1992) European brown hare syndrome in the U.K.; a calicivirus related to but distinct from that of viral haemorrhagic disease in rabbits. *Archives of Virology* 124, 363–370.

Coman, B.J. (1996) *Environmental Impact Associated with the Proposed Use of Rabbit Calicivirus Disease for Integrated Rabbit Control in Australia*. Prepared for the Australia and New Zealand Rabbit Calicivirus Program, 96 pp.

Coman, B. (1997) *Rabbit Control and Rabbit Calicivirus Disease. A Field Handbook for Land Managers in Australia*. Gillingham, Adelaide, 20 pp.

Cooke, B.D. (1990) Rabbit burrows as environments for European rabbit fleas, *Spilopsyllus cuniculi* (Dale) in arid South Australia. *Australian Journal of Zoology* 38, 317–325.

Cooke, B.D. (1996) Analysis of the spread of rabbit calicivirus from Wardang Island through mainland Australia, 25 pp. Unpublished data. Report for the Meat Research Foundation, as part of Project CS.236, 'Field Evaluation of RCD under Quarantine'. Quoted with the author's permission.

Cooke, B.D. (1997) Insects as potential vectors for rabbit calicivirus disease (RCD), 15 pp. Unpublished data. Report to Meat Research Corporation on Project RCD.003, 'Epidemiology of Rabbit Calicivirus'. Quoted with the author's permission.

Cooke, B.D., McColl, K.A. and Amos, N. (1997) Epidemiology of rabbit calicivirus disease in experimental populations of wild rabbits, *Oryctolagus cuniculus* (L.), on Wardang Island, South Australia. Unpublished data. Final Report to the Proponent Committee of the Australian and New Zealand Rabbit Calicivirus Program, 26 pp. Quoted with the author's permission.

Drollette, D. (1996) Australia fends off critics of plan to eradicate rabbits. *Science* 272, 191–192.

Feng-Yi, Z. (1990) The rabbit industry in China. *Journal of Applied Rabbit Research* 12, 278–279.

Foster, M. and Telford, R. (1996) *Structure of the Australian Rabbit Industry: a Preliminary Analysis*. Australian Bureau of Agriculture and Resource Economics, Canberra, 33 pp.

Fuchs, A. and Weissenbock, H. (1992) Comparative histopathological study of rabbit haemorrhagic disease (RHD) and European brown hare syndrome (EBHS). *Journal of Comparative Pathology* 107, 103–113.

Gall, G. le, Huguet, S., Vende, P., Vautherot, J.F. and Rasschaert, D. (1996) European brown hare syndrome virus: molecular cloning and sequencing of the genome. *Journal of General Virology* 77, 1693–1697.

Gavier-Widén, D. and Mörner, T. (1991) Epidemiology and diagnosis of the European brown hare syndrome in Scandinavian countries: a review. *Revue Scientifique et Technique de l'Office International des Épizooties* 10, 453–458.

Gould, A.R., Kattenbelt, J.A., Lenghaus, C., Morrissy, C., Chamberlain, T., Collins, B.J. and Westbury H.A. (1997) The complete nucleotide sequence of rabbit haemorrhagic disease virus (Czech strain V351): use of the polymerase chain reaction to detect replication in Australian vertebrates and analysis of viral population sequence variation. *Virus Research* 47, 7–17.

Gregg, D.A., House, C., Meyer, R. and Berninger, M. (1991) Viral haemorrhagic disease of rabbits in Mexico: epidemiology and viral characterization. *Revue Scientifique et Technique de l'Office International des Épizooties* 10, 435–451.

Guittré, C., Baginski, I., Le Gall, G., Prave, M., Trépo, C. and Cova, L. (1995) Detection of rabbit haemorrhagic disease virus isolates and sequence comparison of the N-terminus of the capsid protein gene by the polymerase chain reaction. *Research in Veterinary Science* 58, 128–132.

Hetzel, B.S. (1978) The changing nutrition of Aborigines in the ecosystem of Central Australia. In: Hetzel, B.S. and Frith, H.J. (eds) *The Nutrition of Aborigines in Relation to the Ecosystem of Central Australia.* CSIRO, Melbourne, pp. 39–47.

Huang, H.-B. (1991) Vaccination against and immune response to viral haemorrhagic disease of rabbits: a review of research in the People's Republic of China. *Revue Scientifique et Technique de l'Office International des Épizooties* 10, 481–498.

Ji, C.-Y., Du, N.-X. and Xu, W.-Y. (1991) Adaptation of the viral haemorrhagic disease of rabbits to the DJRK cell strain. *Revue Scientifique et Technique de l'Office International des Épizooties* 10, 337–345.

Kapikian, A.Z., Estes, M.K. and Chanock, R.M. (1995) Norwalk group of viruses. In: Fields, B.N., Knipe, D.M., Howley, P.M., Chanock, R.M., Melnick, J.L., Monath, T.P., Roizman, B. and Straus, S.E. (eds) *Field's Virology*, 3rd edn. Lippincott-Raven, Philadelphia, pp. 783–810.

König, M., Thiel, H.-J. and Meyers, G. (1998) Detection of viral proteins after infection of cultured hepatocytes with rabbit hemorrhagic disease virus. *Journal of Virology* 72, 4492–4497.

Lavazza, A., Sciscluna, M.T. and Capucci, L. (1996) Susceptibility of hares and rabbits to the European brown hare syndrome virus (EBHSV) and Rabbit haemorrhagic disease virus (RHDV) under experimental conditions. *Journal of Veterinary Medicine* Series B 43, 401–410.

Lenghaus, C., Westbury, H., Collins, B., Ratnamohan, M. and Morrissy, C. (1994) Overview of the RHD project in Australia. In: Munro, R.K. and Williams, R.T. (eds) *Rabbit Haemorrhagic Disease: Issues in Assessment for Biological Control.* Australian Government Publishing Service, Canberra, pp. 104–129.

Liu, S.J., Xue, H.P., Pu, B.Q. and Quian, N.H. (1984) A new viral disease in rabbits. [In Chinese] *Animal Husbandry and Veterinary Medicine* 16, 352–255. Abstract in *Veterinary Bulletin* (1985) 55, 5600.

Lugton, I.W. (1999) A cross-sectional study of risk factors affecting the outcome of rabbit haemorrhagic disease releases in New South Wales. *Australian Veterinary Journal* 77 (in press).

Marcato, P.S., Benazzi, C., Vecchi, G., Galeotti, M., Della Salsa, L., Sarli, G. and Lucidi, P. (1991) Clinical and pathological features of viral haemorrhagic disease of rabbits and the European brown hare syndrome. *Revue Scientifique et Technique de l'Office International des Épizooties* 10, 371–392.

Meyers, G., Wirblich, C. and Thiel H.-J. (1991) Rabbit haemorrhagic disease virus – molecular cloning and nucleotide sequencing of a calicivirus genome. *Virology* 184, 664–676.

Morisse, J.-P., Le Gall, G. and Boilletot, E. (1991) Hepatitis of viral origin in Leporidae: introduction and aetiological hypotheses. *Revue Scientifique et Technique de l'Office International des Épizooties* 10, 283–295.

Moussa, A., Chasey, D., Lavazza, A., Capucci, L., Smid, B., Meyers, G., Rossi, C., Thiel, H.-J., Vlasak, R., Ronsholt, L., Nowotny, N., McCullough, K. and Gavier-Widen, D. (1992) Epidemiology and molecular biology of the calicivirus aetiological agent of the haemorrhagic disease of lagomorphs. *Veterinary Technology Newsletter* 2, 97–101.

Munro, R.K. and Williams, R.T. (eds) (1994) *Rabbit Haemorrhagic Disease: Issues in Assessment for Biological Control.* Australian Government Publishing Service, Canberra, 168 pp.

Mutze, G., Cooke, B. and Alexander, P. (1998a) The initial impact of rabbit hemorrhagic disease on European rabbit populations in South Australia. *Journal of Wildlife Diseases* 34, 221–227.

Mutze, G., Linton, V. and Greenfield, B. (1998b) Impact of rabbit calicivirus disease on the flora and fauna of the Flinders Ranges, South Australia, pp. 153–157. In: *Proceedings of the 11th Australian Vertebrate Pest Conference, Bunbury 1998.*

Nawwar, A.A.M., Mousa, H.A.A., Amin, D.S. and Et-Aziz (1996) Studies on rapid diagnostic methods for rabbit haemorrhagic viral disease (RHVD). *Veterinary Medical Journal Giza* 44, 601–611.

Nowotny, N., Ros Bascuñana, C., Balligi-Pordány, A., Gavier-Widén, D., Uhlén, M. and Belák, S. (1997) Phylogenetic analysis of rabbit haemorrhagic disease and European brown hare syndrome viruses by comparison of sequences from the capsid protein gene. *Archives of Virology* 42, 657–673.

Ohlinger, V.F. and Thiel, H.-J. (1991) Identification of the viral haemorrhagic disease virus of rabbits as a calicivirus. *Revue Scientifique et Technique de l'Office International des Épizooties* 10, 311–323.

Ohlinger, V.F., Haas, B., Meyers, G., Weiland, F. and Thiel, H.-J. (1990) Identification and characterization of the virus causing rabbit haemorrhagic disease. *Journal of Virology* 64, 3331–3336.

Ohlinger, V.F., Haas, B. and Thiel, H.-J. (1993) Rabbit haemorrhagic disease (RHD): characterization of the causative calicivirus. *Veterinary Research* 24, 103–115.
OIE (1991) Viral haemorrhagic disease of rabbits and the European brown hare syndrome. *Revue Scientifique et Technique de l'Office International des Épizooties*, Volume 10.
O'Keefe, J.S., Tempero, J.E., Atkinson, P.H., Pacciarini, M.L., Fallacara, F., Horner, G. and Motha, J. (1998) Typing of rabbit haemorrhagic disease virus from New Zealand wild rabbits. *New Zealand Veterinary Journal* 46, 41–42.
Park J.H., Kida, H., Ueda, K., Ochiai, K., Goryo, M. and Itakura, C. (1991) Etiology of rabbit haemorrhagic disease spontaneously occurring in Korea. *Journal of Veterinary Medicine* B38, 749–754.
Parra, F. and Prieto, M. (1990) Purification and characterization of a calicivirus as the causative agent of a lethal haemorrhagic disease in rabbits. *Journal of Virology* 64, 4013–4015.
Plana-Duran, J., Bastons, M., Rodriguez, M.J., Climent, I., Cortes, E., Vela, C. and Casal, I. (1996) Oral immunization of rabbits with VP60 particles confer protection against rabbit haemorrhagic disease. *Archives of Virology* 141, 1423–1436.
Rasschaert, D., Huguet, S., Madelaien, M.-F. and Vautherot, J.-F. (1995) Sequence and genome organization of a rabbit haemorrhagic disease virus isolated from a wild rabbit. *Virus Genes* 9, 121–132.
Robinson, A.J. and Westbury, H.A. (1997) The Australian and New Zealand Rabbit Calicivirus Disease Program. In: Chasey, D., Gaskell, R.M. and Clarke, I.N. (eds) *Proceedings of the First International Symposium on Caliciviruses.* European Society for Veterinary Virology and the Central Veterinary Library, Weybridge, pp. 144–150.
Rodak, L., Smid, B., Valicek, L., Stepanek, J., Hampel, J. and Jurak, E. (1990) Enzyme-linked immunosorbent assay of antibodies to rabbit haemorrhagic disease virus and determination of its major structural proteins. *Journal of General Virology* 71, 1075–1080.
Rodak, L., Smid, B. and Valicek, L. (1991) Application of control measures against viral haemorrhagic disease of rabbits in the Czech and Slovak Republics. *Revue Scientifique et Technique de l'Office International des Épizooties* 10, 513–524.
Smid, B., Valicek, L., Rodak, L., Stepanek, J. and Jurak, E. (1991) Rabbit haemorrhagic disease: an investigation of some properties of the virus and evaluation of an inactivated vaccine. *Veterinary Microbiology* 26, 77–85.
Smith, A.W. and Boyt, P.M. (1990) Caliciviruses of ocean origin: a review. *Journal of Zoo Wildlife Research* 21, 3–23.
Smith, A.W., Skilling, D.E., Cherry, N., Mead, J.H. and Matson, D.O. (1998) Calilivirus emergence from ocean reservoirs: zoonotic and interspecies movements. *Emerging Virus Diseases* 4, 13–20.
Smyth, R.E., Cooke, B.D. and Newsome, A.E. (1997) The spread of rabbit calicivirus in relation to environmental factors. Unpublished data, 37 pp. Report prepared for the Anti-Rabbit Research Foundation June 1997, as part of the Meat Research Corporation Project RCD.003. Quoted with the authors' permission.
Thompson, J. and Clark, G. (1997) Rabbit calicivirus now established in New Zealand. *Surveillance* 24, 5–6.
Villafuerte, R., Calvette, C., Gortazar, C. and Moreno, S. (1994) First epizootic of rabbit haemorrhagic disease in free living populations of *Oryctolagus cuniculus* at Donana National Park, Spain. *Journal of Wildlife Diseases* 30, 176–179.
Wardhaugh, K. and Rochester, W. (1996) Wardang Island. A retrospective analysis of weather conditions in respect to insect activity and displacement. Unpublished data, 47 pp. CSIRO Division of Entomology, Report to the Meat Research Corporation. Quoted with the authors' permission.
Westbury, H. (1996) Field evaluation of RCD under quarantine. Final Report of Project CS 236 to the Meat Research Corporation.
Williams, K., Parer, I., Coman, B., Burley, J. and Braysher, M. (1995) *Managing Vertebrate Pests: Rabbits.* Australian Government Publishing Service, Canberra, 284 pp.
Wilson, G., McNee, A. and Platts, P. (1992) *Wild Animal Resources – Their use by Aboriginal Communities.* Australian Government Publishing Service, Canberra, pp. 30–36.
Wyatt, R.G., Greenberg, H.G., Dalgard, D.W., Allwood, W.P., Sly, D.L., Thorhill, T.S., Chanock, R.M. and Kapikian, A.Z. (1978) Experimental infection of chimpanzees with the Norwalk agent of epidemic viral gastroenteritis. *Journal of Medical Virology* 2, 89–96.
Xu, W. (1991) Viral hemorrhagic disease of rabbits in the People's Republic of China: epidemiology and virus characterization. *Revue Scientifique et Technique de l'Office International des Épizooties* 10, 393–408.

12

Ecological and Environmental Effects of Biological Control of Rabbits

Overview

Rabbits have been recognized as an important agricultural pest in Australia since the 1880s and in Britain since the First World War. When myxomatosis reduced the large wild rabbit populations in the early 1950s it became apparent that rabbits also had important ecological and environmental effects. In temperate Australia, regeneration of mulga and cypress pine, sheoaks and ti-trees then occurred such as had not been seen before. The effects were particularly pronounced in the semi-arid grasslands, where rabbits drove acacias to extinction by ringbarking small trees during droughts. They also destroyed many species of grass. After myxomatosis, it was feared that in some areas shrubs would become woody weeds. By their destruction of many different species of plants, especially when a severe drought occurred in rangelands where there had previously been several good seasons, rabbits played a major role in enhancing wind and water erosion.

The occurrence of myxomatosis also showed that rabbits were having direct adverse effects on burrowing native mammals such as bilbies and bettongs. Deaths due to myxomatosis sometimes caused foxes and cats to intensify their predation on native animals, and after the early outbreaks there was increased predation of lambs by foxes and eagles, but the predator populations soon adjusted to the lower rabbit numbers.

The reduction of rabbits by more than 95% during the first few years of myxomatosis in Britain led to dramatic changes in many landscapes, especially the chalk downlands, which reverted to woodlands. In some areas there were spectacular increases in palatable grasses and in downland flowers. In France, a luxurious growth of grasses on areas of moorland was followed by an increase in shrubs and the appearance of oak and beech seedlings. Initially there was an increase in hare numbers in some areas, and there was increased competition between predators such as stoats, weasels, foxes and raptors.

Introduction

From the latter half of the 19th century rabbits had been regarded as Australia's major agricultural pest and existing methods of control had not proved effectual, hence the great pressure to introduce biological control, an idea which dated from the days of the Intercolonial Commission in 1888 (see p. 54). However, it was not until myxomatosis (and later rabbit haemorrhagic disease) were introduced that the extent of the ecological and environmental effects of rabbits became apparent. The study sites of the CSIRO Wildlife Survey Section, and those of State pest control authorities, were initially in agricultural land in temperate Australia, but later they extended to the rangelands (see Fig. 8.1, p. 183). The majority of the intensive study sites for rabbit haemorrhagic disease were located in semi-arid areas (see Fig. 11.10, p. 261).

The picture in the agricultural land of south-eastern Australia shortly after myxomatosis was introduced was summed up by Ratcliffe in 1956; it is worth quoting his views at that time[1]:

> Immediately prior to the liberation of myxomatosis, a combination of circumstances had led to the build-up of rabbits to very high levels over much or most of their range; and the situation in many areas could only be described as desperate ... The change, because good seasons have enabled the vegetation to take full advantage of the relief from the pressure of rabbit grazing, has been almost miraculous: the landscape in some areas has been virtually transfigured. Hills that had been grazed to the soil for decades, and whose slopes appeared grey and red on the horizon, are now clothed in grass. The broad margins of the country roads, lying outside the boundary fences of grazing properties, tended to carry dense rabbit populations and as often as not showed it in the poverty of their ground cover. It is now usual to see tall grass and herbage to the road's edge. Looking over the fences, it is now very rare to see a paddock without a dense and healthy pasture.

Initially the Wildlife Survey Section was preoccupied with field studies of myxomatosis and its vectors (see Chapter 7). In the mid-1950s its studies of rabbit biology were broadened, encompassing population biology and social behaviour (Mykytowycz, 1958; Myers and Poole, 1959). In 1962 the Section was expanded to a Division, which began a comprehensive investigation of the native fauna of Australia. In parallel, botanists in other divisions of CSIRO, in State departments and in the universities expanded their studies of the native flora and of the effects of myxomatosis in reducing rabbit grazing. In consequence, by the 1980s there was a much better appreciation of the damage rabbits did to the environment, especially in the rangelands, which constitute some 70% of mainland Australia (Wood, 1984). The environmental and ecological consequences of the persisting high rabbit numbers found in the rangelands in the 1980s have been comprehensively reviewed by Williams *et al.* (1995).

In Europe there was already a detailed knowledge of the flora and fauna and some appreciation of the damage that rabbits did. Rabbits had been regarded as an agricultural pest in Britain since the First World War, but the extreme reduction of rabbit numbers during the first few years after the introduction of myxomatosis in 1953 brought home to all who travelled through the countryside the extent of the ecological effects of rabbits.

Ecological and Environmental Effects of Myxomatosis in Australia

Pastoralists had arrived in the rangelands of New South Wales with flocks of sheep in the 1860s and 1870s, sheep populations reaching a peak of 15.5 million in the Western Division of that State in 1891. Rabbits arrived in these areas in about 1880 and, assisted by the environmental changes already produced by the sheep, soon built up to plague proportions, subject to periodic population crashes due to droughts. Serious environmental degradation was produced by the combination of drought, fire and grazing, principally by sheep and rabbits and to some extent by kangaroos. In suitable soils rabbit warrens were large and numerous, and rabbits often dug shallow scratch holes which were devoid of vegetation. The rabbits played a crucial role because when the grasses and herbs dried out in hot dry summers and in droughts, they ringbarked the woody perennial shrubs and small trees in search of water. They also sought the moister roots of plants, either killing them or lowering their ability to withstand drought. Most of the dominant shrubs and trees in the rangelands were long-lived species, but many gradually disappeared because rabbits consumed the seedlings.

The destruction of vegetation by rabbits prior to myxomatosis had led to serious land degradation, with gully erosion along streams and sheet erosion where the damage by rabbits followed by drought had destroyed most of the vegetation. It took several years after the introduction of myxomatosis to reverse this trend, but in many areas revegetation provided protection against further erosion. For example, the advent of myxomatosis had a dramatic effect on decreasing the rate of siltation in Burrinjuck Dam, in the southern tablelands of New South Wales (Clark, 1990).

By the 1980s it was also realized that rabbits, foxes and cats played a critical role in the depletion or even extinction of some native mammals, and that there were complex interactions between predators and their prey (Newsome, 1993a).

The effects of greatly reduced rabbit numbers on vegetation

On the southern tablelands of New South Wales, rabbits selectively grazed on legumes, greatly reducing important pasture plants such as subterranean clover. Removal of rabbits by myxomatosis in the early 1950s had a dramatic effect, since regeneration was rapid. The reduction of grazing and browsing pressure on seedlings of shrubs and trees following myxomatosis had remarkable effects on the appearance of many outback landscapes. Prior to myxomatosis, the Forestry Commission could ensure regeneration of its *Callitris* reserves only after costly rabbit control measures were in place (Fenner and Ratcliffe, 1965; Fig. 12.1). The widespread and continuing regeneration of mulga (*Acacia aneura*) and cypress pine (*Callitris* spp.) in western New South Wales was so dense that some landholders began to regard the seedlings as noxious weeds. In Tasmania, prior to myxomatosis, most *Pinus radiata* seedlings had to be protected by rabbit-proof netting; as in Europe (see p. 213) the reduction of rabbits by myxomatosis made this unnecessary. In a subalpine environment on mainland Australia, Leigh *et al.* (1987) found that rabbits reduced both the biomass and the diversity of forb species, the combination of burning and rabbits being particularly destructive.

Fig. 12.1. View of pastoral country in the Riverina in 1950, before the spread of myxomatosis, showing the huge population of rabbits among a few aging *Callitris* (native pine) trees. A few years after myxomatosis *Callitris* seedlings became so dense that some landholders began to regard them almost as noxious weeds. Photograph from the *Sydney Morning Herald* about 1950.

The damaging effects of rabbit grazing on vegetation in the Victorian Mallee was demonstrated in an experiment in which Cochrane and McDonald (1966) destroyed warrens in an area of 4 hectares of dune in the Hattah Lakes National Park, then enclosed the area to exclude rabbits. After 21 months, there were 39 species of ground plants inside but only 22 outside, and a fourfold difference in biomass. Of particular interest was the discovery of a species of *Sida* that had not been described scientifically.

The Coorong National Park lies within a coastal region in South Australia south of the mouth of the River Murray, where the average annual rainfall is about 500 mm. Using enclosure experiments and aerial photographs from past surveys, Cooke (1987) showed that two of the dominant trees, sheoaks and saltwater ti-trees, which regenerate poorly unless protected from grazing mammals, regenerated vigorously in the 1950s when myxomatosis kept rabbit numbers low. Unprotected seedlings were quickly found and eaten by rabbits, especially if alternative foods were scarce.

Rabbits have been even more destructive in the arid rangelands, in which the sparse vegetation of perennial grasses, trees and shrubs, many of the last-named with lifespans of 100–400 years, are no longer regenerating because rabbits selectively graze the seedlings that appear when rains come after times of drought. Exclosure experiments in arid rangelands in South Australia in 1979 showed that even after a quarter century of myxomatosis (Fig. 12.2), rabbits were continuing the course to extinction of various *Acacia* species that had begun in the 19th century (Lange and Graham, 1983). Because rabbits can ringbark shrubs and small trees, even an apparently successful germination can be destroyed by rabbits years after the event. Rabbit grazing was more important than grazing by kangaroos in destroying grass species in semi-arid grasslands of central New South Wales, especially within 50 metres of burrows (Leigh *et al.*, 1989). From 1986 onwards, officials of those departments within the South Australian Government that had responsibility for a large part of the rabbit-infested arid country south of the Tropic of Capricorn brought pressure on the Council of Nature Conservation Ministers to investigate other methods of biological control of rabbits in these regions. As a result Dr B.D. Cooke went to Spain to investigate the possibility of obtaining fleas better adapted to hot, dry country than the English variety of *Spilopsyllus cuniculi* (see p. 188). Later, after seeing the effects of a new infectious disease among wild rabbits in Spain, he took active steps to promote the investigation of rabbit haemorrhagic disease virus (see p. 247).

Possible increases in woody weeds

Because of the hot, dry summers, myxomatosis had not been as effective in controlling rabbits in the semi-arid rangelands as it had been in temperate Australia. It was hoped that rabbit haemorrhagic disease would work effectively in such areas, where trees and shrubs are sometimes regarded as 'woody weeds' because they encroach on pasture lands, thereby reducing their capacity for sustaining livestock. Since rabbits graze on such plants, there was some concern that its spread might enhance the spread of woody weeds. However, the consensus of opinion was that this problem could be managed in a better way than by relying on grazing by rabbits, and that the reduction in rabbit grazing would be of great advantage to the overall environment.

The effects of rabbits on native and feral animals

The complex interactions between rabbits and the native fauna and feral animals were comprehensively reviewed in a report prepared as a submission to committees considering the release of rabbit haemorrhagic disease virus (Newsome *et al.*, 1997). Because the situation is complex, it is simplest to consider separately the interactions involving rabbits and native herbivores and those involving predators.

Fig. 12.2. The environmental impact of rabbits in different parts of South Australia. (A) Manunda sheep station, near Yunta (see Fig. 11.8, p. 257), before rabbit control. (B) The same site after control of rabbits by warren ripping. (C) Area in the Coorong National Park (see Fig. 8.1, p. 183) before construction of rabbit-proof fences. (D) The same site eight years after exclusion of rabbits on the right-hand side of the fence. Courtesy of the Anti-Rabbit Research Foundation of Australia.

Interactions between rabbits and native herbivores

In the 1950s attention was focused on the direct economic effects of rabbits and the improvements that followed myxomatosis. In more recent years, with greater environmental awareness, there has been concern that the damaging effects of competition for food and the destruction of shelter by rabbits was reducing the numbers and range, and threatening the local extinction of a number of small native animals. Precise data are hard to come by, but rabbits are suspected of evicting burrowing bettongs (*Bettongia lesueur*) from their burrows and excluding other small marsupials from the best feeding areas. The bilby (*Macrotis lagotis*) is a burrowing animal which was formerly common over a wide area of southern Australia. There is a strong correlation between the spread of the rabbit and the disappearance of the bilby, with a lag period of about 15 years. The bilby is now found only in areas north of the rabbits' distribution or where rabbits and foxes are rare (Southgate, 1990). Likewise the spectacled hare-wallaby (*Lagorchestes conspicillatus*) competes directly with rabbits for seasonally scarce green grass, and it has disappeared from those parts of central Australia where there are rabbits (Ingleby, 1991).

Larger native mammals have also suffered from competition with rabbits. Cooke (1997) reviewed historical records of the contraction of the range in south-eastern

South Australia of the Common Wombat (*Vombatus ursinus*), a large marsupial with an adult weight of up to 40 kg, on which foxes had little impact. Their range contracted some hundreds of kilometres southwards within 20 years of the arrival of rabbits in 1881. Examination of their biology in the Coorong National Park showed that the bulk of the wombat's diet consisted of perennial grasses and sedges. Rabbits are currently the major herbivores in the Park, and experiments with rabbit-proof enclosures showed that rabbit grazing led to a replacement of the native perennial grasses with introduced annual grasses which die off in summer. In consequence, wombats had to retreat to wetter areas where perennial grasses persisted despite grazing by rabbits. The Southern hairy-nosed wombat (*Lasiorhinus latifrons*) and the Western Grey Kangaroo (*Macropus fuliginosus*) appear to have undergone similar reductions in response to rabbit grazing, with resurgences when myxomatosis led to falls in rabbit numbers.

Interactions between rabbits and predators

The extent to which foxes and feral cats control rabbit numbers was demonstrated in a field experiment by Newsome *et al.* (1989; Fig. 12.3). There was a great irruption of rabbits in central New South Wales in 1979, such that spotlight counts along an 18 km transect reached a peak of 5580. During the hot dry summer the counts crashed to about 50, and remained low on two control sites (80 and 50 km^2 in size) for the next 18 months, in spite of good pasture growth. In contrast, there was an accelerated increase in rabbit numbers on another site of 70 km^2 on which foxes and feral cats were persistently shot (Fig. 12.4). After two breeding seasons rabbit counts were more than four times higher where predators were removed than in the control areas, and the peak spotlight counts reached a level comparable to that measured just before the preceding irruption, only to crash because of a severe drought. The authors concluded that carnivores did control the rabbits, but only after rabbit numbers had crashed in the drought. As we noted in Chapter 2, predation on the rabbit in Australia is like a poor handbrake on a car, which will hold the vehicle on a gentle

Fig. 12.3. Alan Newsome (1935–). After graduating with a BSc from the University of Queensland in 1957, Newsome joined the Animal Industry Board of the Northern Territory Administration. After studying the interactions between red kangaroos and the cattle industry in central Australia between 1958 and 1965, he joined the CSIRO Division of Wildlife Research in 1965, where he is now a Senior Principal Research Scientist. He obtained a PhD degree from the University of Adelaide in 1968 for work on the origins of irruptions of house mice in farming land, and then led a 10-year-long study of dingoes in central and south-eastern Australia. Experience at the University of California at Berkeley in 1972 encouraged the development of a sustained interest in predator–prey relationships. This led to the recognition of the role of foxes and cats in regulating rabbit numbers in arid Australia and the effects of population crashes among rabbits, due to drought or rabbit haemorrhagic disease, in switching the attention of these predators to native animals. He was awarded the degree of DSc by the University of Queensland in 1984, and was a Regents Professor in the University of California at Irvine in 1985.

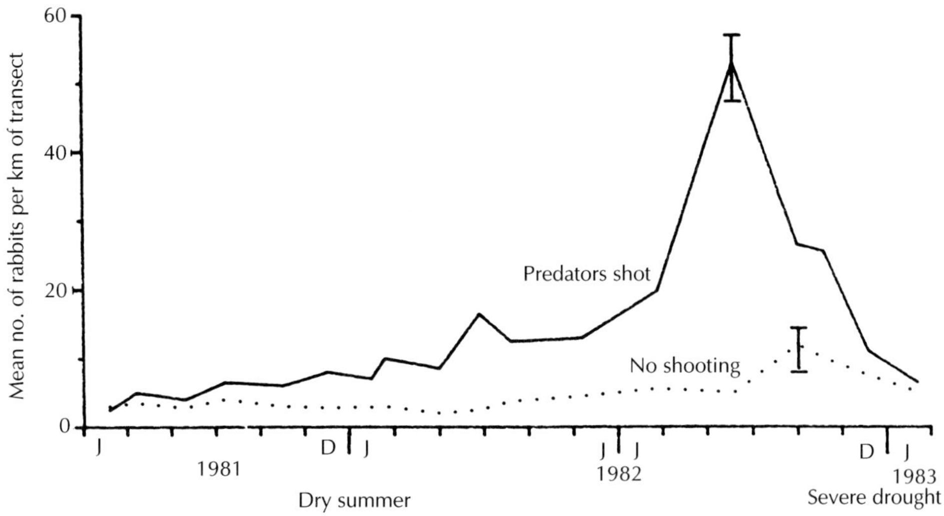

Fig. 12.4. Accelerated increase in rabbits at field sites in semi-arid Australia with continued removal of foxes and cats. The error bars indicate one standard error on either side of the mean. Data from Newsome *et al.* (1989), with permission.

slope, but becomes less and less effective as the car starts to move and gathers momentum.

Prior to myxomatosis the common predators, foxes and some raptors such as the wedge-tailed eagle (*Audax audax*), appear to have had a surfeit of food except during droughts, so that rabbit numbers were not greatly affected by predation but fluctuated in relation to the food available to the rabbits. The immediate effect of myxomatosis was to increase the efficiency of predation, partly because sick rabbits were easy to catch. However, for the first few years after the introduction of myxomatosis, the scarcity of rabbits caused increased attacks on lambs, leading to greater efforts to control predation. In southern Queensland bonuses paid on eagle 'scalps' increased from $5000 in 1950 to $10,000 in 1951 and $12,000 in 1952 and 1953. Over the same period, bonuses on fox scalps increased from $14,000 to $35,000 (Fenner and Myers, 1978). Later the predator population adjusted to the decreased numbers of rabbits, and attacks on lambs ceased to be a major problem.

The question of the ecological and environmental effects of rabbits, especially in the semi-arid rangelands, came to prominence during the investigations embarked on prior to the introduction of rabbit haemorrhagic disease virus. Five species of raptors have benefited by the introduction of rabbits to Australia and were thought likely to be affected by a sustained widescale decline in rabbit numbers such as might be caused when rabbit haemorrhagic disease virus was introduced. It was difficult to predict the outcome, but since raptors are mobile it was suggested that they might be less affected than would be expected from the proportion of rabbits normally in their diet.

Interactions between rabbits, predators and native fauna

In Australia foxes are the principal predators of rabbits and rabbits are the principal prey of foxes. When rabbit populations crash because of drought or myxomatosis, the fox and feral cat populations also collapse, after a lag period (see Fig. 12.5). During the lag period foxes (to a greater extent than cats) are believed to prey heavily on native fauna (Saunders *et al.*, 1995). There is some concern that a sudden decrease of rabbits as a food source in arid Australia following rabbit haemorrhagic disease might cause foxes and cats to prey on small native mammals to a greater extent than previously, so the situation is being monitored. Several of these animals are already under threat because of competition with rabbits for food and shelter (Newsome *et al.*, 1997).

The complexity of the interactions is illustrated by some observations in south-western Australia (Newsome, 1993b). Grazing marsupials in this region were found to have a natural resistance to the poison sodium fluoroacetate ('1080'), because they had evolved in a place where several common under-shrubs contain relatively high concentrations of that chemical, and National Parks there carried flourishing populations of six species of medium-sized marsupials that were rare elsewhere. From the mid-1950s until the early 1970s, 1080 had been widely used for rabbit control in the farmland surrounding the Parks. The type of bait used meant that one rabbit killed by 1080 would contain enough of the poison to kill several adult foxes. Since foxes fed on the dead rabbits, they had declined 'to the point of rarity' in parallel with the fall in rabbit numbers (King *et al.*, 1981), leading to increases in the populations of endangered marsupials. Then in the early 1970s the introduction of the European rabbit flea increased the incidence of myxomatosis to such an extent that the poisoning programme was discontinued (Fig. 12.5). Fox numbers rose again, and because of the paucity of rabbits they preyed on the native fauna to an increased extent, with disastrous consequences. Today, with aerial baiting of foxes over vast areas of the south-western forests of Western Australia (Saunders *et al.*, 1995), marsupial populations are flourishing there once more.

During the investigations preceding the introduction of rabbit haemorrhagic disease virus, it was pointed out that the greatest risk of a precipitate decline in rabbit numbers due to this virus would be its effects on foxes, which might switch to native mammals that were rare or endangered. However, this is a situation that has long existed in the arid and semi-arid parts of Australia, when rabbit and fox populations build up to high numbers in good seasons, only to crash when the inevitable prolonged drought occurs. It is believed by experts that the overall environmental effects of the occurrence of rabbit haemorrhagic disease among wild rabbits in the rangelands of Australia would be favourable, being dominated by the increased growth of native plants and

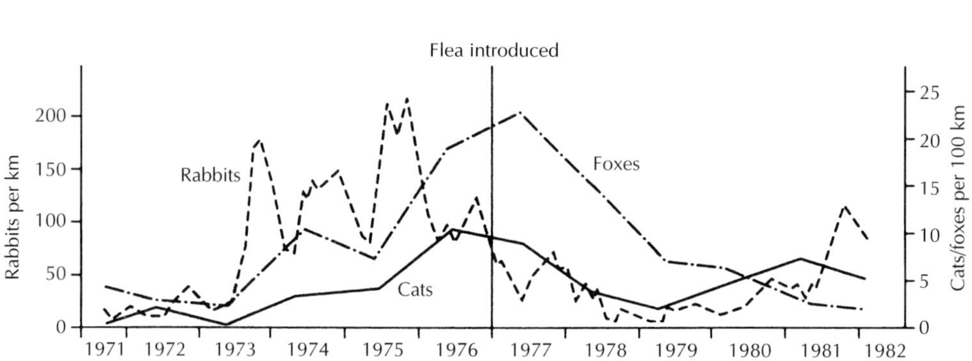

Fig. 12.5. Variation in fox and feral cat numbers in relation to changes in rabbit numbers in south-western Western Australia. There was an increased incidence of myxomatosis following the introduction of the European rabbit flea, and after a lag of some months the numbers of predators fell. From Saunders *et al.* (1995), based on data from King and Wheeler (1985), with permission.

decreased erosion that would follow the destruction of rabbits. The possible adverse effects on the native fauna were discussed in the Biological Control Act report (Bureau of Resource Sciences, 1996) and were examined in detail by Newsome *et al.* (1997).

Ecological and Environmental Effects of Myxomatosis in Europe

In contrast to Australia, where large-scale reductions in the numbers of wild rabbits occurred in the arid rangelands every ten years or so because of droughts, in England and France the effects of myxomatosis were unprecedented. The large-scale destruction of wild rabbits during the years 1952–1955 had dramatic ecological and environmental effects, on vegetation, on other animals and on agricultural production. Because of the more numerous observers and the better monitoring of the changes, the ecological effects of myxomatosis were more adequately described in Europe than was possible in the early years of myxomatosis in Australia. Although the numbers of wild rabbits slowly increased after 1956, many of the environmental changes were permanent.

The effects of declining rabbit numbers on vegetation

Sumption and Flowerdew (1985) and later Thompson (1994) have reviewed British data on the ecological effects of the reduction in rabbit numbers by myxomatosis. Although rabbits had been recognized as an agricultural pest in the 1940s, the extent to which their presence affected the flora of Britain was only realized when myxomatosis reduced the population by some 99% by about 1955. The complex ecological situation created by virtual removal of rabbits from the ecosystem in Britain is summarized in Fig. 12.6. By 1955 grasslands had improved and there was some indication of a natural regeneration of forest trees (Advisory Committee, 1955). Thomas (1960, 1963) recorded changes of species composition in chalk grassland and downland heath, scrubs, and woods of various kinds. Spectacular changes were seen on every site. There was a marked increase in woody plants, in palatable grasses, and in numerous downland flowers and orchids. In 1955 there was a vivid flowering of *Senecio,* normally suppressed by rabbit grazing, as a yellow sheet covering the hills, but it gradually disappeared as the downlands acquired new agricultural and botanical values. One of the authors (F.F.) has vivid memories of wandering with Thomas over one of his sites on the chalk downlands in 1962, and seeing the thousands of yew seedlings and other woody plants springing up everywhere. The Advisory Committee (1955) suggested that the steep chalk downlands that had been maintained by rabbit grazing should be restocked with sheep or cattle. However, in many areas the post-myxomatosis invasions by woody plants has continued unchecked, and many chalk downs, heathlands and sand dunes have become woodlands.

Crompton and Sheail (1975) studied the ecological history of Lackenheath Warren in Suffolk, which was the last rabbit warren to survive out of the many that were established in that area during the 13th century. The effects of centuries of rabbit grazing were obvious. Prior to myxomatosis the warren supported 4.4 rabbits per hectare, compared with 19.8 per hectare on nearby unfenced grasslands. Rabbits in the warren bred only once, in the spring; in nearby areas there was a much longer breeding season. Clearly, 600 years of rabbit grazing had created a system with the rabbit depending on a single spring burst of growth of annual plants for survival. Myxomatosis removed the rabbits, and by 1970 almost half the warren was overgrown by shrubs and seedling pines. Rare flowering plants known to be preferentially grazed by rabbits appeared for the first time, whereas others associated with disturbed soil disappeared.

Similar trends have been described in France, notably in regions where the rabbit density had previously been highest (Morel, 1956). Rabbit grazing had converted oak woodlands into birch, aspen

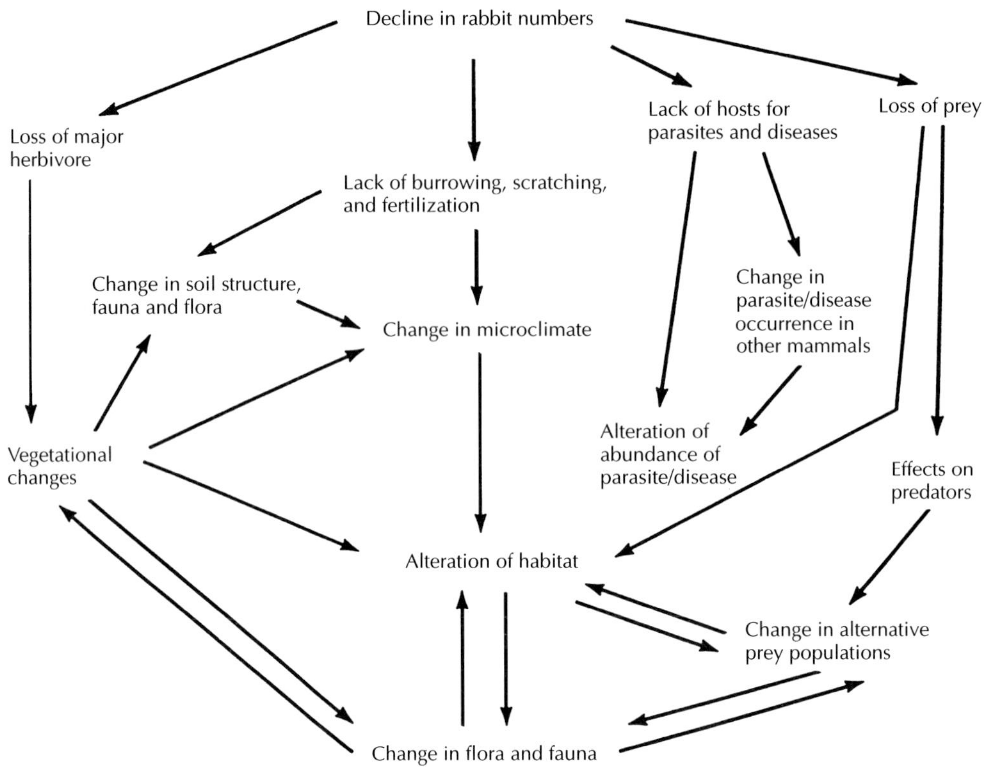

Fig. 12.6. The complex ecological effects of the decline in rabbit numbers due to myxomatosis in Britain. From Sumption and Flowerdew (1985), with permission.

and *Erica* moors, and the abundance of rabbits rendered attempts at reafforestation useless. After myxomatosis there was at first a luxuriant growth of grasses, followed by an increase in the shrub understorey and then by a widespread appearance of oak and beech seedlings. As in Britain, the French countryside was transformed. In Holland, dunes on the North Sea coast developed a rich and luxuriant flowering of various grasses (van Leeuwen and van der Maarel, 1971), and in other areas woody plants reappeared, including oaks and birches. In north Belgium reafforestation became possible and in middle Belgium regeneration of shrub layers occurred in high forest and coastal dunes became stable under better growth of various grasses (Bourlière 1956).

Effects on other animals

The impact of lower rabbit numbers on its herbivorous competitors has not been adequately studied. Although it was found that hares were more common in 1955 than in previous years (Rothschild and Marsh, 1956; Siriez, 1960), in a wider study Barnes and Tapper (1986) found no evidence of an inverse correlation between numbers of rabbits and hares pre- or post-myxomatosis, nor of competition between them. They considered that the decline in numbers of hares that had occurred since the early 1960s was unrelated to the recovery of the rabbit population, and was probably due to the intensification of arable farming methods.

Two predators, foxes and buzzards, were obviously affected by the sudden shortage of rabbits in the early 1950s. Where rabbits

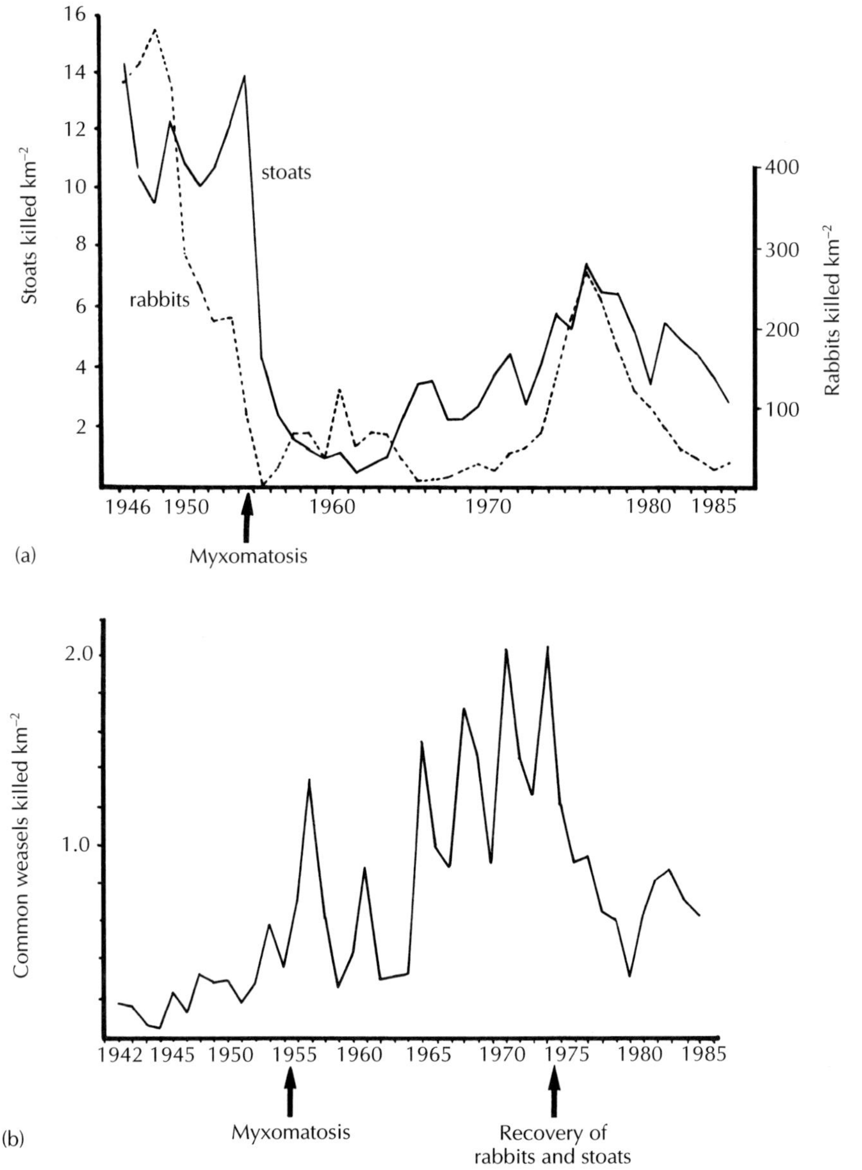

Fig. 12.7. Linked changes in the numbers of rabbits and of stoats (a) and weasels (b). The numbers of stoats was closely linked to the numbers of rabbits. The numbers of weasels increased suddenly after 1953, since they preyed on small rodents whose numbers increased greatly with the great growth of herbage that followed the removal of rabbits by myxomatosis. Modified from Thompson (1994), with permission.

were abundant they had provided 50% or more of the food of the fox in Britain (Southern and Watson, 1941), but after myxomatosis foxes turned to field voles, brown rats and possibly poultry, lambs and game birds (Lever, 1959; Englund, 1965). In the Camargue they altered their diet to birds and fish, in Seine-et-Marne to field mice. The intensified predation on small mammals caused a food shortage among other predators, for example, in one locality near Oxford tawny owls did not breed in 1955 (Southern, 1956). The shortage of rabbits caused a great reduction in the breeding of buzzards, for which young rabbits are a very important food item, especially when they have nestlings to feed (Moore, 1956). Later the buzzard stabilized its population at a new, somewhat lower level (Moore, 1957).

For several centuries the rabbit has been a major item in the diet of the stoat (*Mustela ermina*). However, it was not until myxomatosis occurred that it was realized how dependent they were on rabbits (Southern, 1956). Forced to compete with weasels for small rodents and with foxes, feral cats and raptors for larger animals, stoat numbers dropped precipitately. From the early 1970s increases in rabbit numbers were followed by increases in the numbers of stoats, but with every outbreak of myxomatosis they fell again (King, 1980, 1989; Fig. 12.7). The weasel (*Mustela nivalis*), on the other hand, specializes in feeding on small rodents, and the unprecedented growth of herbage after myxomatosis resulted in a glut of bank voles, wood voles and wood mice. The weasel populations soared, only to decline as the rabbits partially recovered their numbers from the early 1970s and their grazing pressure increased. The numbers of stoats increased concurrently with rabbit numbers (Tapper, 1980).

There is little doubt that the kinds of changes observed in wildlife populations act at all levels of the ecosystem, including soil and grassland animal communities. One interesting example was reported by Le Gac (1966). The destruction of rabbits by myxomatosis on the French Mediterranean coast was followed by a drop in the prevalence of human cases of tick-borne typhus, caused by *Rickettsia conorii*. Deprived of its infected rabbit hosts, the tick *Rhipicephalus sanguineus* could play no further part in transmission of the disease to humans, although the rickettsia–vole cycle was maintained by other vectors.

Endnote

[1]Basser Library Archives MS 143/25/5A. The ecological consequences of myxomatosis in Australia. Report by F.N. Ratcliffe, 21 March 1956.

References

Advisory Committee on Myxomatosis (1955) *Myxomatosis, Second Report of the Advisory Committee on Myxomatosis*. Her Majesty's Stationery Office, London, p. 5.

Barnes, R.F.W. and Tapper, S.C. (1986) Consequences of the myxomatosis epidemic in Britain's rabbit (*Oryctolagus cuniculus*) population on the numbers of brown hares (*Lepus europaeus*). *Mammal Review* 16, 111–116.

Bourlière, F (1956) Biological consequences due to the presence of myxomatosis. *Terre et Vie* 103, 123–136.

Bureau of Resource Sciences (1996) *Rabbit Calicivirus Disease: a Report under the Biological Control Act 1984*. Bureau of Resource Sciences, Canberra, 127 pp.

Clark, R.L. (1990) Ecological history for environmental management. *Proceedings, Ecological Society of Australia* 16, 1–21.

Cochrane, G.R. and McDonald, N.H. (1966) A regeneration study in the Victorian mallee. *Victorian Naturalist* 83, 220–226.

Cooke, B.D. (1987) The effects of rabbit grazing on regeneration of sheoaks, *Allocasuarina verticilliata* and saltwater ti-trees, *Melaleuca halmaturorum*, in the Coorong National Park, South Australia. *Australian Journal of Ecology* 13, 11–20.

Cooke, B.D. (1999) Did European rabbits, *Oryctolagus cuniculus* (L.) displace Common Wombats, *Vombatus ursinus* (Shaw) from part of their range in South Australia? In: Wells, R.T. and Pridmore, P.A. (eds) *Wombats*. Surrey Beatty and Sons in conjunction with Royal Zoological Society Inc., South Australia, pp. 262–270.

Crompton, G. and Sheail, J. (1975) The historical ecology of Lakenheath Warren in Suffolk, England: a case study. *Biological Conservation* 8, 299–313.
Englund, J. (1965) The diet of foxes (*Vulpes vulpes*) on the island of Gotland since myxomatosis. *Viltrevy* 3, 507–530.
Fenner, F. and Myers, K. (1978) Myxoma virus and myxomatosis in retrospect: the first quarter century of a new disease. In: Kurstak, E. and Maramorosch, K. (eds) *Viruses and Environment.* Academic Press, New York, pp. 539–570.
Fenner, F. and Ratcliffe, F.N. (1965) *Myxomatosis.* Cambridge University Press, Cambridge, p. 307.
Ingleby, S. (1991) Distribution and status of the spectacled hare-wallaby (*Lagorchestes conspicillatus*) in Northern Territory and Western Australia. *Wildlife Research* 18, 501–519.
King, C.M. (1980) Population biology of the weasel *Mustela nivalis* on British game estates. *Holarctic Ecology* 3, 160–168.
King, C. (1989) *The Natural History of Weasels and Stoats.* Christopher Helm, London, 253 pp.
King, D.R. and Wheeler, S.H. (1985) The European rabbit in south-western Australia. I. Study sites and population dynamics. *Australian Wildlife Research* 12, 183–196.
King, D.R., Oliver, A. and Mead, R.J. (1981) Bettongia and fluoroacetate: a role for 1080 in fauna management. *Australian Wildlife Research* 8, 529–536.
Lange, R.T. and Graham, C.R. (1983) Rabbits and the failure of regeneration in Australian arid zone *Acacia. Australian Journal of Ecology* 8, 377–381.
Le Gac, P. (1966) Répercussion de la myxomatose sur la fièvre exanthématique boutonneuse méditerranéenne. *Bulletin of the World Health Organization* 35, 143–147.
Leigh, J.H., Wimbush, D.J., Wood, D.H., Holgate, M.D., Slee, A.V., Stanger, M.G. and Forrester, R.I. (1987) Effects of rabbit grazing and fire on a subalpine environment. I. Herbaceous and shrubby vegetation. *Australian Journal of Botany* 35, 433–464.
Leigh, J.H., Wood, D.H., Holgate, M.D., Slee, A. and Stanger, M.G. (1989) Effects of rabbit and kangaroo grazing on two semi-arid grassland communities in central-western New South Wales. *Australian Journal of Botany* 37, 375–396.
Lever, R.J.A.W. (1959) The diet of the fox since myxomatosis. *Journal of Animal Ecology* 28, 359–375.
Moore, N.W. (1956) Rabbits, buzzards and hares: two studies on the indirect effects of myxomatosis. *Terre et la Vie* 103, 220–225.
Moore, N.W. (1957) The past and present status of the buzzard in the British Isles. *British Birds* 50, 173–197.
Morel, A. (1956) Influence de l'épidemie de myxomatose sur la flore Française. *Terre et la Vie* 103, 226–238.
Myers, K. and Poole, W.E. (1959) A study of the biology of the wild rabbit (*Oryctolagus cuniculus*) in confined populations. I. The effects of density on home range and the formation of breeding groups. *CSIRO Wildlife Research* 4, 14–46.
Mykytowycz, R. (1958) Social behaviour of an experimental colony of wild rabbits (*Oryctolagus cuniculus* (L.)). I. Establishmment of the colony. *CSIRO Wildlife Research* 3, 7–25.
Newsome, A. (1993a) Ecological interactions. In: Cooke, B.D. (ed.) *Australian Rabbit Control Conference, 2–3 April 1993.* Anti-Rabbit Research Foundation of Australia, Adelaide, pp. 23–25.
Newsome, A. (1993b) Wildlife conservation and feral animals: the Procustes factor. In: Moritz, C. and Kikkawa, J. (eds) *Conservation Biology in Australia and Oceania.* Surrey Beatty and Sons, Chipping Norton, pp. 141–148.
Newsome, A., Parer, I. and Catling, P.C. (1989) Prolonged prey suppression by carnivores – predator-removal experiments. *Oecologia* 78, 458–467.
Newsome, A., Pech, R., Smyth, R., Banks, P. and Dickman, C. (1997) *Potential Impacts on Australian Native Fauna of Rabbit Calicivirus Disease.* Biodiversity Group, Environment Australia, 130 pp.
Rothschild, M. and Marsh, H. (1956) Increase of hares (*Lepus europaeus* Pallas) at Ashton Wold, with a note on the reduction in numbers of the brown rat (*Rattus norvegicus* Berkenhout). *Proceedings of the Zoological Society of London* 127, 441–445.
Saunders, G., Coman, B., Kinnear, J. and Braysher, M. (1995) *Managing Vertebrate Pests: Foxes.* Australian Government Publishing Service, Canberra, 140 pp.
Siriez, H. (1960) *Lapins et Myxomatose. L'Evolution de la Maladie de 1956 à 1960 et Quelques Compléments à une Précédente Étude.* Société des Editions Pharmaceutique, Paris, 24 pp.

Southern, H.N. (1956) Myxomatosis and the balance of nature. *Agriculture* 63, 10–13.
Southern, H.N. and Watson, J.S. (1941) Summer food of the red fox (*Vulpes vulpes*) in Great Britain: a preliminary report. *Journal of Animal Ecology* 10, 1–11.
Southgate, R.I. (1990) Distribution and abundance of the greater bilby. In: Seebeck, J.B., Brown, P.R., Wallace, R.I. and Kemper, C.M. (eds) *Bandicoots and Bilbies.* Surrey Beatty and Sons, Sydney, pp. 293–302.
Sumption, K.J. and Flowerdew, J.R. (1985) The ecological effects of the decline in rabbits (*Oryctolagus cuniculus*) due to myxomatosis. *Mammal Review* 15, 151–186.
Tapper, S.C. (1980) The status of some predatory mammals. *Game Conservancy Review*, 1979, 48–54.
Thomas, A.S. (1960) Changes in vegetation since the advent of myxomatosis. *Journal of Ecology* 48, 287–304.
Thomas, A.S. (1963) Further changes in vegetation since the advent of myxomatosis. *Journal of Ecology* 51, 151–186.
Thompson, H.V. (1994) The rabbit in Britain. In: Thompson, H.V. and King, C.M. (eds) *The European Rabbit.* Oxford University Press, Oxford, pp. 64–107.
van Leeuwen, C.G. and van der Maarel, E. (1971) Pattern and process in dune vegetations. *Acta botanica Neerland* 20, 191–198.
Williams, K., Parer, I., Coman, B., Burley, J. and Braysher, M. (1995) *Managing Vertebrate Pests: Rabbits.* Australian Government Publishing Services, Canberra, 284 pp.
Wood, D.H. (1984) The rabbit (*Oryctolagus cuniculus* L.) as an element in the arid biome of Australia. In: Cogger, H.G. and Cameron, E.E. (eds) *Arid Australia.* Australian Museum, Sydney, pp. 273–287.

13

Theoretical Aspects of Microbial Control of Vertebrate Pests

Overview

In this chapter and the next we review some of the interesting theoretical questions that have arisen in the consideration of myxomatosis and rabbit haemorrhagic disease and try to put them into a general biological perspective. Since each of the items discussed is essentially a review of matters raised elsewhere in the book, no overview will be given; the 'Contents' list sets out the topics discussed.

The Concept of Emerging and Re-emerging Infectious Diseases

In the mid-1970s it was widely believed that infectious diseases had in large part been conquered in the wealthy industrialized countries, by a combination of improved sanitation and nutrition, vaccination, and the use of effective antibiotics. It was realized that new pandemics of human influenza could occur, caused by essentially 'new' recombinant viruses, but it was not until the AIDS epidemic was well under way, in the mid-1980s, that the medical profession and the public became aware that infectious diseases still constituted a major threat to human health. As well as being a greatly feared if slowly progressive disease, it was soon realized that the immunosuppression induced by AIDS allowed the clinical reappearance of many other infections, both bacterial and viral. In addition, more attention was paid to other infections first noticed when humans moved into new ecological situations, for example Lassa fever, Lyme disease and Crimean–Congo haemorrhagic fever. In 1989 the first international conference on what were called 'emerging' viruses was convened (Morse, 1993), followed shortly after by a report by the Institute of Medicine of the United States (Lederberg *et al.*, 1992) and a conference on historical aspects of emerging infections (Fantini, 1993). The term re-emergence was coined to accommodate the fact that the agents of some diseases which had previously been effectively treated with antibiotics or other drugs had become resistant to them, or that immunosuppression due to AIDS or chemotherapy had produced an increased incidence of some infections previously regarded as being under control. We will not consider re-emergent diseases any further, but concentrate on emerging viral diseases.

While human infections have attracted most attention (Fig. 13.1), many of the same considerations apply to infections of animals. Equine and swine influenza may occur anywhere in the world, since they derive from highly mobile avian hosts, and vector-transmitted diseases may affect livestock derived from Europe when they are introduced into Africa or Australia, from reservoirs in indigenous wild animals.

In the context of this book, myxomatosis, as a disease of European rabbits, was such

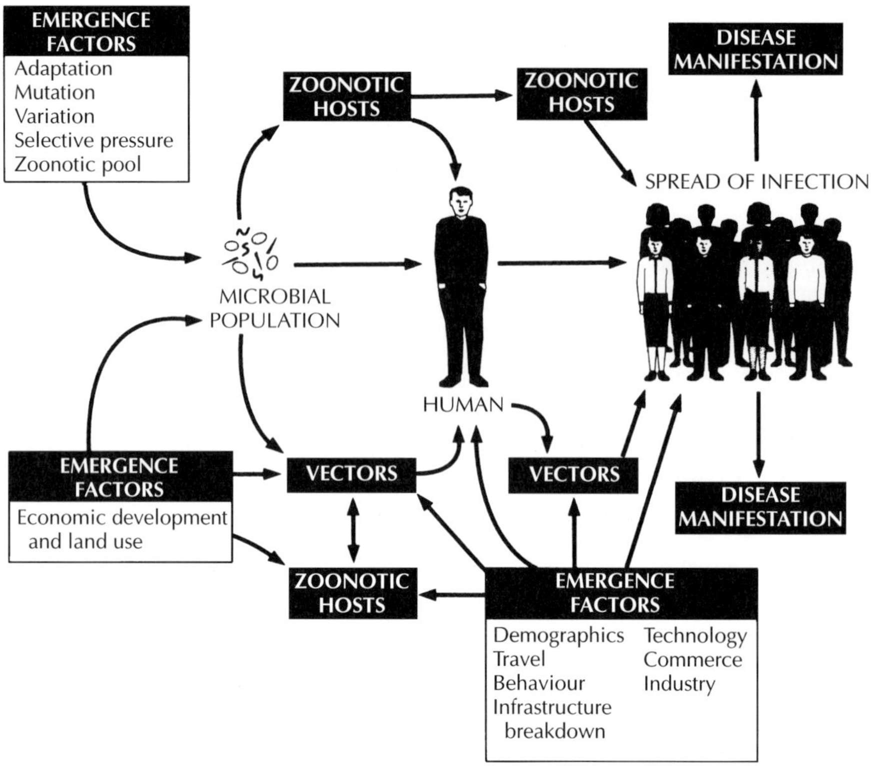

Fig. 13.1. Schematic diagram of the factors influencing the emergence of infectious diseases of humans. Many of the same factors operate with domestic animals and pets, and occasionally with wild animals. From Lederberg *et al.* (1992), with permission.

an emerging disease, first recognized when mosquitoes transferred myxoma virus mechanically from infected *Sylvilagus brasiliensis* in Brazil and from infected *S. bachmani* in California to an imported animal, the European rabbit (*Oryctolagus cuniculus*). As with so many emerging zoonotic viral diseases (e.g. hantavirus infections, most mosquito-transmitted arbovirus infections), the viruses concerned caused mild symptoms in their natural hosts, with which they had co-evolved for millennia, but severe disease in their new hosts. Rabbit haemorrhagic disease virus and the calicivirus that causes the European brown hare syndrome represent an unusual type of emerging infection, in that they appear to be virulent mutants of pre-existing, innocuous and unrecognized caliciviruses, each highly host-specific.

Koch's Postulates as Applied to Viruses

Jacob Henle (1840), one of Koch's teachers, first presented what came to be called 'Koch's (or Henle–Koch) postulates'. In the chapter entitled *Von den Miasmen und Kontagien,* he said 'I will now adduce the reasons which prove that the matter of contagions is not only organic, but also animate, indeed, endowed with independent life, and that it can be thought of as a parasitic organism in the diseased body'. He also recognized the importance of what were later called cultural procedures: 'In order to prove that they are really the causal material, it would be necessary to isolate the ... contagious organism ... and then observe the power of each one of these to see if they corresponded. This is an experiment that must no doubt be renounced'. Koch's contribution was to demonstrate that such experiments *could* be performed. It was not until 1884 that he published the postulates in the form we know today (Koch, 1884), although they first appeared in their present-day form in a paper by Loeffler (1884):

1. The organism must be shown to be constantly present in characteristic form and arrangement in the diseased tissue.
2. The organism which from its behaviour appears to be responsible for the disease must be isolated and grown in pure culture.
3. The pure culture must be shown to induce the disease experimentally.

The two postulates that proved most difficult to fulfil, even with bacteria, were growth in culture and passage in experimental animals, and from an early date (Pettenhofer's experiment with *Vibrio cholerae;* Evans, 1973) until quite recently with *Helicobacter pylori* (Warren and Marshall, 1983), recourse was had to human experimentation. Viruses, rickettsiae and chlamydiae provided an even greater challenge, because until the development of the electron microscope in the 1940s they were invisible, they cannot be cultivated on media used for the cultivation of bacteria but only in living cells, and cell culture did not become a well-established practice until the early 1950s. Further, many viruses are highly host-specific, and can only replicate in the animal in which they cause disease. Most human caliciviruses, for example, do not infect laboratory animals (except chimpanzees) and cannot be grown in tissue culture, even in human cells (Kapikian *et al.*, 1996).

From time to time various authors have updated the Henle–Koch postulates for viruses, for example Rivers (1937), Evans (1976) and more recently zur Hausen (1991) and Barranti-Brodano *et al.* (1998). Myxomatosis was first described at the dawn-time of virology – 1898 – the year in which Loeffler and Frosch described the filterability of foot-and-mouth disease virus and Beijerinck, in studies with tobacco mosaic virus, developed his concept of a *contagium vivum fluidum* (see p. 66). Sanarelli described the causative agent of myxomatosis as an 'invisible virus'. He passed it in rabbits, but using a Chamberland filter he was not able to demonstrate its filterability; this was achieved by Moses a few years later using the slightly coarser Berkefeld filter.

The lagomorph caliciviruses also present problems in fulfilling the original Henle–Koch postulates, in that neither of them has been grown in culture or in any animals except the species of lagomorph from which they were isolated. However, after initial confusion in some laboratories, in which it was claimed that the causative agent of rabbit haemorrhagic disease virus was a parvovirus (see p. 238), their role in the causation of rabbit haemorrhagic disease or European brown hare disease is now in no doubt.

Problems of Host Range – Breadth or 'Switching'

Many microorganisms, especially viruses, are highly host-specific in that in nature they infect only a single species of animal. On the other hand some have a broad host range and can infect and spread between animals of several species. One of the major problems facing scientists who wish to introduce organisms to control pests is to ensure their specificity for the pest concerned. This requirement applies to all types of biological control, but is especially important in the case of microbial control of vertebrate pests. Since there is an overriding concern for the health of humans and their domestic animals, and for native wildlife, any such agent must be carefully tested to ensure that it does not infect any animal other than the designated pest. Both myxoma virus and rabbit haemorrhagic disease virus were subjected to extensive testing for host specificity before their introduction. However, this did not allay the concern of some scientists, especially in the more recent introduction of the calicivirus that causes rabbit haemorrhagic disease, because there is always a chance that a mutation may occur which allows a virus to change its host range.

Some of the concern arises from a confusion of host range, which is the intrinsic ability of a species of virus to infect several different species of animal, and host switching, due to the appearance of mutants able to infect a previous insusceptible species of animal. The existence of virus species with either a narrow or a wide host range can be illustrated by the comparison of different species of the genus *Orthopoxvirus* (Fenner *et al.*, 1989). Cowpox virus has a very wide host range. Its natural hosts are various species of wild rodents living in various parts of Europe and western parts of the former USSR. Serological or other evidence of infection has been found in seven species of wild rodents (Feore *et al.*, 1997), and the virus causes lesions in all species of commonly used laboratory animals. Natural infections have been reported in humans, domestic cats, cows and various captive mammals in zoological collections. On the other hand smallpox virus has a very narrow host range. Natural infections occur only in humans, but apart from rabbits (infected with difficulty), suckling mice and chick embryos, the only other susceptible animals are non-human primates. Monkeys get a rash, and Noble and Rich (1969) showed that serial contact infection occurred for six passages in cynomolgus monkeys before it died out.

Until molecular biological techniques became available it was difficult to produce valid evidence of host switching, i.e. the infection of a species not previously susceptible with a virus believed to have a narrow host range. Even then, it is often difficult to decide, when an apparently novel host is recognized, whether the event is opportunistic infection by an unchanged virus, i.e. due to a wider host range than had been suspected, or the result of a specific mutation which led to the capacity to infect and spread in a novel host. For example, the natural hosts of influenza A virus are aquatic birds of various kinds. Interspecies transmission of influenza A virus has been observed several times, usually from aquatic birds to seal, whales, and mink, all of which have sustained self-limited epidemics, and in 1989 an avian virus caused an epidemic of equine influenza in northern China (Murphy and Webster, 1996). Sometimes avian viruses directly infect humans, as happened in Hong Kong in 1997. However, that virus was not able to spread from person to person and so did not produce a pandemic. Infections of humans with new pandemic strains of influenza virus are usually due to the emergence of reassortant viruses, with genes from avian and human viruses.

The best example of a change in host range due to mutation in a previously highly host-specific virus occurred in 1978, when a new disease occurred among dogs in many countries around the world. Retrospective serological studies showed that it was first present in domestic dogs in Europe in the mid-1970s. The widespread outbreaks and global spread of the virus did not occur until 1978, possibly because of some change in the virus at that time that gave it an epidemiological advantage. It was shown to be due to a parvovirus, which was designated canine parvovirus (CPV) type 2. Subsequently minor sequence changes occurred which led to the emergence worldwide of CPV type 2a, which was later replaced by CPV type 2b. Sequence studies of the capsid protein showed that CPV was very closely related to feline parvovirus, which had long been known to cause disease in cats (Parrish, 1990, 1994). An obvious mechanism for changing specificity would be a change in cell receptor binding, but in fact this does not appear to have occurred; rather changes in the processing of the capsids during cell entry and infection appear to be responsible for the ability of CPV to infect dogs. For any particular virus it is impossible to predict whether or not such a change will occur.

Of microorganisms that have been used for biological control of vertebrate pests, the chicken cholera bacillus, which Pasteur proposed to use for rabbit control (see p. 55) and *Salmonella enteritidis* var. *danysz* and *S. typhimurium*, which have been used for the control of rodents (see p. 49) adversely affect other vertebrate species, including, for *S. typhimurium*, humans. These microorganisms were first suggested or introduced many years ago and were never adequately tested. The only viruses that have been extensively used for the control of vertebrate pests are myxoma virus and the rabbit haemorrhagic disease virus. Although in 1951 there was concern that myxoma virus might cause encephalitis in humans (see p. 141), this proved not to be the case. In spite of its occurrence on a very large scale for almost 50 years in Australia and Europe, the only species other than the rabbit in which natural infections with myxoma viruses are known to have occurred is the European hare (*Lepus europaeus*), in which infections are exceedingly rare.

Because of the lack of a 'proof-reading' mechanism such as is found in eukaryotic cells and in DNA viruses, mutations occur at a very much higher rate in RNA viruses than in DNA viruses, and caliciviruses have an RNA genome of less than 8000 nucleotides. When the proposal to use a calicivirus for rabbit control in Australia first came to public notice, considerable concern was expressed by two American virologists with expertise in the family *Caliciviridae*, D.O. Matson and A.W. Smith, who wrote to the scientific press and to senior officials in Australia and New Zealand protesting against the proposal. Smith's concerns were coloured by his experience with a calicivirus with a wide host range, so-called vesicular exanthem of swine virus, which has now been eradicated from swine but causes infections in many species of pinnepeds (seals and sealions). The Australian decision to proceed with the investigation of the virus for biological control was based on the fact that the vast majority of caliciviruses, and all of those closely related to rabbit haemorrhagic disease virus, are very highly host-specific – so much so that they are impossible to grow in other laboratory animals and even in tissue culture. In addition, rabbit haemorrhagic disease virus (and its avirulent precursor, rabbit calicivirus) had been spreading between domestic rabbits in Europe for many years without having caused recognizable infection in humans or any other domestic animals. Tests in a wide range of domestic animals had failed to find any other susceptible species, but before its release further tests were carried out in a high security laboratory on domestic and companion animals and on a range of native animals of Australia and New Zealand.

Host specificity, French and German schools

Before the discoveries of Pasteur and Koch it was difficult to apply the concept of specificity to infectious diseases, especially those characterized by diarrhoea or respiratory symptoms, because classical medicine and epidemiology showed that epidemics were usually strikingly unselective, affecting old and young, rich and poor alike. As soon as Pasteur became convinced that living microorganisms were the primary and specific causes of fermentation and putrefaction, he applied the concept to the causation of infectious diseases. Thus he was able (Pasteur and Joubert, 1877) to differentiate true anthrax from the septicaemic diseases that had confused the observations of Brauell, Leplat and Jaillard. He further established that the hay bacillus (*B. subtilis*), an organism very similar to Davaine's rod, and like it capable of producing spores, did not cause anthrax when injected into animals.

Concerning specificity, there was at that time a profound difference between the French 'microbiology' and the German 'bacteriology'. In Germany, the germ theory of diseases was based on the specificity and permanence of the biological characteristics of microbial species. For Ferdinand Cohn and Robert Koch, the bacterium had a precise and unvarying morphology, a well-defined life cycle and it produced a well-defined disease. As expressed by Koch (1878): 'A distinct bacterial form corresponds to each disease, and this form always remains the same, however often the disease is transmitted from one animal to another. Further, when we succeed in reproducing the same disease *de novo* by the injection of the putrid substances, only the same bacterial form occurs which was before found to be specific for that disease'. For the German school, one parasite was linked to a particular disease in a particular host animal.

On the other hand, Pasteur believed that the phenomena of variation were of great importance in the epidemiology of infectious diseases. He suggested that epidemics might arise because of an increase in the virulence of a particular microorganism, and that sometimes a microorganism might acquire virulence for a new animal species. For that reason, the French school did not pay much attention to morphology and nomenclature and they often used descriptive names instead of taxonomic nomenclature. Accepting the concept that a given parasite can produce different forms of disease in different host species, the French school developed the idea of preparing vaccines by serial passages of a microbe through a host animal other than its 'natural' host. Pasteur went further, and developed the idea of biological control of vertebrate pests by the used of pathogenic microbes, hence his suggestion of using the chicken cholera bacillus to control rabbits in Australia (see p. 55).

Variability among Myxoma Virus Strains in the Americas

It is not surprising, in view of the probable long periods of evolution in the Americas, that strains of myxoma virus isolated from *Sylvilagus* rabbits in different parts of the Americas differ in virulence and in composition, as judged by virulence tests in rabbits, gel diffusion tests and restriction fragment length polymorphism. This was apparent from the early days, when Martin (1936) found that the strain sent to him by Aragão was less virulent than the Moses strain, also isolated in Brazil from *Sylvilagus brasiliensis* but passaged many times in European rabbits. On the other hand, the Lausanne strain, isolated in Brazil by Bouvier (1954) and subjected to very few passages in laboratory rabbits, was considerably more virulent than the Moses strain, a difference more clearly demonstrated when tests were carried out in genetically resistant rabbits. Among other isolates presumably derived from *Sylvilagus brasiliensis*, crude comparisons of antigenic structure revealed differences between strains isolated in Brazil, Colombia and Panama.

It is not surprising that myxoma virus strains derived from *Sylvilagus bachmani,* a species which is restricted to the west coast of the United States, differed markedly from Brazilian strains in gel diffusion patterns and in the symptomatology of the disease produced in laboratory rabbits. Although no direct comparisons have been made, it appears that strains derived from European rabbits infected in different parts of the range of *Sylvilagus bachmani* differ in virulence; the symptomatology of disease produced by strains isolated in Portland, Oregon (Patton and Holmes, 1977) differs from that produced by strains isolated in California (Fenner and Marshall, 1957; Marshall *et al.,* 1963). It would be of interest to carry out long-distance PCR (polymerase chain reaction) followed by restriction fragment length polymorphism assays, as recently developed for orthopoxviruses by Dr J.J. Esposito (personal communication, 1988) on a range of myxoma virus strains from different parts of the Americas.

Innate Resistance versus Acquired Immunity

Bacteria are by far the oldest and most numerous organisms on planet Earth. To protect themselves from destruction by microbes, systems of innate resistance have evolved in all eukaryotic organisms. In addition to the innate, non-specific systems found in invertebrates, vertebrates have developed a system of specific adaptive responses which are encapsulated in the terms 'immune response' and 'acquired immunity'. To distinguish this from genetic and other types of innate resistance, we shall restrict the term 'immunity' to this class of adaptive responses, using 'resistance' to encompass other types of protection.

Innate resistance

Innate resistance is a complex subject that cannot be fully developed here. It has two components, an innate lack of susceptibility to infection, and in species that can be infected with a virus that causes diseases in most animals of that species, genetic resistance resulting from Darwinian natural selection. Some species of animal are totally resistant to particular infections because they lack the receptors to which particular viruses attach. For example, mice are resistant to infection with polioviruses because they lack the receptors for the virus, but they can be infected with poliovirus RNA and replication of the virus then occurs. At another level, a certain level of innate resistance is provided by various white cells that occur in the bloodstream. Polymorphonuclear and eosinophil leucocytes and 'non-killer' lymphocytes, and other factors like cytokines and interferons, may act to control replication early in the course of infection; these are non-specific responses to infection.

Most animals other than certain members of the family *Leporidae* have an innate resistance to both myxoma virus and the rabbit caliciviruses, hence these viruses are said to be specific to leporids. The other component of innate resistance is the genetic resistance which develops as a consequence of natural selection in animals that are susceptible to infection. Myxoma virus produces infections characterized by small fibromas in the skin in the two species of *Sylvilagus* rabbits which are its natural hosts, but a severe, generalized and usually lethal disease in laboratory and wild European rabbits (*Oryctolagus cuniculus*), whose forebears have not previously been exposed to the virus. After several generations of exposure to strains of virus that did not kill all unselected rabbits, *Oryctolagus cuniculus* developed a high degree of genetic resistance to the effects of infection, even with strains that were highly virulent for unselected (laboratory) rabbits, the result of stringent selection for genotypes that were more resistant to the pathogenic effects of the virus. Natural selection for more resistant genotypes was undoubtedly responsible for most (over 75%) of the resistance that has been observed both in breeding experiments and in the field. However, some 20–25% of the innate resistance found in the progeny of rabbits that have been exposed to myxomatosis is thought to be due to a 'sire effect' (see below).

Acquired immunity

Rabbits which survive infection with myxoma virus, or after infection with the related Shope's fibroma virus, develop acquired immunity to reinfection. The existence of prolonged immunity to reinfection in animals that have recovered from myxomatosis (Fenner *et al.*, 1953) and of short-lived passive immunity among the progeny of immune mothers (Fenner and Marshall, 1954) are well-documented and accord with what is seen after other generalized infections. The time course of immunity to infection of mice with influenza virus, which is representative of acute viral infections in general, is illustrated in Fig. 13.2. After intranasal inoculation, virus replicates in the lung (curve A), reaching maximum titre by day 6 and then decreasing so that no virus remains after day 12. Curve B represents the activity of cytotoxic T lymphocytes, which peaks a few days after the viral titre peaks and then disappears some days after virus can no longer be detected. Antibody-secreting (B) cells (curve C) begin to appear at about the time cytotoxic T lymphocyte activity disappears and serum antibody appears a few days later and slowly rises and declines in titre. At about 3 weeks and 3 months T (cytotoxic lymphocyte) memory cells and B (antibody-producing) memory cells reach their maximum levels; they persist for the life of the mouse. The initial immune responses control (or fail to control) the first infection; the memory cells lead to the secondary responses of

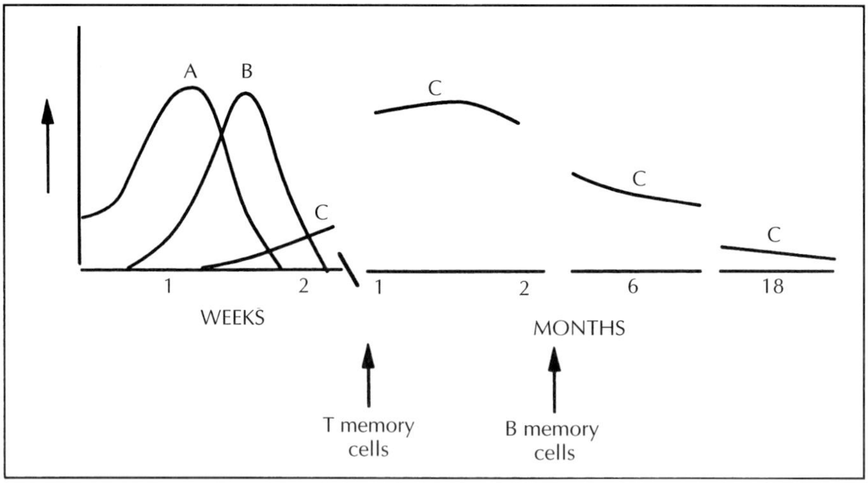

Fig. 13.2. The time sequence of the infectious process and subsequent immune responses in a mouse infected with influenza virus. The ordinate represents the rise and fall of infectivity and effector cells and the abscissa the time after the initiation of infection. Curve A represents the growth and disappearance of virus; Curve B the appearance and disappearance of effector T lymphocytes and curve C the appearance and continuing presence of antibody-secreting cells. The arrows indicate the time of presence of maximum numbers of memory T and B cells. From Ada (1990), with permission.

both cytotoxic T lymphocytes and antibodies that rapidly control a second infection with the same virus.

This adaptive immunity is specific, in that it is effective only against the infecting virus or closely related viruses, and the response is endowed with memory, in that a second exposure to the immunizing virus produces an accelerated response. Thus recovery from prior infection with myxoma virus gives life-long protection from severe disease due to that virus. It is not inheritable, in that after passive immunity (due to antibodies passed across the placenta from mother to fetus) has decayed, progeny animals are fully susceptible. Acquired immunity is the basis of vaccination as a method of protection against infectious diseases. Vaccination with a related virus, such as fibroma virus, gives less complete and less durable immunity than recovery from an acute infection or vaccination with an attenuated strain of myxoma virus.

The puzzling phenomenon of the 'sire effect'

There is no doubt about the reality of the enhanced genetic resistance to myxomatosis that developed within a few years of its introduction into wild rabbit populations and in Sobey's breeding experiments. However, three sets of observations in Australia described in Chapter 8 (p. 198) (Sobey and Conolly, 1986; Williams and Moore, 1991; Parer *et al.,* 1995), suggest that besides genetic resistance and active or passive immunity, there may be another source of enhanced resistance to myxomatosis, namely what has been called the 'sire effect' or 'paternal immunity'. These terms are used to encapsulate the observation that the progeny of recovered male rabbits which are mated to either uninfected or recovered does within seven months of recovering from myxomatosis are significantly more resistant than expected to challenge infection with myxoma virus. There is no such effect when uninfected males are mated to recovered does. The survival advantage for offspring whose sire had recovered from a recent infection with myxoma virus was 20–25%; the other 75–80% of the enhanced resistance, found also in the offspring of recovered does mated to uninfected males, was genetic, i.e. due to neo-Darwinian natural selection.

Noting that the testis is always severely affected in myxomatosis and that in contrast to the production of ova, the production of sperm continues throughout life, Parer *et al.* (1995) speculated that 'the factor responsible for the sire effect, perhaps viral DNA, becomes incorporated into the sperm during their formation'. All observations suggesting the existence of a sire effect so far described, one on domestic rabbits and two on wild rabbits, have been made during the analysis of observations collected for other purposes. It would be desirable to test it in a deliberately planned experiment, incorporating examination of sperm and semen for virus and viral DNA as well as testing progeny for resistance. This would be expensive, in that to have suitable test animals substantial numbers of rabbits would have to be maintained for long periods under conditions in which some lines would be regularly infected with myxoma virus and others maintained free of infection, and the tests themselves would need to incorporate sophisticated laboratory studies as well as using large numbers of animals. However, the phenomenon is of great scientific interest at a time when the common occurrence of reverse transcription in eukaryotic cells has reopened the possibility of Lamarkian inheritance (Steele *et al.*, 1998).

Immunosuppression by Myxoma Virus

It has long been known that infection with viruses that replicate in cells of the immune system, such as macrophages or lymphocytes, may suppress the capacity of the immune system to respond to other microbial infections. Such immunosuppression must be distinguished from immuno evasion (discussed in Chapter 14), which is a mechanism by which an infecting virus may evade control by the host's immune system, but does not necessarily interfere with the capacity of the immune system to deal with other pathogens. Infection with measles virus, which multiplies in monocytes and lymphocytes, can cause reactivation of tuberculosis, and infection of adult mice with lymphocytic choriomeningitis virus also produces a transient immunosuppression. Infection with human immunodeficiency virus, which multiplies in macrophages and lymphocytes, produces a profound and progressive suppression of the cellular immune response, and in the late stages of infection, of the antibody response.

Strayer and his colleagues (Strayer, 1992; Strayer *et al.*, 1983) have investigated the immunosuppression produced by a myxoma-fibroma recombinant virus, 90% of whose genome came from the myxoma virus parent. Infected rabbits suffered from a generalized infection very like myxomatosis, but usually developed 'snuffles' about 10 days after infection and died on the 14th day. The pathology in rabbits infected with the recombinant resembled that found in myxoma-infected rabbits more than the localized fibroma found in infections with fibroma virus. However, there was much more virus in the skin overlying lesions, in the conjunctiva and nasal epithelium, and in the liver, spleen, kidney, lung and lymph nodes in rabbits infected with myxoma virus than in those infected with the myxoma–fibroma recombinant. Extrapolating from these findings, it can be assumed that the immunological dysfunction that Strayer investigated in infections with the recombinant virus would also occur in rabbits infected with myxoma virus. The key findings were an initial immune suppression, probably due to growth of the virus in lymphocytes and the production of immunosuppressive cytokines by T lymphocytes, and recovery of immunological function in the late stages of the disease. However, the depressed immune function allowed the normally commensal *Pasteurella multocida* to multiply in the conjunctiva, nasal passages and lungs; Strayer considered that this was often the cause of death.

In early investigations of the immune response in rabbits infected with a highly virulent and a highly attenuated strain of myxoma virus, Fenner and Woodroofe (1954) found that neutralizing antibody appeared on the tenth day and continued to rise in rabbits infected with the attenuated strain. However, rabbits infected with the virulent strain died on the eleventh day, without developing snuffles. Rabbits infected with Californian strains died even earlier (7–9 days after infection), with minimal signs of disease (Fenner and Marshall, 1957; Marshall *et al.,* 1963). However, in later studies, when most strains collected from the field were less virulent, infected laboratory rabbits almost always developed snuffles. Presumably because the causative commensal bacteria were much less common among wild rabbits, snuffles was rarely seen in the field, in either Australia (J.W. Edmonds and B.D. Cooke, personal communication) or Britain (J. Ross, personal communication), even when rabbits lived longer because of less virulent viruses or greater rabbit resistance.

Effects of Age of Host on Severity of Disease

It is a common observation in viral infections that immature animals and old animals are more likely to suffer severe disease than are mature animals. No studies have been made of the severity of myxomatosis in very old rabbits, and in the wild the matter is of minor importance. Young rabbits (9–28 days old), if not protected by maternal antibody, invariably die after infection with virulent strains of myxoma virus, usually within 5–6 days (Fenner and Marshall, 1954), whereas the mean survival time of mature rabbits was 11 days. In older rabbits, Sobey *et al.* (1970) found that the mean survival time in wild and domestic rabbits with some degree of genetic resistance to myxomatosis fell between 10 and 30 weeks of age and was unchanged for the next 20 weeks.

The behaviour of rabbit haemorrhagic disease in young rabbits is in striking contrast to this, since young rabbits are more resistant than are mature rabbits. Both European and Australian observers found that baby rabbits (<10 days old) can be infected but that virus failed to replicate extensively in the internal organs, although enough virus was excreted to infect sentinel rabbits. Some juvenile rabbits survived infection, and rabbits over 10 weeks of age were fully susceptible. The reason for this unusual response is unknown. Rabbits that recover after having been infected as kittens are immune to reinfection.

Effects of Temperature on Severity of Disease

Apart from the epidemiological influences of environmental temperature (see p. 152), ambient temperatures sometimes affect the physiological response to infectious diseases. Thus it was observed many years ago that if rabbits infected with myxoma virus were maintained at a high temperature they suffered a much less severe disease than controls maintained at ambient temperature (Parker and Thompson, 1942). These results were confirmed and extended by Marshall (1959), who also found that exposure to lower temperatures than normal exacerbated the severity of the disease (see p. 108). The clinical observations were paralleled by high level viraemia and no antibodies in 'cold-room' rabbits and the reverse in 'hot-room' rabbits. Sobey *et al.* (1968) confirmed the ameliorating effects of high temperatures, and in breeding for genetic resistance he exposed rabbits inoculated with virulent virus to a high temperature for 24 hours 3–4 days after inoculation, so as to ensure enough survivors for the next round of breeding.

Marshall had planned his experiments so as to mimic conditions that might occur in the field, rabbits being held at high or low temperatures for 18 hours and at ambient temperature for the other eight hours each day. Field observations at Snowy Plains, a subalpine region of south-eastern Australia, suggested that environmental exposure of wild rabbits to unusually high or low temperatures had similar effects in ameliorating or enhancing the severity of the disease, causing lower or higher mortalities than expected (Dunsmore *et al.*, 1971; Dunsmore and Price, 1972).

Cooke (1998) found that holding rabbits at temperatures occurring in rabbit burrows in arid rangelands in winter and summer (13°C and 27°C) did not affect their response to infection with rabbit haemorrhagic disease virus.

Molecular Aspects of Virulence

For agents used for killing pests, virulence is equated with lethality. This is the definition of virulence used throughout this book for both myxoma virus and rabbit caliciviruses. These are two very different viruses, the one a very large virus with a large double-stranded DNA genome and the other a small virus with a small single-stranded RNA genome. Few molecular biological studies relevant to viral virulence have been carried out yet on either virus; the following account summarizes results so far and looks at future proopects.

Myxoma virus

Virulent strains of myxoma virus, such as those isolated from domestic rabbits that have been infected from *Sylvilagus brasiliensis*, kill the vast majority of rabbits infected with a small dose of virus within 12 days. Some Californian strains are even more rapidly lethal, killing rabbits within 8–9 days. A large number of genes contribute to their virulence. The only genes that have been studied, by McFadden and his colleagues, are those which show homologies with several different immunomodulatory genes that occur in vertebrates. Most of these genes are concerned with cytokine production or receptors and are thought to contribute to virulence by subverting the host's immune system (see p. 309). However, these are by no means the only genes involved in virulence, as demonstrated by the fact that all of these genes that were tested were intact in four attenuated strains recovered from Australian wild rabbits (P.J. Kerr, personal communication, 1996).

Rabbit haemorrhagic disease virus

Rabbit haemorrhagic disease virus was the first calicivirus for which the genome was completely sequenced; it is a single-stranded positive sense RNA molecule of 7437 nucleotides. With the discovery of the avirulent precursor rabbit calicivirus (see p. 238) the opportunity exists for unravelling the genetics of virulence of rabbit haemorrhagic disease virus.

Is Mean Survival Time a Good Surrogate for Lethality?

There are many definitions of viral virulence, but for a biological control agent which exerts its effect by killing its host the most useful definition is one which equates virulence with lethality. The ideal way to test viruses for their lethality is to infect substantial numbers of the host animal with small doses of the virus by a route which mimics natural infection, and determine what percentage of animals die. The problem with myxomatosis was that laboratory rabbits, which had to be used because they were the only animals which were available and had not been subjected to natural selection for resistance, were too expensive to use this test for more than a few key strains. In the early days of the work on the evolution of myxomatosis it was thought necessary to test large numbers of strains of virus, gathered from widely separated areas, so some alternative method was sought. Investigations showed that for viruses that killed more than 50% of a substantial number of rabbits (20 or more), the mean survival time in groups of six rabbits provided a surrogate measure for virulence. Although it was realized that after a few years the strains of virus obtained from rabbits in different parts of the continent were likely to have undergone different mutations, the results of such tests would be unmanageable unless viruses of similar lethality (as judged by survival times) were grouped together, hence the notion of 'virulence grades'.

Many years later, going over the results of large numbers of tests carried out by Edmonds and Shepherd on groups of six rabbits, Parer (1995) queried the reliability of using survival times in this way. He found that in many such small groups all rabbits had died, even though they had been placed in different virulence grades on the basis of their mean survival times. With the recognition that the changes in virulence of myxoma virus have occurred all over Australia and each of the countries of Europe, experimenters should abandon the testing of large numbers of strains and instead carry out annual or biennial tests with a few strains selected from different locations, each in 20–30 rabbits. To ensure comparability with earlier results, genetically unselected laboratory rabbits should be used; however, the use of genetically resistant wild rabbits would give a better impression of what was happening in the wild. One way of obtaining large numbers of reasonably uniformly genetically resistant rabbits would be to use the offspring of laboratory females impregnated with stored semen from resistant wild rabbits. Care would need to be taken to use semen from uninfected wild rabbits so as to avoid the sire effect described earlier.

The Interplay between Virulence and Transmissibility

Infection of leporids with myxoma virus provides some fascinating contrasts in virulence (lethality) and transmissibility. In its natural hosts in the Americas, myxoma virus infection is a benign disease, producing a local lesion in the skin (a fibroma) at the site of infection by insect bite but not making the host sick. The infection can persist in such populations because there is a high concentration of virus in the skin over the lesion, and because lesions in immature animals may persist as infectious lesions for months. When transmitted from *Sylvilagus* to European rabbits the virus causes a generalized disease that is almost always lethal. In the Americas this causes problems for pet owners and those who farm European rabbits, who have to vaccinate their animals. However, the occurrence of enzootic infection of *Sylvilagus* rabbits in California and many countries in South America with strains of myxoma virus that are highly lethal for *Oryctolagus cuniculus* has prevented this species from becoming established, and an agricultural pest, in these countries.

In Australia, where myxomatosis has become an enzootic disease in the wild European rabbit population, virus and host have coevolved (see Chapter 14). At this stage of this coevolution, almost fifty years after its introduction, myxomatosis is still a generalized disease; it has not yet become a benign infection of the kind seen in its natural hosts in the Americas. Initially it was thought that the virus was so virulent that it would kill out rabbits wherever it was introduced and die out in the process. However, it was clear within a few years of its introduction that mutations towards slightly reduced virulence had occurred in many places. Because transmission was primarily due to mosquitoes, which became much less common during the winter, cases of infection by these slightly attenuated strains, which lived for much longer, or survived, provided a mechanism for overwintering, thus providing a source for repeated spread of the disease when mosquitoes once again became common. Since experimental studies many years before had shown that it was relatively easy to produce very attenuated strains of myxoma virus ('neuromyxoma' of Hurst, 1937), it is likely that such variants also arose among wild rabbits. However, the amount of virus in their lesions was such that they would not have been infectious for mosquitoes and therefore would have died out.

The selective advantage of the slightly attenuated strains was such that they became the dominant kinds of virus. Since some 10% of rabbits infected with these strains survived, a mechanism then existed for the rapid selection of resistance in the rabbit population. Over the years, rabbits became so resistant that the slightly attenuated strains produced lesions like those of neuromyxoma, and there was a back-selection for strains that were highly virulent, as judged by tests in genetically unselected laboratory rabbits. This process is discussed at length in the next chapter.

Comparison of Biological and Mechanical Transmission by Arthropods

Many animal viruses, belonging to several families, are transmitted by arthropods. In most cases the virus is ingested by the arthropod during a blood meal on a viraemic vertebrate, multiplies in the cells of its gut, moves to the salivary glands and is transmitted when saliva is injected into the vertebrate host at a subsequent blood feed. This is called 'biological' or 'propagative' transmission and the viruses thus transmitted are called 'arboviruses' (*ar*thropod-*bo*rne *viruses*). The period between ingestion of the virus and the capacity to infect during a subsequent feed is the 'extrinsic incubation period', and is followed by the 'intrinsic incubation period', which is the period between the infective bite and the appearance of symptoms. The length of the extrinsic incubation period, which is the time required for the virus to replicate in the vector and reach the salivary glands, is affected by the ambient temperature and is commonly as long as a week.

A few viruses are transmitted in another way: the arthropod contaminates its mouthparts with the virus in question during its probing through a skin lesion caused by the virus, or rarely during a blood meal or when feeding on an organ containing high concentrations of virus. Table 13.1 sets out the essential differences between these modes of transmission. One important feature of arthropod-borne viruses of both types is that the virus can be readily carried between different host species and for long distances by flying arthropods – for very long distances if these insects are carried in air currents.

By coincidence, the two very different viruses with which this book is concerned, myxoma virus and rabbit haemorrhagic disease virus, are usually mechanically transmitted by arthropods, although they can be transmitted by close contact (usually via the respiratory or intestinal tract). In commercial rabbitries in Europe strains of myxoma virus evolved which were passed by close contact (so-called

Table 13.1. Comparison of mechanical and propagative transmission by arthropods.

Feature	Mechanical transmission	Propagative transmission
Replication in vector	No	Yes
Usual source of virus	Skin lesions ? Exposed organs (RHDV) Blood (eq. inf. anaem. virus)	Blood
Vector specificity	Nil, except for feeding habits	High
Extrinsic incubation period	No	Yes
Interrupted feeding	Yes, highly effective	No
Viruses transmitted	Rabbit papilloma virus Leporipoxviruses Fowlpox virus Rabbit haemorrhagic disease virus Equine infectious anaemia virus (a lentivirus)	Alphaviruses Bunyaviruses (most) Flaviviruses Rhabdoviruses (some) Orbiviruses

amyxomatous or respiratory myxomatosis, see p. 104 and p. 216). Myxoma virus is ordinarily transmitted by mosquitoes or other biting arthropods when they probe through skin lesions that contain high concentrations of virus in the superficial layers. In its natural hosts, *Sylvilagus brasiliensis* and *S. bachmani,* these lesions are usually single, localized tumours in the skin which are produced at the site of the infecting probe or bite in an otherwise healthy animal. In the European rabbit also there is a local lesion, but the disease then becomes generalized and infectious skin lesions develop in many parts of the body, especially on the ears and eyelids. The related leporipoxvirus, Shope's fibroma virus, produces a benign tumour at the site of the bite in the skin of its natural host, *Sylvilagus floridanus,* and in European rabbits as well. These differ in that there is abundant virus in the superficial layers of the lesions in *S. floridanus,* but very little in those in European rabbits. For this reason Shope's fibroma cannot become established as an enzootic disease in European rabbits in the way myxomatosis did in Australia and Europe.

In commercial rabbitries in Europe and China, rabbit haemorrhagic disease virus is usually transmitted by contact. When it was experimentally released in Australia, it initially spread poorly between rabbits in the same enclosed areas, but was then carried outside the trial areas and for hundreds of kilometres by insects, probably flies, which were carried for long distances by air currents. Subsequent investigations showed that in areas where the disease was occurring many species of flies were contaminated, and that both fly faeces and 'flyspots' contained virus. The exact mode of contamination of the flies is unknown, but feeding on the conjunctivae of sick rabbits or the internal organs of dead animals appear likely mechanisms.

Overwintering of Myxoma Virus

When myxomatosis spread through the Murray–Darling Basin in the summer of 1950–51, spread by mosquitoes, the question in everyone's mind was: will it survive the winter? In fact it did survive and spread again, aided by widespread inoculation campaigns. At Lake Urana, where careful field observations were conducted after the first outbreak there in the summer of 1951, sick rabbits were observed in the very small population of surviving rabbits each month between successive outbreaks except for June, August and September (Myers *et al.*, 1954). This fact, and the outbreak of an epidemic caused by a virus substantially less virulent than that used the previous summer as soon as vector mosquitoes became common, suggests that the virus could survive through the winter in temperate Australia by sustained low level transmission. Judging from the 1950–51 outbreak, survival in a few areas would be sufficient to allow further spread of the virus by mosquitoes during the spring and summer.

Since mosquitoes had been the important vectors in Australia and France, attention was initially focused on mosquitoes when myxomatosis broke out in Britain in 1953. *Anopheles labranchiae atroparvus* was incriminated as a vector, and Andrewes *et al.* (1956) showed that some mosquitoes infected by feeding on a rabbit with acute myxomatosis and maintained in a semi-hibernating state at winter temperatures were still infective 149 days and 220 days later.

Subsequently it was shown that the European rabbit flea (*Spilopsyllus cuniculi*) was a much more important vector than mosquitoes in Britain and that both mosquitoes and fleas were important vectors in France (see Chapters 9 and 10). Fleas are present on rabbits throughout the year, and overwintering of myxomatosis is not a problem in Europe except in hotter parts of Spain and France. Myxoma virus was found on quiescent rabbit fleas recovered from soil scrapings from deep burrows that had been abandoned by rabbits ten weeks earlier (Brown *et al.*, 1956; Joubert *et al.*, 1969). Chapple and Lewis (1965) showed that fleas that had fed through lesions of a rabbit with myxomatosis and were then buried in the ground in glass tubes were infective for as long as 112 days. Thus in Europe overwintering could occur either on fleas or mosquitoes. After rabbit fleas were introduced into Australia, the epidemiology in some places changed from summer outbreaks to prolonged low-level transmission in winter and early spring, sometimes followed by short mosquito-borne epizootics in the summer. Where fleas are present in cool temperate parts of Australia they probably play an important part in the overwintering of myxoma virus.

Does myxoma virus cause latent infection? The foregoing observations serve to explain how myxomatosis could persist through the winter, both where rabbit fleas were present, and in their absence if a few mosquitoes could maintain low level transmission. However, observations during outbreaks of myxomatosis in an isolated mountain area and among rabbits at the CSIRO Wildlife Division in Canberra (see p. 200) aroused suspicions in the minds of the investigators that latent infection with myxoma virus might sometimes occur, and that it might be reactivated in such a way as to produce infectious lesions. Experiments by Williams *et al.* (1972) in which recovered rabbits were subjected to heat and hormonal stress provided support for this hypothesis. One of twenty of these rabbits showed eye and genital lesions and material from the eye discharge produced myxomatosis when injected into susceptible animals. These observations and experiments have profound implications for the epidemiology of myxomatosis. They do not accord with what is known about poxvirus infections in general. Although long-lasting infectious lesions may occur in *Sylvilagus* rabbits infected with fibroma virus (Kilham and Dalmat, 1955; Dalmat and Stanton, 1959), and in the unrelated molluscum contagiosum virus (Porter, 1994), latency and reactivation of the kind found in herpesvirus infections have not previously been described for poxvirus infections. The evidence assembled is persuasive, but should be confirmed by an independent study.

Eradication or Control

The usual response of national veterinary authorities to the introduction of exotic diseases of livestock is to eradicate the disease by prompt quarantine and slaughter. This approach has had many successes when diseases such as foot-and-mouth disease, rinderpest and avian influenza have broken out in the United States of America or Australia. It was attempted, but was unsuccessful, when myxomatosis was first recognized in the United Kingdom, and rabbit haemorrhagic disease was successfully eradicated from Mexico shortly after it broke out there. The principal difference of myxomatosis from all the other diseases mentioned is that it was a disease of wild rabbits; the others, including rabbit haemorrhagic disease in Mexico, were diseases of domestic animals.

In recent years the concept of eradication has also been applied to imported animals which have escaped from farms to become a potential pest in the wild. It has been successful with coypus and muskrats in the United Kingdom, where the areas over which the animals had spread was limited. Eradication of rabbits by myxomatosis has also been successful on a few small islands. Except in rare instances, eradication of pest animals is a fantasy.

With both myxomatosis and rabbit haemorrhagic disease, control of the disease in wild rabbits proved impossible. It was not sought in Australia, where the rabbit was a major pest and the virus was deliberately introduced. After the initial attempt at eradication, control of myxomatosis was not pursued in the United Kingdom, where the rabbit was also regarded as a pest, although deliberate spread of the disease was made illegal. However, in parts of France and Spain where wild rabbits were greatly valued for hunting, continued efforts were made to control myxomatosis by vaccination of rabbits and by the importation of genetically resistant animals or even of *Sylvilagus* species. In many countries of Europe domestic rabbits are important as farmed animals and as pets, and in these circumstances vaccination is widely used, initially with fibroma virus and later with an attenuated strain of myxoma virus. Vaccination with a killed rabbit haemorrhagic disease virus is widely used to protect domestic rabbits, and recently attenuated myxoma virus/RHDV-protein recombinants have been introduced to protect domestic rabbits against both diseases. Especially valuable animals, such as laboratory rabbits and breeding stud, can be protected by enclosing their cages within fly-proof wire mesh.

Effectiveness of Biological Control of Vertebrate Pests

Biological control of some plant pests, such as control of prickly pear by *Cactoblastis,* has been dramatically and permanently successful. In general, when methods for biological control of insect pests can be discovered, they prove to be of longer lasting efficacy than chemical control of such pests. However, because of the overwhelming importance of host specificity when biological methods of control are sought for vertebrate pests, there are very few examples of their use. Indeed, apart from experiences on a few small islands, the rabbit is the only pest animal on which they have been successful, and for this animal, surprisingly, two different viruses have been sufficiently lethal to produce very effective control in some places for some decades. However, as Matthams (1921) observed many years ago, in commenting on the unsuccessful outcome of the work of the Intercolonial Commission, there is:

> no evidence to warrant the belief that any known disease can be so employed as to exterminate rabbits. Probably many diseases will be found useful in reducing the rabbit plague ... but even when fuller information concerning these diseases shall have been obtained it will still be necessary to continue the methods of suppressing the pest, which are now generally adopted, subject to such improvements in detail as may from time to time be discovered.

Ratcliffe, who was responsible for introducing myxomatosis into Australia in 1950, had a keen appreciation of this fact, but the initial extraordinary virulence and effective spread of myxomatosis led many landholders to ignore the advice repeatedly given by him and his colleagues. His reports and early correspondence, examples of which are provided in Chapter 6, show that he was repeatedly urging farmers to take advantage of the great reductions in rabbit numbers caused by myxomatosis to clean up their properties by destruction of burrows and fencing of properties, but it was difficult to convince farmers that myxomatosis was not ‘the answer’. By the time Williams *et al.* (1995) were producing their comprehensive book, *Control of Vertebrate Pests: Rabbits,* those responsible for rabbit control accepted myxomatosis as a feature of the life of rabbits in Australia. Although it might well kill some 50% of rabbits in some areas, this was only a partial brake on their enormous fecundity. Like all other vertebrate pests, control could only be achieved by a sustained programme of integrated pest management.

The lesson was even more obvious for rabbit haemorrhagic disease, for experience had demonstrated that immature rabbits were resistant to infection and the overall death rate, before any genetic resistance had developed, was much lower than that during the first years of myxomatosis. The booklets *Rabbit Control and Rabbit Calicivirus Disease* (Coman, 1997), of which 100,000 copies were produced for distribution to land managers in Australia, ends with the admonition ‘Destroying rabbits without destroying their homes or ‘harbour’ often gives only short-term control. In many cases, it is little better than a harvesting operation’. This is followed by a listing of what land managers should do to capitalize on the reduction of rabbits that might follow outbreaks of rabbit haemorrhagic disease (Table 13.2).

Table 13.2. General options for rabbit control in various classes of country, to be used following outbreaks of myxomatosis or rabbit haemorrhagic disease[a].

Situation	Poison	Ripping	Fumigation	Vegetation control[b]	Other
Rangelands or other arable country accessible to machinery	Y	Y	Y	NA	Shooting[c]
Steep or rocky terrain	Y	NA[d]	Y	NA	Shooting
Dense scrub or bush margin	Y	NA	NA	Y	
Erosion-prone country, creek banks	Y	NA	Y	Y	Shooting
Semi-urban areas or around farmhouses	?[e]	NA	Y	Y	

(Y = yes; NA = not applicable.)
[a]After Coman (1997), with permission.
[b]Vegetation control is recommended only in the case of pest plants such as gorse, blackberry or boxthorn.
[c]Shooting is recommended only as a 'mop-up' technique when rabbit numbers have already been reduced to low numbers.
[d]Use of explosives by trained operators is an option for destroying warrens in some situations.
[e]It may be possible to use pindone as a poison in these situations.

References

Ada, G.L. (1990) The immune response to antigens: the immunological principles of vaccination. *Lancet* 335, 523–526.

Andrewes, C.H., Muirhead-Thompson, R.C. and Stevenson, J.P. (1956) Laboratory studies of *Anopheles atroparvus* in relation to myxomatosis. *Journal of Hygiene* 54, 478–486.

Barranti-Brodano, G., Martini, F., De Mattei, M., Lazzarin, L. and Tognon, M. (1998) BK and JC human polyomaviruses and simian SV40: natural history of infection in humans, experimental oncogenicity, and association with human tumors. *Advances in Virus Research* 50, 69–99.

Bouvier, G. (1954) Quelque remarques sur la myxomatose. *Bulletin de l'Office International des Epizooties* 46, 76–77.

Brown, P.W., Allan, R.M. and Shanks, P.L. (1956) Rabbits and myxomatosis in the N.E. of Scotland. *Scottish Agriculture* 35, 204–207.

Chapple, P.L. and Lewis, N.D. (1965) Myxomatosis and the rabbit flea. *Nature* 207, 388–389.

Coman, B. (1997) *Rabbit Control and Rabbit Calicivirus Disease. A Field Handbook for Land Managers in Australia.* Meat Research Corporation, Sydney, 20 pp.

Cooke, B.D. (1998) Recent research on RHD: Insects as vectors. Unpublished data, 4 pp. Quoted with the author's permission.

Dalmat, H.T. and Stanton, M.F. (1959) A comparative study of the Shope fibroma in rabbits in relation to transmissibility by mosquitoes. *Journal of the National Cancer Institute* 22, 595–615.

Dunsmore, J.D. and Price, W.J. (1972) A non-winter epizootic of myxomatosis in subalpine south-eastern Australia. *Australian Journal of Zoology* 20, 405–409.

Dunsmore, J.D., Williams, R.T. and Price, W.J. (1971) Winter epizootic of myxomatosis in subalpine south-eastern Australia. *Australian Journal of Zoology* 19, 275–286.

Evans, A.S. (1973) Pettenkofer revisited. The life and contributions of Max von Pettenkofer (1818–1901). *The Yale Journal of Biology and Medicine* 46, 161–176.

Evans, A.S. (1976) Causation and disease. The Henle–Koch postulates revisited. *The Yale Journal of Biology and Medicine* 49, 175–195.

Fantini, B. (ed.) (1993) Emerging infectious diseases: historical perspectives. *History and Philosophy of Life Sciences* 15, 281–487.

Fenner, F. and Marshall, I.D. (1954) Passive immunity in myxomatosis of the European rabbit (*Oryctolagus cuniculus*): the protection conferred on kittens born by immune does. *Journal of Hygiene* 52, 321–336.

Fenner, F. and Marshall, I. D. (1957) A comparison of the virulence for European rabbits (*Oryctolagus cuniculus*) of strains of myxoma virus recovered in the field in Australia, Europe and America. *Journal of Hygiene* 55, 149–191.

Fenner, F. and Woodroofe, G.M. (1954) Protection of laboratory rabbits against myxomatosis by vaccination with fibroma virus. *Australian Journal of Experimental Biology and Medical Science* 32, 653–668.

Fenner, F., Marshall, I.D. and Woodroofe, G.M. (1953) Studies in the epidemiology of infectious myxomatosis of rabbits, I. Recovery of Australian wild rabbits (*Oryctolagus cuniculus*) from myxomatosis under field conditions. *Journal of Hygiene* 51, 225–244.

Fenner, F., Wittek, R. and Dumbell, W.R. (1989) *The Orthopoxviruses.* Academic Press, San Diego, 432 pp.

Feore, S.M., Bennett, M., Chantrey, J., Jones, T., Baxby, D. and Begon, M. (1997) The effect of cowpox virus on fecundity in bank voles and wood mice. *Proceedings of the Royal Society, Series B* 264, 1457–1461.

Henle, J. (1840) Von den Miasmen und Contagien und von den miasmatisch-contagiosen Krankheiten. In: *Pathologische Untersuchungen*, Berlin, pp. 1–82. (Translated into English and with an introduction by G. Rosen, Baltimore, Johns Hopkins University Press, 1938).

Hurst, E.W. (1937) Myxoma and the Shope fibroma. II. The effect of intracerebral passage on the myxoma virus. *British Journal of Experimental Pathology* 18, 15–22.

Joubert, L., Chippaux, A., Mouchet, J. and Oudar, J. (1969) Entretien hiverno-vernal du virus myxomateux dans les terriers. Myxomatose d'inoculation par la puce du lapin et myxomatose du fouissement. *Bulletin de l'Académie vétérinaire de France* 42, 93–101.

Kapikian, A.Z., Estes, M.K. and Chanock, R.M. (1996) Norwalk group of viruses. In: Fields, B.N., Knipe, D.M., Howley, P.M., Chanock, R.M., Melnick, J.L., Monath, T.P., Roizman, B. and Straus, S.E. (eds) *Fields Virology*, 3rd edn. Lippincott-Raven, Philadelphia, pp. 783–810.

Kilham, L. and Dalmat, H.T. (1955) Host–virus–mosquito relations of Shope fibromas in cottontail rabbits. *American Journal of Hygiene* 61, 45–54.

Koch, R. (1878) *Untersuchungen über die Aetiologie der Wundinfectionkrankheiten.* F.C.W Vogel, Leipzig.

Koch, R. (1884) Die Aetiologie der Tuberkulose. *Mittheilungen aus den Kaiserliche Gesundheitsamte* 2, 1–88. (Translated into English by S. Boyd. In: Cheyne, W.W. (ed.) (1886) *Recent Essays by Various Authors on Bacteria in Relation to Disease.* London, pp. 65–201.

Koch, R. (1890) Ueber bakteriologische Forschung. *Deutsche medizinische Wochenschrifte* 16, 756–757.

Lederberg, J., Shope, R.E. and Oaks, S.C. Jr (eds) (1992) *Emerging Infections. Microbial Threats to Health in the United States.* National Academy Press, Washington, 294 pp.

Loeffler, F. (1884) Untersuchungen über die Bedeutung der Mikrorganismen für die Entstehung der Diphterie beim Menschen, bei der Taube und beim Kalbe. *Mittheilungen aus den Kaiserliche Gesundheitsamte* 2, 421.

Marshall, I.D. (1959) The influence of ambient temperature on the course of myxomatosis in rabbits. *Journal of Hygiene* 57, 484–497.

Marshall, I.D., Regnery, D.C. and Grodhaus, G. (1963) Studies in the epidemiology of myxomatosis in California. I. Observations on two outbreaks of myxomatosis in coastal California and the recovery of myxoma virus from a brush rabbit (*Sylvilagus bachmani*). *American Journal of Hygiene* 77, 195–204.

Martin, C.J. (1936) Observations on Myxomatosis cuniculi (Sanarelli) made with a view to the use of the virus in the control of rabbit plagues. *Bulletin of the Council for Scientific and Industrial Research, Australia*, No. 96, 28 pp.

Matthams, J. (1921) *The Rabbit Pest in Australia.* The Specialty Press, Melbourne, 264 pp.

Morse, S.S. (ed.) (1993) *Emerging Viruses.* Oxford University Press, New York, 317 pp.

Murphy, B.R. and Webster, R.G. (1996) Orthomyxoviruses. In: Fields, B.N., Knipe, D.M., Howley, P.M., Chanock, R.M., Melnick, J.L., Monath, T.P., Roizman, B. and Straus, S.E. (eds) *Fields Virology*, 3rd edn. Lippincott-Raven, Philadelphia, pp. 1397–1445.

Myers, K., Marshall, I.D. and Fenner, F. (1954) Studies in the epidemiology of myxomatosis of rabbits. III. Observations on two successive epizootics in Australian wild rabbits on the riverine plain of south-eastern Australia 1951–1953. *Journal of Hygiene* 52, 337–360.

Noble, J, Jr. and Rich, J.A. (1969) Transmission of smallpox by contact and by aerosol routes in *Macacus iris. Bulletin of the World Health Organization* 40, 279–286.

Parer, I. (1995) Relationship between survival rate and survival time of rabbits, *Oryctolagus cuniculus* (L.), challenged with myxoma virus. *Australian Journal of Zoology* 43, 303–311.
Parer, I., Sobey, W.R., Conolly, D. and Morton, R. (1995) Sire transmission of acquired resistance to myxomatosis. *Australian Journal of Zoology* 43, 459–465.
Parker, R.F. and Thompson, R.L. (1942) The effect of external temperature on the course of infectious myxomatosis of rabbits. *Journal of Experimental Medicine* 75, 567–573.
Parrish, C.R. (1990) Emergence, natural history and variation of canine, mink and feline parvoviruses. *Advances in Virus Research* 38, 403–450.
Parrish, C.R. (1994) The emergence and evolution of canine parvovirus – an example of recent host range mutation. *Seminars in Virology* 5, 121–132.
Pasteur, L. (1881) Le vaccin du charbon. *Comptes Rendus de l'Académie des Sciences de Paris* 92, 666–668.
Pasteur, L. and Joubert, J.F. (1877) Charbon et septicémie. *Comptes Rendus de l'Académie de Sciences de Paris* 85, 101–115.
Patton, N.M. and Holmes, H.T. (1977) Myxomatosis in domestic rabbits in Oregon. *Journal of the American Veterinary Medical Association* 171, 560–562.
Porter, C.D. (1994) Molluscum contagiosum virus. In: Webster, R.G. and Granoff, A. (eds) *Encyclopedia of Virology*, Volume 2, 848–853.
Rivers, T.M. (1937) Viruses and Koch's postulates. *Journal of Bacteriology* 33, 1–6.
Sobey, W.R. and Conolly, D. (1986) Myxomatosis: non-genetic aspects of resistance to myxomatosis in the rabbit *Oryctolagus cuniculus. Australian Wildlife Research* 13, 177–187.
Sobey, W.R., Menzies, W., Conolly, D. and Adams, K.M. (1968) Myxomatosis: the effect of raised temperature on survival time. *Australian Journal of Science* 30, 322.
Sobey, W.R., Conolly, D., Haycock, P. and Edmonds, J.W. (1970) Myxomatosis. The effect of age upon survival of wild and domestic rabbits (*Oryctolagus cuniculus*) with a degree of genetic resistance and unselected domestic rabbits infected with myxomatosis. *Journal of Hygiene* 68, 137–149.
Steele, E.J., Lindley, R.A. and Blanden, R.V. (1998) *Lamarck's Signature. How Retrogenes are Changing Darwin's Natural Selection Paradigm.* Allen & Unwin, Sydney, 286 pp.
Strayer, D.S. (1992) Determinants of virus-related suppression of immune responses as observed during infection with an oncogenic poxvirus. *Progress in Medical Virology* 39, 228–255.
Strayer, D.S., Skaletsky, E., Cabirac, G.F., Sharp, P.A., Corbeil, L.B., Sell, S. and Leibowitz, J.L. (1983) Malignant rabbit fibroma virus causes secondary immunosuppression in rabbits. *Journal of Immunology* 130, 339–404.
Warren, J.R. and Marshall, B. (1983) Unidentified curved bacilli on gastric epithelium in active chronic gastritis. *Lancet* 2, 1273–1275.
Williams, C.K. and Moore, R.J. (1991) Inheritance of acquired immunity to myxomatosis. *Australian Journal of Zoology* 39, 307–311.
Williams, K., Parer, I., Coman, B., Burley, J. and Braysher, M. (1995) *Managing Vertebrate Pests: Rabbits.* Australian Government Printing Service, Canberra, 284 pp.
Williams, R.T., Dunsmore, J.D. and Parer, I. (1972) Evidence for the existence of latent myxoma virus in rabbits (*Oryctolagus cuniculus* (L.)). *Nature* 238, 99–101.
zur Hausen, H. (1991) Papilloma virus/host cell interactions in the pathogenesis of anogenital cancer. In: Brugge, J., Curran, T., Harlow, E. and McCormick, F. (eds) *Origins of Human Cancer. A Comprehensive Review.* Cold Spring Harbor Press, Cold Spring Harbor, New York, pp. 695–705.

14

Coevolution of Parasites and Hosts

Overview

The introduction of the extremely virulent myxoma virus into very large and highly susceptible populations of European rabbits in Australia and Europe in the early 1950s provided an unparalleled opportunity to see how the disease, the virus and the host would change (coevolve) over time. Before considering these changes, the concept of coevolution is briefly discussed and examples of the changes in the host–parasite balance in several important human diseases are reviewed. The critical importance of transmissibility in relation to changes in virulence is discussed.

Looking over available studies on evolutionary changes in myxoma virus and its leporid hosts, the remarkable adaptation of the Californian strain of myxoma virus to its natural host, and its failure to produce transmissible lesions in several other species of rabbit native to North America, are described. Data from myxomatosis in Australia between the early 1950s and the early 1980s show that after about 20 years the virulence of the virus and the resistance of the rabbit both seemed to have reached plateaus. However, limited studies in the 1990s suggest that evolution of both host and parasite continued, so that the rabbits have become highly resistant and most strains of the virus are highly virulent (as judged by tests in unselected rabbits). The data from myxomatosis has provided useful material for mathematical modellers, and their work is briefly discussed.

With the great advances in molecular biology over the past decade, it has been possible to demonstrate that large viruses like myxoma virus contain many genes that are homologues of cellular genes which modulate the immune response, an example of coevolution at the cellular level. The chapter concludes with two other examples of remarkable coevolution discovered in the course of studies on myxomatosis: that of the European rabbit flea (*Spilopsyllus cuniculi*) and the rabbit, and of the presence of fluoroacetate in the native flora of south-western Australia and the resistance of the local herbivorous native animals to the poison, sodium fluoroacetate (1080).

General Considerations on Coevolution

Ever since the publication of *The Origin of Species* (Darwin, 1859), evolutionary biologists have recognized that organisms evolve in relation not only in response to different physical environments, but also through specialized relationships with other species. Darwin called this phenomenon ‘mutual relations’, and emphasized that biological diversity and its evolution had two major components, the diversity within a species and the diversity of interactions between species. Both contribute to the formation of biological communities, which Darwin called the 'entangled bank ... of these elaborately constructed forms, so different from each other, and dependent on each other in so complex a manner ... all produced by laws acting around us'. In the context of this book, the new interactions that are established when a species is introduced into a new environment are particularly important, as for example when European rabbits were introduced into Australia and became a far worse agricultural pest than they were in Europe, and when myxoma virus was transferred from its natural hosts (*Sylvilagus* spp. in the Americas) into European rabbits.

According to Futuyama and Slatkin (1983), the word coevolution was first used by Ehrlich and Raven (1964) in a discussion of the evolutionary interactions that occur between plants and the insects that feed on them. Elsewhere (Flexner, 1987), coevolution is described as 'evolution involving a series of reciprocal changes in two or more non-interbreeding populations that have a close evolutionary relationship and act as agents of natural selection for each other'. Coevolution is a large topic; over a dozen books and 1500 papers including that word in their titles have been published in the last 15 years. Using the term in a broad sense, examples are legion; between plants and their pollinators, between herbivores and food plants, between predatory and prey animals, between parasites and their hosts, plant and animal. As Thompson (1998) has pointed out, coevolution may occur much more rapidly than on the geological timescale in which evolutionary biologists used to think; it is also necessary to discuss it as an ecological process.

For most of the discussion that follows we will use the concept of coevolution in a 'restrictive' sense, i.e. 'a trait of one species has evolved in response to a trait of another species, which trait itself has evolved [or will evolve] in response to the trait in the first'. One of the best-known examples of restrictive coevolution in action is myxomatosis in the European rabbit. In this chapter we will comment on a few other examples in the field of infectious diseases and then discuss the data on the coevolution of virus and host in myxomatosis that have been presented in Chapters 2, 7, 8, 9 and 10.

Resistance of Humans to Infectious Diseases

The notion that infectious diseases had played a major role in recent human evolution was first proposed by Haldane (1949a,b). He was impressed with the high incidence in certain populations of genes which in homozygotes produced severe or lethal diseases, such as those for sickle cell anaemia and thalassaemia. Such high levels could be maintained, he reasoned, only if persons heterozygous for such genes had a high survival advantage under the conditions which led to the common presence of these genes. We now know that these genes have evolved to high frequencies in parts of the world where falciparum malaria was common, and that individuals heterozygous for such genes have a much higher resistance to malaria (Vogel and Motulsky, 1986). While malaria and the haemoglobinopathies provide the classical model, we now know that a great many other genes are major determinants of susceptibility to infectious diseases in humans (Hill, 1998). Twin studies have shown that many humoral and cellular immune responses to antigens of various pathogens have high heritabilities, with most of the genetic components mapping outside the major histocompatibility complex. Genes implicated include those responsible for several cytokines and cytokine receptors and for chemokines and chemokine receptors. Susceptibility to most microorganisms is probably determined by a large number of polymorphic genes, few of which have yet been identified.

Human history abounds with instances of the susceptibility of peoples long isolated from contact with infectious diseases of other human groups to such diseases when they were first exposed to them (McNeill,

1976). For example, not only did centuries of exposure of the peoples of Europe select for genetic resistance to diseases like measles and smallpox, but the pattern of childhood infection and lifelong immunity that developed in urban populations assured that adult Europeans were immune to these diseases. On the other hand, when the inhabitants of distant continents acquired these diseases during European invasions, the whole indigenous population, adults and children alike, were both genetically and immunologically susceptible. The effects of the resulting epidemics were exacerbated by massive social breakdown, so that neglect and starvation accentuated the effects of infectious diseases. This is best illustrated by the history of smallpox in the Americas; measles caused similar devastation when first introduced into Fiji. Europeans were highly susceptible to syphilis when it appeared amongst them in the late 15th century, and malaria provides the classical example of selection for resistance against a human infectious disease.

Smallpox

Smallpox was an acute infection transmitted mainly via the respiratory route, which remained a highly lethal disease for thousands of years in India and China and for hundreds of years after it was introduced into England. It is widely held that outbreaks of smallpox of great severity were a major reason for the conquest of the Aztec and Inca empires by the Spaniards and Portuguese in the early years of the 16th century, and for the relative ease of the European occupation of North America a century later (Hopkins, 1983). The virus was first brought to the American continent in 1520, following an outbreak on the island of Hispaniola due to importations with slaves from West Africa (Fenner, 1993). Although it was a serious disease in West Africa, especially during the 16th century, it was devastating to the Aztecs (Foster, 1950).

> it became so great a pestilence among them throughout the land that in most provinces more than half the population died. ... Many others died of starvation because, as they were all taken sick at once, they could not care for each other, nor was there anyone to give them bread or anything else.

Although the disease was usually classified as 'variola major' (with case-fatality rates of 10–25% in unvaccinated adults) or 'variola minor' (with case-fatality rates of 1–2%), it seems certain that there were a number of strains of differing virulence in each of these categories. Variola minor virus seems to have arisen independently on at least three occasions, once in North America and twice in Africa (Fenner *et al.*, 1988). The strains of variola minor virus that arose in South and Central Africa coexisted with virulent variola major until both were eliminated during the Smallpox Eradication Programme. In the Americas variola minor (alastrim) persisted after variola major had been eliminated, but this was probably due to the much more energetic vaccination programmes that were undertaken against outbreaks of variola major (Fenner *et al.*, 1988). There is little support for the view that in North America variola virus was evolving toward avirulence in a way comparable with the changes in the virulence of myxoma virus in European rabbits, as suggested by Johnson (1986).

Syphilis

There is still argument about the origins of syphilis, contending hypotheses being that it was brought back to Europe from the Americas by early Spanish explorers and that it arose as a mutation from *Treponema pertenue,* the causal agent of yaws. Whatever its origin, it appeared in Europe as a new disease at the very end of the 15th century, 'more horrifying than leprosy or plague, because of its novelty, its profusion of symptoms, its extreme contagiousness, ... and the fact that (in the early years at least) it was often fatal' (Quétel, 1990). Associated with the wars then occurring, it rapidly spread throughout Europe as 'the great pox' through the 16th century. However, as early as 1531 Fracastoro noted in his classical poem *Syphilis, sive Morbus Gallicus* (Wynne-Finch, 1935) that the disease had become less aggressive. As the years passed, the acute infection gradually decreased in virulence but late manifestations (general paralysis of the insane and tabes dorsalis) and congenital infections became more common. Partly because of the prudery associated with sexually transmitted diseases and concern about the late effects, fear of syphilis reached a peak in the first half of the 20th century, then dissolved with the discovery of penicillin and the sexual revolution of the 1970s. Over the course of five centuries syphilis has changed from being a frightening and lethal acute disease, associated with sexual transmission from an early date and therefore evoking strong moral and religious overtones, to a non-lethal disease for which effective therapy is available.

Malaria

Humans suffer from infections with four species of malaria parasite, for all of which humans are the only host. The commonest is *Plasmodium falciparum,* which causes malignant tertian malaria, essentially a disease of tropical and warm temperate regions associated with high mortality, and *Plasmodium vivax,* the cause of benign tertian malaria, which occurs in tropical and temperate regions and has a low mortality. *P. vivax* is closely related to the parasite of simian malaria (*P. fragile*) and has probably coevolved with primates, being able to survive in small populations because of repeated episodes of disease associated with recurrent infectivity for mosquitoes (Brooks, 1986). On the other hand, molecular studies suggest that *P. falciparum* probably originated from one of the plasmodia of birds (Waters *et al.*, 1991). Since it does not cause recurrent disease in humans, its maintenance requires a high and constant rate of infection of both mosquitoes and humans. It was probably established in humans only after human and *Anopheles gambiae* populations burgeoned in sub-Saharan Africa with the introduction of the Malaysian agricultural complex less than 2000 years ago (Weisenfeld, 1967).

As mentioned earlier, falciparum malaria is of interest because it provides the best evidence available of genetic changes in the host which enhance resistance to an infectious disease (Livingstone, 1971; Vogel and Motulsky, 1986). The first abnormality to be associated with resistance to malaria was the gene for haemoglobin S, the sickle cell gene, which produces increased resistance to falciparum malaria in both homozygotes and heterozygotes, and anaemia in homozygotes (Brain, 1952; Allison, 1954). In spite of the severe disease in homozygotes, malignant tertian malaria was so lethal that there was strong selection for the sickle cell gene in most places where it was endemic, especially in West and Central Africa (Hill and Wainscoat, 1986). Other red cell traits that have been associated with resistance to falciparum malaria are thalassaemia and glucose-6-phosphate dehydrogenase deficiency.

Immune Evasion: Coevolution of Virus and Cell at the Molecular Level

In the previous chapter (p. 295) we discussed immune suppression: the way in which some viruses interfere with the immune response in a general way, so that the animal is less well protected against a variety of other pathogens. Recent studies of viruses with large genomes, like the poxviruses, show that their genomes contain a number of genes that are homologues of cellular genes and help the virus to evade the immune system (reviews: McFadden, 1995; Spriggs, 1996). These genes appear to have been acquired by recombination between viral and cellular genes by mechanisms that are not understood; suffice it to say that there is increasing evidence that viral RNAs (in addition to those of retroviruses) may be transcribed into cellular genomes by reverse transcription (Klenerman *et al.*, 1997). A similar mechanism may transcribe cellular messenger RNAs into viral genomes. This is clearly coevolution at the molecular level.

The most detailed studies of viral genes that have homologues in mammalian cells are those carried out with the 'model' poxvirus, vaccinia virus (review: Niemialtowski *et al.*, 1997); the extent of the homology of some of these viral genes with cellular genes is shown in Table 14.1. These genes help viruses to evade the immune response, either by mimicking cytokine receptors, disarming cellular cytokine regulatory responses or interfering with the way cytotoxic lymphocytes induce the death of virus-infected cells. Several such genes have been recognized in myxoma virus (Table 14.2; McFadden *et al.*, 1995). Some of these genes, M-T2, M-T5 and M11L in myxoma virus, also affect apoptosis (programmed cell death). Delayed apoptosis allows viral replication to be completed; initiation of apoptosis then accelerates the release of mature virus particles (McFadden and Barry, 1998). Independent disruption of each of the genes listed in Table 14.2 leads to attenuation of myxoma virus. However, investigations in McFadden's laboratory by Kerr (P.J. Kerr, personal communication, 1996), showed that virulence was much more complicated than might be assumed from such results. He examined seven Australian strains, three of which were highly virulent and four of which were attenuated, for the presence of the genes M-T2, M-T7, Serp 1 and M11L. In all seven strains all four of these genes were intact. Other studies showed that both virulent and attenuated viruses down-regulated CD4 and MHC 1 molecules on the surface of infected cells to the same degree. Modulation of the immune response and of apoptosis by such genes, rather than being associated with virulence, may be a precondition for the survival of these viruses in nature.

Table 14.1. Products of some vaccinia virus genes that are cellular gene homologues[a].

Action of gene product	Amino acid homology
Gene products that inhibit host defence mechanisms	
1. Inhibitors of interferon (IFN) action	
IFN-γ-receptor homologue	25%
2. Inhibitors of cytokine responses	
Tumour necrosis factor receptor homologue	24%
Interleukin-1 type II receptor homologue	30%
3. Inhibitors of complement pathway	
Vaccinia virus complement control protein	40%
Gene product that stimulates poxvirus replication	
Epidermal growth factor homologue	38%

[a]Data from Niemialtowski *et al.* (1997).

Table 14.2. Products of some myxoma virus genes that are cellular gene homologues[a].

	Gene	Cellular homologue	Localization of viral protein	Function
Virokines	SERP-1	Serpin	Secreted	Inactivates multiple host serine proteases (inhibits inflammatory response)
Viroceptors	M-T2	TNF receptor	Secreted	Binds and inhibits TNF-α,β
	M-T7	IFN-γ receptor	Secreted	Binds and inhibits IFN-γ
	M11L	?	Cell surface	Inhibits inflammation

[a]Based on McFadden *et al.* (1995).
TNF = tumour necrosis factor; IFN = interferon.

The Relationship between Resistance, Virulence and Transmissibility

A key element in the coevolution of viruses and their hosts is the relationship between host resistance, viral virulence and the efficiency of transmission between infected and susceptible host animals, a topic recently reviewed by Levin (1996). We will restrict this discussion to viruses that cause acute, self-limited infections, since myxomatosis and rabbit haemorrhagic disease are such diseases. Different considerations enter into the evolution of viruses that cause persistent infections, especially if these are associated with recurrent infectivity, such as infections with herpesviruses.

It is often stated that 'successful' viruses evolve to become harmless (or at least non-lethal) to their animal hosts. This is probably true in most instances, but it may be that such adaptations sometimes occur very slowly. For example, yellow fever virus is readily attenuated by passage in developing chicken embryos or in cultured cells. Natural infections of African monkeys, which circulate virus at titres sufficient to infect mosquitoes, rarely cause disease. On the other hand, most species of South American monkeys, to which it was first introduced in the 1640s, still often suffer fatal infections. Although it has been known for centuries as a disease that can be maintained by person-to-person transmission, yellow fever remains a highly lethal disease in humans, in whom the high level viraemia necessary for infecting mosquitoes is attained by wild-type but not by attenuated strains of the virus. Presumably the difference between African monkeys on the one hand and humans and South American monkeys on the other reflects the fact that the virus has been circulating amongst African monkeys for millennia and the other hosts only for centuries.

Rabies virus has persisted in populations of foxes, dogs, raccoons, bats and other animals for thousands of years, yet it is thought to be almost invariably fatal in each of these animal species. The key to the survival of any particular mutant of an animal virus is its transmissibility. It had been thought that maintenance of rabies virus in animal populations was achieved by a combination of a variable incubation period, sometimes very long, followed by centrifugal spread along neuronal routes to the brain and to the salivary glands, whence it was transferred to other hosts by licking or biting. However, the nature of the disease is such that detailed studies of its epidemiology have been difficult. Recent investigations in the Serengeti have suggested that another mechanism may be operative. Although many species of wild animals there are affected, Cleaveland and Dye (1995) believe that domestic dogs are the reservoir host. Such a state could be achieved in large populations of dogs by serial transmission of lethal, short incubation period, infections. However, the dog density in some districts where the disease was endemic was only about five dogs per square kilometre, which is much lower than the density required for persistence by this mechanism. The dog is the only species for which there is unequivocal evidence of infectious carriers, although they are very rare, and serological studies showed that a substantial proportion of healthy dogs in the Serengeti had detectable levels of rabies-specific antibody. Mathematical modelling of the results suggested that persistence of the disease was due to dogs that were infectious carriers, rather than dogs in which the disease had an abnormally long incubation period. Nevertheless, rabies remains a lethal disease; the virus transmitted from such carriers was almost invariably lethal in dogs as well as other animals.

If long-term carriage and persistent or recurrent viral infectivity can occur as a very rare feature of a disease like rabies, may it also occur in other diseases that are ordinarily regarded as self-limited, acute infections? Such a phenomenon could not have occurred in smallpox; if it had, the reduction of incidence from 20 million cases in 1967 to zero at the end of 1977, and maintenance of zero incidence for the next 20 years, would have been impossible. The possibility was discussed in the previous chapter (see p. 301) that in very rare cases latent infection and recurrent infectivity might occur in rabbits infected with myxoma virus.

The most neglected factor in discussions of the evolution of the virulence of infectious agents is their transmissibility. This factor is particularly important in diseases transmitted by insect vectors, such as malaria and arbovirus infections. The interaction can be expressed graphically as an interplay between three factors:

Transmissibility ⇄ Virulence of the causative agent

⤡ ⤢

Resistance of the host

The complexity of these interactions is obvious with organisms that have to multiply in both insects and vertebrate hosts; it has long been assumed that they were less important in the much rarer cases in which the common mode of transmission was mechanical. However, as we shall demonstrate, evidence produced over the last few decades suggests that natural survival of myxoma virus in populations of wild European rabbits is due to a constantly evolving balance between viral virulence, host resistance and infectivity for insect vectors. It has yet to be determined how this balance will evolve in rabbit haemorrhagic disease.

Myxoma virus was introduced in Australia for rabbit control because it was highly host-specific and extremely lethal. In the years preceding its introduction there were serious doubts it was not sufficiently contagious to spread through wild rabbit populations (see Chapter 6), but in fact it is readily transmitted by any insect that will bite two rabbits in succession, and mosquitoes answering this requirement were common in Australia. There was concern from the outset about how long myxomatosis would continue to be highly lethal. Evidence presented below shows that slightly less virulent strains had a survival advantage even in fully susceptible rabbits, because these strains were better maintained through the winter, when mosquitoes still occurred, but were uncommon. These slightly less virulent strains allowed some 10% of rabbits to survive, in contrast to the 1% or fewer that survived infection with the original highly virulent virus. This provided enough breeding animals to allow for the selection of increasingly resistant rabbits. In the conditions under which most infectious diseases have to be studied, i.e. where the virulence of the viruses occurring at any particular time is determined by their lethality for their natural hosts at that time, it could be said that myxomatosis has been evolving towards relatively low virulence. However, in myxomatosis it was possible to continue to test the virulence of the virus in laboratory rabbits, which had not been subjected to selection for resistance. Tests in these hosts showed that after some 40 years there was a substantial increase in viral virulence, because these more virulent virus strains were able to produce lesions in the more resistant wild rabbits that were more readily transmitted by mosquitoes than the less virulent strains.

Coevolution of Leporipoxviruses and *Sylvilagus* spp. in the Americas

Three of the four known members of the genus *Leporipoxvirus* occur in the Americas: myxoma virus, which is known to occur in *Sylvilagus brasiliensis* and *S. bachmani*, fibroma virus, which has been recovered from *S. floridanus*, and squirrel fibroma virus, which produces fibromas in the eastern grey squirrel, *Sciurus carolinensis*. A fourth possible American member is the poxvirus found in skin lesions of western grey squirrels (*Sciurus griseus griseus*) by Regnery (1975), but this has not yet been adequately characterized. The only known non-American leporipoxvirus is the virus which produces fibromas in the European hare (*Lepus europaeus*) but this also is inadequately characterized, and there have been no reports of its occurrence in hares in Europe since the early 1960s. Interesting studies demonstrating the coevolution of myxoma viruses and North American leporids were carried out some years ago by Regnery, Marshall and their collaborators.

Californian myxoma virus and Sylvilagus bachmani

As described in Chapter 4, the strains of myxoma virus isolated from *Sylvilagus bachmani* differ in many respects from those presumed to have been derived from *S. brasiliensis* – 'presumed' because the only isolation from *S. brasiliensis* was that made by Aragão (1943), all other strains being recoveries from European rabbits in places where the principal or only leporid was *S. brasiliensis*.

Comparisons of the behaviour of the Californian and Brazilian strains of myxoma virus in several Californian leporids show that the Californian strain is exquisitely adapted to survive in *S. bachmani* (Regnery and Marshall, 1971). Using fibromas produced in *S. bachmani* by Californian myxoma virus as sites for mosquito probing, infectivity was tested in five species of *Sylvilagus* and in *Oryctolagus cuniculus*. If a local lesion was produced in the recipient rabbit, mosquitoes were induced to probe on this at various times from the first appearance of the tumour until it scabbed, and then tested for infectivity on the most susceptible host, *Oryctolagus cuniculus*. Localized tumours, but no secondary lesions, developed in each of five species of *Sylvilagus* (*S. audubonii, S. bachmani, S. floridanus, S. idahoensis* and *S. nuttallii*), but not in *S. brasiliensis* or the blacktailed jack rabbit (*Lepus californicus*). The skin lesions on *S. audubonii* resembled those produced in *Oryctolagus cuniculus*; those in *S. idahoensis* and *S. nuttallii* appeared later and were smaller. However, further transfer by mosquito bite from these tumours to *Oryctolagus cuniculus* was successful only from lesions in *S. bachmani* (Table 14.3). The success with lesions of *S. bachmani* as a source of virus and the failures with *S. audubonii, S. floridanus* and *S. idahoensis* parallelled the titres of virus in the superficial layers of the skin. On the other hand, although the Brazilian strain of virus produced skin tumours in *S. bachmani*, these did not contain sufficient virus for further mosquito transmission (Marshall and Regnery, 1963). These results suggest a high degree of coevolution between the Californian strain of myxoma virus and its natural host, the brush rabbit, *S. bachmani*.

Table 14.3. Transmission of Californian myxoma virus to several American leporids by mosquitoes probing lesions in *S. bachmani*, and failure to transmit it from skin lesions of all recipient species except *S. bachmani*[a].

Donor host	Mosquito transfer	Recipient host	Skin lesion titre[b]	Second mosquito transfer	Second recipient host
S. bachmani	26/54[c]	*S. bachmani*	$10^{8.2}$	153/253[d,e]	*O. cuniculus*
S. bachmani	16/20	*S. audubonii*	$10^{7.3}$	0/27	*O. cuniculus*
S. bachmani	4/7	*S. floridanus*	$10^{4.5}$	0/21	*O. cuniculus*
S. bachmani	3/3	*S. idahoensis*	—	0/15	*O. cuniculus*
S. bachmani	7/12	*S. nuttallii*	$10^{6.8}$	0/18	*O. cuniculus*

[a]Data from Regnery and Marshall (1971) and Marshall and Regnery (1963).
[b]Titres of rabbit-infectious particles per gram of skin slice of tumours at time of their maximum growth.
[c]Numerator, positive results on attempted transfers from *S. bachmani* to recipient host indicated; denominator, numbers of transfers attempted.
[d]Numerator, positive results on attempted transfers from recipient host to *O. cuniculus*; denominator, numbers of transfers attempted.
[e]The rather low efficiency of transfers from *S. bachmani* (first line) is due to inclusion of many attempts made at unfavourable times.

Brazilian myxoma virus in North American* Sylvilagus *spp.

In contrast to the situation with Californian myxoma virus, the Brazilian strain of myxoma virus produces transmissible tumours in *S. audubonii*, *S. floridanus* and *S. nuttallii* (Regnery, 1971; Table 14.4), those in *S. floridanus* being small and flat, those in *S. audubonii* larger and bulbous. In addition to a large tumour at the site of infection, *S. nuttallii* suffered a generalized disease reminiscent of that seen in *Oryctolagus cuniculus*. In contrast to the reactions of these species to infection with the Californian strain (Table 14.3), mosquito transfer from the lesions in *S. audubonii* and *S. nuttallii* to *Oryctolagus* was successful with much the same frequency as found after mosquitoes had probed through lesions in *Oryctolagus*; attempted transfers from *S. floridanus* were less successful. Although leporiviruses are not known to occur naturally in either *S. audubonii* or *S. nuttallii*, Regnery concluded that the Brazilian strain of myxoma virus, if introduced, could well become established in populations of *S. nuttallii* and possibly in *S. audubonii*.

Coevolution of Host Resistance and Viral Virulence in Myxoma Virus Infection of *Oryctolagus cuniculus*

From the time of its initial spread through the Murray–Darling basin in the summer of 1951, informal conversations between F.M. Burnet and Fenner at the Walter and Eliza Hall Institute, and with L.B. Bull at the nearby CSIRO Division of Animal Health, had focused on speculations about possible evolutionary changes in myxoma virus,

Table 14.4. Tumour production in four *Sylvilagus* species infected with the Brazilian (Lausanne) strain of myxoma virus by mosquitoes that had probed through tumours in the skin of *Oryctolagus cuniculi*, and the results of attempts at transfers from these tumours[a].

Recipient host	Mosquito transfers	Longevity of tumours (days)	Mosquito transfers to *Oryctolagus*: Day after infection	Mosquito transfers to *Oryctolagus*: Result
S. audubonii	3/3[b]	2–27	7	2/6[c]
			11	3/4
			15	3/5
			19	4/6
			23	1/6
S. floridanus	—[d]	5–26	8	0/4
			14	0/4
			17	1/3
S. nuttallii	1/2	6–22,	10	1/1
		then died	12	1/1
S. bachmani	1/2	5–28	8	0/4
			9	0/1
			11	0/2
			12	0/1
			14	0/2
			15	0/4
			18	0/6

[a]Data from Regnery (1971).
[b]Numerator, positive results on attempted transfers from *O. cuniculus* to recipient host indicated; denominator, numbers of transfers attempted.
[c]Numerator, positive results on attempted transfers from recipient host to *O. cuniculus*; denominator, numbers of transfers attempted.
[d]Infected by injection of tumour tissue suspension.

and after what then was thought likely to be a much longer time, the development of resistance in the rabbit. Burnet was a member of the CSIRO Advisory Committee on Myxomatosis that had been set up early in 1951, and as he noted some years later (Burnet, 1971) '[I took] some pride in having correctly predicted at the first meeting [on 13 February 1951] that myxomatosis would not exterminate the rabbit, that rabbits would develop an inheritable resistance and that it would be to the evolutionary advantage of the virus if it came down to a lower level of virulence'.

For his part Fenner (1952), addressing the Australian and New Zealand Association for the Advancement of Science in August 1952, had listed as last among five possible explanations of the lower mortality rates then being reported in some parts of Australia: 'there may have been a mutation in the virus so that forms transmitted in certain areas are less virulent than the classical strain which caused such a high mortality in places like Lake Urana'. With an additional year of experience, he noted (Fenner, 1953) that 'virus variants of reduced virulence ... have been isolated from several localities' and that 'variation in the innate resistance of the European rabbit to myxomatosis may have already occurred, but its demonstration is difficult'.

Because of the theoretical interest of such evolutionary changes in the virus and its host and their great practical importance, laboratory work in Australia during the next 25 years was focused on following changes in viral virulence and in the innate resistance of rabbits. Most of the tests of viruses and host animals were carried out by Fenner's team in Canberra, by Edmonds and Shepherd at the Keith Turnbull Institute near Melbourne, and by Sobey at the CSIRO Division of Animal Genetics in Sydney. After a period of some 15 years during which no studies of this kind were carried out, in 1996 P.J. Kerr, of the CSIRO Division of Wildlife and Ecology, tested the virulence of a few field strains and the resistance of local wild rabbits to highly virulent strains of virus. Similar but less extensive studies were undertaken in Britain, where a different strain of virus had been introduced and the principal vector was the European rabbit flea, rather than mosquitoes.

Information on the coevolution of the European rabbit and myxoma virus differs from that available for any other disease of vertebrates in that it has been possible to assay the virulence of the virus by tests in essentially unchanging host animals, namely genetically unselected and serologically negative laboratory rabbits, as well as in wild rabbits as they occurred at the time the viruses were isolated. In other infections it has been possible only to observe historical changes in the severity or lethality of disease. To make possible the assay of the lethality of large numbers of strains of virus, obtained over many years, in animals as expensive and difficult to house as rabbits (compared, for example, with mice), virulence was tested in groups of five or six laboratory rabbits and the results expressed in terms of virulence grades, as determined by survival times (see p. 94). Measurement of changes in the resistance of wild rabbits was by comparison somewhat easier, although it involved the periodic collection of reasonable numbers of uninfected wild rabbits, and their maintenance in mosquito-proof quarters until the tests were completed (see p. 174). Using genetically unselected and serologically negative laboratory rabbits as controls, resistance was then tested with strains of virus of appropriate virulence, using more virulent viruses as the rabbits became more resistant.

Changes in viral virulence and rabbit resistance in Australia

The changes in the virulence of the strains of myxoma virus recovered from Australian wild rabbits after 1951 are described in some detail in Chapters 7 (p. 172) and 8 (p. 193). To put the results of the virulence tests in the context of coevolution, all available data are summarized in Table 14.5. The data on changes in the resistance of wild rabbits over time are summarized in Table 14.6.

Table 14.5. The virulence of strains of myxoma virus recovered from the field between 1951 and 1994 (expressed as percentages)[a].

Virulence grade	I	II	III	IV	V
Mean survival time (days)	9–13	14–16	17–28	29–50	—
Presumed case-fatality rate (%)	>99	95–99	70–95	50–70	<50
1950–51	>99				
1952–55	13.3	20.0	53.3	13.3	0
1956–58	0.7	5.3	54.6	24.1	15.5
1959–63	1.8	11.1	60.5	21.8	4.7
1964–66	0.6	0.6	63.4	34.0	1.3
1967–69	0	0	63.0	35.7	1.6
1970–74	0.6	4.6	74.1	20.7	0
1975–81	1.9	3.3	67.0	27.8	0
1984–85	0	17.6	76.5	5.9	0
1992–94	83.3	0	0	16.7	0

[a]Consolidated data from Tables 8.1 and 8.4.

Table 14.6. Case-fatality rates of non-immune rabbits challenged with four different strains of myxoma virus.

	Case-fatality rate (%)			
	KM 13 strain (Grade III)	Standard Laboratory Strain (Grade I)	Glenfield strain (Grade I)	Lausanne strain (Grade I)
Selective breeding experiments (laboratory rabbits)[a]				
grade 0 (no selection)	95	99		
grade 4	80	88		
grade 6.5		79		
Wild rabbits[b]				
1952–53	90			
1956–57	33	87		
1961		73	97	
1961–66		68	98	
1967–71		66	94	
1972–75		67	96	100
1976–81		60	91	98
1996				44

[a]Data from Sobey (1969).
[b]Data from various sources (Urana rabbits, Mallee rabbits, Canberra rabbits).

The first isolations of strains of reduced virulence were made in 1952 and early in 1953. These strains were characterized by longer survival times of infected rabbits and the survival of about 10% of genetically unselected wild and laboratory rabbits. Strains of reduced virulence soon appeared in many widely separated areas, all over Australia. Although myxoma virus may be carried for long distances by infected mosquitoes, experience in Tasmania and Western Australia was so similar to that in south-eastern Australia that it is clear that such changes were occurring repeatedly and independently. Strains of Grade III virulence, with presumed case-fatality rates of 70–90%, became established as the dominant group by about 1955. In spite of the regular introduction of strains of Grade I virulence (Standard Laboratory Strain, Glenfield strain and Lausanne strain) by means of inoculation campaigns, strains of very high virulence (Grades I and II) were very rarely recovered from wild rabbits between the late 1950s and 1981, when systematic testing ceased. However, although strains of Grade III virulence remained dominant, by the 1970s strains of Grade I and Grade II virulence were being recovered from about 8% of wild rabbits in the Mallee region of Victoria, where the most resistant rabbits occurred, and about 5% over Victoria as a whole (see Table 8.1, p. 194), rising to about 17% by 1985. No systematic tests were conducted between 1985 and 1996, but the majority of the few strains collected in the 1990s and tested in 1996 were of Grade I virulence. They produced a clinical picture in genetically unselected laboratory rabbits very like that produced by the original Standard Laboratory Strain.

If there had been no attenuation of the virus and only about 1% of wild rabbits had survived, the disease would often have died out at the end of summer, and emergence of resistant rabbits would have been extremely slow. However, since strains that allowed some 10% of rabbits to survive emerged so soon, it was not surprising that in the face of the very stringent selection, genetically resistant rabbits quickly became dominant (Table 14.6). Over the next 25 years there was a steady increase in rabbit resistance, such that by the late 1970s the Standard Laboratory Strain, which had initially killed over 99% of the wild rabbits, was lethal for only 60%. Such high levels of resistance enabled three strains of Grade I virus to be differentiated. The Glenfield strain, isolated from a wild rabbit in Dubbo, New South Wales, in 1951, was clearly more virulent than the Standard Laboratory strain from which it evolved, and other tests showed that the Lausanne strain (a different isolate from Brazilian rabbits), which had been used to introduce myxomatosis into Europe, was even more virulent. There was no systematic testing of rabbit resistance between 1981 and 1996, but limited tests in 1996 showed that less than half the wild rabbits infected with the highly virulent Lausanne strain died of myxomatosis, after survival times that were usually longer than 25 days.

The explanation for these progressive and interactive changes was provided in early investigations of the mechanism of transmission (see p. 80). It was shown that the usual mode of transmission of myxomatosis was mechanical transfer by insect vectors, usually by mosquitoes but if they were present, by European fleas as well. These vectors contaminated their mouthparts with myxoma virus only when they probed through the skin overlying myxoma lesions. The epidermal cells over lesions produced in laboratory rabbits by a highly attenuated 'laboratory' strain, neuromyxoma virus (and presumably of other, naturally occurring, very attenuated strains), contained very few virus particles and the infected animal quickly recovered, so infection by probing or biting vectors was rare. On the other hand viruses of Grade I and Grade III virulence produced highly infectious skin lesions. Since genetically unselected rabbits infected with Grade I virus died within 4 or 5 days of these lesions becoming infectious, whereas lesions produced by Grade III strains persisted throughout the longer life of the rabbit, and for as long as a month in animals which survived, such strains had a great survival advantage during the winter, when mosquitoes were rare.

Under the influence of this intense selection pressure, wild rabbits became steadily more resistant. For many years, viruses of intermediate virulence (Grade III) appeared to produce the kinds of lesions that allowed viruses to overwinter. However, by the early 1990s the majority of strains recovered from rabbits in temperate parts of Australia were as virulent as the Standard Laboratory strain. Because of the large size of the genome of myxoma virus, it has not been practicable to sequence its genome, and the genetic basis of the changes from high to intermediate levels of virulence, and later back to high virulence, has not been determined. Given the appearance of the Glenfield strain in 1951, it is reasonable to assume that the high virulence of current field strains is not due to 'reversion' to the genetic structure of the Standard Laboratory strain, but to other mutations, probably in a number of different genes. No studies have been made of the genetic changes underlying increased rabbit resistance.

Changes in viral virulence and rabbit resistance in Britain and France

Myxomatosis in Europe was initiated by a different strain of virus from that used in Australia, the Lausanne strain, and inoculation of wild rabbits was prohibited in all European countries. Early investigations in Australia showed that the Lausanne strain caused distinctive protuberant lesions, but in laboratory rabbits the case-fatality rate and the mean survival time were the same as those in animals infected with the Standard Laboratory strain – both were classed as of Grade I virulence. Subsequent studies in resistant Australian wild rabbits (Table 14.6) showed that the Lausanne strain was substantially more virulent than the Standard Laboratory Strain, producing case-fatality rates of 98–100% at times when the Standard Laboratory strain was recording rates of 60–67%. It was also thought that flea transmission, which is the dominant mode in Britain, might favour viruses that killed rabbits, since it was thought that fleas would move away from dead rabbits more readily than from live rabbits (Fenner and Ratcliffe, 1965). This assumption was later shown to be only partially true; fleas abandoned dead rabbits, but also moved readily between live rabbits. Among wild rabbits in Britain fleas are by far the most important vectors; in France, fleas are important year-round vectors and mosquitoes are the most important vectors during the summer.

The greater virulence of the Lausanne strain, or transmission by fleas, appears to have delayed somewhat the emergence and dominance of moderately attenuated strains of virus (Table 14.7). In France and Britain, some 11% and 4% of field strains examined nine years after introduction of the virus appeared to be of Grade I virulence and 19% and 18% of Grade II virulence; equivalent figures in Australia were 0.5% (Grade I) and 20% (Grade II). Later figures, for Britain, were 2% and 0% Grade I strains and 26% and 36% Grade II strains 22 and 28 years after the virus was introduced; equivalent figures in Australia were 1% and 3% (Grade I) and 7% and 6% (Grade II). The only European data on rabbit resistance come from tests on rabbits from Norfolk, in Britain, which showed a steady fall in case-fatality rates from 90% to 21%, with a virus causing a 98% case-fatality rate in control wild rabbits, over the period 1966 to 1976. No more recent data are available from Europe.

Modelling of Coevolution in Myxomatosis in *Oryctolagus cuniculus*

Ecologists often use mathematical models to provide logically consistent explanations of biological phenomena and to predict the

Table 14.7. The virulence of strains of myxoma virus recovered from wild rabbits in Britain and France between 1955 and 1981 (expressed as percentages)[a].

Virulence grade	I	II	III	IV	V
Mean survival time (days)	9–13	14–16	17–28	29–50	—
Presumed case-fatality rate (%)	>99	95–99	70–95	50–70	<50
1953–55 (Britain and France)[b]	99.0				
1962 Britain)	4.1	17.6	63.6	14.0	0.9
1962 (France)	11.0	19.3	55.4	13.5	0.8
1968 (France)	2.0	4.1	35.1	58.8	4.3
1975 (Britain)	1.6	25.8	66.4	5.5	0.8
1981 (Britain)	0	35.8	62.6	1.6	0.8

[a]Except where indicated, based on tests in laboratory rabbits.
[b]Based on field observations.

outcome of interactions between antagonists, such as predators and prey or hosts and their parasites. For host–pathogen systems, one of the most important conclusions from these models is that coevolution can lead to many different outcomes; the commonly held view that diseases evolve toward a commensal state does not always hold and sometimes the endpoint of a coevolutionary process is a system in which the pathogen is highly virulent. Studies of the virulence of myxoma virus in wild rabbits (which are subjected to selection for resistance), compared with their lethality in laboratory rabbits which have not been subjected to such selection, provide data that has guided the development of some important coevolutionary models; indeed myxomatosis is one of the most important case studies in the ecological and epidemiological literature.

The first critical analysis of coevolution in the rabbit–myxoma virus system was produced by Rendel (1971), who used quantitative genetics models to estimate the selection pressure on rabbits for resistance to myxomatosis. He argued that the emergence of less virulent viruses would accelerate the evolution of resistant rabbits by increasing the selection pressure for resistance. Using the data on changing levels of viral virulence and rabbit resistance then available, he concluded that by 1980 the rabbit population would have acquired a level of resistance that would enable some 70% to recover from infection with the Standard Laboratory Strain and 20–25% from the Glenfield strain. The only data available to check this prediction are those produced in the period 1976 to 1981 by Edmonds and Shepherd (quoted in Fenner, 1983). Over that time-span they found that in the Mallee region, where rabbits were genetically more resistant than elsewhere in Victoria (see Table 8.3, p. 197), the survival rate was 40% for the Standard Laboratory Strain and 9% for the more virulent Glenfield strain. In Gippsland, where rabbits were less resistant, the corresponding figures were 21% and 5%.

A decade later Anderson and May, in a series of influential papers (summarized in Anderson and May, 1991), focussed attention on the 'basic reproductive rate', R_0, as a means to compare the success of different strains of virus competing within a host population. This quantity is the average number of secondary infections produced in a population of susceptible individuals during the infectious period of a single infected host. The infectious period is the inverse of the rate of recovery and death, and so R_0 is often written in the form

$$R_0 = \frac{\text{new infections per infected host}}{\text{rate of death and recovery}}$$

For an epidemic to arise in a population, R_0 must be greater than one. Moreover, Bremermann and Thieme (1989) showed that, under a variety of conditions, pathogens that maximize R_0 can competitively exclude other strains that are competing within the same host population. In a coevolutionary system, the equation shows that pathogens have two paths to success: a high R_0 may result if a pathogen is highly transmissible (the numerator is large), or if it has low virulence (the denominator is small). Trade-offs between virulence and transmissibility imply that there is some kind of relationship between the numerator and the denominator.

There may also be a relationship between virulence and recovery rates. Using data on the survival times of rabbits infected with different strains of myxomatosis, Anderson and May (1982) investigated the relationship between death rates (virulence) and recovery rates, and predicted that the myxoma virus would evolve to an intermediate virulence grade. Their prediction qualitatively matches the observed virulence grade, but ignores any relationship between virulence and transmissibility, i.e. any relationship between the numerator and denominator of the expression for R_0. This motivated Massad (1987) to refine the model of Anderson and May to include a relationship between the death rate and transmissibility, resulting in a prediction that more closely matches the virulence observed in Australia, at least for the period 1975–1981 (Fenner, 1983).

Neither of the above models, however, allows for the probability that rabbits will evolve to become more resistant, assuming that evolutionary change in the virus is rapid enough to track any slow change in the resistance of the rabbit. More recently, Dwyer *et al.* (1990) used a detailed model – including, for example, explicit consideration of virus titres in the skin of infected rabbits and realistic rabbit demography – to address the question of whether rabbits and the myxoma virus will continue to evolve antagonistically in a coevolutionary 'arms race' (*sensu* Dawkins and Krebs, 1979), and whether the virus will continue to control the rabbit population. They concluded that there was insufficient information on the relationship between increasing resistance and virulence to make strong predictions, but that the virus should continue to control the rabbit in the short term. An arms race is likely in progress, as evidenced by the data on rabbit resistance available to Dwyer *et al.* (1990) at the time. This included the 1975–1981 data of Edmonds and Shepherd (quoted earlier) and a small sample for 1984–1985 (J.W. Edmonds, personal communication, 1998), both of which show a slight swing back to more virulent strains. Since then, the majority of the few strains obtained from wild rabbits over the period 1991 to 1994 (see Table 8.2) behave as Grade I strains in unselected laboratory rabbits (P.J. Kerr, personal communication, 1998), suggesting that an arms race is indeed in progress.

Recent theoretical models of coevolution have begun to address questions of how variation in the number and susceptibility of hosts (in space and time) affects host–pathogen systems. For example, Lipsitch *et al.* (1995) show that populations within which pairs of hosts have repeated opportunities for transmission favour the evolution of pathogens with lower virulence than populations with less spatial or temporal structuring. On a finer scale, models have recently been developed to predict the outcome of the evolutionary processes that can occur within a host as a result of the interaction of a population of pathogens with the immune system (e.g. Anderson, 1995; May and Nowak, 1995). Selection within the host may actually produce pathogens with characteristics that are not optimal for transmission between hosts.

Such thinking is also reflected in recent models of the rabbit–myxoma virus system. Using British data, Seymour (1992) modelled the interaction of viral virulence, rabbit resistance and resource limitation. Although Seymour's model assumes a homogeneous population of rabbits, his central conclusion has implications for spatially structured populations. He concluded that, when fleas are the only vectors, the interaction of virulence and resistance is highly dependent on resource limitation, i.e. the capacity of the environment to support a substantial rabbit population. In resource-poor environments, viruses of low virulence are likely to die out for the lack of susceptible hosts, while epizootics with viruses of high virulence – that can cause drastic population crashes in good environments – are unlikely to occur. In a nutshell, poor environments act as disease-free refuges for the rabbit. Seymour's argument is less easily sustained in Australia and France, where highly mobile mosquito vectors occur, so that only in rare situations would poor environments act as refuge areas. However, in all countries where myxomatosis occurs, there is enormous variation in the distribution of rabbits and insect vectors, providing selection pressure for viruses with particular virulence and transmission characteristics. In turn, this must select for more resistance in some areas than others. Developing a general theory to predict and explain patterns of virulence remains a major challenge for models of coevolution in host–parasite systems.

Coevolution of the *Spilopsyllus cuniculi* and *Oryctolagus cuniculus*

During the investigations on myxomatosis in Britain a remarkable example of the coevolution of a vertebrate and one of its ectoparasites was discovered. The European rabbit (*Oryctolagus cuniculus*) harbours four species of fleas: *Spilopsyllus cuniculi*, which occurs throughout its range in Europe, and *Xenopsylla cunicularis, Caenopsylla lactaevi* and *Odontopsyllus quirosi,* which occur only in France and Spain (including Morocco for *Xenopsylla*) (Beaucournu, 1980). Since it was an efficient vector of myxomatosis which promised to be useful in situations in which mosquitoes were uncommon, attempts were made to introduce *S. cuniculi* into Australia in the early 1950s (see p. 171) but it could not be induced to breed. The reason for this was discovered by Mead-Briggs and Rothschild a few years later (see p. 85), and represents a remarkable example of the coevolution of an ectoparasite and its host (Fig. 14.1).

Investigators in Britain and Australia had found that fleas released on domestic or wild rabbits kept in hutches failed to produce eggs. Rothschild (1957) had reported similar results with rabbits in hutches, but found that the fleas bred successfully if the rabbits were kept in semi-wild conditions. Mead-Briggs and Rudge (1960) then showed that if the fleas were placed on pregnant does, their ovaries matured and eggs were laid shortly after the young rabbits were born, whereas the ovaries of fleas kept on non-pregnant rabbits did not develop. Subsequent work, summarized by Rothschild and Ford (1972a) and Mead-Briggs (1977), showed that the mammalian hormones of the adult rabbit control the maturation of both sexes of *S. cuniculi*, and demonstrated the complexity of the interactions between the fleas and their hosts. This relationship does not hold for other fleas that parasitize *O. cuniculus*, such as *X. cunicularis,* which Cooke had no difficulty in breeding in the laboratory before and after he brought it to Australia (see p. 189). However, the flea, *Cediopsylla simplex,* which parasitizes the eastern cottontail (*Sylvilagus floridanus*), a North American leporid, has a life cycle essentially similar to that of the European rabbit flea, suggesting that the evolution of this relationship is of considerable antiquity (Rothschild and Ford, 1972b).

Coevolution of Plants Containing Fluoroacetate and Native Animals in Western Australia

The south-western corner of Western Australia has a unique native flora and fauna, with many species that are found nowhere else in Australia. European settlement on the Swan River, in that area, began in 1829, and in 1901 the colony became the State of Western Australia. Life was difficult for the early settlers. Not only was the environment harsh and the soil infertile, but when they moved their flocks and herds inland from the coastal plain many of the sheep and cattle sickened and died. After some ten years it was realized that the stock were dying because they had been eating poisonous pea-flowered legumes, now known to belong to the genus *Gastrolobium*. This puzzled the settlers; they knew that legumes were nutritious and they had seen the local bronze-wing pigeons eating the seeds with impunity. The solution to this puzzle came over a century later (Mead *et al.,* 1985; King and Kinnear, 1991).

Sodium monofluoroacetate ('1080') is one of the most toxic substances known. After being patented as a moth-proofing agent in the 1920s, it was first used as a vertebrate pesticide in the 1940s, for the control of rodents (Kalmbach, 1945). It was introduced into Australia to control rabbits in the early 1950s and is now the most widely used poison for rabbits, foxes and other vertebrate pests (McIlroy, 1981; Saunders *et al.,* 1995; Williams *et al.,* 1995). In the mid-1960s it was found that the toxin in *Gastrolobium* was fluoroacetate, and it is now known that this poison occurs naturally in five genera of plants, three of which, *Acacia, Gastrolobium* and *Oxylobium,* occur only in Australia. Thirty-three species belonging to the genera *Gastrolobium* and *Oxylobium* occur only in south-western Australia. Tests on native animals in that region showed that they were relatively resistant to fluoroacetate, compared with introduced animals and similar native species occurring in other parts of Australia (Oliver *et al.,* 1977; King, 1990; review: Twigg and King, 1991).

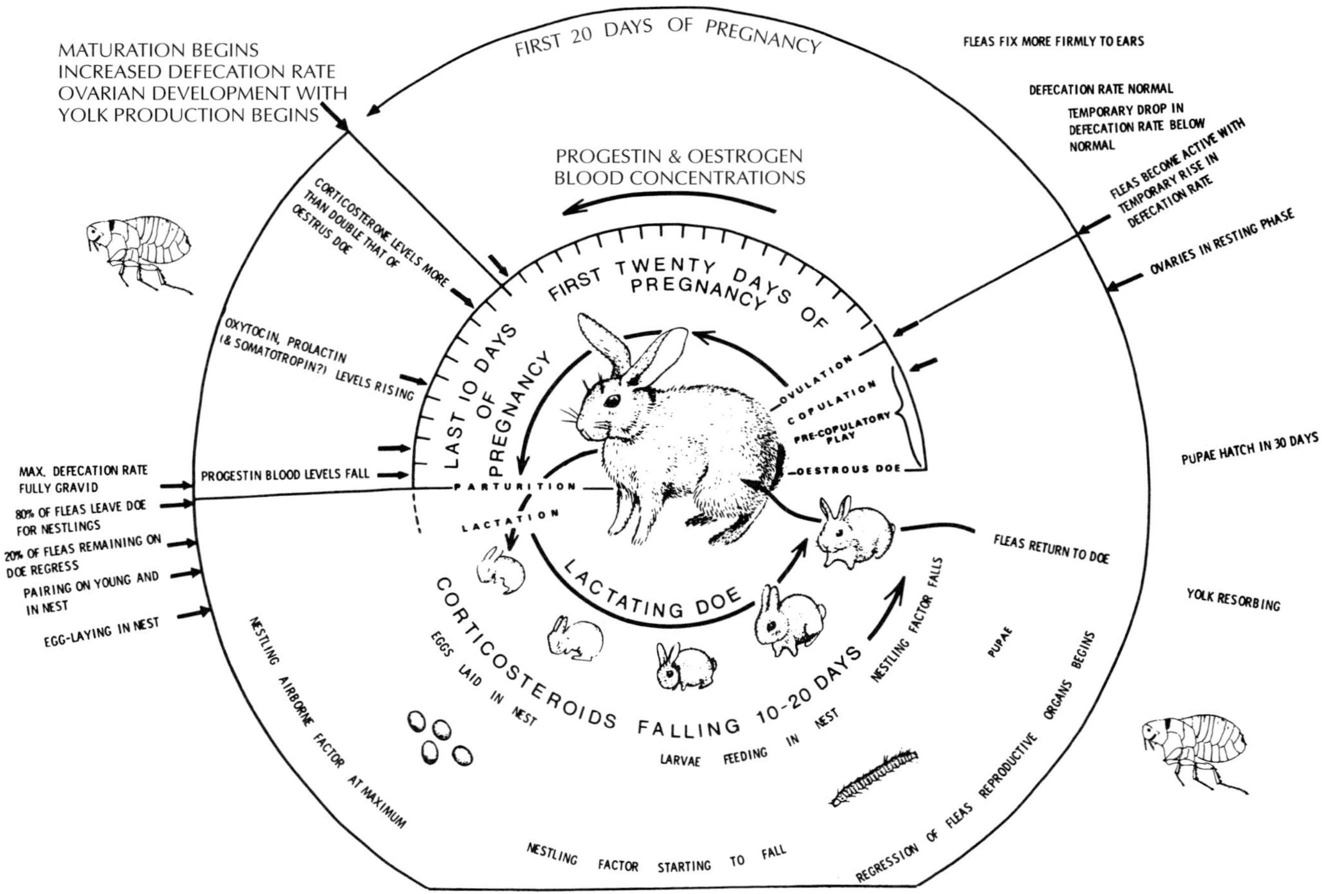

Fig. 14.1. Diagram illustrating the synchronization of the breeding cycles of *Oryctolagus cuniculus* and *Spilopsyllus cuniculi* and the complex interactions between host hormonal activity and the breeding of the flea. Modified from Rothschild and Ford (1972a), with permission.

Rabbits, foxes, cats and dingoes, which are the principal vertebrate pests in that part of Australia, are very sensitive to fluoroacetate, with lethal doses (LD_{50}) of 0.2–0.4 mg kg^{-1}, whereas the brush-tailed bettong (*Bettongia penicillata*), which is native to south-western Western Australia, is very resistant, with an LD_{50} of 100 mg kg^{-1}. Further, brush-tailed possums (*Trichosurus vulpecula*) from south-west Western Australia are over 100 times more resistant than the same species from parts of eastern Australia where plants containing fluoroacetate do not occur (LD_{50} of 125 mg kg^{-1} compared with 0.75 mg kg^{-1}). The local marsupial carnivores are also more resistant than their eastern colleagues, and much more resistant than introduced carnivores (Table 14.8).

From 1955 until the early 1970s 1080 was widely used for rabbit control in the farmland surrounding the National Parks in south-western Western Australia. The type of bait used meant that one rabbit killed by 1080 would contain enough poison to kill several adult foxes. Since foxes fed on the dead rabbits, they had declined in parallel with the fall in rabbit numbers. The removal of their competitors (rabbits) and their major predators (foxes) led to increases in the populations of endangered marsupials. In 1969 the European rabbit flea was introduced for the control of rabbits and the incidence of deaths from myxomatosis rose to the point that in 1970 the rabbit poisoning programme was greatly curtailed (King *et al.*, 1981). Freed from the risk of secondary poisoning, fox numbers rose again, and because of the paucity of rabbits, they preyed on the native fauna to an increased extent, with disastrous consequences. To remedy the situation, in 1978 1080 baiting of foxes was introduced over large areas of the south-western forests of Western Australia, initially at ground level and since 1994 by air, using baits that would be unlikely to be taken by other animals. Small nature reserves surrounded by farmland were baited at intervals varying from monthly to four times a year. With the reduction of competition with rabbits and the removal of their major predator, bettongs and other marsupials are flourishing once more.

Table 14.8. Lethality of '1080' poison for different groups of native animals which evolved in south-western Australia (where fluoroacetate occurs in native vegetation) and for introduced animals[a].

	Known exposure		No exposure	
Groups of animals	Number of species	Mean LD_{50} (mg kg^{-1})[b]	Number of species	Mean LD_{50} (mg kg^{-1})
Introduced carnivores	—	—	3	0.2
Marsupial carnivores	12	8.3	9	2.7
Introduced herbivores	—	—	5	0.4
Marsupial herbivores	10	42.0	10	0.3

[a]Based on McIlroy (1992).
[b]The LD_{50} represents the number of milligrams of 1080, per kilogram of body weight, which will kill 50% of the test animals.

References

Allison, A.C. (1954) Protection afforded by sickle cell trait against subtertian malarial infection. *British Medical Journal* 1, 290–294.

Anderson, R.M. (1995) Evolutionary pressures in the spread and persistence of infectious agents in vertebrate populations. *Parasitology* 111, S15–S31.

Anderson, R.M. and May, R.M. (1982) Coevolution of hosts and parasites. *Parasitology* 85, 411–426.

Anderson, R.M. and May, R.M. (1991) *Infectious Diseases of Humans. Dynamics and Controls.* Oxford University Press, Oxford, pp. 17–23.

Aragão, H.B. (1943) O virus do mixoma no coelho do mato (*Sylvilagus minensis*), sua transmissão pelos *Aedes scapularis* e *aegypti*. *Memorias do Instituto Oswaldo Cruz* 38, 93–99.
Beaucournu, J.C. (1980) Les puces du lapin de garenne. In: Traub, R. and Starcke, H. (eds) *First International Conference on Fleas*. Balkema, Rotterdam, pp. 383–389.
Brain, P. (1952) The sickle cell trait. Its clinical significance. *South African Medical Journal* 26, 925–928.
Bremermann, H.J. and Thieme, H.R. (1989) A competitive exclusion principle for pathogen virulence. *Journal of Mathematical Biology* 27, 179–190.
Brooks, D.R. (1986) Analysis of host–parasite coevolution. In: Howell, M.J. (ed.) *Parasitology – quo vadit?* Australian Academy of Science, Canberra, pp. 291–297.
Burnet, M. (1971) *Walter and Eliza Hall Institute 1915–1965*. Melbourne University Press, Melbourne, p. 140.
Cleaveland, S. and Dye, C. (1995) Maintenance of a microparasite infecting several host species: rabies in the Serengeti. *Parasitology* 111, S33–S47.
Darwin, C. (1859) *The Origin of Species by Means of Natural Selection, or the Preservation of Favoured Races in the Struggle for Life*. J. Murray, London, p. 489.
Dawkins, R. and Krebs, J.R. (1979) Arms races between and within species. *Proceedings of the Royal Society* B205, 259–285.
Dwyer, G., Levin, S.A. and Buttel, L. (1990) A simulation model of the population dynamics and evolution of myxomatosis. *Ecological Monographs* 60, 423–447.
Ehrlich, P.R. and Raven, P.H. (1964) Butterflies and plants: a study in coevolution. *Evolution* 18, 586–608.
Fenner, F. (1952) Myxomatosis: the virus and the disease it causes. *Australian Journal of Science* 15, 81–85.
Fenner, F. (1953) Host parasite relationships in myxomatosis of the Australian wild rabbit. *Cold Spring Harbor Symposia on Quantitative Biology* 18, 291–294.
Fenner, F. (1983) Biological control, as exemplified by smallpox eradication and myxomatosis. *Proceedings of the Royal Society of London* B 218, 259–285.
Fenner, F. (1993) Smallpox: emergence, global spread, and eradication. *History and Philosophy of the Life Sciences* 15, 397–420.
Fenner, F. and Ratcliffe, F.N. (1965) *Myxomatosis*. Cambridge University Press, Cambridge, pp. 343–344.
Fenner, F., Henderson, D.A., Arita, I., Jezek, Z. and Ladnyi, I.D. (1988) *Smallpox and its Eradication*. World Health Organization, Geneva, pp. 332–335.
Flexner, S.B. (ed.) (1987) *The Random House Dictionary of the English Language*, 2nd edn. Random House, New York.
Foster, E.A. (transl. and ed.) (1950) *Motolinia's History of the Indians of New Spain*. The Cortes Society, Berkeley. (Reprinted in 1973 by Greenwood Press, Westport, Conn.)
Futuyama, D.J. and Slatkin, M. (1983) *Coevolution*. Sinauer Associates, Sunderland, Mass., 555 pp.
Haldane, J.B.S. (1949a) The rate of mutations of human genes. In: *Proceedings of the 8th International Congress of Genetics (Stockholm, 1948), Heriditas (Suppl.)*, p. 267.
Haldane, J.B.S. (1949b) Diseases and evolution. In *Symposium sui fattori ecologici e genetici della speciazione negli animali (Pallanza, 31 July–2 August, 1948)*. Rome, pp. 68–76.
Hill, A.V.S. (1998) The immunogenetics of human infectious diseases. *Annual Review of Immunology* 16, 593–617.
Hill, A.V. and Wainscoat, J.S. (1986) The evolution of the alpha- and beta-globin gene clusters in human populations. *Human Genetics* 74, 16–23.
Hopkins, D.R. (1983) *Princes and Peasants. Smallpox in History*. University of Chicago Press, Chicago, 380 pp.
Johnson, R. (1986) The group selection of variola minor in the Americas: an anthropnootic analogue for the myxomatosis epizootic in Australia. *Medical Hypotheses* 19, 341–344.
Kalmbach, E.R. (1945) 'Ten eighty'. A war-produced rodenticide. *Science* 102, 232–233.
King, D. and Kinnear, J. (1991) 1080: the toxic paradox. *Landscope* 6, 14–19.
King, D.R. (1990) 1080 and the Australian fauna. *Agriculture Protection Board of Western Australia, Technical Series No. 8*.
King, D.R., Oliver, A. and Mead, R.J. (1981) Bettongia and fluoroacetate: a role for 1080 in fauna management. *Australian Wildlife Research* 8, 529–536.

Klenerman, P., Hengarner, H. and Zinkernagel, R.M. (1997) A non-retroviral RNA virus persists in DNA form. *Nature* 390, 298–301.

Levin, B.R. (1996) The evolution and maintenance of virulence in microparasites. *Emerging Infectious Diseases* 2, 93–102.

Lipsitch, M., Herre, E.A. and Nowak, M.A. (1995) Host population structure and the evolution of virulence – a law of diminishing returns. *Evolution* 49, 743–748.

Livingstone, F.B. (1971) Malaria and human polymorphisms. *Annual Review of Genetics* 5, 33–64.

Marshall, I.D. and Regnery, D.C. (1963) Studies in the epidemiology of myxomatosis in California. I. Observations on two outbreaks of myxomatosis in coastal California and the recovery of myxoma virus from a brush rabbit (*Sylvilagus bachmani*). *American Journal of Hygiene* 77, 195–204.

Massad, E. (1987) Transmission rates and the evolution of pathogenicity. *Evolution* 41, 1127–1140.

May, R.M. and Nowak, M.A. (1995) Coinfection and the evolution of parasite virulence. *Proceedings of the Royal Society of London* B 261, 209–215.

McFadden, G. (ed.) (1995) *Viroceptors, Virokines and Related Immune Modulators Encoded by DNA Viruses.* R.G. Landes, Austin, 218 pp.

McFadden, G. and Barry, M. (1998) How poxviruses oppose apoptosis. *Seminars in Virology* 8, 429–442.

McFadden, G., Graham, K., Ellion, K., Barry, M., Macen, J., Schreiber, M., Mossman, K., Nash, P., Lalani, A. and Everett, H. (1995) Interruption of cytokine networks by poxviruses: Lessons from myxoma virus. *Journal of Leukocyte Biology* 57, 731–738.

McIlroy, J.C. (1981) The sensitivity of Australian animals to 1080 poison. I. Intraspecific variation and factors affecting acute toxicity. *Australian Wildlife Research* 8, 369–383.

McIlroy, J.C. (1992) The effect on Australian animals of 1080-poisoning campaigns. In: J.E. Borroco and R.E. Marsh (eds) *Proceedings of the 15th Vertebrate Pest Conference*, California, pp. 356–359.

McNeill, W.H. (1976) *Plagues and People.* Blackwell, Oxford, 369 pp.

Mead, R.J., Oliver, A.J., Fing, D.R. and Huback, P.H. (1985) The co-evolutionary role of fluoroacetate in plant–animal interactions in Australia. *Oikos* 44, 55–60.

Mead-Briggs, A.R. (1977) The European rabbit, the European rabbit flea and myxomatosis. *Applied Biology* 2, 183–261.

Mead-Briggs, A.R. and Rudge, A.J.B. (1960) Breeding of the rabbit flea, *Spilopsyllus cuniculi* (Dale); requirement of a 'factor' from a pregnant rabbit for ovarian maturation. *Nature* 187, 1136–1137.

Niemialtowski, M.G., Toka, F.N., Malicka, E., Spohr de Faundez, I., Gierynska, M. and Schollenberger, A. (1997) Orthopoxviruses and their immune escape. *Reviews in Medical Virology* 7, 35–47.

Oliver, A.J., King, D.R. and Mead, R.J. (1977) The evolution of resistance to fluoroacetate intoxication in mammals. *Search* 8, 130–132.

Quétel, L. (1990) *History of Syphilis.* Translated from the French by J. Braddock and B. Pike. Polity Press, Cambridge, 342 pp.

Regnery, D.C. (1971) The epidemic potential of Brazilian myxoma virus (Lausanne strain) for three species of North American cottontails. *American Journal of Epidemiology* 94, 514–519.

Regnery, D.C. and Marshall, I.D. (1971) Studies in the epidemiology of myxomatosis in California. IV. The susceptibility of six leporid species to Californian myxoma virus and the relative infectivity of their tumors for mosquitoes. *American Journal of Epidemiology* 94, 508–513.

Regnery, R.L. (1975) Preliminary studies on an unusual poxvirus of the western grey squirrel (*Sciurus griseus griseus*) of North America. *Intervirology* 5, 364–366.

Rendel, J.M. (1971) Myxomatosis in the Australian rabbit population. *Search* 2, 89–94.

Rothschild, M. (1957) Breeding the rabbit flea (*Spilopsyllus cuniculi* (Dale)) in captivity. *The Entomologist* 90, 304–305.

Rothschild, M. and Ford, B. (1972a) Factors influencing the breeding of the rabbit flea (*Spilopsyllus cuniculi*): A spring-time accelerator and a kairomone in nestling rabbit urine with notes on *Cediopsylla simplex*, another "hormone bound" species. *Journal of Zoology* 170, 87–137.

Rothschild, M. and Ford, B. (1972b) Breeding cycle of the flea *Cediopsylla simplex* is controlled by breeding cycle of host. *Science* 178, 625–626.

Saunders, G., Coman, B., Kinnear, J. and Braysher, M. (1995) *Managing Vertebrate Pests: Foxes.* Australian Government Printing Service, Canberra, 141 pp.

Seymour, R.M. (1992) A study of virulence, resistance and resource limitation in a model of myxomatosis mediated by the European rabbit flea *Spilopsyllus cuniculi* (Dale). *Ecological Modelling* 60, 281–308.

Sobey, W.R. (1969) Selection for resistance to myxomatosis in domestic rabbits (*Oryctolagus cuniculus*). *Journal of Hygiene* 67, 743–754.
Spriggs, M.K. (1996) One step ahead of the game: Viral immunomodulatory molecules. *Annual Review of Immunology* 16, 101–130.
Thompson, J.N. (1998) Rapid evolution as an ecological process. *Trends in Ecology and Evolution* 13, 329–332.
Twigg, L.E. and King, D.R. (1991) The impact of fluoroacetate-bearing vegetation on native Australian fauna: a review. *Oikos* 61, 412–430.
Vogel, F. and Motulsky, A.G. (1986) *Human Genetics. Problems and Approaches*, 2nd edn. Springer-Verlag, New York, pp. 456–472.
Waters, A.P., Higgins, D.G. and McCutchan, T.F. (1991) *Plasmodium falciparum* appears to have arisen as a result of lateral transfer between avian and human hosts. *Proceedings of the National Academy of Sciences* 88, 3140–3144.
Weisenfeld, S.L. (1967) Sickle-cell trait in human biological and cultural evolution. *Science* 157, 1134–1140.
Williams, K., Parer, I., Coman, B., Burley, J. and Braysher, M. (1995) *Managing Vertebrate Pests: Rabbits.* Australian Government Printing Service, Canberra, 284 pp.
Wynne-Finch, H. (1935) *Fracastor: Syphilis or the French Disease. A Poem in Latin Hexameters by Girolamo Fracastoro. With Translation, Notes and Appendix.* Heinemann, London, 253 pp.

Glossary

Definitions are focused on the way in which the words are used in this book.

acclimatization The habituation of animals or plants to a new environment, commonly applied when moved by colonists from Europe to other continents.
antibody Specialized serum protein (immunoglobulin) produced in response to an antigen, which has the ability to combine specifically with that antigen (virus).
antibody response (primary) Production of antibody on first exposure to a particular antigen (virus).
antibody response (secondary) Production of antibody on a second exposure to a particular antigen (virus).
antigen Substance that can induce an immune response when introduced into an animal and which binds to the corresponding antibody *in vitro.*
arbovirus *Ar*thropod-*bo*rne; a virus that replicates in an arthropod and is transmitted by bite to a vertebrate host in which it also replicates.
arthropod Invertebrate animal with jointed body and limbs; the phylum Arthropoda includes insects, ticks, mites and other invertebrates.
attenuated Weakened in virulence (compared with original virus).
biological control Control of a pest or weed by means of another living agent.
biological transmission Transmission of a virus to a vertebrate by an arthropod after replication in that arthropod.
calicivirus Member of the viral family *Caliciviridae,* which includes rabbit haemorrhagic disease virus.
case-fatality rate Proportion of deaths to number of cases of a disease.
cell-mediated immunity Immunity produced by the action of cytotoxic lymphocytes produced after exposure to an antigen, rather than antibody.
chorioallantoic membrane Membrane surrounding a developing chick embryo, on which it is possible to grow viruses.
coevolution Evolution involving a series of reciprocal changes between a virus and its host, each acting as an agent of selection for the other.
double-stranded nucleic acid Genome of a virus that consists of a double strand of either DNA or RNA. The genomes of all organisms higher than viruses consist of double-stranded DNA.
ecology Branch of biology dealing with the relations of living organisms with each other and with their environment.
ecosystem The system which includes the physical environment and all living organisms that live within that particular environment.

ectoparasite A parasite that resides on the outside of the body of another living organism (its host).

emerging disease A previously unknown infectious disease which comes to attention because it affects humans or animals in which humans are interested.

enzootic An infectious disease continously present in a population of animals without introduction from elsewhere.

epidemiology The scientific study of the factors related to the spread of infectious diseases in human and/or animal populations.

epidemic (epizootic) An outbreak of an infectious disease previously absent or uncommon in a population of humans (animals), usually after introduction from elsewhere.

eradication Literally, tearing out by the roots; totally destroying populations of an organism (animal, microorganism, plant) in a particular area.

fibroma A non-malignant tumour consisting largely of fibroblasts that occurs in the skin (or in certain internal organs) of an animal.

field experiment A scientific experiment, with controls, set up in the field, as distinct from the laboratory.

gene Segment of the genome of an organism that codes for a particular protein.

genetic resistance Resistance to the ill-effects of an infectious disease due to mutations in the genes of the host organism.

genome Complete set of genes of a virus or organism.

haemorrhagic disease Infectious disease associated with internal or external bleeding.

health hazard Potential of an infectious agent occurring in an animal to affect the health of humans.

host Organism that supports the life of a parasitic microorganism.

host range Variety of species of animals which a particular parasite can infect.

host resistance Resistance of an animal to infection by a particular parasite.

host specificity Extent to which the host range of a particular parasite is restricted to one or a few species of animal.

immune response, primary Antibody production and cell-mediated immunity following the first exposure of a vertebrate to an antigen or virus.

immune response, secondary Antibody production and cell-mediated immunity following a second or subsequent exposure of a vertebrate to an antigen or virus.

immunity, active Adaptive response associated with antibody production and cell-mediated immunity following the exposure of a vertebrate to an infectious agent.

immunity, passive Transfer of protective antibody by inoculation of serum or from mother to foetus or newborn via the placenta or in the milk or colostrum.

immunization Production of active immunity by the deliberate inoculation of a vaccine.

immunocontraception Prevention of conception by inoculation of material which produces an immune response that prevents the production of sperm or the fertilization or implantation of eggs.

incubation period The interval between entry of an infectious microorganism into the animal body and the appearance of symptoms of disease.

infectivity The extent to which a preparation of an infectious agent can be diluted and still cause infection. Commonly measured by counting plaques or pocks produced by particular dilutions of a suspension, or as the 50% infectious dose (ID_{50}) when a number of animals are infected with an appropriate dilution.

inoculation Introduction of material (virus, vaccine) into the body by penetrating the skin.

Koch's postulates A set of rules developed by Robert Koch for determining whether a particular microorganism is responsible for a particular disease.

latent infection Persistent infection in which little or no infectious virus is detectable, despite the continued presence of the viral genome.

leporid Member of the family Leporidae, the rabbits and hares.

leporipoxvirus A virus belonging to the genus *Leporipoxvirus,* which belongs to the family *Poxviridae.* Includes myxoma and Shope fibroma viruses.

lethality Extent to which infection with a particular microorganism causes death.

live virus vaccines Material used for immunization which consists of virus which multiplies in the inoculated animal.

mean survival time The average number of days of survival among a group of animals inoculated on the same day with a virus that is lethal for most animals.

mechanical transmission Transmission of a virus mechanically, by introduction beneath the skin by a pin or on the mouthparts of an arthropod.

microorganism An organism (including virus) which is too small to be seen by the naked eye.

morbidity Sickness caused by a particular infectious agent.

mucus Slimy substance secreted by mucous membranes or which exudes from the cut surface of skin lesions of myxomatosis.

mustelid Member of the family Mustelidae, which includes ferrets, stoats and weasels.

mutant A microorganism or organism whose genome has undergone a mutation.

mutation A heritable change in the nucleotide sequence of the genome of an organism.

parasite An organism (microorganism, virus) that depends directly on another organism for its sustenance or survival.

pathogenesis The process by which an infectious agent produces disease in its host.

pests Animals, plants or microorganisms which interfere with the comfort, health or well-being of humans.

plaque Area of destruction produced by a virus in a monolayer of susceptible cells. Counts of plaques produced by suitable dilutions of virus are used to determine the infectivity of a virus suspension.

Pleistocene Glacial and post-glacial geological epoch following the Tertiary era.

pock Area of infiltration of the chorioallantoic membrane produced by a virus.

poxvirus A virus belonging to the family *Poxviridae.*

reactivation Process leading to the recovery of viable virus.

reservoir host Host animal in which a virus responsible for disease in another animal (or human) survives, even in the absence of the other animal.

resistance (genetic resistance; acquired resistance) Capacity of an animal to resist the ill-effects of an infectious agent. Genetic resistance is resistance associated with the host's genome; acquired resistance (immunity) is that conferred by vaccination or previous infection.

single-stranded nucleic acid Genome of a virus consists of a single strand of either DNA or RNA.

strain (of laboratory animal) A line of animals that has been inbred to the extent that it is genetically uniform.

strain (of virus) Preparations of virus that have been conserved in the laboratory and are associated with particular characteristics.

surveillance Careful oversight of populations of animals or humans to determine whether they are infected with particular viruses.

survival rate Proportion of animals in a group which survive a particular infection.

susceptibility Potential of a particular animal or species of animal to be infected with a particular virus.

transmissibility The likelihood that a particular virus or strain of virus will be transmitted from one animal to others.

vector **1.** Insect involved in transmitting a virus from one animal to another. **2.** Virus used to incorporate foreign genetic material so that the related protein is expressed when an animal is infected with that virus.

viraemia Occurrence of virus in blood; it may occur in plasma or leucocytes.
virulence Capacity of virus to produce severe disease or death in particular host species. Initially, with myxoma and RHD viruses, equated with lethality.
virulence grade An artificial measure designed to obtain a measure of the lethality of myxoma virus using both proportion of a small group that were killed and their survival time as a measure of virulence.
virus Infectious agents that are extremely small (filterable), can only replicate within living cells, and contain only one kind of nucleic acid, DNA or RNA.
zoogeography The science of the distribution of animals on the earth.

Index of Names

Names in endnotes or references (in text and at ends of chapters) are excluded from this index. Numbers in bold indicate that the person is the subject of a figure and biographical legend on that page.

Subject Index